Biomedical Library

Queen's University Belfast

Tel: 028 9097 2710

E-mail: biomed.info@qub.ac.uk

For due dates and renewals:

QUB borrowers see 'MY ACCOUNT' at

http://library.qub.ac.uk/qcat

or go to the Library Home Page

HPSS borrowers see 'MY ACCOUNT' at

www.honni.qub.ac.uk/qcat

This book must be returned not later than its due
date, but is subject to recall if in demand

Fines are imposed on overdue books

Methods of
Enzymatic Analysis

Methods of Enzymatic Analysis

Third Edition

Editor-in-Chief: Hans Ulrich Bergmeyer
Editors: Jürgen Bergmeyer and Marianne Graßl

verlag chemie

Weinheim · Deerfield Beach, Florida · Basel

Methods of Enzymatic Analysis

Third Edition

Editor-in-Chief: Hans Ulrich Bergmeyer
Editors: Jürgen Bergmeyer and Marianne Graßl

Volume II
Samples, Reagents,
Assessment of Results

verlag
chemie

Weinheim · Deerfield Beach, Florida · Basel

Editor-in-Chief:
Prof. Dr. rer. nat. Hans Ulrich Bergmeyer
Hauptstraße 88
D-8132 Tutzing
Federal Republic of Germany

Language Editor:
Prof. Donald W. Moss, Ph. D., D. Sc.
Appletrees 42
Greenways Hinchlay Wood
Surrey, KT10 OQD, United Kingdom

Editors:
Dr. rer. nat. Jürgen Bergmeyer
In der Neckarhelle 168
D-6900 Heidelberg
Federal Republic of Germany

Dr. rer. nat. Marianne Graßl
Frauenchiemsee-Straße 20
D-8000 München 80
Federal Republic of Germany

Note

The methods published in this book have not been checked experimentally by the editors. Sole responsibility for the accuracy of the contents of the contributions and the literature cited rests with the authors. Readers are therefore requested to direct all enquiries to the authors (addresses are listed on pp. XV – XVI).

Previous editions of "Methods of Enzymatic Analysis":
1st Edition 1963, one volume
  2nd printing, revised, 1965
  3rd printing, 1968
  4th printing, 1971

2nd Edition 1974, four volumes
  2nd printing, 1977
  3rd printing, 1981

Previous editions of "Methoden der enzymatischen Analyse":
1. Auflage 1962, one volume
2. neubearbeitete und erweiterte Auflage 1970, two volumes
3. neubearbeitete und erweiterte Auflage 1974, two volumes

Deutsche Bibliothek, Cataloguing-in-Publication Data
Methods of enzymatic analysis / ed.-in-chief: Hans Ulrich Bergmeyer.
Ed.: Jürgen Bergmeyer and Marianne Graßl. –
Weinheim; Deerfield Beach, Florida; Basel: Verlag Chemie
  Dt. Ausg. u.d.T.: Methoden der enzymatischen Analyse
NE: Bergmeyer, Hans Ulrich [Hrsg.]
Vol. II. Samples, reagents, assessment of results. – 3. ed. – 1983.
  ISBN 3-527-26042-0 (Weinheim)
  ISBN 0-89573-232-7 (Deerfield Beach)

Production manager: Heidi Lenz
Composition: Krebs-Gehlen Druckerei, D-6944 Hemsbach
Printing: Hans Rappold Offsetdruck GmbH, D-6720 Speyer
Bookbinding: Josef Spinner, D-7583 Ottersweier
Printed in the Federal Republic of Germany

Preface to the Series

"Methods of Enzymatic Analysis" appeared for the first time in 1962 as a one-volume treatise in German. Several updated and improved editions in English and German have been published since then. The latest English edition appeared in 1974.

In the meantime, enzymatic analysis has continued to find new applications, refinements and extensions at a pace that justifies – indeed, demands – the preparation of a new and completely revised edition. However, the field has grown so enormously that it can no longer be surveyed adequately by one person. Fortunately, therefore, I am supported in this new enterprise by Dr. M. Graßl, who is highly experienced in biochemical analysis, and Dr. J. Bergmeyer, who represents the younger generation of biochemists.

With the 1974 edition of "Methods of Enzymatic Analysis" as a starting point for our work towards the new edition, it soon became obvious that many chapters had to be eliminated, re-written or added. Moreover, the increased number of analytes that can now be determined enzymatically and of enzymes regularly requiring analysis, especially in the clinical laboratory, together with the emergence of an entirely new field of application through the technique of the enzyme-immunoassay, demanded a new arrangement and subdivision of the contents, if the vast range of material was to be dealt with properly and lucidly.

The result is the plan of the work printed on the page opposite the title page of this volume. Of course, it would be impossible to publish a whole series such as this at one moment and still maintain an equal degree of topicality for all contributions. Therefore, we decided to produce the series at a pace of several volumes per year. The volumes will not necessarily appear in their numerical order, but will be made available as they can be planned and completed.

As before, the purpose of the work is to provide reliable descriptions of well-developed procedures of enzymatic analysis in the broadest sense of the term. Special efforts are being made to arrange every chapter, and to coordinate the contents of all chapters, in such a way that the volumes are useful as laboratory manuals for daily work.

Internationally-agreed enzyme nomenclature as well as quantities and units correlating with the "Système International d'Unités" are used wherever possible in order to make statements and data unambiguous and comparable over time and space.

All contributions are and will be written in English: however, contributors come from all over the world and their manuscripts naturally show various versions of English. These have to be harmonized in style and spelling in order to achieve uniformity throughout the series without, we hope, entirely eliminating each author's personal approach. Professor Donald W. Moss has kindly agreed to undertake this task. We agreed with him to use modern English spelling, but to try to minimize differences between British and American practice. We hope that this will be considered as a fair solution and one which will make the series accessible to as wide a readership as possible.

Thanks are due to the authors in the first place for responding so readily to our invitations, for writing their chapters so diligently within a short time and for communicating their experience and expertise. We are also indebted to all colleagues who gave their advice and to Professor Moss for accepting the task of language editor. Finally I wish to record my gratitude to Verlag Chemie for the fruitful and excellent cooperation during all stages from planning to production.

Tutzing, February 1983   Hans Ulrich Bergmeyer

Preface to Volume II

This volume makes the transition from theory to practice. It covers the three steps in obtaining results from laboratory: the pre-analytical phase, the post-analytical phase, and the analysis itself. It describes the handling of reagents as well as their characteristic.

For the "Pre-analytical Phase", methods of obtaining and processing specimens and samples have been indicated in detail for the clinical laboratory and for laboratories engaged in food, cosmetic and pharmaceutical chemistry. A chapter is included with descriptions and validations of the various methods for protein determination that occupy such a central place in biological analysis.

A variety of new points of view are stressed under "Handling of Reagents for Enzymatic Analysis". Not only are "Quality Requirements" described in general, but the philosophy of the use of reagent kits and the eventful developments in the field of "Reagents on Carriers", are also discussed. The step-by-step improvement of methods, facilitating the experimenter's work, may initiate further progress in this direction.

The list of biochemical reagents comprises 143 enzymes, each with short but comprehensive instructions for measurement, and presents all the data needed for application in enzymatic analysis. Characteristic data for 85 coenzymes, metabolites and other biochemical reagents are also listed, and so is − for the first time − the huge variety of standard and reference materials now offered by various authorities. Manufacturers and dealers are not named, since biochemicals are now available almost everywhere.

In compiling the list, we surveyed Volumes III to VII and described the reagents needed for the methods presented there. Only an approximate forecast can be given at present of the methods to be covered in Volumes VIII to X. It is for this reason that the list is called "Biochemical Reagents *for General Use*". Volumes VIII to X will include parts requiring special reagents, separate lists of which will be compiled.

To the benefit of the reader, the chapter on "The Quality of Experimental Results" has been kept short. The editors were guided by the view that the central problem is the *control* of the quality of experimental results and the way in which this should be achieved. Statistics need only be described to the extent necessary for their use as tools for quality control.

We hope that readers will find this volume useful for their daily work in setting-up and performing enzymatic analysis. With its 267-page *collection of data* on biochemical reagents it is intended to be an indispensable part of laboratory equipment.

The editors are grateful to all colleagues and especially to all authors for their assistance, for numerous suggestions, and for their willingness to co-operate so readily.

Tutzing, November 1983   Hans Ulrich Bergmeyer

Contents

Contents of Volumes I – X

(Chapter Headings only)

Volume V
Enzymes 3: Peptidases, Proteinases and Their Inhibitors

1 Peptidases and Their Inhibitors
2 Proteinases and Their Inhibitors
3 Blood Coagulation Enzymes
4 Complement Enzymes

Volume VI
Metabolites 1: Carbohydrates

1 Poly-, Oligo- and Disaccharides
2 Monosaccharides and Derivatives
3 Three-Carbon Compounds
4 Two- and One-Carbon Compounds

Volume VII
Metabolites 2: Tri- and Dicarboxylic Acids, Purine and Pyrimidine Derivatives, Coenzymes, Inorganic Compounds

1 Citric Acid Cycle Compounds
2 Purines, Pyrimidines, Nucleosides
3 Nucleotides, Coenzymes and Related Compounds
4 Nucleoside Diphosphate Sugars and Derivatives
5 Inorganic Compounds

Volume VIII
Metabolites 3: Lipids, Steroids, Drugs

1 Lipids
2 Steroids
3 Drugs

Volume IX
Metabolites 4: Proteins, Peptides, Amino Acids

1 Proteins
2 Peptides and Peptide Hormones
3 Amino Acids and Related Compounds

Volume X
Molecular Biologicals (tentative)

Contributors

Bergmeyer, Hans Ulrich
Boehringer Mannheim GmbH
Biochemical Research Centre
Bahnhofstr. 9 – 15
D-8132 Tutzing/Obb.
Federal Republic of Germany

p. 20, 102, 126, 442

Beutler, Hans-Otto
Boehringer Mannheim GmbH
Biochemical Research Centre
Bahnhofstr. 9 – 15
D-8132 Tutzing/Obb.
Federal Republic of Germany

p. 328

Brand, Karl
Institut für Physiol. Chemie
der Universität Erlangen-Nürnberg
Fahrstraße 17
D-8520 Erlangen
Federal Republic of Germany

p. 26, 30

Brettschneider, Horst
Boehringer Mannheim GmbH
Werk Penzberg
Nonnenwald 2
D-8122 Penzberg
Federal Republic of Germany

p. 459, 477

Glocke, Manfred
Boehringer Mannheim GmbH
Sandhofer Str. 116
D-6800 Mannheim 31
Federal Republic of Germany

p. 459, 477

Gottschalk, Gerhard
Institut für Mikrobiologie
der Universität Göttingen
Griesebachstr. 8
D-3400 Göttingen
Federal Republic of Germany

p. 66

Graßl, Marianne
Boehringer Mannheim GmbH
Werk Penzberg
Nonnenwald 2
D-8122 Penzberg
Federal Republic of Germany

p. 102, 126, 442

Guder, Walter G.
Städt. Krankenhaus Schwabing
Klinisch-Chemisches Institut
Kölner Platz 1
D-8000 München 40
Federal Republic of Germany

p. 2

Henniger, Günther
Boehringer Mannheim GmbH
Biochemical Research Centre
Bahnhofstr. 9 – 15
D-8132 Tutzing/Obb.
Federal Republic of Germany

p. 20

Hess, Benno
Max-Planck-Institut
für Ernährungsphysiologie
Rheinlanddamm 201
D-4600 Dortmund
Federal Republic of Germany

p. 26, 30

Kresze, Georg-Burkhard
Boehringer Mannheim GmbH
Biochemical Research Centre
Bahnhofstr. 9 – 15
D-8132 Tutzing/Obb.
Federal Republic of Germany

p. 84

Kula, Maria-Regina
Gesellschaft für Biotechnologische
Forschung mbH
Macheroder Weg 1
D-3300 Braunschweig
Federal Republic of Germany

p. 66

Peters, Timothy J.
Division of Clinical Cell Biology
MRC Clinical Research Centre
Watford Road
Harrow, Middlesex Ha1 3UJ
United Kingdom

p. 49

Portenhauser, Rudolf
Boehringer Mannheim GmbH
Biochemical Research Centre
Bahnhofstr. 9 – 15
D-8132 Tutzing/Obb.
Federal Republic of Germany

p. 393

Supp, Martin
Boehringer Mannheim GmbH
Werk Penzberg
Nonnenwald 2
D-8122 Penzberg
Federal Republic of Germany

p. 328

Theimer, Roland R.
Botanisches Institut der Universität
Menzinger Straße 67
D-8000 München 19
Federal Republic of Germany

p. 73

Wahlefeld, August Wilhelm
Boehringer Mannheim GmbH
Biochemical Research Centre
Bahnhofstr. 9 – 15
D-8132 Tutzing/Obb.
Federal Republic of Germany

p. 2

Walter, Hans-Elmar
Boehringer Mannheim GmbH
Werk Penzberg
Nonnenwald 2
D-8122 Penzberg
Federal Republic of Germany

p. 126

1 Specimens and Samples

1.1   Preparation and Processing of Samples

Enzymatic analysis of biological material is performed to evaluate the substrate concentration and enzymatic activity of the material, e.g. tissue, cells, fluid, or foodstuff. The results are interpreted as indicators of the composition of the organism or biological material from which the sample was taken. However, this can only be assured if the sample is representative and does not undergo changes in composition before enzymatic analysis. A typical example demonstrating the problems of pre-analytical influences is the use of enzymatic analysis of human blood samples in the diagnostic evaluation of disease. The complex composition of this material provides well documented examples to illustrate nearly all possible pitfalls and analytical interferences in one material.

In food chemistry and in analysis of cosmetics and pharmaceutical preparations, collection of the specimen and the specific pretreatment of the specimen and samples are particularly important because the kinds of sample are so different. No general rules can be given; the procedures are all dependent on the type of analyte and sample material. In contrast to the application of chemical methods, in which isolation and separation methods are necessary to yield satisfactory results, the use of enzymatic methods does not need any elaborate pre-treatment: only procedures which are easy to perform and which facilitate the measurement are needed in the preparation of the sample.

1.1.1   Specimens and Samples in Clinical Laboratory Sciences

Walter G. Guder and August-Wilhelm Wahlefeld

Although similar principles apply for all biological materials (tissues, urine, cerebrospinal fluid, liquor, cell cultures, etc.), the following short overview concentrates on blood analysis, since the complex composition of this material provides enough well documented examples to illustrate nearly all possible pitfalls and analytical interferences which may be encountered in sample material.

1.1.1.1   The Pre-analytical Phase

The pre-analytical phase consists of several steps:

 − preparation of subject to be investigated
 − collection of specimen

- separation of sample from specimen
- transport of specimen and/or sample
- storage of specimen and/or sample
- pre-treatment of samples for enzymatic analysis

The *specimen* is defined as that part of the subject which is taken as representative for analysis. The *sample* is the material which is actually analyzed. It can be either a part of the specimen or is derived or prepared from the specimen. Thus a tissue biopsy is a specimen, which is homogenized for measurement of enzyme activities. Consequently, the aliquot of homogenate which is analyzed is the sample. Only under certain conditions is the sample identical with the specimen.

Possible changes in analyte concentration (activity) and interfering factors can arise at all steps of the pre-analytical phase. For practical reasons factors which lead to non-representative results have been divided into biological influence factors (Einfluß-größen) and interference factors (Störfaktoren) [1, 2, 3].

Biological influence factors lead to changes of the analyte *in vivo*. They are by definition independent of the analytical method used. They can be either variable or invariable. Only influence factors which can be changed by the analyst can be the subject of standardization [1].

Interference factors, on the other hand, include all factors that alter the result after the specimen has been collected. They can be divided into two groups:

a) interference factors which *in vitro* lead to changes of the parameter to be measured (e.g. decrease in enzyme activity caused by inappropriate cooling of sample),

b) interference factors that are different from the parameter to be measured, but interfere with the analytical procedure (e.g. apparent decrease in glucose measured enzymatically with glucose oxidase procedures in the presence of ascorbic acid).

The latter can be eliminated by increasing the specificity of the analytical procedure, whereas the former can only be controlled by appropriate standardization of the pre-analytical phase.

A typical example demonstrating the problems of pre-analytical influences is the use of enzymatic analysis of human blood samples [1, 2]. For other types of material the reader is referred to the respective chapters of this series: 1.1.2, this volume; and the individual procedures for the determination of analytes in Vols. V – X. Special care must be taken with urine [5, 6], cerebrospinal fluid [7] and other body fluids (cf. Vol. III, chapter 1.2). The preparation of tissue for enzymatic analysis is described in the following chapters 1.2.2 and 1.2.3.

1.1.1.2 Specimens

Specimen collection

Before a specimen is taken, several questions should be answered to select an adequate technique and avoid later misinterpretations of results.

– What will be measured in the sample?
– When will this be measured?
– Are interfering factors known, which need standardization of the subject (material) before sampling?
– Are all materials available for optimal processing of specimen?
– Are additions needed: if yes, in what amounts?

Either of these points can cause inadequate sampling if not answered properly in advance.

Preparation for specimen collection

Specimen collection not only includes the preparation of the technical material needed to draw and process the material to be analyzed: the preparation of the subject from whom the specimen is taken seems to be even more important. Preparation of the subject includes several factors which may influence the composition of the sample if they are not standardized properly. Among these are dietary habits, type and degree of muscular work, interfering drugs, other diagnostic and/or therapeutic procedures, and the position of the body before and during specimen collection. Table 1 summarizes some influence factors which lead to changes of blood constituents due to improper preparation.

Table 1. Variable factors influencing human blood composition, which can be standardized by preparation for specimen collection (for additional references see [2, 3, 8]).

Influence factor	Examples of blood constituents affected	Recommendation to standardize specimen collection
Food	urea, triglycerides, uric acid, glucose, phosphorus	sampling after 12 h fasting
Prolonged fasting	clotting factors, cholinesterase, oral glucose tolerance	3 days feeding before sampling
Rhythms	iron, oral glucose tolerance, cortisol, acid phosphatase [9]	sampling between 7 and 9 a.m.
Muscular activity	creatine kinase, lactate dehydrogenase, glucose, clotting factors, ASAT*	no exhausting muscular activity 3 days before sampling
Ethanol	glucose, uric acid, γ-GT** lactate, amylase	control of alcohol consumption
Drugs	uric acid, cholinesterase, alkaline phosphatase, γ-GT**	control of drug effects use of interference list [10]
Position	proteins, lipids, enzymes	15 min supine position before sampling
Smoking	lipase, amylase after secretin [11], cholesterol, glucose [12]	control of smoking
Coffee	glycerol, fatty acids, cortisol [1]	sampling after 12 h fasting

* L-Aspartate: 2-oxoglutarate aminotransferase, EC 2.6.1.1.
** (5-Glutamyl)-peptide: amino-acid 5-glutamyltransferase, EC 2.3.2.2.

The time of specimen collection has to be standardized if circadian rhythms may alter the metabolite or enzyme to be measured. With regard to human blood samples, several national and international recommendations have been published which include standardization of the preparation of patients before sampling [13 – 15]. Blood drawing in the morning after an overnight fast and after 10 – 15 minutes in a supine position is proposed [13].

Anatomical sites of specimen collection

The site of blood sampling depends on the question raised, the analyte to be measured, the amount of sample needed and the anatomical situation. In contrast to arterial blood, the composition of which is the same in all accessible vessels, venous blood not only represents whole-body blood composition but also is largely influenced by the metabolism and haemodynamics of the region which it drains. Capillary blood, on the other hand, is a mixture of blood derived from arterioles, venules and capillaries as well as interstitial and intracellular fluids [14], with relative differences between different capillary beds.

Because of their ease of accessibility, superficial veins of the palmar surface of the forearm are the preferred site of venous blood specimen collection in adult patients, whereas superficial head veins are preferred in young children. Capillary blood is taken from the outer plantar surface of the heel in young children [14], the lower ear lobe and the fingertips, although the last is less recommended because of the more painful procedure. In general, venepuncture is suggested, unless specific reasons favour the analysis of capillary blood. One of these reasons is the metabolism of the forearm, which uses glucose and produces lactate depending on blood flow and muscular work. This leads to concentration differences between capillary and venous blood samples.

Techniques of blood collection

Capillary blood

According to the recommendations published recently by the National Committee of Laboratory Standards [14, 15], capillary samples may be obtained after warming and cleaning the puncture site to avoid repeated puncture and contamination of sample. Silicone may be used to improve blood flow. The puncture lancette should be big enough to provide sufficient blood flow but should be prevented from entering deeper tissues in order to avoid artery and bone puncture. In children under the age of 1 year the puncture should not exceed 2.4 mm in depth [14]. The first drop of blood must be discarded, because it contains excess tissue fluid. Pressure on the wound, wiping with alcohol and repeated puncturing at the same site should be avoided because of the danger of haemolysis, excess tissue fluid and prolonged wound healing. Up to 0.5 ml blood can be obtained from a single puncture site.

Venepuncture

Although venepuncture is the technique used most often to obtain human blood samples, no generally accepted standard procedure exists which would cover all influencing aspects. In spite of this fact many published experiences have led to some standardization which may be the basis for future recommendations [16]. Possible objects of standardization and sources of erroneous results are:

– time and degree of tourniquet application
– – before puncture
– – at puncture and during sampling
– cleaning of puncture site
– type of needle (gauge, material)
– duration of venepuncture
– pressure (vacuum, free flow)
– specimen containers, stoppers
– additives
– – type (anticoagulant, stabilizers, stimulators of coagulation)
– – dosage
– labelling and coding

Several types of errors in sample composition can be caused by improper vene-puncture and can be prevented by following the standard procedure. Table 2 summarizes some mechanisms, which may lead to changes in blood sample composition by non-standardized venepuncture.

Table 2. Mechanisms which lead to changes in blood sample composition by improper venepuncture technique (for references see [2, 12, 16])

Mechanism	caused by	analytes changed
Decrease of blood water space by redistribution into extravasal space	prolonged or forced tourniquet application	proteins (enzymes), lipids, cells
Contamination with tissue constituents	hypoxia, repeated injection, pressure	potassium, phosphorus, LDH*
Haemolysis	improper cleaning of skin (iso-propanol) or sampling material (detergents) needle gauge too small, vacuum too strong	potassium, phosphorus, LDH, acid phosphatase, ASAT
Contaminations	exogenous impurity	ethanol (from skin cleaning) amylase (from saliva) ammonia (urine)
	additives	ions, water
	infusions	glucose, ions, proteins
Thrombolysis, coagulation	sampling too slow, inappropriate anticoagulant, inadequate mixing	fibrinogen degradation products, coagulation factors

* Lactate dehydrogenase, EC 1.1.1.27

As a standard procedure, a tourniquet not exceeding systolic blood pressure should be applied not longer than one minute before puncturing the vein with a No. 1 needle or a higher gauge size. After the needle is in the right position in the direction of blood flow, the tourniquet should be released and blood obtained by either controlled vacuum or free flow into a clean sample tube. Cleaning of the venepuncture site should avoid contaminating or haemolyzing substances. The skin should be dried before puncturing.

Identification of specimen

Before a specimen is taken, the tube should be labelled with appropriate information. In the case of a patient's blood these should include at least the name of the patient, the number or name of the ward or place where the sample was drawn, and the time of day if more than one sample is drawn on the same day. In some instances it is necessary to give additional information if this is not given on a separate request form. This includes additives, site of puncture, date and possible interfering factors included in the sample. Additives should be coded according to given standards [15].

Stability of metabolites and enzymes in the pre-analytical phase

Specimen collection is the first, but not the only procedure which may cause sample composition changes. Each of the pre-analytical steps bears several dangers which may be sources of interferences and cause significant changes. Fig. 1 summarizes these pre-analytical factors and their possible dangers and provides methods to prevent changes.

Sources of changes in specimen composition in vitro

For most metabolites the analyst can only minimize changes, since artefacts due to improper specimen collection, transport, sample separation and storage are nearly unavoidable. Likewise, he has clearly to define the optimal procedure for a given sample to be processed. Optimal conditions for one type of analysis may be unfavourable for others.

In the case of blood, clotting, additives, time and temperature before and after separation of blood cells, light, interfering metabolites and enzymes are possible sources of errors (Fig. 1). These factors have only been partially described in recommendations given by the National Committee for Clinical Laboratory Standards [15]. If *serum* is the desired sample to be analyzed, at least 30 minutes are needed at room temperature for optimal clotting, before the blood can be centrifuged. Cooling of whole blood and use of plastic tubes without additives prolongs this time. On the other hand, *plasma*, the material more relevant to the intravasal fluid, can only be preserved by the addition of anticoagulants none of which is ideally suitable for all

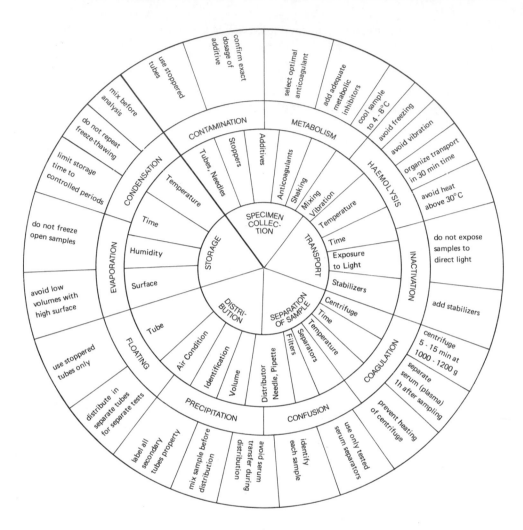

Fig. 1. The pre-analytical phase in enzymatic analysis of human blood. The inner circle gives the various steps of the pre-analytical phase, surrounded by possible sources of disturbing factors. The machanisms which lead to changes in sample composition are given in the third circle with recommendations for preventing changes in the outer segments.

blood constituents. Addition of *anticoagulants* not only preserves clotting factors, but decreases haemolysis, shortens the time needed before centrifugation and thereby increases the similarity between analytical results and *in vivo* conditions. However, each of the anticoagulants in use bears its own problems, limiting the general use of one substance. Table 3 summarizes the limitations in applicability for some plasma constituents [17]. Thus in analyzing serum or plasma a compromise has always to be accepted between the analytical aims and the proposed procedure of sample processing.

Table 3. Interference of anticoagulants with enzyme and metabolite determinations in human plasma

Measured metabolite/ enzyme	Ammonium heparinate 0.75 mg/ml	Li, K, Na heparinate 0.75 mg/ml	EDTA 1 mg/ml	Citrate 5 mg/ml	Oxalate 2 mg/ml	NaF 2 mg/ml
Acid phosphatase (*Hillmann*'s method)	+	+	+	−	+	+
Alkaline phosphatase	−	−	+	+	+	+
Ammonia (enzymatic)	+	+	−	+	+	+
Ethanol (enzymatic)	−	−	−	−	−	−
Cholesterol (CHOD-PAP)	−	−	−	−	−	−
Cholinesterase	−	−	−	−	−	+
Creatine kinase NAC, CK-MB (immunol.)	−	−	−	+	+	+
Glucose (hexokinase) deproteinized by HClO$_4$	−	−	−	−	−	−
Glucose (GOD) deproteinized by a) perchlorid acid	−	−	−			−
b) uranyl acetate	−	−	−	+	+	−
Glutamate dehydrogenase						⏐
Glutamate-oxalactate transaminase (ASAT)	−	−	−	−	+	−
Glutamate-pyruvate transaminase (ALAT)	−	−	−	−	+	−
γ-Glutamyltransferase (Glucana) a) sample start	+	+	−	+	+	+
b) substrate start	−	−	−	+	+	+
Leucine amino peptidase	+	+	+	+	+	+
Lactate dehydrogenase	−	−	−	−	+	−
Triglycerides (enzymatic)	−	−	−	+	+	+
Urea (enzymatic)	+	−	−	−	−	−
Uric acid a) uricase	−	−	+	+	+	+
b) uricase PAP	−	−	−	−	−	−

+ determination interfered with by anticoagulant
− no influence
All enzyme activities were measured according to the standard procedures of the German Soc. Clin. Chem. [18].

Procedures to increase stability

Several metabolites and enzymes are unstable even under optimal conditions, due to either the presence of metabolizing enzymes or bacteria in the sample or chemical instability [17]. Cooling the sample often decreases the metabolism to an acceptable degree. However, in the case of blood, this prolongs clotting time, increases haemolysis and can therefore only be recommended for anticoagulated blood samples [13, 16]. After separation of serum or plasma, cooling is an effective procedure to increase stability. $+4°C$ is recommended for most serum or plasma constituents, since freezing can lead to inactivation (lactate dehydrogenase), precipitation of insoluble constituents (calcium, uric acid) and condensation. Detailed experiences with plasma (serum) constituents have been published [17] and the reader is referred to the respective chapters of these volumes.

Some constituents need additional precautions to assure stability over a few hours. Glucose is the best known example, since it is rapidly used up by the metabolizing capacity present in blood cells. This can be inhibited either by *deproteinization* before transport of specimen or by addition of *inhibitors of glycolysis* to the whole blood. To keep the concentration stable in whole blood sodium fluoride in a concentration of 2 mg/ml is used [19].

Examples of *enzyme stabilizers* are SH-reagents for creatine kinase and acidification to stabilize acid phosphatase. *Protease inhibitors* can prevent enzymatic degradation of hormonal peptides such as glucagon and ACTH. *Complexing calcium* by citrate and other anticoagulants prevents activation of most coagulation proteases. Stability of serum or plasma constituents is likewise increased by *using only stoppered cups* or tubes for storage. This not only prevents evaporation (which would lead to increased concentration of non-volatile constituents) but also prevents bacterial and chemical contamination, as well as condensation of water during storage below 4°C. Evaporation is a serious problem in open, flat sample containers with a relatively large surface area. Dependent on temperature, humidity and air conditions, evaporation can cause significant loss of sample even in a few hours. Another cause of possible changes in concentration can be inhomogeneity of constituents in the sample. This may be serious in lipid analysis and with some other constituents upon prolonged storage. Mixing of samples before distribution and before analysis at room temperature seems sufficient to prevent this source of error. Direct light must be excluded if light sensitive constituents are to be measured (e.g. creatine kinase [20], bilirubin, porphyrins [2]).

In general, stability cannot be assured in pathological sera even if it has been investigated in large series of normal and abnormal samples. This is due to the higher enzyme activities present in pathological samples, by the absence of physiological protease inhibitors, and by the presence of drugs and other disturbing factors. Thus fatty acids can be measured accurately in normal samples after cold storage, whereas they increase rapidly if lipase activities, triglycerides or both are increased. Therefore, recommendations should be given for the worst condition observed in order to obtain optimal results.

1.1.1.3   Preparation of Samples

Separation of sample from specimen

Blood, like most biological materials, cannot be analyzed directly but must be processed to obtain the optimal sample for enzymatic analysis. Several treatments are obligatory for blood in order to obtain stable and reproducible analytical results:

- Separation of blood cells
- Separation of fibrin (to obtain serum)
- Deproteinization
- Extraction, chromatography, precipitation
- Delipidation

Separation of serum and plasma from whole blood

In most instances serum or plasma is the sample analyzed. Its separation from whole blood by centrifugation is the procedure most often used. In spite of its general use, no standard recommendations exist regarding centrifugal force, time and temperature limits. The generally accepted experience, suggesting 3000 rpm for 10 min, seems to apply for standard rotors and serum samples of 10 ml at room temperature. The main purpose of this procedure is to get rid of blood cells containing enzyme activities which by far exceed those of the supernatant serum or plasma. The most critical cells are thrombocytes, which can invisibly contaminate the lower plasma phase if not centrifuged sufficiently. Fibrin aggregates are spun down with the blood cells if blood is not anticoagulated. Insufficient clotting time and low temperature prevent clear separation of the clot.

Several devices have been described which help to overcome the problems of serum separation. Polystyrene beads, which separate serum from blood clot by their specific weight, and plastic serum separators are added to the tube before centrifugation of the blood. However, only the latter provide a clear separation, because the separator forms an impermeable stopper on top of the deposit of fibrin and blood cells upon centrifugation [21]. After centrifugation serum or plasma can thus be separated into a container stoppered into the same tube, thus preventing confusion of samples. In most instances, however, serum is decanted into a secondary container either by hand, pipette or an automatic sample distributor.

Deproteinization

Inactivation of metabolic activities by deproteinization of whole blood, serum or plasma is recommended in all cases, since even at room temperature rapid changes in metabolite concentrations take place.

Methods

Only a few procedures are applied in well-established routine methods for the preparation of protein-free samples: in acidic deproteinization, perchloric acid is usually used; in neutral deproteinization, cationic precipitants are used. There is no universal protein precipitant available yet for less widely-determined analytes or those for which a new procedure has to be developed. For every analytical problem the appropriate protein precipitant should be selected by experiment. In most instances, when preparing protein-free samples for the determination of acid-stable analytes, perchloric acid is preferred as precipitant.

Neuberg [22] introduced this procedure. 1 volume blood, serum or plasma is mixed with 9 volumes of $HClO_4$, 1 mol/l, the mixture is shaken, allowed to stand for 10 min at room temperature, then centrifuged 10 min at 3000 g. The supernatant has a pH of nearly 0. Usually one obtains optically clear supernatants.

The excess of perchloric acid can be removed from the supernatant by neutralization with KOH, K_2CO_3 or K-acetate. The resulting $KClO_4$ precipitates in the cold (ice-bath) and can be filtered off. This procedure is of advantage when perchlorate anions or acidic pH values interfere with the assay procedure. Perchloric acid also does not adsorb ultraviolet light, as compared with trichloroacetic acid.

Other anionic precipitants are tungstic acid and trichloroacetic acid.

Folin & Wu [23] recommended precipitation by an acidic tungstic acid solution in which 1 volume 10% w/v sodium tungstate, $Na_2WO_4 \cdot 2\ H_2O$ and 1 volume H_2SO_4, 0.66 mol/l, are mixed with 1 volume serum/plasma diluted with 7 volumes of water. This precipitant results in a crystal-clear filtrate, but when applied to blood some floating material can be observed making it difficult to remove clear samples.

Greenwald [24] introduced the use of 5% (v/v) trichloroacetic acid solution (1 volume sample + 9 volumes precipitant). The mixture should stand for 20 min at room temperature prior to centrifugation for 10 min at 3000 g. The final pH is about 1.

The precipitant is more voluminous than with the *Folin-Wu*-precipitant, but sometimes the protein-free supernatant is not always optically clear. This drawback has to be taken into account when the assay procedure involves measurement at 339 nm. Another problem may arise from the decomposition of trichloroacetic acid solution to form chloroform and carbon dioxide during storage (avoid higher temperature). Also, an incomplete precipitation of proteins in urine has been reported [25].

Cationic precipitants generally are used for acid labile analytes.

Several cations have been proposed for deproteinization procedures: Hg^{++}, Cd^{++}, Zn^{++} or combinations of $Hg^{++}/Ba^{++}/Zn^{++}$ and Ba^{++}/Zn^{++}. Uranyl cations are also used.

Somogyi [26] recommended the use of zinc sulphate solution (5% w/v, $ZnSO_4 \cdot 7\ H_2O$) in combination with barium hydroxide solution (0.3 mol/l). 5 Volumes water are added to 1 volume of blood, followed by 2 volumes barium hydroxide, then mixed. Two volumes zinc sulphate solution are added, mixed and centrifuged for 10 min at 3000 g.

This deproteinization procedure has been recommended for the reference method for glucose in serum or plasma [27]. It should be noted, however, that this procedure is not applicable to all analytes: e.g. uric acid adsorbs to the precipitate and is co-precipitated, resulting in too low values with any assay method.

Uranyl acetate [28] in isotonic saline solution has been introduced especially for deproteinization of whole blood without disrupting erythrocytes. This procedure is advantageous for glucose determinations according to the glucose oxidase/peroxidase method, since without disrupting red blood cells no gluthathione (which interferes with the hydrogen peroxide in the peroxidase reaction) is released to the supernatant.

1 volume of blood is mixed with 10 volumes of 0.16% (w/v) uranyl acetate in 0.9% (w/v) saline solution, mixed and after 10 min at room temperature centrifuged for 10 min at 3000 g.

Care has to be taken that the protein concentration in the sample should not be below 1 g/100 ml, since otherwise the yellowish uranyl cation is not completely bound by proteins present and colours the supernatant, presenting a problem in photometric measurements at about 450 nm.

Volume displacement effect

In general, it should always be recognized that preparing protein-free supernatants not only results in dilution of the sample, but also in change of concentration due to the so-called "volume displacement effect". Its influence on serum and plasma specimens has been investigated.

Van Slyke [29] was the first to note in 1927 differences in results of analyses with and without deproteinization. This effect was defined in 1936 by Ball & Sadusk [30]. However, Buergi [31, 32] was the first to consider the theory in detail and proposed that it applied to all low-molecular-weight compounds in serum or plasma. A certain total volume consists to a large extent of fluid volume, and to a lesser extent of the space occupied by the protein molecules. This corresponds to the so-called partial specific volume and is the reciprocal of the specific gravity of the protein, i.e., a characteristic property of each protein. The partial specific volume for plasma proteins is around 0.7 – 0.75. If a certain volume of plasma is pipetted, this contains a fraction of the volume as protein. After deproteinization, the low-molecular-weight compounds are somewhat more concentrated. The following equation permits calculation of the difference in the concentrations of a substance in plasma and in a protein-free extract:

$$\Delta c = c_1 \left[\frac{100}{100 - \bar{V} \times \rho_p} - 1 \right]$$

Where is

c_1 concentration of substance in plasma
c_2 concentration of substance in protein-free supernatant
Δc the difference $c_2 - c_1$
ρ_p concentration of protein in g/100 ml
$\bar{V}$ partial specific volume of the proteins.

An experimental value of ca. 5% has been found for serum with regard to the acetate determination [33]. It shows increasing overestimation of the analyte with increasing protein concentration. In our laboratories, with serum or plasma, values have been observed for glucose or inorganic phosphate higher by about 5 to 6% than those found without deproteinization. According to our experience, when triglycerides from serum or plasma are chemically hydrolyzed we find an overestimation of 3%; therefore, we recommend for the most accurate results the investigation of the volume displacement effect for the proposed assay procedures.

Special treatments of samples

In some instances, the separated serum or plasma sample needs additional treatment to assure stability of the analyte and/or to separate the analyte from interfering substances of serum. The procedures are closely linked to the analytical procedures and the special conditions of the sample.

Samples from heparinized patients often exhibit prolonged coagulation after centrifugation without contraction of the fibrin clot. This can prevent pipetting of sample and needs additional separation steps, either by *high force centrifugation* (above $10^4 g$), *filtration* or *deproteinization*. Similar problems can arise from cryoglobulins and other monoclonal immunoglobins. The latter interferences may be met by *warming* or *dilution* of samples, respectively. Similarly, high turbidity of the sample caused by elevated triglycerides can be a serious problem in enzymatic analysis. Appropriate blanking is sufficient, if turbidity is not outside the photometer's range. Otherwise the sample must be delipidated. This can be done by either ultracentrifugation, lipoprotein precipitation [34] or lipid extraction with fluorocarbons [35].

Separation from interfering substances is less often needed in enzymatic analysis than in chemical procedures. With regard to blood samples, the interfering substance is most often unknown to the analyst, if it is not visible or indicated on the specimen. This can be a normal or pathological blood constituent, a drug or drug metabolite, which interferes with a nonspecific reaction of the analytical procedure or is an inhibitor of an enzyme to be measured.

In some cases interference can be prevented by sample dilution, chromatography, dialysis, or specific adsorption procedures. The reader is referred to the excellent computer list of *Young et al.* [10] for detailed information.

The sum of all non-measured constituents of the sample constitutes the sample matrix, which may be the cause of analytical interferences if not regarded properly. The matrix has always to be included in the description of the analytical procedure, when applied to blood serum or plasma.

1.1.1.4   Stability of Analytes in Specimens and Samples

The following tables summarize current knowledge about the stability of important analytes in specimens and samples [17].

The general conditions are described above but, especially for metabolically active substances, it is important to discuss quantitatively the changes in concentration which result when proper precautions are, or are not, taken [36].

Glucose is rapidly metabolized by the erythrocyte enzyme system. The process can be effectively stopped by adding inhibitors to the enzymes of the glycolytic pathway; e.g. the use of maleinimide is well established.

Table 4. Stability of metabolites in blood and deproteinized solutions.

Substance	Material	Storage temperature	Percentage of starting value after days				
			1	2	3	5	8
Glucose	blood	room temperature	30	6		0	
		+4 °C	80	32		20	
	blood + HClO$_4$ (without centrifugation)	room temperature	110*	100		115*	
		+4 °C	99	102		108*	
	blood + HClO$_4$ (supernatant after centrifugation)	room temperature	99	100		102	
		+4 °C	99	100		100	
		−25 °C			99		104
	plasma	room temperature	99	99	99		97
		+4 °C	101	101	98		99
		−20 °C	99	101	99		77
Lactate	blood	room temperature	341	425			
		+4 °C	168	210			
	blood + HClO$_4$ (supernatant after centrifugation)	room temperature	99	102	98		
		+4 °C	101	105	102		
		−25 °C			102		
Pyruvate	blood	room temperature	856	1065			
		+4 °C	101	155			
	blood + HClO$_4$ (supernatant after centrifugation)	room temperature	101		95		
		−25 °C			96		
ATP	blood	room temperature	95	74			
		+4 °C	62	50			
	blood + HClO$_4$ (supernatant after centrifugation)	room temperature	78	62			32
		+4 °C	93	85	87		76
ADP	blood	room temperature	131				
		+4 °C	112	100			
	blood + HClO$_4$ (supernatant after centrifugation neutralized)	room temperature	97	100	93		
		+4 °C	98	95	92		

Substance	Material	Storage temperature	Percentage of starting value after days				
			1	2	3	5	8
AMP	blood	room temperature	248				
		+4°C	317	346			
	blood + HClO$_4$ (supernatant after centrifugation neutralized)	room temperature	102	100			
		+4°C	100	100			
2-Oxo-glutarate	blood	room temperature	135	205			
	blood + HClO$_4$ (supernatant after centrifugation neutralized)	room temperature	80			77	
		+4°C	95	91			

* The increase of glucose concentrations is caused by acidic hydrolysis of glucose-containing substances of erythrocytes.

Lactate or pyruvate, on the other hand, increase dramatically by the same mechanism, if the blood specimen is not deproteinized at once. Alternatively, when fluoride plus EDTA are added to the blood specimen, plasma can be used.

The changes in the concentrations of individual metabolites are very variable due to the effects of specific reactions: ATP concentration in blood decreases more slowly at room temperature – due to high glycolytic activity – than at +4°C, when the degradation reaction becomes predominant.

The products of ATP degradation, ADP and AMP, increase as expected from the assumptions made above, but AMP increases more rapidly at +4°C than at room temperature, whereas ADP increases more slowly. These effects demonstrate the need to establish the individual conditions for proper storage of samples by experiment, rather than from theoretical considerations.

As a rule, analytes of considerable metabolic activity should be quantitatively determined only after deproteinization of samples, or in samples in which the metabolizing enzyme reactions have been blocked by specific inhibitors.

The stabilities of metabolites important in clinical chemical laboratories are collected in Table 5 (cf. also [33, 37, 38]).

The stability of enzymes in specimens or samples is also influenced by a number of general factors, such as pH values, exposure to higher temperature, and the presence of detergents or organic solvents, but also by freezing and thawing. The situation in tissue extracts is usually very different for enzymes; the extracting agents or proteases present sometimes decrease stability. Therefore, it is not possible to give stability data for enzymes in homogenates or extracts.

However, because human serum is a fairly uniform matrix and is the most common sample in the clinical laboratory, extensive data about stability of enzymes in serum have been collected [38 – 41].

Table 5. Stability of metabolites in serum or plasma samples.

	Storage temperature (always use well-stoppered containers)	
	2 – 8 °C	20 – 25 °C
Ammonia	EDTA plasma:	
	2 h	
Bilirubin, total	use only fresh serum or plasma avoid heat and direct sunlight	
Cholesterol, total	serum:	serum:
	max. 6 days	max. 6 days
HDL-cholesterol	serum:	
	max. 6 days	
	supernatant after precipitation:	
	7 days	7 days
Creatinine	serum:	
	24 h	
Copper	serum:	serum:
	max. 14 days	max. 14 days
Glucose	whole blood: immediate deproteinization plasma (separated from cells)	
	7 days	3 days
Galactose	whole blood: immediate deproteinization supernatant	
	max. 3 days	
Iron	serum:	serum:
	max. 7 days	max. 4 days
Inorganic Phosphate	serum:	serum:
	max. 7 days	max. 2 days
Total protein	serum:	serum:
	max. 6 days	max. 6 days
Triglycerides	serum:	serum:
	max. 3 days	storage not recommended, formation of free glycerol
Urea	serum:	serum:
	max. 3 days	24 h (sterile!)
Uric acid	serum:	serum:
	up to 5 days	

Table 6. Stability of enzymes in serum samples.

	Storage temperature (always use well-stoppered containers)	
	2 − 8 °C	20 − 25 °C
Aldolase	loss of activity:	
	5 days 8%	5 days 15%
Alk. Phosphatase	max. 7 days	loss of activity: 7 days 10%
Acid Phosphatase	when stabilized (5 mg NaHSO$_4$/ml) for 7 days: no loss of activity	
α-Amylase	for 5 days: no loss of activity	
Cholinesterase	for 7 days: no loss of activity	
Creatine kinase (NAC activated)	loss of activity:	
	7 days 2%	24 h 2%
CK-MB isoenzyme (NAC activated)	24 h <10%	1 h <10%
GlDH	loss of activity:	
	3 days 5%	3 days 15%
GOT (ASAT)	loss of activity:	
	3 days 8%	3 days 10%
GPT (ALAT)	loss of activity:	
	3 days 10%	3 days 17%
γ-GT	for 7 days: no loss of activity	
LAP	for 7 days: no loss of activity	
LDH	loss of activity:	
	3 days 8%	3 days 2%
LDH-1-isoenzyme (α-HBDH)	7 days 5%	7 days 5%
Lipase	for 5 days	for 24 h
	no loss of activity	

In conclusion, the stability of the different analytes should be established according to the conditions used. Problems arise when different analytes are to be determined in one sample. It is recommended that the sample should be analyzed as soon as possible: changes in composition may then remain within acceptable limits.

References

[1] *D. S. Young,* Biological Variability, in: Chemical Diagnosis of Disease, Elsevier/North Holland Biomedical Press, Amsterdam 1979, pp. 1 – 113.

[2] *W. G. Guder,* Einflußgrößen und Störfaktoren bei klinisch-chemischen Untersuchungen, Internist *21,* 533 – 543 (1980).

[3] *D. Stamm,* A New Concept for Quality Control of Clinical Laboratory Investigations in the Light of Clinical Requirements and Based on the Reference Matter to Values, J. Clin. Chem. Clin. Biochem., in press (1982).

[4] *W. G. Guder,* Standardisierung von Verfahren der Probennahme und Probenvorbereitung bei Enzymaktivitätsbestimmungen im Blut, Lab. Med. *5,* A + B 233 – 235 (1981).

[5] *U. Schmidt, U. C. Dubach,* Diagnostic Significance of Enzymes and Proteins in Urine, Huber, Bern (1979).

[6] *L. Hallmann,* Klinische Chemie und Mikroskopie, Thieme, Stuttgart 1979, pp. 142 – 213.

[7] *L. Hallmann,* Klinische Chemie und Mikroskopie, Thieme, Stuttgart 1979, pp. 214 – 220.

[8] *H. Keller,* Einflüsse auf klinisch-chemische Meßgrößen, in: *H. Lang, W. Rick, H. Büttner* (eds.), Validität Klinisch-chemischer Befunde, Springer, Heidelberg 1980, pp. 25 – 49.

[9] *H. Wisser, E. Knoll,* Tageszeitliche Änderungen klinisch-chemischer Meßgrößen, Ärztl. Lab. *28,* 99 – 108 (1982).

[10] *D. S. Young, L. C. Pestander, V. Gibbermann,* Effects of Drugs on Clinical Laboratory Tests, Clin. Chem. *21,* 1D – 423D (1975).

[11] *G. Balldin, A. Borgström, A. Eddeland, S. Genell, L. Hagberg, K. Ohlson,* Elevated Serum Levels of Pancreatic Secretory Proteins in Cigarette Smokers after Secretin Stimulation, J. Clin. Invest. *66,* 159 – 162 (1980).

[12] *B. E. Statland, P. Winkel,* Effects of Pre-analytical Factors on the Intraindividual Variation of Analytes in the Blood of Healthy Subjects. Consideration of Preparations of the Subject and Time of Venipuncture, Crit. Rev. Clin. Lab. Sci. *8,* 105 – 144 (1977).

[13] *B. Junge, H. Hoffmeister, H. M. Feddersen, L. Röcker,* Standardisierung der Blutentnahme, Dtsch. med. Wochenschr. *103,* 160 – 265 (1978).

[14] Standard Procedures for the Collection of Skin Puncture Blood Specimen, Natl. Com. Clin. Lab. Stand. (NCCLS), Vol. 2, No. 5, 132 – 149 (1982).

[15] Eur. Comm. for Clin. Lab. Stand. (ECCLS) *1,* 47 – 56 (1982).

[16] Standard Procedures for the Collection of Diagnostic Blood Specimens by Venipuncture, Natl. Comm. Clin. Lab. Stand. (NCCLS) ASH3 1 – 27 (1977).

[17] Probennahme, Probenvorbereitung, Probenverwahrung, Boehringer Mannheim 1981.

[18] Klinische Chemie. Eine Dokumentation der Deutschen Gesellschaft für Klin. Chemie, 1982.

[19] *W. Guder, J. Kruse-Jarres,* Is Glucose a Reliable Index of Carbohydrate Metabolism? J. Clin. Chem. Clin. Biochem. *20,* 135 – 140 (1982).

[20] *D. F. Davidson, I. R. Hainsworth, R. Rowan, G. Kousourou, M. Colgan,* Possible effect of bilirubin concentration on the in-vitro lability of creatine kinase during storage, Ann. Clin. Biochem. *18,* 185 – 186 (1981).

[21] *R. H. Laessig, J. O. Westgard, R. N. Carey, D. J. Hassemer, Th. Schwartz, D. H. Feldbruegge,* Assessment of a Serum Separator Device for Obtaining Serum Specimens Suitable for Clinical Analysis, Clin. Chem. *22,* 235 – 239 (1976).

[22] *C. Neuberg, E. Strauss, L. E. Lipkin,* Convenient Method for Deproteinization, Arch. Biochem. *4,* 101 – 104 (1944).

[23] *O. Folin, H. Wu,* A System of Blood Analysis, J. Biol. Chem. *38,* 81 – 110 (1919).

[24] *T. Greenwald,* The Estimation of Non-Protein Nitrogen in Blood, J. Biol. Chem. *34,* 97 – 101 (1918).

[25] *W. W. Beckman, A. Hiller, T. Shedlovsky, R. M. Archibald,* The Occurrence in Urine of a Protein Soluble in Trichloroacetic Acid, J. Biol. Chem. *148,* 247 – 248 (1943).

[26] *M. Somogyi,* Notes on Sugar Determination, J. Biol. Chem. *195,* 19 – 23 (1952).

[27] An FDA Medical Device Standards Publication, Technical Report: Proposed Performance Standard for in Vitro Diagnostic Devices Used in the Quantitative Measurement of Glucose in Serum of Plasma, May 1, 1980, US Dept. of Health, Education, and Welfare.

[28] *E. Bernt, H.-U. Bergmeyer,* Glukose-Bestimmungsmethoden, Ärztl. Lab. *13,* 472 – 475 (1967).
[29] *D. D. van Slike, A. Hiller, K. C. Berthelsen,* A Gasometric Micro Method for Determination of Iodates and Sulfates, and its Application to the Estimation of Total Base in Blood Serum, J. Biol. Chem. *74,* 659 – 675 (1927).
[30] *Z. E. Ball. J. F. Sadusk,* A Study of the Estimation of Sodium in Blood Serum, J. Biol. Chem. *113,* 661 – 665 (1936).
[31] *W. Bürgi, R. Richterich, M. L. Mittelholzer,* Der Einfluß der Enteiweißung auf die Resultate von Serum und Plasma-Analysen, Klin. Wochenschr. *45,* 83 – 86 (1967).
[32] *W. Bürgi,* The Volume Displacement Effect in Quantitative Analyses of Red Blood Cell Constituents, Z. Klin. Chem. Klin. Biochem. *7,* 458 – 460 (1969).
[33] *H. U. Bergmeyer, H. Möllering,* Enzymatische Bestimmung von Acetat, Biochem. Z. *344,* 167 – 189 (1966).
[34] *J. Freise, P. Magerstedt, E. Schmidt,* Enzymdiagnostik im lipämischen Serum vor und nach Polianionenpräzipitation mit Heparin und Magnesiumchlorid. J. Clin. Chem. Clin. Biochem. *15,* 485 – 488 (1977).
[35] *H. W. Voigt,* Klärung lipämischer Seren durch neues Verfahren, Laboratoriumsblätter *27,* 168 – 172 (1977).
[36] *H. U. Bergmeyer, E. Bernt, K. Gawehn, G. Michal,* The Sample: Stability of Metabolites and Enzymes in the Sample, in: *H. U. Bergmeyer* (ed.), Principles of Enzymatic Analysis, Verlag Chemie, Weinheim, New York 1978, pp. 111 – 114.
[37] *H. Hoffmeister, B. Junge,* Über die Haltbarkeit von Serumbestandteilen und die Zuverlässigkeit ihrer Bestimmung im AutoAnalyzer SMA 12/30 – Survey, J. Clin. Chem. Clin. Biochem. *8,* 613 – 617 (1970).
[38] *F. Knüchel,* Haltbarkeit von Fermenten und Substraten im Serum, Med. Welt *31,* 525 (1980).
[39] *H. U. Bergmeyer,* Standardization of Enzyme Assays, Clin. Chem. *18,* 1305 – 1311 (1972).
[40] *G. Szasz, W. Gerhardt, W. Gruber,* Creatine Kinase in Serum: 5. Effect of Thiols on Isoenzyme Activity during Storage at Various Temperatures, Clin. Chem. *24,* 1557 – 1563 (1978).
[41] *E. and. F. W. Schmidt,* Aspekte der Enzym-Diagnostik, Med. Welt *21,* 805 – 816 (1970).

1.1.2   Specimens and Samples in Food Chemistry, Cosmetics, Pharmacy

Günther Henniger and Hans Ulrich Bergmeyer

In general, the same rules are valid in the preparation and processing of samples in food chemistry, cosmetics and pharmacy (cf. e.g. [1, 2]) as in clinical laboratory sciences. Sample preparation is particularly simple for enzymatic analysis. Only a few sources of error exist. The fewer the number of steps during sample preparation, the smaller is the danger of introducing interfering substances or of partially removing the analyte [3].

Photometric methods are mainly used in enzymatic analysis, and, for these methods, almost clear and colourless sample solutions have to be prepared. Therefore, the material to be investigated has to be transformed into a suitable aqueous solution by special treatments which may consist of the following steps:

- disintegration
- homogenization
- extraction
- de-fatting
- de-gassing

- deproteinization
- decolorization
- filtration, centrifugation
- neutralization
- dilution

Sample preparation and sample processing are the first important steps in analysis. They must be carried out correctly to avoid erroneous results.

One should not forget that samples can change during storage, either

- by taking up water or by desiccation,
- by oxidation, or
- due to micro-organisms such as bacteria or fungi.

The stability of the sample solution is limited; storage in the frozen state may be effective.

1.1.2.1   Solid Specimens

Especially in food chemistry, the size of specimen is critical in many instances in obtaining representative results: e.g. it is inadequate to use 100 mg of black pudding containing large particles of meat, in this case several grams, at least, are necessary. On the other hand, 50 mg of a homogeneous sour cream are absolutely sufficient for analysis.

Insufficient homogeneity of the sample material is a serious source of error, because the recovery is unsatisfactory and the results are too high or too low. It has even been observed that, in sausages hanging in a butcher's shop for a period of time, the content of fat and water has shifted from bottom to top and from top to bottom, respectively, due to the force of gravity. Poor disintegration of plant cells yields results that are too low since the extraction of the analyte is incomplete (cf. chapter 1.2.4).

The treatment of a solid specimen yields in every case an extract consisting of a solution of the analyte. This sample solution must then be treated like a liquid specimen.

Disintegration, homogenization

Solid samples can be ground in a mortar, minced in a blender or ground in a mill. If the material is of soft consistency a mixer can be used, or even a homogenizer (cf. p. 33).

All these steps can be applied to the dry material, or in the presence of extraction media or reagents for simultaneous deproteinization.

Extraction

In general, extraction is performed with re-purified water in a volumetric flask. The extraction temperature depends on the specimen and the analyte: for measurement of volatile compounds, e.g. ethanol or acetic acid, extraction is performed at room temperature; if higher temperature is necessary, a condenser must be applied. For the extraction of pasty foodstuffs, paper or tobacco a 60 °C or 80 °C water bath or a heated magnetic stirrer is used. The extract can be used for analysis after removal of residues by filtration, centrifugation or by clarification with *Carrez*'s reagents. Extraction and clarification by *Carrez*'s reagents may also be combined (cf. p. 24).

For the preparation of samples containing fat the extraction temperature must be higher than the melting point of fat; otherwise the analyte may be located within fat particles (triglycerides) so that the extraction is incomplete. As examples, citrate is extracted from cremes with boiling water, glycerol from soap with boiling diluted hydrochloric acid, the fatty acids separating on the surface of the liquid in the cold.

In some cases treatment with organic solvents is indicated; e.g. for the determination of esters in emulgators the sample is extracted with chloroform.

Examples of application: meat products, wheat germs, bread, cakes, candies, chocolate, jam, coffee, tea, tobacco.

De-fatting

Fats of the sample are already removed in most cases during filtration of the cold extract. If the solid sample contains high amounts of fat extraction in the Soxhlet apparatus is recommended. Care must be taken in the differentiation between analytes soluble in fat or insoluble in fat.

1.1.2.2 Liquid Specimens

The goal of pre-treatment of liquid specimens is to achieve a reliable measuring signal. For photometry, it is necessary to get as clear and colourless solutions as possible. Nevertheless, even slightly turbid or opalescent solutions as well as slightly coloured ones can be used, if the initial absorption allows proper measurements.

Dilution, neutralization, de-gassing

Practically clear and colourless specimens with nearly neutral pH can be analyzed directly without pre-treatment.

It is often necessary to dilute the specimen to get a useful signal in photometric measurement. The degree of dilution of the liquid specimen depends on the concentration of the analyte. Its concentration in the assay mixture should be high enough to yield a measured signal between $\Delta A = 0.050$ and $\Delta A = 0.800$. This corresponds in NAD(P) dependent reactions to a concentration of the analyte in the diluted sample of approximately 0.25 to 3.5 mmol/l or, for a molecular weight of 100, approximately 25 to 350 mg/l, measured at 339 nm.

Dilution is performed with water, buffer, or a salt solution. The sample is pipetted into a 100 ml volumetric flask containing some diluent and the volume is made up to 100 ml with diluent.

In Table 1 the upper limits of the concentration of various analytes are presented for undiluted specimens.

If acidic samples such as acetic acid or lemon juice are used undiluted they must be neutralized, because the buffer capacity of the assay system might not be sufficient to maintain the optimum pH necessary for the enzyme reaction. Sodium or potassium hydroxide is recommended for this purpose. If buffers are used, the same buffer as that used in the assay mixture should be chosen since the enzyme activity may be severely impaired by certain ions; e.g., β-galactosidase exhibits nearly no activity in Tris or triethanolamine buffer, so that the determination of lactose is not possible in presence of these buffers.

Table 1. Upper limits of mass concentration* of various analytes measurable in undiluted samples (in order of increasing concentrations).

Analyte	up to g/l	Analyte	up to g/l
Ethanol	0.06	Cholesterol	0.4
Glutamic acid	0.07	Starch	0.4
Ammonia	0.1	Glycerol	0.5
Sorbitol, xylitol	0.1	Glucose + fructose,	
Urea	0.15	Lactose + galactose, isocitric acid	0.5
Ascorbic acid	0.2	Gluconic acid	0.6
Malic, lactic acid	0.2	Saccharose + glucose	0.8
Acetaldehyde, formic acid	0.1	Lecithin	1.0
Acetic acid	0.15	Raffinose	1.3

* For measurements at Hg 365 nm approx. twice these concentrations can be measured.

Examples of application: clear fruit and vegetable juices, wine, after-shave lotion, Eau de Cologne, infusion solutions.

Beverages containing carbon dioxide, e.g. beer, champagne, lemonades, must be freed from gas in order to be able to pipet the sample properly. CO_2 can be bound by alkalization of the sample. The simplest way is filtration through filter paper.

Deproteinization, de-fatting

The solubility of proteins depends on the pH and ionic strength of the solution among other factors. Thus, clear sample solutions, even in the smallest quantities, can cause turbidity when they are introduced into the assay mixture. Furthermore, when plant or animal tissues are extracted with non-deproteinizing agents, enzymes will also be extracted which may interfere with the enzymatic assay procedure. Specimens containing protein must be deproteinized even if clear liquid specimens appear to contain only small amounts of protein.

Most conveniently perchloric acid is used; after centrifugation the clear supernatant is neutralized with KOH which causes nearly complete removal of $HClO_4$ precipitating as $KClO_4$ in an ice bath. Emulsions such as milk can be transformed into a solution by this method of deproteinization.

Perchloric acid deproteinization cannot be applied to acid labile substances such as saccharose, maltose and other disaccharides. In these cases clarification by *Carrez*'s reagents is recommended (see below).

Trichloroacetic acid is less appropriate for deproteinization because its excess must be removed with ether, unless such small concentrations are chosen that no interference with the assay system will occur. In this case, however, the sample solutions are not always clear.

De-fatting of liquid solutions directly or after deproteinization can be achieved by cooling the extract or liquid sample in an ice-bath for 20 minutes and filtering off the cold solution. For high amounts of fat see p. 22.

Decolorization, clarification

With liquid and coloured specimens, the dyes may absorb light at the measuring wavelength, thus exceeding the optimal measuring range of the photometer.

If the concentration of the analyte is such that the sample can be diluted to a certain extent, no photometric problems arise, so that it is unnecessary to decolorize the sample. If the concentration of the analyte is too low the sample solution must be decolorized, e.g. by removal of the dye with a suitable adsorber. The choice of adsorber depends on the dye to be adsorbed and the substance to be determined. The dyes must be adsorbed as completely as possible without simultaneously adsorbing the analyte.

Commercially available adsorbers such as polyamide, gelatine, bentonite or polyvinylpyrrolidone can be used. Charcoal is not suitable in many cases. The adsorber usually is stirred into the liquid sample and the suspension is filtered. However, these procedures must be standardized with respect to amount of adsorber (e.g. 1%) and time of action (e.g. 1 min), for each single analyte to avoid its removal from the solution. Appropriate conditions are determined by recovery experiments.

In many cases inhibitors such as tannins in wine are also removed in the decolorization step. This means that there are no further interferences in the enzymatic determination after decolorization.

Examples of application: Analysis of red wine, fruit juices, blackberries, red beets.

Decolorization and clarification by *Carrez*'s reagent is routinely applied in sugar analysis for the removal of turbidities, dyestuffs and proteins, and for breaking of emulsions.

Procedure:
Carrez's solution I. $K_4[Fe(CN)_6] \cdot 3 H_2O$, 85 mmol/l = 3.60 g/100 ml.
Carrez's solution II. $ZnSO_4 \cdot 7 H_2O$, 250 mmol/l = 7.20 g/100 ml.
In a 100 ml volumetric flask add to the liquid sample 5 ml *Carrez*'s I solution, shake vigorously, add 5 ml *Carrez*'s II solution, shake vigorously. The precipitate of $Zn_2[Fe(CN)_6]$ adsorbs dispersed and coloured compounds as well as proteins. Add 10 ml NaOH (0.1 mol/l) for precipitation of exceeding Zn^{2+} ions. Shake vigorously, adjust with water at 100 ml, mix and filter.

Examples of application: turbid juices, desserts, lotions, extracts of foodstuff, cosmetics and tobacco.

1.1.2.3   Special Techniques

Special techniques [4] must be used in sample preparation in some cases, depending e.g. on the analyte, on the sample, and on the assay method applied.

Ascorbic acid: due to the instability of ascorbic acid, especially in solutions, m-phosphoric acid, pH 3.5 − 4.0, must be used as a stabilizer for extraction and as solvent [5]. The pH value of 3.5 − 4.0 is very important for the assay procedure.

Starch is a high molecular weight compound and insoluble in water. To dissolve starch a solution of dimethylsulphoxide and HCl is used [6]; there is no decomposition during sample preparation.

Lecithin: this substance also is insoluble in water. t-Butanol can be used as solvent, or a suspension can be prepared by ultrasonication. In some other cases the test principle requires a special sample preparation [7].

Glutamic acid or ascorbic acid, and sorbitol/xylitol: reducing substances interfere with the indicator reaction usually yielding a formazan. The reducing substances can be removed e.g. be H_2O_2 in alkaline solution [4].

Saccharose and fructose: a large excess of glucose (more than 5 − 10 times higher than the sucrose or fructose concentration) is responsible for a bad precision of the measurement. In these cases it is recommended to oxidize glucose with glucose oxidase and to remove the H_2O_2 formed in the reaction by catalase [4].

Esters, e.g. of lactic acid, citric acid, acetic acid and glycerol must be saponified with alkaline solutions in the analysis of emulsifiers. The hydrolysis is carried out, e.g., with alcoholic potassium hydroxide solution by boiling under a reflux condenser.

References

[1] *G. Henniger,* Enzymatische Lebensmittelanalytik, Z. Lebensmittel-Technologie und -Verfahrenstechnik *30,* 137 − 144 and 182 − 185 (1979).
[2] *G. Henniger, H. Boos,* Anwendung der enzymatischen Analyse bei der Untersuchung kosmetischer Präparate − dargestellt an einigen Beispielen, Seifen-Öle-Fette-Wachse *104,* 159 − 164 (1979).

[3] *P. Kohler,* Probenvorbehandlung – Überprüfung des Testansatzes, Alimenta *20*, 3 – 6 (1981).
[4] Methods of Enzymatic Food Analysis 1980, Boehringer Mannheim, D-6800 Mannheim 31.
[5] *G. Henniger,* Enzymatische Bestimmung von L-Ascorbinsäure in Lebensmitteln, Pharmazeutika und biologischen Flüssigkeiten, Alimenta *20*, 12 – 14 (1981).
[6] *H.-O. Beutler,* Enzymatische Bestimmung von Stärke in Lebensmitteln mit Hilfe der Hexokinase-Methode, Starch/Stärke *30*, 309 – 312 (1978).
[7] *H.-O. Beutler, G. Henniger,* Enzymatische Bestimmung von Lecithin, Swiss Food *3*, 27 – 29 (1981).

1.2   Cell and Tissue Disintegration

1.2.1   General*

Benno Hess and Karl Brand

One aim of enzymatic analysis is to obtain information on the concentration and localization of metabolites in the living cell**. It is therefore desirable to prepare the living material in a form suitable for the measurements, but without altering the structure or the relative amounts of the substances to be analyzed. However, two difficulties stand in the way of this ideal situation:

1. the destruction of the physiological state is accompanied by a change in the physicochemical state and the concentration of the metabolites ("operational isomers"),
2. when a tissue is disintegrated the compartmental concentration gradients may be disturbed (non-linear change of the compartmental conditions).

These two complications are inter-dependent and in consequence the analytical results provide a more or less distorted picture of the conditions in the living cell.

Every assay of metabolites in biological material is therefore an "operational test", i.e. the results are influenced by the property tested for and by the method used. A comparison of the results of different workers is only possible if the methods used are well defined.

Apart from these main obstacles, it is important to consider the possible instability of the metabolites during the disintegration of biological material. They may be acid-

---

* For general and detailed discussions of the subject, see [1 – 3].
** For a review and definition of the terms concentration, reacting concentration, content and level, activity of metabolites, cellular concentration, tissue concentration, status *in vivo* and altered status, see [4 – 6].

labile (ATP, ADP, NADH, NADPH), alkali-labile (triose phosphate, NAD, NADP), oxidizable or easily denaturated (proteins). They may also be transformed enzymatically during the relatively long disintegration process. For example, the cooling of guinea pig kidney from 38 °C to 0 °C with the "quick-freeze" tongs takes 90 ms [7]. However, 22 ms is sufficient for a 10% change in the steady state concentration of FADH and FAD in intact ascites tumour cells [8]. Usually the steady state levels of low molecular weight metabolites undergo an exothermic shift during the disintegration towards the true position of equilibrium. Anaerobiosis, aerobiosis and dilution may be contributory factors in changing the amounts of substrate (e.g. displacement of the aldolase equilibrium, displacement of the ATP/ADP or lactate/pyruvate ratios on disintegration with acid; decomposition of enzyme-substrate complexes).

The disintegration conditions must therefore be governed by two considerations:

1. preservation of the chemical structure of the metabolites. For this, the temperature, pH, ionic strength, time, etc., used in the fixation and extraction of the tissue must be compatible with the decay constants and the biological half-life times of the compounds to be determined,

2. preservation of the localization of the compounds in the tissue and cell compartments. The tissue is divided according to its morphological characteristics; i.e. into structures visible with the light and electron microscopes.

Generally, distinction is made between disintegration methods used for the assay of high molecular weight compounds (e.g. enzymes) and those used for low molecular weight metabolites. If enzyme activity is to be determined, the speed with which the tissue is fixed, disintegrated and extracted is less important. To preserve the structure of enzyme proteins the extraction is carried out at low temperature with distilled water, salt solutions [1], sucrose [1], dextran or albumin solutions. To avoid denaturation of protein by the acid produced in tissue with high glycolytic rates, the solutions used for extraction should be buffered at a suitable pH. The addition of chelating agents to prevent oxidation, or of cysteine to protect SH groups, is also recommended. The tissue is disintegrated by careful homogenization, sometimes with the addition of compounds which act on the membranes (e.g. digitonin, proteinases) leading to a solubilization of cytosolic enzymes. In contrast, the disintegration of tissue for the analysis of low molecular weight metabolites requires the fixation of the cellular state within seconds or even milliseconds. For this, the sample is frozen and/or metabolic inhibitors or denaturing agents are used.

Rapid cooling (as in the "quick-freeze" method) brings about nearly ideal fixation of the tissue. It has the advantage that more time is available for the subsequent operations (deproteinization, extraction). Direct spectrophotometric measurements on deep-frozen tissue samples (e.g. in liquid air) are particularly elegant. Inhibitors prevent certain metabolic reactions and so fix the concentrations of the metabolites (inhibition of glycolysis by addition of fluoride, iodoacetic acid, cyanide or hydrazine; production of anaerobic conditions by the addition of hydrocyanic acid, sodium azide, or sodium sulphide or by gassing with nitrogen).

Deproteinizing agents include: acids (trichloroacetic acid, perchloric acid, etc.), alkalis and organic solvents. Metabolic processes can also be stopped by rapid heating (e.g. exposing tissue or cells to hot alcohol).

Generally, it is best to use deproteinizing agents, metabolic inhibitors and fixation by low temperature in combination with one another and in conjunction with the extraction or homogenization of the tissue.

Occasionally it is possible to make use of the lability of certain metabolites. For example, the instability of NAD in alkali is utilized to destroy this coenzyme in a mixture of NAD and NADH. Triose phosphates are hydrolyzed by alkali and the phosphate liberated can be determined. NADH is sensitive to acid. The acid degradation of NADPH gives phosphorylated ADP-ribose, which can be detected chromatographically.

If the metabolites or enzyme proteins are stable or can be stabilized artificially, the tissue can be processed with the cell topography in mind. The cell components are separated from one another by homogenization and subsequent differential centrifugation (perhaps in a continuous flow centrifuge). Each fraction is then analyzed.

When a tissue is being quantitatively processed it is important to check the content of the substance being assayed after every step. This type of check should show any degradative processes or changes in the limits of the cell compartments. Stable properties of biological material such as dry or fresh weight [1], cell count, haematocrit, DNA content or protein content serve as reference standards. The concentrations of stable enzymes can also serve as reference standards (cytochrome a in mitochondria or intact cells [8], glyceraldehyde-3-phosphate dehydrogenase for glycolytic enzymes and soluble cytoplasm [9].

The extracellular fluid mixes with the cell contents on extraction and therefore distorts the true concentration ratios. The measured values must therefore be corrected for the volume and the metabolite concentration of the extracellular fluid. To correct for blood content the oxyhaemoglobin concentration is determined [6, 10, 11]. The size of the intercellular space can be calculated from a chloride determination. This calculation is based on the assumption that the intracellular chloride concentration is low, while the chloride concentrations in the intercellular tissue fluid and plasma are roughly equal [6]. The distribution of metabolites or other compounds in the intracellular compartments is obtained from the analysis of the cell fractions. The amount of protein, DNA, or the amount of cytochrome c or a (in mitochondria) as well as the weight of the tissue extracted can serve as reference standards for the calculation of the metabolite concentrations. The size of the intracellular space cannot be determined exactly. Usually a particular method of tissue disintegration cannot satisfy all the requirements. Only a few methods are optimal. To improve the outcome, it is better to use several complementary methods of extraction for comparison. In view of the heterogeneity of biological structures practically every tissue requires a different treatment, even though the same compound is being estimated.

Special care is required in the treatment of the organisms whose tissue is to be examined. The nutritional and functional state of experimental animals must be considered. Any periods of hunger, thirst, darkness or exercise (continuous or at regular intervals) before the experiment should be reported, as well as the conditions of narcosis and the removal of blood (arterial or venous, with or without pressure) or organs.

The experimental result can naturally only give the value for what is present in the assay material at the time of analysis. Despite strict observance of correct techniques, however, experimental results for biological material are not always reproducible. The quantitative and topographic compositions of some cells and tissues exhibit significant variations with the time of day and even with the time of year; sexual and racial differences have also been observed, as well as relationships with the body weight (refer e. g. to [12 – 15]). This is true both of metabolites (e. g. glycogen) and of enzyme activities (e. g. esterases, aminotransferases).

In animal experiments, the fact that anaesthesia has been carried out and the anaesthetic used are as important as the manner in which the animals are sacrified. According to *Tarnowski* et al. [16, 17] the physiological metabolite content of rat liver may be drastically changed both by the degree of excitation of the animals and by the method of killing. The depth and duration of anaesthesia and the nature of the anaesthetic may falsify the result of the assay [17]. These findings no doubt also apply to other types of animals and other organs. Anoxia of the tissue, even for a short time, changes in particular the concentrations of the energy-rich nucleotides and their degradation products as well as the concentrations of the redox couples and other labile metabolic intermediates [17]. Instead of collecting the tissue *in situ* with freeze-clamps, the excised organs or even whole animals are still often frozen in liquid nitrogen or liquid air. This method should be avoided in the measurement of metabolite contents because the tissue samples are insulated for 30 s or more by the gas envelope of the evaporating freezing agent and thus prevented from complete freezing.

With growing knowledge of these relationships, there is an increasing awareness of the pressing need to standardize the conditions under which biological samples are collected and/or to state these conditions when the experimental results are reported.

References

[1] *K. A. C. Elliott,* Tissue Slice Technique, in: *S. P. Colowick, N. O. Kaplan* (eds.), Methods in Enzymology, Vol. *I*, Academic Press, New York 1955, pp. 3 – 9.
[2] *P. P. Cohen, H. Beinert,* Methods of Preparing Animal Tissues, in: *W. W. Umbreit, R. H. Burris, J. F. Stauffer* (eds.), Manometric Techniques, 3rd edit., Burgess Publishing Co., Minneapolis 1959, pp. 135 – 150.
[3] *H. A. Krebs, W. Bartley, D. E. Griffiths, L. A. Stocken,* Tissue Preparations in Vitro, in: *H. M. Rauen* (ed.), Biochemisches Taschenbuch, Vol. *II*, Springer-Verlag, Heidelberg 1964, pp. 522 – 541.
[4] *Th. Bücher, M. Klingenberg,* Wege des Wasserstoffs in der lebendigen Organisation, Angew. Chem. *70*, 552 – 570 (1958).
[5] *B. Hess, B. Chance,* Über zelluläre Regulationsmechanismen und ihr mathematisches Modell, Naturwissenschaften *46*, 248 – 257 (1959).
[6] *H. J. Hohorst, F. H. Kreutz, Th. Bücher,* Über Metabolitgehalte und Metabolit-Konzentrationen in der Leber der Ratte, Biochem. Z. *332*, 18 – 46 (1959).

[7] *A. Wollenberger, O. Ristau, G. Schoffa,* Eine einfache Technik der extrem schnellen Abkühlung größerer Gewebestücke, Pflügers Archiv ges. Physiol. Menschen, Tiere *270,* 399 – 412 (1960).
[8] *B. Chance, B. Hess,* Metabolic Control Mechanisms, J. Biol. Chem. *234,* 2404 – 2412 (1959).
[9] *W. Vogell, F. R. Bishai, Th. Bücher, M. Klingenberg, D. Pette, E. Zebe,* Über strukturelle und enzymatische Muster in Muskeln von Locusta migratoria, Biochem. Z. *332,* 81 – 117 (1959).
[10] *B. Chance,* The Kinetics and Inhibition of Cytochrome Components of Succinic Oxidase System, J. Biol. Chem. *197,* 557 – 565 (1952).
[11] *H. Holzer, G. Sedlmayr, M. Kiese,* Bestimmung des Blutgehaltes von Leberproben zur Korrektur biochemischer Analysen, Biochem. Z. *328,* 176 – 186 (1956).
[12] *P. Yap,* The Cellular Aspects of Biorhythms; VIIIth Intern. Congr. of Anatomy, Wiesbaden 1965, p. 143.
[13] *H. v. Mayersbach,* Biochemisch und histochemisch faßbare Zirkadianschwankungen in Organen standardisierter Versuchstiere, Verh. dtsch. Ges. inn. Med. 73. Kongr. 1967, p. 942 – 960.
[14] *R. Leske, H. v. Mayersbach,* The Role of Histochemical and Biochemical Preparation Methods for the Detection of Glycogen, J. Histochem. Cytochem. *17,* 527 – 538 (1969).
[15] *H. v. Mayersbach,* Die Bedeutung normaler biologischer Faktoren für die topochemische Zell- und Gewerbeforschung, Hoppe-Seyler's Z. physiol. Chem. *350,* 1169 – 1170 (1969).
[16] *R. P. Faupel, H. J. Seitz, W. Tarnowski, V. Thiemann, Ch. Weiss,* The Problem of Tissue Sampling from Experimental Animals with Respect to Freezing Technique, Anoxia, Stress and Narcosis, Arch. Biochem. Biophys. *148,* 509 – 522 (1972).
[17] *H. J. Seitz, R. P. Faupel, C. S. Kampf, W. Tarnowski,* Influence of Different Types of Narcosis and of Neck Fracture on the Concentration of Glycolytic Intermediates and Related Compounds in Rat Liver, Arch. Biochem. Biophys. *158,* 12 – 18 (1973).

1.2.2   Methods for Animal Tissue

Karl Brand and Benno Hess

1.2.2.1   Surviving Tissue

The metabolic function of organs can be studied with a spectrum of organ preparations ranging from the intact organ *in vivo,* through perfusion systems, tissue slices, isolated cells, homogenates, cell fractions to purified enzymes. Each preparation has its special advantages and disadvantages. The principal advantage of the perfused organ as a complex metabolic model resides in the possibility of investigating steady states in an open metabolic system, in contrast to closed metabolic systems such as isolated cell incubations. On the other hand the latter are more suitable e.g. for following changes of intracellular metabolite concentrations or in studying hormone interactions with receptors and transport processes.

Tissue slices

Tissue slices are organized, surviving tissue without a blood supply, but allowing free diffusion of oxygen and a variety of metabolites. The use of tissue slices was introduced by *O. Warburg* [1] as a method to study *in vitro* metabolism of tissue. They are usually prepared with a microtome (for detailed description see [2]). The critical thickness of a tissue slice for the diffusion of oxygen is 0.2 mm (for the calculation of the critical thickness see [2, 3]). The free diffusion of metabolites varies and can be limiting (e.g. sodium glutamate) [4]. Tissue slices are suitable for the assay of enzymes (directly or after fractional extraction), for the determination of metabolites after deproteinization and for metabolic studies [5].

Types of instrument: Stadie-Riggs tissue slicer [6], Tissue slicer from *Arthur H. Thomas Co.,* Philadelphia, USA.

Heterogeneous cell suspensions

Heterogeneous cell suspensions, such as blood, exudates, transudates (ascites, pleural fluids, etc.) or cell cultures, are best collected with a syringe, preferably under paraffin. In the collection of venous blood the effect of stasis should be taken into account. For the rapid deproteinization of blood, it should be led directly from the vein (or by use of a syringe) into a weighed centrifuge tube containing ice-cold perchloric acid.

Haemolysis of erythrocytes with digitonin [7] and hypotonic haemolysis have become popular; rupture of the cells also occurs when they are cooled to $-90\,°C$ (e.g. in acetone-dry ice). Heparin, oxalate, citrate, EDTA or siliconized glassware is used to prevent the coagulation of blood. The addition of metabolic inhibitors such as fluoride or iodoacetate to inhibit glycolysis is also recommended. The effect of these substances on enzyme activities and metabolite concentrations must be checked.

Homogeneous cell suspensions

Suspensions of isolated cells from various organs and tissues are now being used in an increasing number of biochemical and pharmacological investigations, including studies on metabolic regulation, hormone action, drug metabolism, biosynthetic processes and many aspects of differentiation and ageing.

Although most of the techniques currently employed to obtain isolated cells involve perfusion of the organ with digestive enzymes such as collagenase and hyaluronidase, attempts have also been made to avoid the perfusion step and achieve cell isolation by incubating cut pieces e.g. of the liver in enzyme containing solutions [8, 9].

In Table 1 a list of references is given for approved methods to isolate viable cells from various organs and tissues.

1.2.2.2   Tissue Fixation

To fix tissue rapidly, flat tongs with aluminium jaws are used, which can be pre-cooled with liquid nitrogen ("quick-freeze" tongs or clamps [27, 28]). The tissue gripped between the cold jaws is compressed to form a thin layer, and protruding tissue is broken off. A test of this method showed that 1.6 g of guinea pig kidney was

Table 1. List of References for Approved Methods to Isolate Cells from Various Organs and Tissues

Type of Cells	Author(s)	Reference
Hepatocytes	*M. N. Berry, D. S. Friend*	10
	P. Moldèus et al.	11
	J. R. Fry et al.	9
Kupffer Cells	*D. L. Knook, E. Ch. Sleyster*	12
Renal Cells	*D. P. Jones et al.*	13
Kidney tubules	*P. Vinay et al.*	14
	W. G. Guder	15
Adipocytes	*M. Rodbell*	16
Lung cells	*Th. R. Devereux, J. R. Fouts*	17
Enterocytes	*M. M. Weiser*	18
	J. W. Porteous et al.	19
Neurons and glial cells	*S. Sinha, S. P. R. Rose*	20
Cerebral neurons	*W. B. Huttner et al.*	21
Pancreatic islets	*D. W. Scharp et al.*	22
B-Lymphocytes	*F. Indiveri et al.*	23
Thymocytes	*J. Segal, S. H. Ingbar*	24
	C. A. Pupar, G. M. W. Cook	25
Macrophages	*R. T. Dean*	26

cooled from 38 to $0\,°C$ in 0.09 s, from 38 to $-40\,°C$ in 0.15 s, and to $-160\,°C$ in 0.5 s [28]. The tissue is compressed to a layer 0.7 mm thick. The method is particularly suitable for fixing tissue containing labile metabolites (liver, kidney, brain, muscle, nerves [29]).

Procedure according to [27]:

Grip exposed liver of liver lobes with two forceps, lift up slightly and press between two light metal blocks (provided with wooden handles; block dimensions 50 mm × 20 mm × 8 mm) which have been pre-cooled in liquid air. Separate the frozen piece of liver from the excess tissue with a pair of scissors, break off any pieces of incompletely frozen tissue protruding over the edges of the metal blocks and immerse the tissue contained in the blocks in liquid air. With frequent additions of liquid air, powder the piece of liver in a metal or porcelain mortar pre-cooled with liquid air. The tissue should not be allowed to thaw during the process. It is therefore best to work in a cold room. Also the absorption of water by the cold tissue, which is considerable in the highly humid air of laboratories at normal temperature, is minimized by working in a cold room. Add a portion of the tissue to a weighed amount of ice-cold 6% perchloric acid (about 1 g tissue to 5 ml $HClO_4$) and grind rapidly with a glass pestle. The tissue powder must not be allowed to remain on the walls of the vessel during its addition to the perchloric acid. Weigh the vessel containing the tissue suspension, homogenize the contents for 30 s and centrifuge for 4 min in the cold at $2800\,g$. Decant the supernatant, re-extract the sediment with 3% perchloric acid and centrifuge again. Combine the supernatants.

The authors [27] state that powdering the tissue and extracting the powder is sufficient for the quantitative recovery of metabolites, and that with five metabolites, subsequent homogenization of the final sediment in an Ultra-Turrax (see 1.2.2.3) resulted in no increase in the yields. In spite of this, re-homogenization was advised to ensure the quantitative extraction of metabolites.

Types of instrument: "Quick-freeze" tongs according to *Wollenberger et al.* [28]. "Quick-freeze" tongs according to *Hohorst et al.* [27].

A procedure has been described [30] for the isolation of peptides and proteins from animal tissue that greatly reduces the possibility of proteolytic digestion. In principle, pieces of frozen tissue are pulverized with dry ice in a *Waring* Blender. The still frozen powder is added to a solution of guanidinium chloride, 6 mol/l, the suspension returned to the blender and blended at high speed to shear DNA. Nucleic acids are removed by the addition of 0.5 volumes of ethanol and centrifugation. The supernatant containing the peptides is ultrafiltered successively through *Whatman* No 541 and *Whatman* No 1 filters and pumped through a hollow fibre concentration system (*Amicon* DC 2, HP 10 Cartridge) having a cut-off for substances with a molecular weight of 25 000.

1.2.2.3   Homogenates

Table 2 gives a survey of methods of homogenization. It is followed by a detailed description of the various procedures and types of instrument.

Table 2. Methods for preparation of homogenates and for the disintegration of cells

Type of homogenate	Method	Tissue
Wet homogenates	Mechanical Homogenization: Pestle homogenizer	parenchymatous tissue, brain, heart
	Grinding with sand or glass beads	muscle, frozen tissue
	Glass bead homogenizer	ascites tumour cells, muscle
	Blade homogenizer (blender)	muscle
	Sonic homogenization	animal tissue and cells
	Thermal disintegration freezing and thawing [freezing mixtures (acetone-CO_2; alcohol-CO_2); liquid air; liquid nitrogen]	universal application
	Chemical disintegration with isopentanol; butanol; digitonin; petroleum ether	erythrocytes, mitochondria, muscle
	Biological-enzymatic disintegration autolysis, maceration; bacterial proteases	muscle, heart
Dry homogenates	homogenization by dehydration acetone; lyophilization	various tissues

Wet homogenates

Mechanical homogenization

Pestle homogenizer according to *Potter* and *Elvehjem* [31]

At present the most widely used instrument for the homogenization of tissue (particularly parenchymatous tissue) is the pestle homogenizer. It consists of a tight-fitting pestle (free play ca. 0.2 mm) made of glass or plastic, which is rapidly rotated by a stirrer motor in a thick-walled test tube also made of glass or plastic (see Fig. 1). The tissue is cut up into small pieces with scissors and suspended in a medium. This suspension is poured into the test tube and then by rapidly rotating the pestle (about 1000 revolutions per min) the tissue is homogenized in a few minutes. Fresh tissue (1 − 2 g) is suspended in 8 − 10 ml water to give a 1 : 10 dilution. Usually the test tube is pushed up and down the rotating pestle by hand (see Fig. 1). The test tube is cooled to prevent over-heating during the homogenization. The extent of the homogenization can be varied by using either a loose (e.g. for kidney) or tight-fitting pestle. The use of pestle homogenization includes the preparation of tissue for the subsequent separation of cell fractions by differential centrifugation (cf. part 1.2.2.4).

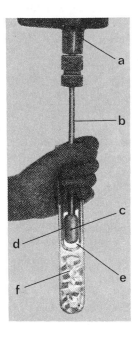

Fig. 1. Pestle homogenizer
a: Stirrer motor
b: Steel shaft
c: Pestle
d: Thick-walled test tube
e: Homogenate
f: Cooling jacket.

Types of instrument: Pestle homogenizers are available in various sizes (5 − 50 ml) and in Pyrex glass or Teflon. The pestles may be of various shapes.

Manufacturers: *Arthur H. Thomas Comp.,* Philadelphia 5, Pa., USA; *B. Braun Melsungen AG,* D-3508 Melsungen; *Glenco Scientific,* Chicago, Ill., USA; IKA-

Rührwerk RW 20 or RW 20 DZM, *Janke and Kunkel GmbH & Co. KG,* D-7813 Staufen.

Grinding in a mortar

Grinding biological material in a cooled mortar with sand, alumina, diatomaceous earth or glass powder is an important preparative method, which may be followed by homogenization. The disintegration of heart muscle for the isolation of sarcosomes by grinding with alumina or sand has proved successful [32, 33]. Pulverization in a metal or porcelain mortar is the method of choice for disintegration of pieces of tissue [27] which have been frozen with liquid air or liquid nitrogen.

Glass bead homogenizers

A tissue or cell sample is disintegrated by mechanical vibration, rotation or shaking in the presence of glass beads. The method is particularly suitable for types of cells which are resistant to disintegration, such as bacteria, yeast, ascites tumour cells, etc. By variation of the diameter of the glass beads and the frequency of the vibration, rotation or shaking, the degree of homogenization can be varied.
    Types of instrument: Vibrogen Cell Mill Vi4 − *E. Bühler,* D-7400 Tübingen; Micro-Dismembrator − *B. Braun Melsungen AG,* D-3508 Melsungen; MSK Homogenizer: Glass bead homogenizer according to *Merkenschlager, Schlossmann & Kurz* [34] (rotation principle). 75 ml Duran glass containers, which can hold, e.g. 20 g yeast (fresh weight). Rotation speed: 2000 − 4000 rpm. Cooling: liquid carbon dioxide (the temperature of the material remains below 5 °C). Time of homogenization: 30 − 120 s for complete homogenization and preparation of cell-free extracts. With shorter homogenization times (10 s) the mitochondria remain intact. Manufacturer: *B. Braun Melsungen AG,* D-3508 Melsungen. The glass beads must be washed with dilute nitric or hydrochloric acid, then rinsed repeatedly with distilled water. Diameter of the glass beads: 0.1 mm for bacteria; 0.5 mm for yeast and ascites tumour cells. Manufacturer: Superbrite Glass Beads, *Minnesota Mining and Manufacturing Comp.,* St. Paul 6, Minn., USA; Ballotini beads, *English Glass Co. Ltd.,* Leicester, England.

Blade homogenizer (Blender)

The tissue is disintegrated in a short time by rapidly rotating blades (10000 − 40000 rpm), resulting in a fine brei. The method is unsuitable for cell fractionation since the intracellular elements, such as mitochondria, are destroyed. It is suitable, however, for the isolation and purification of enzymes and for the assay of stable metabolites after further manipulations e.g. sonication and extraction.
    Types of instruments: Bottom drive homogenizers: for the preparation of large amounts of tissue (100 g to 1 kg). Usually available with several attachments of different sizes. Speed: 6000 to 12000 rpm. Manufacturers and trade names: *Waring Comp.,* New Hartford, Con. USA − *Waring* Commercial Blendor.

Top drive homogenizers: for the preparation of small amounts of tissue (up to 100 g). Available with attachments up to 100 ml. Variable speeds, with some types up to 50000 rpm. Some instruments are available with cooling jackets. Trade names and manufactures: IKA-Ultra-Turrax T18/10 or T45 — *Janke and Kunkel GmbH & Co. KG,* D-7813 Staufen; Bühler-Homogenizer — *E. Bühler,* D-7400 Tübingen; Sorvall-Omnimixer — *Du Pont,* Willington, Del., USA; Polytron Models PT 7, PT 10 S or PT 20 S — *Kinematica,* CH-6010 Kriens/Luzern.

Sonic homogenization

The pressure changes of several thousand atmospheres caused by ultrasonics break cell membranes and cell walls. Acoustic methods are therefore suitable for the preparation of homogenates. Animal tissues (spleen, liver, ascites tumour cells, erythrocytes, kidney, Hela tumour cells, thymus and lymph nodes) can be disintegrated by sonication for $10-90$ s). Many enzymes are liberated. Mitochondria are destroyed. It is usually necessary to cool during the disintegration.

Procedure: prepare a 5% cell or tissue sespension in potassium phosphate buffer (0.1 mol/l, pH 7.5). Stir this suspension for 10 min at 0°C before the sonication to obtain an even suspension of cells, so that no air-containing aggregations can pass into the sonication vessel. Sonicate the suspension in $10-15$ ml portions. Frequency: 19.5 kc/s; output: about 50 ± 5 watts/cm^2. Cool the sonicator head to between -5 and -8°C during sonication at maximum output, with 70% ethanol as the refrigerant, so that the sonicated fluid has a final temperature of 1 to 3°C. At lower intensities of sonication, cool just sufficiently to prevent the suspension from freezing.

Types of instrument: Branson Sonifier B-12 or B-15 — *Branson Sonic Power,* Danburry, Conn., USA; Labsonic 1510 — *B. Braun Melsungen AG,* D-3508 Melsungen.

Thermal disintegration

Repeated freezing and thawing is a successful means of disintegrating tissue and intact cells, particularly erythrocytes and bacteria. As a general rule [35] slow freezing leads first to the intercellular formation of ice nuclei. On more rapid freezing intracellular ice crystals are also formed and these destroy the intracellular structure. On thawing, the cells are ruptured osmotically due to the presence of pure water. If the temperature is lowered very quickly (within seconds), crystallization cannot occur and the tissue becomes vitrified. In this case, if the tissue is thawed quickly to avoid crystallization, living cells and tissue are undamaged. Consequently, if a tissue fixed by the "quick-freeze" method is to be prepared for the analysis of metabolites, the thawing and deproteinization (addition of perchloric acid) of the tissue must be combined. For spectrophotometric measurements in nitrogen de-vitrification of the tissue is obtained by addition of glycerol (e.g. 1 part cell suspension containing 0.3 g fresh cells/ml + 1 part glycerol).

Chemical disintegration

In chemical disintegration methods the cell wall is attacked chemically. The disintegrating agent is allowed to act for the shortest possible time and is usually helped by the use of a mechanical homogenization method (*Potter-Elvehjem* homogenizer or blade homogenizer). Since the chemicals may interfere during the analysis, only those which can be easily removed are used. An example is the preparation of erythrocyte haemolysates with digitonin [7]. This method depends on the destruction of the erythrocyte membrane by the reaction of digitonin with the cholesterol of the cell wall. It has also been used in the disintegration of mitochondria (from liver cell homogenates) for the subsequent extraction of mitochondrial particles [36] (cf. 1.2.2.4). The method is also suitable for the lysis of leukocytes and pletelets.

Procedure according to [37]: mix 5 ml washed erythrocytes, leukocytes or platelets, 2.5 ml triethanolamine buffer (5 mmol/l, pH 7.5), 2.5 ml water distilled in a quartz apparatus, and 1.0 ml saturated, aqueous solution of digitonin. Incubate the mixture for 60 min at 3 °C (until haemolyzed). Centrifuge off the cell stroma (15 min at 3000 rpm). In addition, disintegrate leukocytes and platelets mechanically in a *Potter-Elvehjem* homogenizer.

Biological-enzymatic disintegration

One of the oldest methods of disintegration is the autolytic decomposition of cellular structures by endogenous proteolytic enzymes. It is the preliminary step before subsequent maceration (mostly by autolysis at 35 °C, 2 to 3 hours) and for the preparation of maceration juice capable of carrying out fermentation. The maceration is usually carried out with the addition of toluene, ethyl acetate or sodium sulphide.

Enzymatic lysis of tissue is employed to disintegrate tissue for the preparation of isolated animal cells (cf. 1.2.2.4). Proteolytic enzymes (e.g. proteases from *Bacillus subtilis*) have proved to be of value also for the disintegration of tissue and free cells to prepare cell organelles, e.g. intact mitochondria from muscle (cf. part 1.2.2.4).

Dry homogenates

The classical method for the preparation of dry homogenates is dehydration with acetone (acetone-dried powder) [38]. It is still used for the preparation of enzymes from bacteria, yeast or animal material. However, it is less suitable for the preparation of extracts for the quantitative analysis of enzyme activity. For this purpose as well as for the concentration, storage and distribution of biological materials, vacuum freeze-drying (lyophilization) of tissue homogenates, cell suspensions and cell extracts is widely used. Many biological materials, which will rapidly deteriorate even in frozen solutions, can be kept in a usable state for many years after lyophilization. Samples to be lyophilized should be in aqueous solutions and quickly frozen in a cold bath of acetone and solid carbon dioxyde prior to lyophilization. It is further important to select a lyophilizer of sufficient capacity, equipped with a vacuum pump capable of maintaining a proper vacuum and a cold trap sufficient to handle the high flow of vapour.

Type of instrument: Freeze Dryer Models Alpha, Beta I, Delta I or Delta III, *Martin Christ,* D-3360 Osterode am Harz; Automatic Freeze Dryer Model 10 – 100 NL, *Cenco Instruments B. V.,* Breda, The Netherlands.

1.2.2.4   Cell and Tissue Fractionation

Separation of cell particles and subcellular organelles is generally achieved after cell or tissue disintegration by methods described in 1.2.2.1 (homogeneous cell suspensions) and 1.2.2.3 (homogenates) by differential centrifugation or by density gradient centrifugation. Purification of subcellular organelles on a density gradient should not be performed if purification on a sucrose gradient is feasible, since in order to obtain higher densities, such high concentrations are required that some cell organelles may be damaged by the osmotic conditions. Furthermore, the increase in viscosity of the medium necessitates long centrifugation times. An appropriate reagent for obtaining high densities with small changes in osmolarity and viscosity is the non-ionic sucrose polymer Ficoll (*Pharmacia,* Uppsala, Sweden) which is widely used to construct discontinuous and continuous (5 – 35% in sucrose, 250 mmol/l) gradients.

Criteria to confirm the integrity and purity of the separated subcellular fractions are analyses of enzyme activities and functions (e.g. respiration) specific for the respective cell organelle and electron micrographs. Contamination with other subcellular fractions can best be established by measuring the corresponding specific enzyme activities.

In the following sections a survey of methods for the isolation of cell organelles by tissue and cell fractionation with the appropriate references, is given. Procedure to isolate mitochondria and synaptosomes are described in more detail.

Isolation of plasma membranes

Methods for the preparation of plasma membranes are described for liver cells [39, 40] intestinal brush border cells [41, 42] adipocytes [43, 44] erythrocytes [45] human platelets [46] bladder epithelial cells [47] and ascites tumour cells [48]. A critical step of the isolation procedure is cell rupture which should be as gentle as possible to avoid scrambling and disruption of the plasma membranes. For most mammalian cells, cell rupture with a *Dounce* homogenizer is appropriate; in the case of cells more resistant to breakage (e.g. ascites tumour cells) less gentle techniques (see 1.2.2.3) and even hypotonic conditions have to be employed. Isotonicity and the concentration of divalent metal ions of the buffer solution is of importance for the yield and quality of plasma membranes.

Isolation of *Golgi* apparatus

Methods for the isolation of the *Golgi* apparatus from rat liver cells, mammary gland and rat testes are reviewed by *D. J. Morré* [49].

Isolation of cell nuclei

Approved methods to isolate cell nuclei are described for rat liver [50, 51] and rat hepatoma cells [52]. Cell lysis can be achieved by detergent (Triton X-100, 0.1%) treatment or by allowing the cells to swell in a hypotonic medium. Subsequent homogenization and centrifugation should be carried out after addition of sucrose (0.3 mol/l) as an osmotic stabilizer to minimize the release of lysosomal enzymes.

Preparation of lysosomes

Isolation and characterization of lysosomes has been first achieved by *DeDuve* (for review see [53]). *Wattiaux et al.* [54] have published a method for the separation of rat liver lysosomes on a continuous sucrose gradient based on the finding that lysosomes filled with Triton WR-1339 are rendered lighter than mitochondria. This method can also be used to isolate liver mitochondria (see below) free of lysosomes [55].

Preparation of peroxisomes

A method for the isolation and characterization of peroxisomes from rat liver and kidney involving isopycnic sucrose density gradient centrifugation is described by *Hsieh & Tolbert* [56].

Preparation of ribosomes

Animal and human ribosomes can be prepared from powdered tissue material that has been stored under liquid nitrogen or from fresh tissue. Procedures for the preparation of ribosomes from eukaryotic cells are given by *Martin et al.* [57], from rat liver by *Petermann & Pavlovec* [58] and from mouse liver by *Staehelin & Falvey* [59]. In the last article a method to isolate mammalian ribosomal subunits active in polypeptide synthesis is also described. An approved method for the preparation of derived and native ribosomal subunits from rat liver is reported also by *Sadnik et al.* [60].

Microsomes and active ribosomes can be prepared from animal tissue as follows [60]: all operations are carried out at $0°C-4°C$. 20 g of minced rat liver is homogenized in 50 ml of a solution containing per litre 35 mmol $KHCO_3$, 20 mmol K_2HPO_4, 25 mmol KCl, 4 mmol $MgCl_2$, and 0.35 mol sucrose, pH 7.6. The homogenate is centrifuged at 13 000 g for 30 min. The supernatant is filtered through 4 layers of cheesecloth, and centrifuged in a preparative ultracentrifuge at 100 000 g for 2 h to obtain the microsomes. The sedimented microsomes are re-suspended in about 10 ml of a solution containing in one litre 50 mmol Tris-HCl, 0.5 mol NH_4Cl, 10 mmol $MgCl_2$ and 6 mmol β-mercaptoethanol, at pH 7.6. 1 ml of 15% sodium deoxycholate is added dropwise to the microsomal suspension under stirring which is continued for 15 min. The solution (11 ml) is layered on a discontinuous sucrose gradient, constructed with 10 ml each of sucrose solutions containing 0.5 mol/l (top layer) and 0.1 mol/l (bottom layer) and Tris-HCl, 10 mmol/l, NH_4Cl, 0.5 mmol/l, and β-mercaptoethanol, 6 mmol/l, pH 7.6. Centrifugation is carried out at 22 500 rpm (*Spinco* No 30 rotor) for 20 h. The supernatant is decanted and the precipitated ribosomes are re-suspended in about 5 ml of a solution containing in one litre 50 mmol Tris-HCl, 0.5 mol NH_4Cl, 10 mmol $MgCl_2$ and 6 mmol β-mercaptoethanol, pH 7.6. The discontinuous gradient centrifugation is repeated under the same conditions. The sedimented ribosomes are re-suspended in 2 ml of a medium containing in 1 litre 50 mmol Tris-HCl, 50 mmol KCl or NH_4Cl, 4 mmol $MgCl_2$, 1 mmol dithiothreitol and 0.35 mol sucrose, pH 7.6. After gentle homogenization using a Teflon-glass homogenizer the ribosomal suspension is centrifuged at about 3000 g for 15 min to remove any heavy insoluble material and the supernatant containing the ribosomes is stored in aliquots at $-70°C$.

Preparation of mitochondria

Isolation of mitochondria requires the breakage of intercellular connections and plasma membranes which, dependent on the type of cells or tissue, is usually achieved either by mechanical or enzymatic treatment (cf. 1.2.2.3). All operations have to be performed at $0°C - 4°C$. Mitochondria are generally isolated from the homogenate by differential centrifugation. The resulting mitochondrial pellet should be considered "crude". A further purification based on density can be performed by density gradient centrifugation using a continuous (5 – 35% in sucrose, 250 mmol/l) or discontinuous Ficoll gradient. Because the sedimentation coefficients for mitochondria from different tissues vary greatly, specifications for the values of centrifugation have to be determined for each new preparation. The mitochondria should be suspended in a minimal volume of non-ionic isolation medium or if used inmediately, in incubation medium.

Criteria of integrity: Respiratory Control Ratio (RCR) = (respiration rate in the presence of added substrate and phosphate acceptor ADP)/(respiration rate after consumption of ADP). High values indicate a tight coupling between ADP phosphorylation to ATP and respiration (electron transport).

ADP/O Ratio = (total amount of ADP added)/(amount of oxygen consumed for ADP stimulated respiration).

In the following articles procedures for the isolation of mitochondria from liver, heart muscle and brain are described in detail; for other animal tissues references to approved methods are given (Table 3).

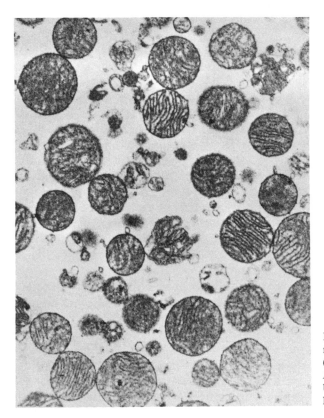

Fig. 2. Electron micrograph of rat heart mitochondria (Courtesy of *Dr. E. Drecoll-Lütjen,* Institute of Anatomy, University of Erlangen-Nürnberg).

Table 3. List of references for approved methods to isolate mitochondria from various organs and tissues

Organ/Tissue	Author(s)	Reference
Adrenal cortex	*D. R. Pfeiffer, T. T. Tchen*	65
	E. R. Simpson, D. L. Williams-Smith	66
Kidney cortex	*M. W. Weiner, H. A. Lardy*	67
	J. G. Ghazarian et al.	68
Striated muscle	*P. H. E. Groot, W. C. Hülsmann*	69
White adipose Tissue	*D. L. Severson et al.*	70
Brown adipose	*K. J. Hittelman et al.*	71
Tissue	*B. Cannon*	72

Liver

The key references are *Schneider* [61] and *Cooper & Lehninger* [62]. After sacrificing the animal remove the liver immediately and place it in about 50 ml of ice-cold isolation medium containing per litre 250 mmol sucrose, 10 mmol Tris-HCl and

0.5 mmol K^+-EDTA, pH 7.4. The weight of the liver is obtained by difference, by weighing the tared beaker.

All operations from this point on are carried out in the cold ($0\,°C - 4\,°C$) either by keeping samples in ice or by working in the cold room. Slice the liver into small pieces with a pair of scissors, transfer them to a 50 ml glass-Teflon homogenizer (*Potter-Elvehjem* type) containing 15 ml of cold isolation medium and homogenize with passes of a pestle rotating at 1000–1200 rpm until no larger pieces remain. If necessary homogenization can be repeated using a tighter pestle. Dilute the fine homogenate to 6 volumes (w/v) with isolation medium, mix thoroughly and centrifuge in 10 ml centrifuge tubes for 5 min at 1500 g. Decant the "postnuclear supernatants" carefully and centrifuge again for 10 min at 10000 g. Discard the supernatants, suspend the mitochondrial pellets thoroughly in a few ml of homogenization medium, combine the suspensions and wash twice by centrifugation for 10 min each at 10000 g. Re-suspend the washed mitochondria in a minimal volume of isolation medium or, if respiration measurements are carried out subsequently, in 2 ml of incubation medium containing in 1 litre 250 mmol sucrose, 20 mmol Tris-HCl, 2.4 mmol K_2HPO_4, and 0.05 mmol K^+-EDTA, pH 7.4.

If necessary the mitochondria can be further purified by density gradient centrifugation at 12000 g for 30 min using a discontinuous Ficoll gradient (3 ml each of 8.16 and 32% Ficoll in sucrose, 250 mmol/l).

Preparation of submitochondrial particles [36]: re-suspend the washed mitochondrial pellet in 8 ml of 1% aqueous cold digitonin solution (dissolve by heating), keep the suspension for 30 min at $0\,°C - 4\,°C$ and spin the suspension for 30 min at 58000 g in an ultracentrifuge. Collect the clear supernatant by decanting in a cold ultracentrifuge tube, add 1% digitonin solution to fill up the tube and centrifuge for 30 min at 105000 g. Discard the supernatant and wash the submitochondrial pellet once by suspending in 1% digitonin solution and centrifugation under the same conditions. Re-suspend the washed submitochondrial particles in a minimal volume of incubation medium. The preparation is active in oxidative phosphorylation only for a few hours.

Heart muscle

The key reference is *Pande & Blanchaer* [63]. After sacrificing the animal quickly remove the heart, cut it into pieces and wash repeatedly with several portions of chilled isolation medium (225 mmol mannitol, 75 mmol sucrose, 0.05 mmol K^+-EDTA and 10 mmol Tris-HCl in one litre, pH 7.4) to free it from external blood. Chop further, suspend in 8 ml of isolation medium containing 2–4 mg bacterial proteinase Nargase (*Serva,* Feinbiochemica GmbH & Co., Heidelberg, FRG) incubate for 8 min at 4°C with occasional stirring, add 12 ml of isolation medium and homogenize using a glass-Teflon homogenizer (*Potter-Elvehjem* type) with a loosely fitting pestle (700 rpm) until no larger pieces remain. Leave the homogenate at ice bath temperature for 8 min, add 15 ml of isolation medium and re-homogenize with a tighter pestle (clearance 0.2 mm) rotating at about 300 rpm until the resistance to homogenization is no longer felt. Centrifuge the resulting homogenate at 400 g for 5 min at 4°C, re-centrifuge the supernatant at 12000 g for 10 min and rinse the pellet obtained with isolation medium to remove any loosely adhering upper layer of white material. Suspend the mitochondrial pellet in 15 ml of isolation medium and re-centrifuge at 12000 g for 10 min. Rinse the pellet obtained again with isolation medium and suspend it uniformly in 0.6 ml of the isolation medium or – if respiration measurements are carried out subsequently – in 0.6 ml of incubation solution containing 230 mmol mannitol, 70 mmol sucrose, 20 mmol Tris-HCl, 5 mmol K_2HPO_4 and 0.02 mmol K^+-EDTA in 1 litre, pH 7.4.

Brain

Brain mitochondria can be prepared according to *Clark & Nicklas* [64] with slight modifications as follows: remove the cerebral hemispheres rapidly from decapitated rats into ice-cold isolation medium (250 mmol sucrose, 10 mmol Tris-HCl and 0.5 mmol K^+-EDTA in 1 litre, pH 7.4). Chop the tissue finely with a pair of scissors while washing it frequently with ice-cold isolation medium. Place the material from four rats in a glass-Teflon homogenizer (*Potter-Elvehjem* type) together with 20 ml of cold isolation medium and homogenize by eight up and down strokes of a tight fitting Teflon pestle rotating at 800 rpm. Add a further 5 ml of ice-cold medium and centrifuge the total homogenate at 0°C – 4°C for 3 min at 2000 g. Re-centrifuge the supernatant from this spin for 8 min at 12500 g and re-suspend the crude mitochondrial pellet thus obtained in a 3% Ficoll medium (3% Ficoll in a solution of 120 mmol mannitol, 30 mmol sucrose and 25 µmol K^+-EDTA per litre, pH 7.4) to a final volume of 5 ml. Layer this suspension carefully onto 10 ml of a 6% Ficoll medium (6% Ficoll in a solution of 240 mmol mannitol, 30 mmol sucrose and 50 µmol K^+-EDTA per litre, pH 7.4) and centrifuge for 30 min at 11500 g. Decant the supernatant from this spin and remove the slight fluffy layer from the pellet with a plastic rod. Re-suspend the mitochondrial pellet in about 5 ml of isolation medium and re-centrifuge for 10 min at 12500 g. Re-suspend the washed mitochondrial pellet in about 1 ml of cold isolation medium or, if respiration measurements are performed subsequently, in 1 ml of incubation medium containing in 1 litre 225 mmol mannitol, 75 mmol sucrose, 15 mmol Tris-HCl, 5 mmol K_2HPO_4 and 0.05 mmol K^+-EDTA, pH 7.4. The average yield per rat brain is between 3 – 4 mg mitochondrial protein.

Other tissues

For the preparation of mitochondria of other tissues references to approved methods are given in Table 3, p. 41.

Separation of mitochondrial membranes

A method [73] that has proved effective is the separation of the outer mitochondrial membrane from the inner membrane plus matrix by swelling mitochondria in potassium phosphate buffer, 20 mmol/l, pH 7.4. Both membranes can then be isolated by sonication and density gradient centrifugation using a three-layer sucrose gradient. Outer membranes of mitochondria, however, cannot be separated by this procedure from lysosomal membranes if membranes of both types are present in the suspension applied to the sucrose gradient.

Isolation of brain synaptosomes

The key references are *Gray & Whittaker* [74], *De Robertis et al.* [75], *Gurd et al.* [76], *Krueger & Greengard* [77] and *Booth & Clark* [78]. From the methods published in [74 – 78] the procedure described by *Krueger & Greengard* [77] with minor modifications [79] is in our experience most suitable to yield highly purified active synaptosomes.

Quickly remove the brains from 6 decapitated rats, hemisect and discard the cerebellum, brain stem mesencephalon and most of the white matter. Carry out all operations from this point at $0°C - 4°C$. Transfer the cerebral cortices (about 4 g from six rats) into 20 ml of ice cold isolation medium containing per litre 320 mmol sucrose and 4 mmol Tris-HCl, pH 7.4 and homogenize using 12 up and down strokes at 900 rpm in a glass-Teflon homogenizer (0.2 mm clearance). Dilute the homogenate with 20 ml of isolation medium and centrifuge at 1000 g for 10 min. Collect the supernatant with a pipette, re-homogenize the pellet using 4 up and down strokes at 900 rpm and centrifuge at 1000 g for 10 min. Combine the two supernatants, centrifuge again and discard the pellet. Centrifuge the supernatant at 12500 g for 15 min and re-suspend the crude synaptosomal pellet (P_2) in 5 ml of isolation medium. Dilute with 15 ml of 19% Ficoll (w/v) dissolved in isolation medium to give a final Ficoll concentration of about 14%. Form a discontinuous density gradient by overlaying 15 ml of this suspension, first with 10 ml of 7.5% (w/v) Ficoll in isolation medium and then with 5 ml of isolation medium alone. Centrifuge this gradient for 90 min at 90000 g in a *Beckman* SW 27 swinging bucket rotor. Harvest the synaptosomes at the 7.5 to 14% interface with a *Pasteur* pipette, dilute to 90 ml with isolation medium and centrifuge at 20000 g for 10 min. Wash the synaptosomal pellet by re-suspending it in 45 ml of oxygenated calcium-free *Krebs-Ringer* phosphate buffer, pH 7.4, containing in one litre 124 mmol NaCl, 4 mmol KCl, 1.3 mmol $MgSO_4$, and 16 mmol Na_2HPO_4 and centrifuge again at 20000 g for 10 min. Re-suspend the synaptosomal pellet in calcium-free *Krebs-Ringer* phosphate buffer to give a final concentration of about 10 mg protein/ml. Fig. 3 shows an electron micrograph of a typical preparation.

Preparation of synaptosomal plasma membrane: For the preparation of the synaptic membranes according to *Burbach et al.* [80] the synaptosomal pellet obtained as described above is lysed by homogenization in deionized water (5 ml/g of tissue) and kept at $0°C$ for 30 min. The supernatant obtained after centrifugation of the suspension at 10000 g for 20 min contains the light synaptic membrane structures.

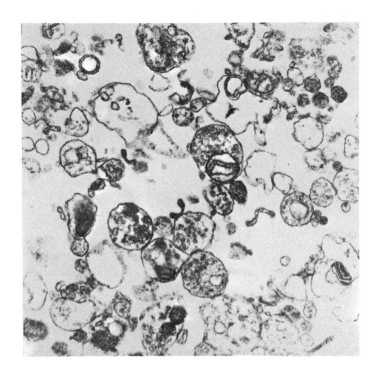

Fig. 3. Electron micrograph of cerebral cortical synaptosomes (Courtesy of *Dr. E. Drecoll-Lütjen,* Institute of Anatomy, University of Erlangen-Nürnberg).

This suspension is diluted 5 times with a sodium chloride solution (155 mmol/l) and the membranes are centrifuged down at 100000 g in 30 min in an ultracentrifuge and washed twice. For further purification the membranes are subjected to density gradient centrifugation using a discontinuous sucrose gradient consisting of 1 mol/l and 0.4 mol/l [80]. The light synaptosomal plasma membrane fraction accumulates at the 1.0 – 0.4 mol/l sucrose interface during swing-out centrifugation at 100000 g for 80 min. The membranes are collected, diluted 5 times with a sodium chloride solution, centrifuged down at 100000 g in 30 min and washed twice.

References

[1] *O. Warburg,* Versuche an überlebendem Carcinomgewebe (Methoden), Biochem. Z. *142*, 317 – 333 (1923).

[2] *K. A. C. Elliott,* Tissue Slice Technique, in: *S. P. Colowick, N. O. Kaplan* (eds.), Methods in Enzymology, Vol. *I*, Academic Press, New York 1955, pp. 3 – 19.

[3] *O. Warburg, F. Kubowitz, W. Christian,* Über die katalytische Wirkung von Methylenblau in lebenden Zellen, Biochem. Z. *227*, 245 – 271 (1930).

[4] *P. P. Cohen, M. Hayano,* The Conversion of Citrulline to Arginine (Transamination) by Tissue Slices and Homogenates, J. Biol. Chem. *166*, 239 – 250 (1946).

[5] *K. Brand,* Metabolism of 2-Oxoacid Analogues of Leucine, Valine and Phenylalanine by Heart Muscle, Brain and Kidney of the Rat, Biochim. Biophys. Acta *677*, 126 – 132 (1981).

[6] *W. C. Stadie, B. C. Riggs,* Microtome for the Preparation of Tissue Slices for Metabolic Studies of Surviving Tissues in Vitro, J. Biol. Chem. *154,* 687 – 690 (1944).

[7] *G. W. Löhr, H. D. Waller, O. Karges,* Quantitative Fermentbestimmung in roten Blutzellen, Klin. Wochenschr. *35,* 871 – 876 (1957).

[8] *R. B. Howard, L. A. Pesch,* Respiratory Activity of Intact, Isolated Parenchymal Cells from Rat Liver, J. Biol. Chem. *243,* 3105 – 3109 (1968).

[9] *J. R. Fry, C. A. Jones, P. Wiebkin, P. Belleman, J. W. Bridges,* The Enzymic Isolation of Adult Rat Hepatocytes in a Functional and Viable State, Anal. Biochem. *71,* 341 – 350 (1976).

[10] *M. N. Berry, D. S. Friend,* High Yield Preparation of Isolated Rat Liver Parenchymal Cells, J. Cell. Biol. *43,* 506 – 520 (1969).

[11] *P. Moldéus, J. Högberg, S. Orrenius,* Isolation and Use of Liver Cells, in: *S. P. Colowick, N. O. Kaplan* (eds.), Methods of Enzymology, Vol. *LII,* Academic Press, New York 1978, pp. 60 – 71.

[12] *D. L. Knook, E. C. Sleyster,* Preparation and Characterization of *Kupffer* Cells from Rat and Mouse Liver, in: *E. Wisse, D. L. Knook* (eds.), *Kupffer* Cells and other Liver Sinusoidal Cells, Elsevier/North Holland, Biomedical Press, Amsterdam 1977, pp. 273 – 288.

[13] *D. P. Jones, G.-B. Sundby, K. Ormstad, S. Orrenius,* Use of Isolated Kidney Cells for Study of Drug Metabolism, Pharmacol. Biochem. *28,* 929 – 935 (1979).

[14] *P. Vinay, A. Gougoux, G. Lemieux,* Isolation of a Pure Suspension of Rat Proximal Tubules, Am. J. Physiol. *241,* F403 – F411 (1981).

[15] *W. G. Guder,* Stimulation of Renal Gluconeogenesis by Angiotensin II, Biochim. Biophys. Acta *584,* 507 – 519 (1979).

[16] *M. Rodbell,* Metabolism of Isolated Fat Cells, J. Biol. Chem. *239,* 375 – 380 (1964).

[17] *Th. R. Devereux, J. R. Fouts,* Isolation of Pulmonary Cells and Use in Studies of Xenobiotic Metabolism, in: *S. P. Colowick, N. O. Kaplan* (eds.), Methods in Enzymology, Vol. *77,* Academic Press, New York 1981, pp. 147 – 151.

[18] *M. M. Weiser,* Intestinal Epithelial Cell Surface Membrane Glycoprotein Synthesis, J. Biol. Chem. *248,* 2536 – 2541 (1973).

[19] *J. W. Porteous, H. M. Furneaux, C. K. Pearson, C. M. Lake, A. Morrison,* Poly (Adenosine Diphosphate Ribose) Synthetase Activity in Nuclei of Dividing and Non-Dividing but Differentiating Intestinal Epithelial Cells, Biochem. J. *180,* 455 – 461 (1979).

[20] *A. K. Sinha, S. P. Rose,* Bulk Separation of Neurones and Glia: a Comparison of Techniques, Brain Res. *33,* 205 – 217 (1971).

[21] *W. B. Huttner, R. Meyermann, V. Neuhoff, H.-H. Althaus,* Neurochemical and Morphological Studies of Bulk Isolated Rat Brain Cells. II. Preparation of Viable Cerebral Neurones Which Retain Synaptic Complexes, Brain Res. *171,* 225 – 237 (1979).

[22] *D. W. Scharp. C. B. Kemp, M. J. Knight, W. F. Ballinger, P. E. Lacy,* The Use of Ficoll in the Preparation of Viable Islets of Langerhans from the Rat Pancreas, Transplantation *16,* 686 – 689 (1973).

[23] *F. Indiveri, M. A. Pellegrino, G. A. Molinaro, V. Quaranta, S. Ferrone,* Rosetting of Human T Lymphocytes with Goat Red Blood Cells: Effect of Treatment with 2-Aminoethylisothiouronium Bromide (AET) and Comparison with AET Treated Sheep Red Blood Cells, J. Immunol. Methods *30,* 317 – 328 (1979).

[24] *J. Segal, S. H. Ingbar,* Stimulation by Trijodothyronine of the in vitro Uptake of Sugars by Rat Thymocytes, J. Clin. Invest. *63,* 507 – 515 (1979).

[25] *C. A. Pupar, G. M. W. Cook,* Glycoprotein Biosynthesis in Quiescent and Stimulated Thymocytes and a T-Cell Lymphoma, Biochem. J. *201,* 377 – 385 (1982).

[26] *R. T. Dean,* Macrophage Protein Turnover, Biochem. J. *180,* 339 – 345 (1979).

[27] *H. J. Hohorst, F. H. Kreutz, Th. Bücher,* Über Metabolitgehalt und Metabolit-Konzentrationen in der Leber der Ratte, Biochem. Z. *332,* 18 – 46 (1959).

[28] *A. Wollenberger, O. Ristau, G. Schoffa,* Eine einfache Technik der extrem schnellen Abkühlung größerer Gewebestücke, Pflügers Archiv ges. Physiol. Menschen, Tiere *270,* 399 – 412 (1960).

[29] *E. Gerlach, H. J. Doring, A. Fleckenstein,* Papierchromatographische Studien über die Adenin-und Guanin-Nukleotide sowie andere säurelösliche Phosphor-Verbindungen des Gehirns bei Narkose, Ischämie und in Abhängigkeit von der Technik der Gewebeentnahme, Pflügers Archiv ges. Physiol. Menschen, Tiere *266,* 266 – 291 (1958).

[30] *E. Hannappel, St. Davoust, B. L. Horecker,* Isolation of Peptides from Calf Thymus, Biochim. Biophys. Res. Commun. *104,* 266 – 272 (1982).

[31] *V. R. Potter, C. A. Elvehjem,* A Modified Method for the Study of Tissue Oxidations, J. Biol. Chem. *114,* 495 – 504 (1936).

[32] *K. W. Cleland, E. C. Slater,* Respiratory Granules of Heart Muscle, Biochem. J. *53,* 547 – 556 (1953).

[33] *B. Chance, M. Baltscheffsky,* Spectroscopic Effects of Adenosine Diphosphate upon the Respiratory Pigments of Rat-Heart-Muscle Sarcomes, Biochem. J. *68,* 283 – 295 (1958).

[34] *M. Merkenschlager, K. Schlossmann, W. Kurz,* Ein mechanischer Zellhomogenisator und seine Anwendbarkeit auf biologische Probleme, Biochem. Z. *329,* 322 – 340 (1957).

[35] *H. T. Meryman,* General Principles of Freezing and Freezing Injury in Cellular Materials, Ann. N.Y. Acad. Sci. *85,* 503 – 509 (1960).

[36] *C. Cooper, T. M. Devlin, A. L. Lehninger,* Oxidative Phosphorylation in a Enzyme Fraction from Mitochondrial Extracts, Biochim. Biophys. Acta *18,* 159 – 160 (1955).

[37] *G. W. Löhr, H. D. Waller,* Zellstoffwechsel und Zellalterung, Klin. Wochenschr. *37,* 833 – 839 (1959).

[38] *I. C. Gunsalus,* Extraction of Enzymes from Microorganisms (Bacteria and Yeast), in: *S. P. Colowick, N. O. Kaplan* (eds.), Methods in Enzymology, Vol. *I,* Academic Press, New York 1955, pp. 51 – 62.

[39] *D. M. Neville, Jr.,* The Isolation of a Cell Membrane Fraction from Rat Liver, J. Biophys. Biochem. Cytol. *8,* 413 – 422 (1960).

[40] *P. Emmelot, C. J. Bos,* Adenosine Triphosphatase in the Cell-Membrane Fraction from Rat Liver, Biochim. Biophys. Acta *58,* 374 – 375 (1962).

[41] *G. G. Forstner, S. M. Sabesin, K. J. Isselbacher,* Rat Intestinal Microvillus Membranes, Biochem. J. *106,* 381 – 390 (1968).

[42] *J. Schmitz, H. Preiser, D. Maestracci, B. K. Gosh, J. J. Cerda, R. K. Crane,* Purification of the Human Intestinal Brush Border Membrane, Biochim. Biophys. Acta *323,* 98 – 112 (1973).

[43] *Y. Giudicelli, R. Pecquery,* β-Adrenergic Receptors and Catecholamine-Sensitive Adenylate Cyclase in Rat Fat-Cell Membranes: Influence of Growth, Cell Size and Aging, Eur. J. Biochem. *90,* 413 – 419 (1978).

[44] *Y. Giudicelli, D. Lacasa, B. Agli,* Alterations Induced by a Prolonged Fasting: Opposite Effects on the β-Adrenergic Receptor-Coupled Adenylate-Cyclase System and on Lipolysis in Fat Cells from Rat, Eur. J. Biochem. *121,* 301 – 308 (1982).

[45] *A. K. Shukla, R. Schauer,* Fluorimetric Determination of Unsubstituted and 9(8)-O-Acetylated Sialic Acids in Erythrocyte Membranes, Hoppe-Seyler's Z. Physiol. Chem. *363,* 255 – 262 (1982).

[46] *A. J. Barber, G. A. Jamieson,* Isolation and Characterization of Plasma Membranes from Human Blood Platelets, J. Biol. Chem. *245,* 6357 – 6365 (1970).

[47] *R. M. Hays, P. Barland,* The Isolation of the Membrane of the Toat Bladder Epithelial Cell, J. Cell. Biol. *31,* 209 – 215 (1966).

[48] *V. B. Kamat, D. F. H. Wallach,* Separation and Partial Purification of Plasma-Membrane Fragments from Ehrlich Ascites Carcinoma Microsomes, Science *148,* 1343 – 1344 (1965).

[49] *D. J. Morré,* Isolation of Golgi Apparatus, in: *S. P. Colowick, N. O. Kaplan* (eds.), Methods in Enzymology, Vol. *XXII,* Academic Press, New York 1971, pp. 130 – 144.

[50] *R. K. Roy, A. S. Lau, H. N. Munro, B. S. Baliga, S. Sarkar,* Release of in Vitro-Synthesized Poly(A)-Containing RNA from Isolated Rat Liver Nuclei: Characterization of the Ribonucleoprotein Particles Involved, Proc. Natl. Acad. Sci. USA *76,* 1751 – 1755 (1979).

[51] *M. D. Dabeva, K. P. Dudov, A. A. Hadjiolov,* Quantitative Analysis of Rat Liver Nucleolar and Nucleoplasmic Ribosomal Ribonucleic Acids, Biochem. J. *171,* 367 – 374 (1978).

[52] *H. Jacobs, G. D. Birnie,* Isolation and Purification of Rat Hepatoma Nuclei Active in the Transport of Messenger RNA in vitro, Eur. J. Biochem. *121,* 597 – 607 (1982).

[53] *C. de Duve, R. Wattiaux,* Function of Lysosomes, Annu. Rev. Physiol. *28,* 435 – 492 (1966).

[54] *R. Wattiaux, M. Wibo, P. Baudhuin,* Influence of the Injection of Triton WR-1339 on the Properties of the Rat-Liver Lysosomes, in: *A. V. S. de Reuck, M. P. Cameron* (eds.), Lysosomes, Ciba Foundation Symposium, Churchill Verlag, London 1963, pp. 176 – 178.

[55] *P. M. Vignais, J. Nachbaur, P. V. Vignais, J. André,* A Critical Approach to the Study of the Localization of Phospholipase-A in Mitochondria, in: *L. Ernster, Z. Drahota* (eds.), Mitochondria-Structure and Function, Fifth FEBS Meeting, Vol. *17*, Academic Press, New York 1969, pp. 43 – 58.

[56] *B. Hsieh, N. E. Tolbert,* Glyoxylate Aminotransferase in Peroxisomes from Rat Liver and Kidney, J. Biol. Chem. *251*, 4408 – 4415 (1976).

[57] *T. E. Martin, F. S. Rolleston, R. B. Low, I. G. Wool,* Dissociation and Reassociation of Skeletal Muscle Ribosomes, J. Mol. Biol. *43*, 135 – 149 (1969).

[58] *M. L. Petermann, A. Pavlovec,* Studies on Ribonucleic Acid from Rat Liver Ribosomes, J. Biol. Chem. *238*, 3717 – 3724 (1963).

[59] *T. Staehelin, A. K. Falvey,* Isolation of Mammalian Ribosomal Subunits Active in Polypeptide Synthesis, in: *S. P. Colowick, N. O. Kaplan* (eds.), Methods in Enzymology, Vol. *XX*, Academic Press, New York 1971, pp. 433 – 446.

[60] *I. Sadnik, F. Herrera, J. McCuiston, H. A. Thompson, K. Moldave,* Studies on Native Ribosomal Subunits from Rat Liver. Evidence for Activities Associated with Native 40S Subunits that Affect the Interaction with Acetylphenylalanyl-tRNA, Methionyl-tRNA$_f$, and 60S Subunits, Biochemistry *14*, 5328 – 5335 (1975).

[61] *W. C. Schneider,* Intracellular Distribution of Enzymes, J. Biol. Chem. *176*, 259 – 266 (1948).

[62] *C. Cooper, A. L. Lehninger,* Oxidative Phosphorylation by an Enzyme Complex from Extracts of Mitochondria, J. Biol. Chem. *219*, 489 – 506 (1956).

[63] *S. V. Pande, M. C. Blanchaer,* Reversible Inhibition of Mitochondrial Adenosine Diphosphate Phosphorylation by Long Chain Acyl Coenzyme A Esters, J. Biol. Chem. *246*, 402 – 411 (1971).

[64] *J. B. Clark, W. J. Nicklas,* The Metabolism of Rat Brain Mitochondria, J. Biol. Chem. *245*, 4724 – 4731 (1970).

[65] *D. R. Pfeiffer, T. T. Tchen,* The Role of Ca^{2+} in Control of Malic Enzyme Activity in Bovine Adrenal Cortex Mitochondria, Biochem. Biophys. Res. Commun. *50*, 807 – 813 (1973).

[66] *E. R. Simpson, D. L. Williams-Smith,* Effect of Calcium(ion) Uptake by Rat Adrenal Mitochondria on Pregnenolone Formation and Spectral Properties of Cytochrome P-450, Biochim. Biophys. Acta *404*, 309 – 320 (1975).

[67] *M. W. Weiner, H. A. Lardy,* Reduction of Pyridine Nucleotides Induced by Adenosine Diphosphate in Kidney Mitochondria, J. Biol. Chem. *248*, 7682 – 7687 (1973).

[68] *J. G. Ghazarian, C. R. Jefcoate, J. C. Knutson, W. H. Orme-Johnson, H. F. DeLuca,* Mitochondrial Cytochrome P$_{450}$, a Component of Chick Kidney 25-Hydroxychole-calciferol-1 α-hydrolase, J. Biol. Chem. *249*, 3026 – 3033 (1974).

[69] *P. H. E. Groot, W. C. Hülsmann,* The Activation and Oxidation of Octanoate and Palmitate by Rat Skeletal Muscle Mitochondria, Biochim. Biophys. Acta *316*, 124 – 135 (1973).

[70] *D. L. Severson, R. M. Denton, B. J. Bridges, P. J. Randle,* Exchangeable and Total Calcium Pools in Mitochondria of Rat Epididymal Fat-Pads and Isolated Fat-Cells, Biochem. J. *154*, 209 – 223 (1976).

[71] *K. J. Hittelman, O. Lindberg, B. Cannon,* Oxidative Phosphorylation and Compartmentation of Fatty Acid Metabolism in Brown Fat Mitochondria, Eur. J. Biochem. *11*, 183 – 192 (1969).

[72] *B. Cannon,* Control of Fatty-Acid Oxidation in Brown-Adipose-Tissue Mitochondria, Eur. J. Biochem. *23*, 125 – 135 (1971).

[73] *C. Godinot, C. Vial, B. Font, D. C. Gautheron,* Régulation de l'Activité Respiratoire des Mitochondries de Coeur de Porc et Transformations des Nucléotides Adényliques et du Phosphate, Eur. J. Biochem. *8*, 385 – 394 (1969).

[74] *E. G. Gray, V. P. Whittaker,* The Isolation of Nerve Endings from Brain: an Electron Microscopic Study of Cell Fragments Derived by Homogenization and Centrifugation, J. Anat. (London) *96*, 79 – 87 (1962).

[75] *E. de Robertis, A. Pellegrino de Iraldi, G. Rodriguez de Lores Arnaiz, L. Salganicoff,* Cholinergic and Non-Cholinergic Nerve Endings in Rat Brain – I: Isolation and Subcellular Distribution of Acetylcholine and Acetylcholinesterase, J. Neurochem. *9*, 23 – 25 (1962).

[76] *J. W. Gurd, L. R. Jones, H. R. Mahler, W. J. Moore,* Isolation and Partial Characterization of Rat Brain Synaptic Plasma Membranes, J. Neurochem. *22*, 281 – 290 (1974).

[77] *B. K. Krueger, P. Greengard,* Depolarization-Induced Phosphorylation of Specific Proteins, Mediated by Calcium Ion Influx, in Rat Brain Synaptosomes, J. Biol. Chem. *252,* 2764 – 2773 (1977).

[78] *R. F. G. Booth, J. B. Clark,* A Rapid Method for the Preparation of Relatively Pure Metabolically Component Synaptosomes from Rat Brain, Biochem. J. *176,* 365 – 370 (1978).

[79] *U. Bauer, K. Brand,* Carbon Balance of Glucose Metabolism in Rat Cerebral Cortical Synaptosomes, J. Neurochem. *39,* 239 – 243 (1982).

[80] *J. P. H. Burbach, E. R. De Kloet, P. Schotman, D. De Wied,* Proteolytic Conversion of β-Endorphin by Brain Synaptic Membranes. Characterization of generated β-Endorphin Fragments and Proposed Metabolic Pathway, J. Biol. Chem. *256,* 12463 – 12469 (1981).

1.2.3   Subcellular Fractionation and Enzymatic Analysis of Tissue Biopsy Specimens

Timothy J. Peters

Analysis of tissue constituents for enzymes, hormones, vitamins, trace metals, etc., requires a consideration of their compartmentalization within the constituent cells of the tissue. There may be subforms of the constituent in different compartments and in disease states there may be selective changes in only one compartment. It is therefore essential in many studies of tissue enzymes to combine enzyme assays with subcellular fractionation procedures.

1.2.3.1   Subcellular Fractionation Techniques

Subcellular fractionation may be either analytical or preparative [1, 2]. *Analytical* subcellular fractionation does not aim to isolate individual organelles to a high purity. Using a variety of methods based on differences in the biophysical or biochemical properties of the individual organelles, they are separated (resolved) from each other. The organelles are identified in the separated fractions mainly by assay of their marker enzymes. Suitable markers are discussed below. Analytical subcellular fractionation is quantitative, in that all the organelles in the tissue sample are accounted for. This is particularly important in the study of diseased tissue where alterations in the properties of certain organelles may occur. If these are the properties that are also important in the isolation procedures, misleading conclusions may be reached if a standardized organelle-isolation procedure is used for both control and pathological tissue. Using the analytical approach, changes in the physical properties of the organelles can be detected and this information can be used to indicate organelle pathology.

Preparative subcellular fractionation aims to isolate specific organelles to a high degree of purity. Much of our current knowledge of cell biology has been obtained by this approach. It employs large amounts of starting material and uses several different fractionation techniques, successively, to isolate the organelles, often with little regard to yield or quantitative recoveries. Morphological methods are frequently used to assess purity. Clearly, this approach has only very limited application to tissue biopsy samples which are usually only available in milligram quantities. In addition, it is usual to investigate several organelles in the same tissue sample, a problem not amenable to the preparative approach. Another difficulty in applying preparative fractionation procedures to pathological tissue is that alterations may occur in the physical properties of organelles. If the properties play a key role in the isolation procedure, markedly differing yields of the organelles may be obtained from the diseased tissue compared with normal tissue. Alternatively, the diseased organelles may be lost during the fractionation procedures.

Several distinct techniques have been used for the study of organelles by both the analytical and preparative approaches. These will be considered in turn and special comments made on their applicability to small tissue biopsy samples.

Centrifugation methods

Centrifugation techniques rely on differences in organelle density or size (sedimentation coefficient) to effect a separation.

Isopycnic centrifugation (Fig. 1)

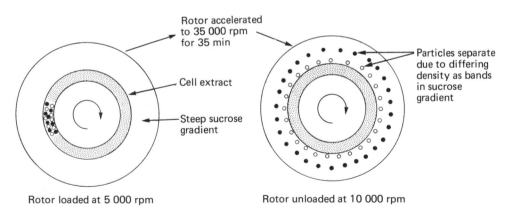

Fig. 1. Separation of organelles by isopycnic centrifugation in a zonal rotor. Reproduced from ref. [3].

Separations based on density differences in organelles involve layering the tissue homogenate or extract in isotonic medium (usually sucrose) onto a linear density gradient and centrifugation for prolonged periods until the organelles reach their equilibrium densities. This is a useful technique, since the individual organelles usually have a narrow density range and are readily resolved from one another; furthermore, providing that equilibrium densities are achieved, the exact centrifugation conditions are not critical. Sucrose is the most commonly used medium but sugar polymers, e.g. Ficoll®, glycogen, Stractan®, have also been used. These high molecular weight polymers are of particular value if the organelles are damaged by hypertonic solutions, for example, of sucrose. A disadvantage of these polymers is that they are highly viscous and prolonged centrifugation times are necessary to achieve equilibrium densities. Fig. 2 illustrates the relationship between viscosity and density of solutions of com-

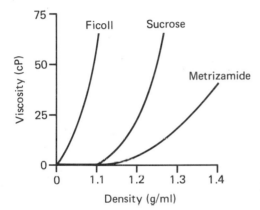

Fig. 2. Viscosity of aqueous solutions of density gradient media. Reproduced from ref. [4].

commonly used centrifugation media. A relatively recent addition to this range of polymers has been the introduction of Percoll®. Percoll forms a self-generating gradient and is also particularly suitable for separating cells. Colloidal silica had previously been used as a density gradient medium but its use had largely been abandoned because of its cytotoxicity. Coating the particles with polyvinylpyrrolidone overcame the problems.

   In order to overcome many of the disadvantages of these high molecular weight polymers, iodinated organic compounds have been introduced. These are of high density and low viscosity, with low toxicity and relatively low osmotic activity. Again, these compounds (e.g. metrizamide) have been frequently used for subcellular fractionation of tissue organelles. Separation of a broad range of organelles in a single tissue sample have been disappointing [5] but the use of iodinated media in the isolation of specific organelles has been useful as part of a multi-step procedure [4].

   Apart from the introduction of new media, there have been other innovations designed to overcome the problems of density gradient centrifugation. One of the disadvantages mentioned above has been that prolonged centrifugation times are necessary to reach equilibrium densities, particularly with the highly viscous media. Two approaches have been used to overcome this problem.

Firstly, the introduction of very high speed centrifuges capable of generating over 500 000 *g* has considerably shortened centrifugation times. However, an unexpected problem with high speed centrifugation has been organelle disruption due to the high hydrostatic pressure generated in the swinging-bucket rotors [6, 7]. This particularly affects mitochondria, but other organelles are also affected. The damage can be reproduced by exposing organelles in isotonic sucrose at 4 °C to high pressures in a pressure cell [8]. The hydrostatic pressure appears to disrupt the inner mitochondrial membrane. This renders it permeable to sucrose so that the equilibrium density increases. The phenomena can be minimized by centrifugation at higher temperatures [9] and slower speeds, the use of protective agents [10] or use of zonal rather than swinging-bucket rotors.

The second and more useful approach to shortening the centrifugation time needed to achieve equilibrium density is to reduce the path-length of organelles in the gradient. This can be achieved with the small volume automatic zonal centrifuge designed by *H. Beaufay* [11]. This rotor, which is unfortunately not commercially available, has a smaller chamber, peripherally placed in the rotor. It allows organelles to reach their equilibrium density in as little as one tenth of the time necessary in conventional swing-out rotors. In addition, the rotor has unique features of automatic unloading, improved filling and temperature control, and absence of hydrostatic organelle damage. In many ways the recently-introduced vertical pocket re-orientating rotor combines the advantages of the *Beaufay* rotor with ability to process several samples simultaneously. In addition, these rotors are commercially available for most ultracentrifuges [12].

Rate zonal and differential centrifugation (Fig. 3)

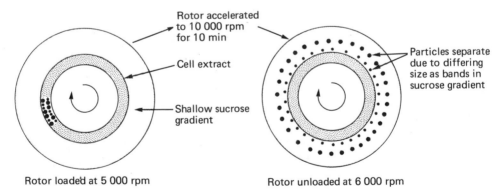

Fig. 3. Separation of organelles by isokinetic (rate zonal) centrifugation in a zonal rotor. Reproduced from ref. [3].

These techniques separate organelles on the basis of differences in their sedimentation coefficients. Zonal centrifugation is performed in either swinging-bucket or zonal rotors. The latter are generally of larger volume and thus are more suitable for preparative fractionation. Zonal rotors minimize wall effects and avoid the disturbance of the separated zones which occurs during stopping the rotor prior to collection of the fractions. However, they are more expensive, and require complex ancillary equipment including pumps, gradient-making devices and various monitors. In this form of centrifugation a shallow gradient is employed to stabilize the sample layer and the separated bands of organelles. Resolution of organelles may be limited by this approach as the size distribution is generally considerably broader than the density distribution of organelles; thus, there is often considerable overlap. Another problem which is sometimes encountered in rate zonal or isokinetic centrifugation is the phenomenon of drop sedimentation. This occurs when highly concentrated tissue extracts are layered onto shallow sucrose gradients. Droplets of the extract sediment *en masse* through the gradient, blurring any organelle separation. Nevertheless, rate zonal techniques can be used to isolate selected organelles and to determine the sedimentation coefficients of various organelles or granules. Provided that sufficiently sensitive marker assays are available, low concentrations of tissue extracts can then be used, thus avoiding drop sedimentation.

Differential pelleting (centrifugation) is a form of separation based on differences in the sedimentation coefficients of the organelles. It is the earliest form of centrifugation technique to be developed and applied to organelle isolation, and still has many applications. It yields limited numbers of fractions, simplifying analysis. Moreover, the composition of the principal fractions for several tissues, particularly liver, are well defined. Differential centrifugation has the particular advantage that throughout the procedure the organelles are exposed only to isotonic isolation media. However, the technique is lengthy and is not particularly suited to the subcellular fractionation of human tissue biopsy samples. The technique is principally used to isolate an enriched fraction of a particular organelle for subsequent purification studies. It can also be used to estimate sedimentation coefficients of organelles [13].

Use of membrane perturbants in conjunction with centrifugation procedures

Many centrifugation procedures achieve only a limited separation of certain organelles. This can be enhanced, often achieving a high degree of purity, by combining centrifugation with selective organelle perturbants. Several approaches have been used. One of the earliest was to use a specific enzyme associated with the organelle to form a product which alters the density of that organelle. Examples include the formation of lead phosphate by the glucose-6-phosphatase of endoplasmic reticulum [14], and the formation of intra-mitochondrial formazans following incubation of these organelles with succinate and tetrazolium compounds [15, 16]. In both instances the organelles show selective increases in density but, although of value in analytical studies, the presence of the reaction product is usually a disadvantage.

More recent examples include the use of pyrophosphate [17, 18] or heparin [19] to release ribosomes selectively from the rough endoplasmic reticulum. Loss of ribosomes leads to a striking decrease in density of the endoplasmic reticulum membranes. Addition of Ca^{2+} or Mg^{2+} can be used to facilitate the isolation of endoplasmic reticulum [20, 21] and mitochondria [22] or, alternatively, can be used to achieve rapid isolation of a highly purified plasma membrane fraction [23].

The chaotropic agent digitonin has also been used in the subcellular fractionation of several tissues. It complexes to accessible cholesterol and, as the molecule contains five sugar residues, it causes a marked increase in membrane density; most typically of the plasma membranes, which are rich in cholesterol [17, 18, 24]. Digitonin also has effects on the other organelles, mainly due to its detergent action, and this is clearly concentration dependent. The lysosomal membrane is most susceptible to the lytic effects of detergent, and at low concentrations, these organelles are disrupted and their contents are released into the cytosolic fraction. At higher concentrations the effects are less selective and solubilization of components of the endoplasmic reticulum, outer mitochondrial membranes and other organelles may occur. The effect of digitonin is highly concentration-dependent and the effects on different tissues and organelles are probably related to their cholesterol/lipid/protein ratios.

The application of the perturbing effect of digitonin on the organelles of a human liver biopsy are illustrated in Fig. 4. There is a marked increase in density of the plasma membrane marker, 5'-nucleotidase. It is clear that γ-glutamyltransferase and portions of acid phosphatase and leucyl-2-naphthylamidase are also localized to this organelle. The endoplasmic reticulum marker shows a small decrease in density due to loss of some ribosomes [24, 25] and it is clear that β-glucuronidase is also located, in part, to the endoplasmic reticulum. The lysosomes are disrupted and their marker enzyme, N-acetyl-β-glucosaminidase, is recovered in the soluble fraction. Mitochondria, as reflected by malate dehydrogenase, are unaffected by the digitonin.

A recent use of this approach has been to treat whole cell preparations with low concentrations of digitonin in an isotonic medium. As the plasma membrane is impermeable to this agent and excess can readily be washed away, digitonin can act solely on this membrane, selectively increasing the equilibrium density of the plasma membrane in a subsequent centrifugation step. This approach has been successfully used with peripheral blood neutrophils [27], but was found to be unsuitable with human peripheral blood lymphocytes as their plasma membranes were apparently permeable to even very low concentrations of digitonin [28].

The use of selective organelle perturbants is likely to find further applications in subcellular fractionation techniques. The use of antibodies directed against surface components of certain organelles are a selective perturbants. If these antibodies are coupled to dense carbohydrates or proteins, e.g. ferritin, they can be used to alter selectively the density of certain organelles [29].

Other methods of fractionation

Affinity chromatography

Like the last example of selective organelle perturbants, this method of organelle separation relies on the presence of certain proteins or glycoproteins on the exposed surface of the organelle. Immobilized antibodies to organelle surface proteins, or

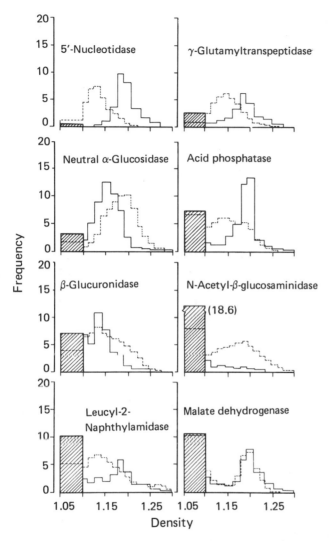

Fig. 4. Comparison of principal marker enzyme distribution after homogenization of human needle biopsy samples of human liver in the absence (interrupted line) and presence (continuous line) of digitonin. Graphs show frequency-density histograms. Activity present in the density span $1.05 - 1.10$ g/ml corresponds to the soluble fraction. Further details see reference [26].

lectins attached to inert column materials, should provide selective isolation of the particular organelles. Problems due to steric hindrance of the ligands, accessibility to organelle constituents and difficulties of eluting the organelles without disrupting them, has so far meant that these potentially elegant techniques have not been fully exploited in the separation or isolation of subcellular organelles [30]. The technique also relies on the formation of vesicles of a consistent "sidedness", e.g. from plasma membrane and endoplasmic reticulum. (A consistent sidedness implies that the vesicles must always close up in such a way that the particular reactive surface component is accessible to the binding agent.)

Gel permeation chromatography

In view of the value of this technique in the separation of many different molecules of biological interest, it is surprising that separation of subcellular organelles by this approach has been attempted to a limited extent only. Polysaccharides including Sepharose 6B and Sepharose 1000M have been used to fractionate microsomal components. In particular, they have been used to separate adsorbed cytosolic proteins from the membraneous components [31, 32]. Their exclusion volume is too small to resolve adequately larger subcellular organelles.

Chromatography on controlled-porosity glass beads has been used to a limited extent in the purification of vesicles, including synaptosomal vesicles [33] and fragments of intestinal brush border vesicles [34]. Separation is largely on the basis of apparent molecular volume but hydrophobic and ionic interactions may play a role. It is likely that, providing suitable matrices can be developed, chromatographic methods for the separation and isolation of intracellular organelles will be of increasing value.

Phase-partition chromatography

Phase-partition, particularly with immiscible organic phases, has been used for the separation of molecules of biological interest for many years. The introduction, by *Albertsson* and colleagues [35], of two phase systems of dextran and polyethylene glycol polymers has enabled this approach to be used for the isolation of biological macromolecules, subcellular organelles, viruses and bacteria, and even intact cells [36]. Mixtures of dextran and polyethylene glycol above their critical concentrations yield an upper polyethylene glycol-rich phase and a lower, dextran-rich phase. Various ions, in particular phosphate, distribute unequally between the two phases and thus there is a charge- and ionic-gradient between the two phases. The principles of counter-current partition are illustrated in Fig. 5.

Many different phase mixtures can be employed, including substituted dextrans and polyethylene glycol derivatives, enhancing ionic and hydrophobic interactions between the two phases. The attachment of ligands to one or other of the phase constituents can further be used to enhance the separation and some remarkably effective isolation techniques have been reported [36].

The disadvantages of the two-phase systems are the relatively high viscosity of the polymers necessitating long settling times of the phase mixtures and the multiple transfers necessary to achieve useful separations. *Albertsson*'s group have used an automatic thin-film counter-current apparatus [37], whereas we have used centrifugation techniques to facilitate the separation of the two phases. This can be performed in either a discrete transfer apparatus [38] or in a continuous toroidal oil centrifuge [39].

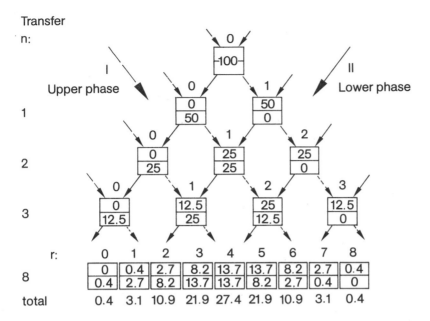

Fig. 5. Principles of countercurrent partition. A substance with a partition coefficient of 1, i.e. is equally distributed between the two phases, is subjected to 8 transfers. After each transfer, the lower phase in tube 0 is partitioned against new upper phase and the upper phase is transferred to fresh lower phase in a new tube. Thus the upper phase is transferred to the right and the lower phase to the left. Substances partitioning selectively between the two phases will be similarly transferred.

Phase separations are very sensitive to alterations in the polymer composition and there are often marked variations in organelle separations using different batches of reagents. Careful standardization of the mixtures, particularly the dextran, is important.

Separation of organelles frequently occurs due to differential partitioning between the polyethylene glycol-rich upper phase and the interface and thus it is sensitive to the centrifugation conditions used to separate the phases. A potential deficiency in counter-current studies is that only limited amounts of material can be processed in a single experiment. However, the study of tissue biopsy organelles by this technique is not subject to these limitations. Nevertheless, a systematic study of rat liver organelles [40] by this technique indicates that resolution of certain organelles is possible. In particular, subfractionation of membranes which are difficult to resolve by centrifugation techniques, e.g. plasma membrane, smooth endoplasmic reticulum, can be achieved (Fig. 6).

Electrophoretic methods

Separation of subcellular organelles by electrophoretic techniques has, like chromatographic procedures, been limited by problems with support matrices. However, the

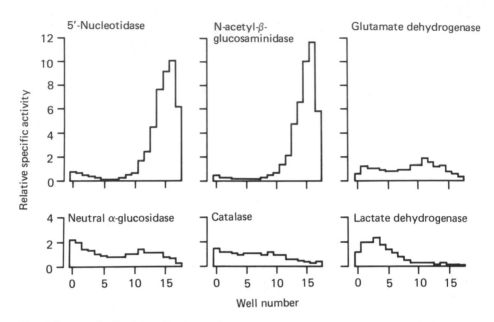

Fig. 6. Enzyme distributions following 17 discrete transfers of rat liver homogenate partitioning between two-phase mixture of 5.8% dextran T500 and 4% poly(ethylene glycol) 6000. Relative specific activity is plotted against well number. Note excellent resolution of plasma membrane (5'-nucleotidase) from endoplasmic reticulum (neutral α-glucosidase). Plasma membrane is however not separated from lysosomes (N-acetyl-β-glucosaminidase). Reproduced from ref. [40].

introduction of carrier-free electrophoresis has overcome these problems and useful separations of intact cells, organelles, bacteria and viruses have been obtained [41, 42]. At present the equipment is costly and is not yet suitable for processing small amounts of tissue.

Enzyme analysis of biopsy samples and density gradient fractions

The approach adopted in, what is in fact, a biochemical approach to tissue pathology, is to assay selected marker enzymes for individual organelles [3, 43]. Where possible, marker enzymes solely confined to a particular organelle are assayed [30, 44]. However, limitations due to low activities or insensitive assays may necessitate the use of markers that are less than ideal. The most appropriate markers differ, in some cases, from one tissue to another, particularly for plasma membrane domains. Table 1 lists the principal marker enzymes used in tissue studies in man, where only milligram quantities of sample are available. Detailed procedures are given in the appropriate references, but the assays rely on radiometric or fluorimetric techniques. Hydrolases are readily determined with 4-methylumbelliferyl or coumarin derivatives. Dehydrogenases can be assayed at high sensitivity with the methods of *Lowry* [45], and radiometric assays are available for most other analyses.

Table 1. Marker enzymes for principal subcellular organelles

Organelle	Marker Enzyme	Ref.
Plasma Membrane	5'-Nucleotidase	46
	Na^+, K^+ ATPase	47, 48
Brush Border (intestine)	γ-Glutamyltransferase	49
	Alkaline phosphatase	50
	α-Glucosidase (pH 6.0)	50
Golgi apparatus	Galactosyl transferase	51
Endoplasmic Reticulum	α-Glucosidase (pH 8.0)	46, 50
Mitochondria	Glutamate dehydrogenase	52
	Malate dehydrogenase	52
	Monoamine oxidase	52
Peroxisomes	Catalase	50
Lysosomes	N-Acetyl-β-glucosaminidase	46
Cytosol	Lactate dehydrogenase	50
Nuclei	DNA	53

Analysis of tissue samples for representative organelle marker enzymes entails many assays, particularly if the sample has been subjected to subcellular fractionation. Thus, analysis of six organelles in a single jejunal biopsy sample may entail over 250 estimations. Clearly, automated analysis should prove particularly valuable. Unfortunately, automated methods for most of the radioassays are not yet available and the complete requirements for fully-automated assay of tissue samples and sucrose density gradient fractions by fluorimetric techniques are not met by

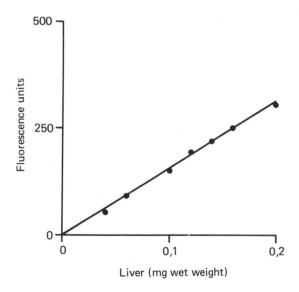

Fig. 7. Calibration curve for liver N-acetyl-*β*-glucosaminidase assayed by an automated version of the method described in ref. [46].

commercially-available instruments. The specific requirements are that the tissue homogenates are kept at 4 °C during sampling and that minimal amounts of material are used for duplicate assays. Because of the low levels of activity of many enzymes, end-point assays with incubation times of up to 1 hour are required. Analysis of gradient fractions which many contain up to 60% (w/w) sucrose or other media also poses a problem for many automatic analysers because of their high viscosity. We have coupled the *Pye-Unicam* AURA® analyzer to a *Perkin Elmer* 1000 M spectro-fluorimeter and a *Hewlett Packard* desk top calculator to provide automatic enzyme analysis of density gradient fractions. Most hydrolases, dehydrogenases and oxidases can be readily assayed. Excellent linear enzymic kinetics are obtained, even in the presence of high sucrose concentrations (Fig. 7), and there is very close agreement between enzyme distribution determined manually and with the automatic enzyme analyzer (Fig. 8).

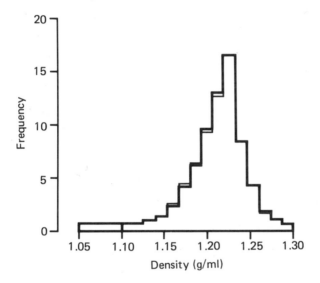

Fig. 8. Comparison of distribution of N-acetyl-β-glucosaminidase assayed in density gradient fractions following subcellular fractionation of rat liver homogenate [18]. Manual assays shown in thin lines; automated assays in thick line.

1.2.3.2   Application of Analytical Subcellular Fractionation Techniques

These micro-techniques have been applied to a wide variety of human and animal samples from both normal and diseased specimens. Tissues investigated in this manner include stomach [54], duodenum [55], jejunum [50], ileum [56] and colon [57], liver [26], kidney [56], cardiac [59] and skeletal muscle [48], lymphocytes [28, 60], polymorphs [27, 61], monocytes [62], macrophages [63], and pituitary [64]. These references should be consulted for specific details. Examples will be provided of the localization of peptide hydrolases in human jejunal biopsy specimens.

The solution to the question of the subcellular localization of peptide hydrolases within the human small intestinal mucosa required, in the first instance, the development of suitably sensitive assays for the determination of a wide variety of peptide hydrolases in sucrose density gradient fractions from milligram quantities of tissue. For several peptide substrates a fluorimetric modification [65] of the L-amino acid oxidase-coupled peptidase assay, as used for micro-assay of disaccharidases [66], was suitable. For glycine-containing peptides, a new assay procedure using a fluorescent method for measuring the released glycine had to be developed [67]. The second problem was the development of a technique for the subcellular fractionation of these small amounts of jejunal mucosa. Single-step sucrose density gradient centrifugation of biopsy homogenates in a *Beaufay* automatic zonal rotor was used in conjunction with fluorimetric assays of marker enzymes.

Fig. 9 shows the distribution of several peptide hydrolases in the density gradient fractions. The two dipeptidases are located exclusively to the cytosolic fractions. The two tripeptidases also show a particulate component at density 1.23 g/ml, whereas the tetra- and penta-peptide hydrolases are almost exclusively located to this fraction,

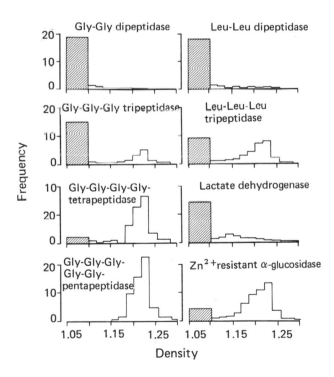

Fig. 9. Subcellular distribution of peptide hydrolases in human jejunal biopsy sample. Frequency-density distribution of 4 peptidases and cytosolic (lactate dehydrogenases) and brush border (α-glucosidase) markers are shown. Activity over density span 1.05 – 1.10 g/ml (hatched region) represents cytosolic fraction. Reproduced from ref. [68].

which corresponds to the brush border membrane as reflected by Zn^{2+}-resistant α-glucosidase. This location was confirmed by repeating the fractionation procedure after homogenization in the presence of digitonin. It is clear from the results shown in Fig. 10 that the particulate oligopeptidase activity is located to the brush border membrane which shows a density shift with digitonin, in contrast to the lysosomal marker N-acetyl-β-glucosaminidase which is now recovered in the cytosolic fraction.

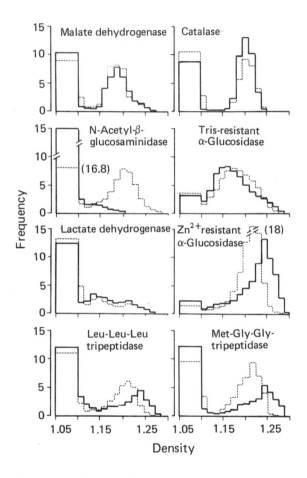

Fig. 10. Subcellular distribution of peptide hydrolases in human jejunal biopsy samples. Distributions of peptidases and organelle marker enzymes after homogenization in presence (thick line) or absence (interrupted line) of digitonin. Reproduced from ref. [67].

Similar studies aimed at answering both physiological and pathological questions, in both experimental animal and human tissue, are clearly amenable to analogous techniques for their solution.

References

[1] *C. De Duve,* Tissue Fractionation Past and Present, J. Cell. Biol. *50,* 20 – 55 D (1971).

[2] *C. De Duve, H. Beaufay,* A Short History of Tissue Fractionation, J. Cell. Biol. *91,* 293 – 299 S (1981).

[3] *T. J. Peters,* Investigation of Tissue Organelles by a Combination of Analytical Subcellular Fractionation and Enzymic Microanalysis: A New Approach to Pathology, J. Clin. Path. *34,* 1 – 12 (1981).

[4] *D. Rickwood,* Biological Separations in Iodinated Density Gradient Media, Information Retrieval Ltd, London 1976, p. 205.

[5] *J. Dawson, M. G. Bryant, S. R. Bloom, T. J. Peters,* Characterization of Gut Hormone Storage Granules from Normal Human Jejunum using Metrizamide Density Gradients, Regulatory Peptides *2,* 305 – 351 (1981).

[6] *R. Wattiaux,* Behaviour of Rat Liver Mitochondria during Centrifugation in a Sucrose Gradient, Mol. Cell. Biochem. *4,* 21 – 29 (1974).

[7] *M. Collot, S. Wattiaux-De Coninck, R. Wattiaux,* Deterioration of Rat Liver Mitochondria during Isopycnic Centrifugation in an Isoosmotic Medium, Eur. J. Biochem. *51,* 603 – 608 (1975).

[8] *M. Bronfman, H. Beaufay,* Alteration of Subcellular Organelles induced by Compression, FEBS Lett. *36,* 163 – 168 (1973).

[9] *S. Wattiaux-De Coninck, M.-F. Ronveaux-Dupal, F. Dubois, R. Wattiaux,* Effect of Temperature on the Behaviour of Rat Liver Mitochondria during Centrifugation in a Sucrose Gradient, Eur. J. Biochem. *39,* 93 – 99 (1973).

[10] *S. Wattiaux-De Coninck, F. Dubois, R. Wattiaux,* Effect of Imipramine on the Behaviour of Rat Liver Mitochondria during Centrifugation in a Sucrose Gradient, Eur. J. Biochem. *48,* 407 – 416 (1974).

[11] *H. Beaufay,* La centrifugation en gradient de densité (Thèse d'Agrégation de l'Enseignement Superieur Université Catholique de Louvain, Louvain, Belgium). Ceuterick S.A., Louvain, Belgium, 1966, p. 132.

[12] *D. Rickwood,* An Assessment of Vertical Rotors, Anal. Biochem. *122,* 33 – 40 (1982).

[13] *E. Slinde, T. Flatmark,* Determination of Sedimentation Coefficients of Subcellular Particles of Rat Liver Homogenates. A Theoretical and Experimental Approach, Anal. Biochem. *56,* 324 – 340 (1973).

[14] *J. A. Lewis, J. R. Tata,* Heterogeneous Distribution of Glucose-6-Phosphatase in Rat Liver Microsomal Fractions as shown by Adaptation of a Cytochemical Technique, Biochem. J. *134,* 69 – 78 (1973).

[15] *G. A. Davis, F. E. Bloom,* Isolation of Synaptic Junctional Complexes from Rat Brain, Brain Res. *62,* 135 – 153 (1973).

[16] *B. R. Holloway, J. M. Thorp, G. D. Smith, T. J. Peters,* Analytical Subcellular Fractionation and Enzymic Analysis of Liver Homogenates from Control and Clofibrate-treated Rats, Mice and Monkeys with Reference to the Fatty Acid Oxidising Enzymes, Ann. N. Y. Acad. Sci. *386,* 453 – 455 (1982).

[17] *A. Amar-Costesec, M. Wibo, D. Thines-Sempoux, H. Beaufay, J. Berthet,* Analytical Study of Microsomes and Isolated Subcellular Membranes from Rat Liver. 4. Biochemical, Physical and Morphological Modifications of Microsomal Components Induced by Digitonin, EDTA and Pyrophosphate, J. Cell. Biol. *62,* 717 – 745 (1974).

[18] *G. D. Smith, T. J. Peters,* Analytical Subcellular Fractionation of Rat Liver with Special Reference to the Localisation of Putative Plasma Membrane Marker Enzymes, Eur. J. Biochem. *104,* 305 – 311 (1980).

[19] *P. J. Freidlin, R. J. Patterson,* Heparin Releases Monosomes and Polysomes from Rough Endoplasmic Reticulum, Biochem. Biophys. Res. Commun. *93,* 521 (1980).

[20] *S. A. Kamath, F. A. Kummerow, K. A. Narayan,* A Simple Procedure for the Isolation of Rat Liver Microsomes, FEBS Lett. *17,* 90 – 122 (1971).

[21] *S. A. Kamath, K. A. Narayan,* Interaction of Ca^{2+} with Endoplasmic Reticulum of Rat Liver: A Standardized Procedure for the Isolation of Rat Liver Microsomes, Anal. Biochem. *48,* 53 – 61 (1972).

[22] *D. Gratecos, M. Knibiehler, V. Benoit, M. Semeriva,* Plasma Membranes from Rat Intestinal Epithelial Cells at Different Stages of Maturation. Preparation and Characterization of Plasma Membrane Subfractions Originating from Crypt and Villous Cells, Biochim. Biophys. Acta *512,* 508 – 524 (1978).

[23] *M. Kessler, O. Aguto, C. Storelli, H. Murer, H. Muller, G. Semenza,* A Modified Procedure for the Rapid Preparation of Efficiently Transporting Vesicles from Small Intestinal Brush Border Membranes, Biochim. Biophys. Acta *506,* 137 – 154 (1975).

[24] *K. A. Mitropoulos, S. Venkatesan, S. Balasubramanian, T. J. Peters,* The Submicrosomal Localization of 3-Hydroxy-3-Methyl Glutaryl Coenzyme A Reductase, Cholesterol 7-α-hydroxylase and Cholesterol in Rat Liver, Eur. J. Biochem. *82,* 419 – 429 (1978).

[25] *S. Balasubramanin, S. Venkatesan, K. A. Mitropoulos, T. J. Peters,* The Submicrosomal Localization of Acyl-Coenzyme A-Cholesterol Acyltransferase and its Substrate and of Cholesterol Esters in Rat Liver, Biochem. J. *174,* 863 – 872 (1978).

[26] *T. J. Peters, C. A. Seymor,* Analytical Subcellular Fractionation of Needle Biopsy Specimens from Human Liver, Biochem. J. *174,* 435 – 446 (1978).

[27] *G. P. Smith, T. J. Peters,* Subcellular Localization and Properties of Adenosine Diphosphatase Activity in Polymorphonuclear Leukocytes, Biochim. Biophys. Acta *673,* 234 – 242 (1981).

[28] *G. P. Smith, T. Shah, A. D. B. Webster, T. J. Peters,* Studies on the Kinetic Properties and Subcellular Localization of Adenine Nucleotide Phosphatases in Peripheral Blood Lymphocytes from Control Subjects and Patients with Common Variable Primary Hypogammaglobulinaemia, Clin. Exp. Immunol. *49,* 393 – 400 (1982).

[29] *A. E. Brown, J. Elovson,* Subfractionation of Liver Membrane Preparations by Specific Ligand-induced Density Perturbation, Biochim. Biophys. Acta *597,* 247 – 262 (1980).

[30] *W. H. Evans,* Subcellular Membrane and Isolated Organelles: Preparative Techniques and Criteria for Purity, Techniques in Lipid and Membrane Biochemistry *B407,* 1 – 46 (1982).

[31] *J. H. Venekamp, C. Kliffen, W. H. M. Mosch,* Purification of Cytoplasmic Ribosomes by Column Chromatography, J. Chromatogr. *87,* 449 – 454 (1973).

[32] *J. A. Higgins, J. E. Mazurkiewicz,* A Rapid Method for the Separation of Free and Membrane-bound Polysomes of Rat Liver by Gel Filtration on Columns of Sepharose 4B, Prep. Biochem. *10,* 317 – 330 (1980).

[33] *A. Nagy, R. R. Baker, S. J. Morris, V. P. Whittaker,* The Preparation and Characterisation of Synaptic Vesicles of High Purity, Brain Res. *109,* 285 – 309 (1976).

[34] *K. Ohsawa, A. Kano, T. Hoshi,* Purification of Intestinal Brush Border Membrane Vesicles by the use of Controlled Pore Glass-Bead Column, Life Sci. *24,* 669 – 678 (1979).

[35] *P.-Å. Albertsson,* Partition of Cell Particles and Macromolecules, Wiley-Interscience, New York 1971, pp. 323.

[36] *D. Fisher,* The Separation of Cells and Organelles by Partitioning in Two-Polymer Aqueous Phases, Biochem. J. *196,* 1 – 10 (1981).

[37] *P.-Å. Albertsson, B. Andersson, C. Larsson, H. E. Akerlund,* Phase Partition – A Method for Purification and Analysis of Cell Organelles and Membrane Vesicles, Methods Biochem. Anal. *28,* 115 – 150 (1982).

[38] *D. G. Pritchard, R. M. Halpern, J. A. Halpern, B. C. Halpern, R. A. Smith,* Fractionation of Mucopolysaccharides by Counter Current Distribution in Aqueous Two-phase System, Biochim. Biophys. Acta, *404,* 289 – 299 (1975).

[39] *Y. Ito,* Countercurrent Chromatography, Trends Biochem. Sci. *7,* 47 – 52 (1982).

[40] *W. B. Morris, T. J. Peters,* Microanalytical Partition of Rat Liver Homogenates by Poly(ethylene glycol)-Dextran Counter-current Distribution, Eur. J. Biochem. *121,* 421 – 426 (1982).

[41] *E. Harms, H. Kern, J. A. Schneider,* Human Lysosomes can be Prepared from Diploid Skin Fibroplasts by Free-flow Electrophoresis, Proc. Natl. Acad. Sci. *77,* 6139 – 6143 (1980).

[42] *S. Menashi, H. Weintraub, N. Crawford,* Characterisation of Human Platelet Surface and Intracellular Membranes Isolated by Free-Flow Electrophoresis, J. Biol. Chem. *256,* 4095 – 4101 (1981).

[43] *T. J. Peters,* Application of Analytical Subcellular Fractionation Techniques and Tissue Enzymic Analysis to the Study of Human Pathology, Clin. Sci. Mol. Med. *53,* 505 – 511 (1977).

[44] D. J. Mooré, G. B. Cline, R. Coleman, W. H. Evans, H. Glaumann, D. R. Headon, E. Reid, G. Siebert, C. C. Widenell, Markers of Membraneous Cell Components, Eur. J. Cell. Biol. 20, 195 – 199 (1979).

[45] O. H. Lowry, J. V. Passionneau, A Flexible System of Enzymatic Analysis, Academic Press, New York 1972, p. 291.

[46] C. A. Seymour, T. J. Peters, Enzyme Levels in Human Liver Biopsies. Assay Methods and Activities of some Lysosomal and Membrane-bound Enzymes in Control Tissue and Serum, Clin. Sci. Mol. Med. 52, 229 – 239 (1977).

[47] F. J. Bloomfield, G. Wells, E. A. Welman, T. J. Peters, Analytical Subcellular Fractionation of Guinea Pig Myocardium, Clin. Sci. Mol. Med. 53, 63 – 74 (1977).

[48] F. C. Martin, A. J. Levi, G. Slavin, T. J. Peters, Analytical Subcellular Fractionation of Normal Human Skeletal Muscle by Sucrose Density Gradient Centrifugation, Eur. J. Clin. Invest. 13, 49 – 56 (1983).

[49] G. D. Smith, J. L. Ding, T. J. Peters, A Sensitive Fluorimetric Assay for γ-Glutamyl Transferase, Anal. Biochem. 100, 136 – 139 (1979).

[50] T. J. Peters, The Analytical Subcellular Fractionation of Jejunal Biopsy Specimens. Methodology and Characterization of the Organelles in Normal Tissue, Clin. Sci. Mol. Med. 51, 557 – 574 (1976).

[51] D. S. Leake, G. E. Lieberman, T. J. Peters, Properties and Subcellular Localisation of Adenosine Diphosphatase in Arterial Smooth Muscle Cells in Culture, Biochim. Biophys. Acta, in press (1983).

[52] W. J. Jenkins, T. J. Peters, Mitochondrial Enzyme Activities in Liver Biopsies from Patients with Alcoholic Liver Disease, Gut 19, 341 – 344 (1978).

[53] J. Kapiscinski, B. Skoczylas, Simple and Rapid Fluorimetric Method for DNA Microassay, Anal. Biochem. 83, 252 – 257 (1977).

[54] J. Dawson, M. G. Bryant, S. R. Bloom, T. J. Peters, Subcellular Fractionation Studies of Human Gastric Antrum. Characterization of Gastrin, Somatostatin and VIP Storage Granules and the Principal Organelles in Normal Tissue, Clin. Sci. 59, 1 – 6 (1980).

[55] D. Stiel, D. J. Murray, T. J. Peters, Mucosal Enzyme Activities, with Special Reference to Enzymes Implicated in Bicarbonate Secretion, in the Duodenum of Rats with Cysteamine-induced Ulcers, Clin. Sci. 64, 341 – 347 (1983).

[56] C. O'Morain, F. Ameglio, R. Tosi, S. Durichio, S. Cucchiara, T. J. Peters, Localisation of HLA and Ia Antigens and β2 Microglobulins in Human Small Intestinal Mucosal Cells, Clin. Sci. 64, 69 P (1983).

[57] J. Dawson, M. G. Bryant, S. R. Bloom, T. J. Peters, Subcellular Fractionation Studies of Human Rectal Mucosa: Localisation of the Mucosal Peptide Hormones, Clin. Sci. 59, 457 – 462 (1980).

[58] U. N. Bhuyan, C. R. B. Welbourn, D. J. Evans, T. J. Peters, Biochemical Studies of the Isolated Rat Glomerulus and the Effects of Puromycin Amino-nucleoside Administration, Brit. J. Exp. Path. 61, 69 – 75 (1980).

[59] D. H. Fitchett, G. Wells, T. J. Peters, Analytical Subcellular Fractionation of Human Heart: A Comparison of Left and Right Ventricle with Hypertrophic Obstructive Myopathic Tissue, Cardiovas. Res. 13, 532 – 540 (1979).

[60] T. Shah, A. D. B. Webster, T. J. Peters, Enzyme Analysis and Subcellular Fractionation of Human Peripheral Blood Lymphocytes with Special Reference to the Localization of Putative Plasma Membrane Enzymes, Cell Biochem. Funct., in press (1983).

[61] A. W. Segal, T. J. Peters, Analytical Subcellular Fractionation of Human Granulocytes with Special Reference to the Localisation of Enzymes involved in Microbicidal Mechanisms, Clin. Sci. Mol. Med. 52, 429 – 442 (1977).

[62] K. T. Hughes, M. Davies, P. W. Andrew, T. J. Peters, Analytical Fractionation of Subcellular Organelles from Human Peripheral Monocytes with Particular Reference to their Neutral Proteinase Content, Biochem. Soc. Trans. 8, 594 – 595 (1980).

[63] D. B. Lowrie, P. W. Andrew, T. J. Peters, Analytical Subcellular Fractionation of Alveolar Macrophages from Normal and BCG-Vaccinated Rabbits with Particular Reference to Heterogeneity of Hydrolase-containing Granules, Biochem. J. 178, 761 – 767 (1979).

[64] *K. Mashiter, T. J. Peters,* Analytical Subcellular Fractionation of Human Pituitary: Characterization of the Organelles and Hormone-Containing Granules from Normal and Adenomatous Tissue, Clin. Sci. Mol. Med. *55,* 13 – 14 P (1978).

[65] *J. A. Nicholson, T. J. Peters,* Fluorometric Assay for Intestinal Peptidases, Anal. Biochem. *87,* 418 – 424 (1978).

[66] *T. J. Peters, R. M. Batt, J. R. Heath, J. Tilleray,* The Microassay of Intestinal Disaccharidases, Biochem. Med. *15,* 145 – 148 (1976).

[67] *J. A. Nicholson, T. J. Peters,* Fluorimetric Assay for Human Intestinal Glycine Peptidases, Clin. Chim. Acta, *91,* 153 – 158 (1979).

[68] *J. A. Nicholson, T. J. Peters,* The Subcellular Localisation of Peptide Hydrolase Activity in the Human Jejunum, Eur. J. Clin. Invest. *9,* 349 – 354 (1979).

1.2.4   Methods for Micro-organisms

Gerhard Gottschalk and Maria-Regina Kula

Bacteria do not contain organelles such as the contractile vacuoles and they are, therefore, unable to maintain osmotic balance by active means. Since bacteria usually live in a hypotonic environment the osmotic pressure of their interior has to be counterbalanced. This is achieved by a thick and rigid cell wall. As a consequence bacterial cells are comparatively difficult to disrupt. The composition of the bacterial cell walls is subject to great variations, and micro-organisms vary in the ease with which they can be disintegrated. In general *Gram*-positive bacteria are more difficult to disrupt than *Gram*-negative bacteria, and rods can be disintegrated more easily than cocci. Physical and chemical methods of breakage have been developed. Reviews of the various techniques that can be used have been presented by *Hughes et al.* [1], *Coakley et al.* [2] and *Dobrogosz* [3].

1.2.4.1   Physical Methods

Alumina grinding

This is the most inexpensive method for the disruption of micro-organisms. Alumina (Polishing Alumina, Grade A-301, *Fisher Scientific Co.*) is washed and dried. One part of alumina is mixed with 1 part of a bacterial cell paste in a pre-cooled mortar. The mixture is ground in a cold room until it gains a claylike consistency. Depending on the bacterial species this may require 15 min. The soluble material is then dissolved in an appropriate buffer. Alumina and cell debris are removed by centrifugation.

Grinding with alumina is a very mild method for cell breakage. However, the yields are often low.

Wet milling

Micro-organisms can be disintegrated by wet milling, which is an extention of the grinding method described above. In wet milling disruption of cells occurs by the combined action of direct impact and high liquid shear; both forces are enhanced by adding an abrasive to the suspension. Difficulties arise because the size of most micro-organisms falls in the range of $1-10$ µm, and the cells walls encountered are rather strong. This means that considerable forces have to be exerted on very small objects in order to obtain an efficient cell disruption. Therefore, high load volumes of glass beads are usually employed in the mill, e.g. 80% of the working volume in agitator mills or equal volumes of suspension and beads in vibratory mills. Reducing the size of the glass beads will increase the number of impacts and the shearing surface, but will lower the kinetic energy of the single bead, so a small optimal size is expected. For the disruption of yeast species, for example, beads of $0.5-1$ mm diameter have been found effective while for bacteria $0.1-0.5$ mm size gave better results. The cell concentration in the suspension is usually set quite high: $40-50$% (w wet cell mass/v suspension). Lower cell concentrations lead to extended periods of treatment for the same degree of cell disruption. However, it should be noted that the release of nucleic acids especially from the cells increases the viscosity of the suspension during homogenization, which may impair energy transfer to the beads and in addition lead to floating of very small beads, thereby abolishing grinding action and strongly damping shear stress. If difficulties arise in the disruption of a given micro-organism it is recommended to check the operating variables: size of beads concentration of cells and duration of treatment. Much heat is also generated during wet milling as an unwanted side effect, which has to be effectively removed if the isolation of labile enzymes is attempted. Two general types of wet mills are currently in use (vibratory mills and agitator mills), which differ primarily in the mode of energy transfer to the glass beads.

Vibratory mills

Vibratory mills are convenient to use for the disintegration of micro-organisms on a small scale and appear especially suitable if cell walls are to be prepared [4]. Cells are suspended in a suitable buffer and filled into a special flask together with the glass beads. The flask is closed and fastened to the mill. Shaking occurs at high frequency, at the same time and excentric motion is imparted.

In the Vibrogen Zellmühle (*E. Bühler*, Tübingen, Germany) the frequency is set to 70 Hz; the Braun MSK mill (*B. Braun Melsungen AG*, Melsungen, Germany) can be operated at 33.3 or 66.7 Hz. In the Braun mill cooling is provided by CO_2 and the temperature in the reaction vessel can be kept $\leq 5\,°C$. Various species of yeast have been disintegrated with 80 – 99% efficiency within 30 – 60 s. Bacteria require longer reaction times to achieve high solubilization of intracellular constituents: 60 – 120 s are usually sufficient [5]. In vibratory mills the rate of disintegration was found to depend also on the liquid load/air ratio in the vessel [1]. Air bubbles contribute significantly to an amplification of the power input, especially near resonance. Foaming may become a problem, and addition of some antifoam agent such as tributyl citrate is recommended.

After completion of the experiment the cell homogenate together with the glass beads is poured from the reaction vessel. The glass beads have to be rinsed in some suitable way in order to ensure high recovery of solubilized protein.

Agitator mills

High speed agitator mills have been successfully employed for the disintegration of yeast [6, 7] and bacteria [8]. Such mills are usually operated in a continuous mode so that disintegration of large amounts of cells is possible. Reports in the biochemical literature about the utilization of an agitator mill are concerned mainly with the isolation of enzymes from various organisms up to pilot-plant scale. A large scale preparation of mitochondria from *Saccharomyces cerevisiae* after disruption in an agitator mill has been described [9].

Recently some small versions of agitator mills for batch operation have become commercially available:

> Pearl-Mill – 50 ml sample volume
> (*VMA-Getzmann Verfahrenstechnik,* D-5226 Reichshof, Germany)
> Minimotor-Mill – 50 ml sample volume
> (*Eiger Engineering Ltd.,* Warrington, Great Britain)
> Dyno-Mill – 85 ml sample volume,
> (*W. A. Bachofen, Maschinenfabrik,* Basel, Switzerland).

In this way smaller quantities of cells can also be handled, and the agitator mill may become a useful alternative for the disruption of cells in the biochemical laboratory. The grinding chamber should be completely filled during operation, so that foaming should present no problem. The glass beads are activated by a stirrer driven by an electric motor and carrying one or several impellers. Tip speeds > 5 m $\times$ s^{-1} are usually required to achieve an efficient disruption within 2 – 10 min. Jacketed grinding chambers are employed and ice water or ethylene glycol-water mixtures at $-10\,°C$ are circulated to remove heat during grinding. Some organisms which are difficult to break by other mechanical treatments, such as mycelia [10] or *Brevibacterium ammoniagenes* [8], have been successfully disintegrated in agitator mills with high recovery of enzymes in solution.

Ultrasonication

Disruption of micro-organisms by ultrasonication has been widely used since it was introduced by *Harvey & Loomis* [11] for the breakage of luminous bacteria. Cell disintegration is caused by cavitational forces which produce shock waves or cavitational micro-streaming [2].

A frozen cell paste is thawed in approximately the same volume of buffer and poured into the sonicator cup which is cooled in ice or by circulating coolant through the jacket. The tip of the ultrasonicator probe is then placed a few millimetres beneath the liquid surface. The generator is set to the recommended power setting and tuned to the resonant frequency. Ultrasonication is continued for about 30 s. The sample is then allowed to cool down to 0 °C again and, if necessary, ultrasonication is resumed. The overall ultrasonication time depends on the kind of micro-organism to be disrupted. Many clostridial species, pseudomonads and *E. coli* are disrupted in 30 to 60 s. For *Streptomyces, Streptococcus* or *Arthrobacter* species several min of ultrasonication are required. To increase the efficiency of ultrasonication, aluminum oxide (ALCOA A-305) in amounts of 1 g per 10 g of cells (wet weight) may be added to the suspension (For optimization of ultrasonication conditions see [12]). It is important to place the tip of the sonicator probe beneath the liquid surface: if it is too close to the surface "cavitation unloading" resulting in large air bubbles will occur, and the rate of cell breakage will be decreased [2].

A great variety of sonicators is commercially available. They operate at 20 kHz and up to 200 W power. Large samples can be subjected to ultrasonication using a continuous flow cell. Such a system has, for instance, been used to sonicate kilograms of Brewer's yeast per hour [13].

Hughes Press

The use of a *Hughes* Press for the disruption of micro-organisms is laborious but very effective with respect to the extent of cell breakage and mild with respect to the recovery of labile enzymes. The principle is that a frozen cell suspension is forced by pressure through a small gap into a receiving chamber [1]. The press consists of a steel body and a steel plunger. The body has a cylindrical bore which is connected to the receiving chamber by means of an annular slit. The press is chilled to -25 to -30 °C. The cell suspension (as concentrated as possible) is poured into the chamber, and the plunger is brought into position. Pressures of 10000 to 15000 N/cm^2 are applied with a hydraulic press. The high pressure forces the frozen material through the slit into the receiving chamber. This brings about a disruption of the cells.

Another press, called X-Press, with quite similar design has been described and results for the disintegration of a variety of organisms have been reported [14].

French Press

The *French* Press was developed for the breakage of *Chlorella* [15]. Later it was applied to bacteria, and it proved to be very effective in disrupting cells from many bacterial species [1]. This press consists of a steel cylinder, a basal seal and a plunger. The steel cylinder has a small orifice with a needle valve. The press is pre-cooled to

0°C, and the piston is pre-set to provide the desired chamber volume. The cell is then mounted upside-down on a tripod stand, and the cell suspension is added with a slight excess, so that some suspension flows out through the open needle valve when the basal seal is brought into position (air has to be completely removed). The valve is then closed, the cell is mounted in the hydraulic press and pressure is applied (10000 to 15000 N/cm^2). The needle valve is carefully opened so that the suspension very slowly seeps out of the orifice into a chilled tube.

The *French* press is available in different sizes from *Aminco, Inc.,* Silver Spring, Md., USA. A refined modification is the Ribi fractionator (*Ivan Sorvall, Inc.,* Norwalk, Con., USA).

For preparative purposes the Manton Gaulin High Pressure Homogenizer is used [16].

The *French* press has been widely employed for the preparation of cell extracts in work on enzymes and on membranes. Cells which are difficult to disrupt by this method may be passed two or three times through the press. In order to employ the *French* press successfully it is important to operate it in such a way that the temperature of the cell suspension is kept below certain limits. These limits depend on the lability of the enzyme under investigation.

1.2.4.2   Chemical Methods

Permeabilization

It is sometimes necessary to determine the activity of a particular enzyme in small samples withdrawn from a growing culture. It is troublesome to prepare extracts from these samples by one of the physical methods, and it may be possible to make the cells permeable and to measure enzyme activity using the permeabilized cell suspension.

Toluene-treated cells of *Escherichia coli* have been used to determine β-galactosidase activity [17]. Culture samples are centrifuged, washed with potassium phosphate buffer, 50 mmol/l, pH 7.5, and suspended in the same buffer. 2 ml of cell suspension and 2 drops of toluene are shaken in test tubes for 30 min at 37°C. The β-galactosidase activity of the cells can then be measured directly. If this procedure is applied to other bacterial species or to other enzyme systems the optimum time of incubation with toluene has to be determined.

The glutamine synthetase activity of a number of micro-organisms has been determined using hexadecyltrimethylammonium bromide (CTAB) to make the cells permeable [18]. 10 ml cell suspension are treated with 0.1 ml of a 1% (w/v) CTAB solution. After 10 min glutamine synthetase activity can be measured. This procedure also allows the determination of the adenylylation state of glutamine synthetase because it is fixed by the treatment of the cells with the detergent.

Osmotic disruption

For osmotic disruption of bacteria it is necessary to weaken or abolish the mechanical strength of the cell wall. If this is done in a hypertonic medium, spheroplasts are formed and proteins outside the cytoplasmic membrane can be recovered. Disruption of spheroplasts and release of the intracellular enzymes occur in a hypotonic medium.

Separation of peripheral membrane proteins and of periplasmic proteins

Heppel and collaborators developed an osmotic shock procedure which results in a release of periplasmic enzymes and peripheral membrane proteins from *E. coli* [19, 20]. Cells from the logarithmic growth phase are centrifuged at 16000 *g* for 20 min and washed twice with Tris-HCl, pH 8.0, 10 mmol/l, and EDTA, 1 mmol/l. Following an incubation for a few min at room temperature, the suspension is centrifuged and the pellet is rapidly dispersed in 80 volumes of icecold $MgCl_2$, 0.5 mmol/l.

This rapid decrease of the osmolarity of the medium results in the release of proteins from the periplasmic space and the membrane. After centrifugation these proteins are present in the supernatant (shock fluid).

Peripheral membrane proteins of *Desulfovibrio vulgaris* have been released as follows [21]: Cells are suspended in a buffer, pH 9.0, containing Tris base, 50 mmol/l, EDTA, 50 mmol/l, and Na_2CO_2, 1.7 mmol/l (10 ml buffer per 1 g of wet weight). The suspension is gently stirred for 30 min at 35 °C and centrifuged at 10000 *g* for 20 min. The pellet is again suspended in the above buffer. After incubation and centrifugation the supernatants are combined; they contain the peripheral membrane proteins.

The following procedure was developed for *Clostridium formicoaceticum* [22]: 2 g (wet weight) of freshly harvested cells are suspended in 10 ml N-2-hydroxyethylpiper-azine-N-2-ethanesulphonic acid (Hepes) buffer, pH 8.0, 20 mmol/l, dithioerythritol, 1 mmol/l, and sucrose, 0.6 mol/l. 1 mg of lysozyme is added per ml, and the suspension is incubated at 37 °C for 30 min. During this incubation the cells are converted to round or oval spheroplasts. Centrifugation at 50000 *g* and 4 °C for 20 min yields a supernatant containing the proteins of the periplasm.

Lysis of cells with lysozyme-EDTA

Incubation of the spheroplasts of *C. formicoaceticum* from above in Hepes buffer, pH 8.0, 20 mmol/l, containing dithioerythritol, 20 mmol/l, $MgCl_2$, 0.5 mmol/l, and DNase, 1 mg/10 ml, results in lysis and release of the intracellular enzymes. A general procedure for lysis of *Gram*-negative bacteria by lysozyme-EDTA is given by *Kaback* [23]: exponentially growing cells are harvested by centrifugation at 16000 *g* for 20 min, and the pellet is washed twice with Tris-HCl, pH 8.0, 10 mmol/l, at 0 °C. After re-suspending the cells in Tris-HCl, pH 8.0, 30 mmol/l, containing 20% (w/v) sucrose (90 ml per 1 g wet weight), potassium EDTA, pH 7.0, and lysozyme are added to give final concentrations of 10 mmol/l and 0.5 mg/ml, respectively. The suspension is incubated for 30 min at room temperature. Spheroplast formation is then complete. The spheroplasts are collected by centrifugation, suspended in potassium phosphate buffer, pH 6.6, 50 mmol/l and incubated at 37 °C. This incubation in a hypotonic medium brings about the lysis of spheroplasts. If the lysozyme-EDTA treatment is done in a hypotonic medium from the beginning, cell lysis occurs concomitantly with cell wall hydrolysis.

Modifications of this lysozyme-EDTA method have been summarized by *Hughes et al.* [1]. The addition of EDTA is not necessary if *Gram*-positive organisms are subjected to a lysozyme treatment. If the cells do not lyse readily, the addition of glycine (1% w/v) to the growth medium one or two generation times before harvest can be tried. If lysis is done for the isolation of DNA from *Gram*-negative bacteria, treatment with sodium dodecylsulphate plus EDTA is a very effective method [24].

Yeast cell walls arc not attacked by the lysozyme methods described, but they can by lysed by an extract from the snail *Helix pomatia* [25, 26].

References

[1] *D. E. Hughes, J. W. T. Wimpenny, D. Lloyd,* The Disintegration of Micro-organisms, in: *J. R. Norris, D. W. Ribbons* (eds.), Methods in Microbiology, Vol. *5 B*, Academic Press, London and New York 1971, pp. 1 – 54.

[2] *W. T. Coakley, A. J. Bater, D. Lloyd,* Disruption of Micro-organisms, in: *A. H. Rose, D. W. Tempest* (eds.), Adv. Microb. Physiol., Vol. *16*, Academic Press, London, New York, San Francisco 1977, pp. 279 – 341.

[3] *W. J. Dobrogosz,* Enzymatic Activity, in: *P. Gerhardt, R. G. E. Murray, R. N. Costilow, E. W. Nester, W. A. Wood, N. R. Krieg, G. B. Phillips* (eds.), Manual of Methods for General Bacteriology, American Society for Microbiology 1981, pp. 365 – 392.

[4] *A. Rodgers, D. E. Hughes,* The Disintegration of Micro-organisms by Shaking with Glass Beads, J. Biochem. Microbiol. Tech. Engng. *2*, 49 – 70 (1960).

[5] *M. Merkenschlager, K. Schlossmann, W. Kurz,* Ein mechanischer Zellhomogenisator und seine Anwendbarkeit auf biologische Probleme, Biochem. Z. *329*, 332 – 340 (1957).

[6] *J. Rehacek, K. Beran, V. Bicik,* Disintegration of Microorganisms and Preparation of Yeast Cell Walls in a New Type of Disintegrator, Appl. Microbiol. *17*, 462 – 466 (1969).

[7] *F. Marffy, M.-R. Kula,* Enzyme Yields from Cells of Brewer's Yeast Disrupted by Treatment in a Horizontal Disintegrator, Biotechnol. Bioeng. *16*, 623 – 634 (1974).

[8] *H. Schütte, K. H. Kroner, H. Hustedt, M.-R. Kula,* Experiences with a 20 l Industrial Bead Mill for the Disruption of Microorganisms, Enzyme Microbial Technol. *5*, 143 – 148 (1983).

[9] *D. Deters, U. Müller, H. Homberger,* Breakage of Yeast Cells: Large Scale Isolation of Yeast Mitochondria with a Continuous-Flow Disintegrator Anal. Biochem. *70*, 263 – 267 (1976).

[10] *K. Zetelaki,* Disruption of Mycelia for Enzymes, Process Biochem. *4* (12), 19 – 24 (1969).

[11] *E. N. Harvey, A. L. Loomis,* The Destruction of Luminous Bacteria by High Frequency Sound Waves, J. Bacteriol. *17*, 373 – 376 (1929).

[12] *W. T. Coacley, R. C. Brown, C. J. James,* Optimization of the Release of Undegraded Material Extracted from Cells by Ultrasound, Biotechnol. Bioeng. *16*, 659 – 673 (1974).

[13] *C. J. James, W. T. Coakley, D. E. Hughes,* Kinetics of Protein Release from Yeast Sonicated in Batch and Flow Systems at 20 kHz, Biotechnol. Bioeng. *14*, 33 – 42 (1972).

[14] *L. Edebo,* A New Press for the Disruption of Microorganisms and other Cells, J. Biochem. Microbiol. Tech. Engng. *2*, 453 – 479 (1960).

[15] *H. W. Milner, N. S. Lawrence, C. S. French,* Colloidal Dispersion of Chloroplast Material, Science *111*, 633 – 634 (1950).

[16] *M. Follows, P. J. Hetherington, P. Dunnill, M. D. Lilly,* Release of Enzymes from Baker's Yeast by Disruption in an Industrial Homogenizer, Biotechnol. Bioeng. *13*, 549 – 560 (1971).

[17] *A. B. Pardee, F. Jacob, J. Monod,* The Genetic Control and Cytoplasmic Expression of "Inducibility" in the Synthesis of β-Galactosidase by *E. coli*, J. Mol. Biol. I, 165 – 178 (1959).

[18] *B. C. Johansson, H. Gest,* Adenylylation/Deadenylylation Control of the Glutamine Synthetase, Eur. J. Biochem. *81,* 365 – 371 (1977).
[19] *H. C. Neu, L. A. Heppel,* The Release of Enzymes from *Escherichia coli* by Osmotic Shock and during the Formation of Spheroplasts, J. Biol. Chem. *240,* 3685 – 3692 (1965).
[20] *N. C. Nossal, L. A. Heppel,* The Release of Enzymes by Osmotic Shock from *Escherichia coli* in Exponential Phase, J. Biol. Chem. *241,* 3055 – 3062 (1966).
[21] *W. Badziong, R. K. Thauer,* Vectorial Electron Transport in *Desulfovibrio vulgaris* (Marburg) Growing on Hydrogen plus Sulfate as Sole Energy Source, Arch. Microbiol. *125,* 167 – 174 (1980).
[22] *M. Dorn, J. R. Andreesen, G. Gottschalk,* Fumarate Reductase of *Clostridium formicoaceticum,* a Peripheral Membrane Protein, Arch. Microbiol. *119,* 7 – 11 (1978).
[23] *H. R. Kaback,* Bacterial Membranes, in: *S. P. Colowick, N. O. Kaplan* (eds.), Methods in Enzymology, Vol. *XXII,* Academic Press, New York and London 1971, pp. 99 – 120.
[24] *J. Marmur,* A Procedure for the Isolation of Deoxyribonucleic Acid from Micro-organisms, J. Mol. Biol. *3,* 208 – 218 (1961).
[25] *A. A. Eddy, D. H. Williamson,* A Method of Isolating Protoplasts from Yeast, Nature, London, *179,* 1252 – 1253 (1957).
[26] *E. A. Duell, S. Inoue, M. F. Utter,* Isolation and Properties of Intact Mitochondria from Spheroplasts of Yeast, J. Bacteriol. *88,* 1762 – 1773 (1964).

1.2.5   Methods for Plant Tissues

Roland R. Theimer

1.2.5.1   Pecularities of Plant Cells and Tissues

Plant cells show several distinctive properties that affect the choice of tissue disintegration procedures. The cells are encased by a rigid cell wall that is a network of cellulose fibres embedded in a matrix of hemicellulose and some proteins. Different cell types possess walls of differing thickness that may be stiffened by incorporated lignin. Frequently, adjacent cells are connected through plasma bridges, called plasmodesmata, that traverse the common cell walls. A second distinct feature is the voluminous central vacuole present in most mature plant cells. It develops by coalescence of numerous small vacuoles that are typical of immature or dividing plant cells. The central vacuole contains, in aqueous solution, inorganic ions, polyphosphates, a variety of organic compounds such as carboxylic and amino acids, phenols, quinones, tannins, flavons, sugars, and alkaloids. In addition, a number of lytic enzymes, e.g. proteinases, nucleases, and phosphatases, are also present in the vacuole [1]. Thus, depending on the plant tissue the vacuolar contents have a more or less detrimental potential when mixed with the cytoplasm. The central vacuole occupies up to 90% of the cellular lumen, thus confining the cytoplasm with its various structures and organelles to a thin peripheral layer. Due to its enormous size the vacuole is inevitably ruptured during tissue disintegration if preventive measures are not taken (see below). Therefore, plant tissues with relatively innocuous vacuolar saps are preferred.

Also, plant tissues consist largely of non-plasma material, mainly cell walls, vacuolar sap, and starch granules. For example, 97.4% of the fresh-weight of a spinach leaf is water, minerals, fat, and carbohydrates. Only 3.2% is extractable protein whereas liver tissue contains 20% protein [2]. Moreover, the composition of the cellular compartments of a given plant tissue may change rather dramatically and rapidly. CAM (crassulacean acid metabolism) plants that accumulate up to 200 μequ. malic acid per g fresh weight in the vacuoles during the dark period and degrade it during the light period (day) represent an extreme example. Thus, towards the end of the dark period the pH value of the cell sap will be below 4 [3]. Also, chloroplasts of all plants accumulate starch grains during the light period which disappear during the subsequent dark period. Starch-laden chloroplasts tend to rupture during tissue homogenization and show different sedimentation characteristics [4].

Most plant tissues consist of a number of different cell types that show rather different cytological and metabolic characteristics leading to a heterogeneous population of cell organelles in the homogenates. It is therefore beneficial to examine a hand-cut section of a selected tissue in the light microscope to obtain an idea of the fine structure, grade of lignification, number of inclusions, and diversity of cell types.

Plants grown under non-sterile conditions in the field, greenhouse or growth chamber may be contaminated considerably on their surface areas by bacteria which will be carried over into the homogenates, leading to misinterpretations of enzyme activities or protein contents if no microscopic assessment of the possible contamination is performed. Plant tissues should be thoroughly rinsed or peeled before disintegration.

1.2.5.2   Media for Tissue Disintegration

The disintegration is performed in an isolation medium that protects the extracted cell components from the deleterious mechanical or chemical stress during homogenization and subsequent isolation. Ideally, the medium should simulate the intracellular milieu. An impressive array of recipes for such isolation media may be found in the literature, each formulated more or less specifically for a distinct plant tissue or cell organelle. For the separation of membrane-bounded cell compartments a general list of ingredients includes a buffered osmoticum, certain ions, chelators, and a sulphydryl compound. Usually sucrose, mannitol, or sorbitol, 0.2 – 0.5 mol/l, are added as osmoticum to give a slightly hypertonic medium, which helps to prevent loss in proteins and ions from membrane-bounded compartments [5].

Addition of a carefully selected, suitable buffer maintains a constant proton concentration usually around pH $7-8$ which is important for the stability and function of many cell constituents, especially when plant tissues with highly vacuolated cells are disintegrated. Its chemical structure may be crucial to the preservation of a given cell component. For example, phosphate buffers were shown to extract cytochromes from membranes leading to rupture of mitochondria [6].

Also, amines such as Tris buffer, uncouple photophosphorylation in chloroplasts, but buffers without free amino groups are readily available [7]. The concentration of the buffer generally ranges between 0.05 to 0.15 mol/l.

To date the specific requirements for ions in the isolation medium are still not sufficiently understood. As a result, the types and quantities given in the different recipes are largely empirical. Mg ions in low concentration (0.1 mmol/l) are generally included in isolation media since they are thought to preserve membrane structures and coupling of oxidative phosphorylation [5] and, in concentrations of at least 3 mmol/l, maintain ribosome binding to ER membranes [8].

However, in higher concentrations, bivalent ions produce aggregation or membranous organelles which hampers the separation of the different cell components [9]. Therefore, the endogenous content of such ions in the plant material should be taken into account. For example, when using cells rich in Ca ions, such as onion tissue, the addition of Ca ions to the isolation medium may be reduced or even omitted [10].

As a paradoxon isolation media contain in most cases up to 10 mmol EDTA or EGTA (ethylene-glycol-bis(β-aminoethylether)-N,N' tetraacetic acid) per litre which are thought to chelate bivalent toxic ions, e.g. heavy metal ions, liberated during tissue disintegration [6]. In the medium used for washing (usually by re-centrifugation steps) of the isolated organelles the concentration of EDTA should be substantially reduced in order to avoid loss of membrane bound Mg and Ca ions.

The addition to the isolation medium of a thiol compound such as dithiothreitol, cysteine, reduced glutathione, or mercaptoethanol that protects SH-groups of proteins, is strongly recommended.

Some recipes include weakly negatively charged macromolecules such as Ficoll, dextran, or gum arabic which aid in preserving cellular particles. *Honda* [11] formulated a complex aqueous extraction medium containing Ficoll and dextran that was routinely used to isolate chloroplasts, mitochondria, nuclei, and lipid bodies from several plant tissues. The medium helps to retain structural integrity of isolated cellular organelles for over 24 h. Also, bovine serum albumin or other proteins, e.g. cleared coconut milk [12], are added to isolation media. They are reputed to have a stabilizing effect on cellular constituents and structures [13]. Besides their buffering action they may scavenge ionized substances that could damage membranes and proteins [13, 14], and emulgate free fatty acids that may disturb integrity of membranes, and even inhibit lipase activity that produces such fatty acids [14]. Some plant tissues show high contents of phenolic compounds that *in vivo* are stored in membrane-bounded compartments. Their deleterious action, especially in the presence of polyphenol oxidases, is abolished by the inclusion of polyvinylpyrrolidone (molecular weight ca. 40000) in the isolation medium [13].

1.2.5.3   Disintegration Procedures

General remarks

Essentially two major strategies have been developed for plant tissue disintegration. The more traditional one is the mechanical disruption of the cells by hand or by power-driven devices. Again, the method of choice for tissue disintegration depends on both the rigidity of the tissue and the cell component to be isolated. Therefore, special techniques will be outlined in section b). For studies on the metabolite contents of cellular compartments, leaching of water soluble substances from the cell organelles may be minimized by disintegrating the plant tissues in non-aqueous extraction media. Such methods are described for nuclei [15], chloroplasts [16], and protein storage bodies [17].

More recently, impressive advances in the techniques for rapid protoplast isolation in sufficient quantities from intact tissues and cell suspension culture cells have provided a new powerful tool for cell fractionation studies [18 – 21]. The intermediate protoplast formation permits gentle rupture of the cells and facilitates the isolation of intact cell components from a great number of tissues [18]. Since vascular and epidermal cells are largely removed, the cell material used for fractionation becomes relatively homogeneous.

Protoplasts are released by incubation of tissue pieces in mixtures of commercially available cell wall-degrading enzymes [18 – 21]. Usually the epidermal layers are peeled or brushed off and/or the enzymes are forced into the intercellular spaces by vacuum infiltration. Both the incubation period and the temperature depend on the type of tissue and enzyme mixture. They vary mainly between 1 h and 20 h at 20 – 37°C [18]. Evidently, prolonged treatment may alter compartmentation and metabolic properties of cells. In addition, most enzyme preparations are of limited purity being contaminated by a wide range of unspecific hydrolases [22].

Hence, protoplasts should be thoroughly washed before disruption. They are separated by sieving the mixture through a net with 50 – 80 μm pores, pelleted, and washed by low speed centrifugation. Disintegration may be achieved either mechanically [23, 24], by osmotic lysis in sugar solutions [25, 26] or in potassium phosphate solution [22, 27], or chemically by incubation in media containing polybases [22, 28, 29].

Special disintegration techniques

Disintegration procedures have to be optimized to suit a specific tissue with its pecularities of toughness and noxious constituents and to give best recovery of the particular cell component desired.

Total protein and lipid contents

Standard procedures for the isolation of total protein or enzyme contents of a tissue usually require disintegration of the cells in a blender in an appropriate solute, e.g.

ice-cold buffer solution (e.g. [30]), that also contains pertinent protectants (see chapter 3.3.4.2). Foaming of the aqueous extract in high-speed blenders should be rigorously avoided to prevent loss in enzyme activities. Alternatively, plant tissue may be quickly frozen by immersing in liquid nitrogen and comminuted with mortar and pestle. The resulting powder is then extracted with a suitable extraction medium [31]. Total lipid content is extracted when tissue pieces are ground or blended in a mixture of chloroform/methanol, ratio 2:1 [32], or with a low-toxicity solvent comprising hexane/isopropanol, ratio 3:2 [33].

The isolation of fragile organelles, however, requires procedures that mostly were worked out by trial and error. Established techniques for specific cell organelles will be listed below along with recent references for a more detailed protocol.

Nuclei

For best success the tissue should consist of cells with thin walls and only few lipid and starch granules [25]. When the occurrence of certain proteins in nuclei is to be examined, the tissue should be prefixed for $10-45$ min in glutardialdehyde solution, $10-20$ g/l, before disintegration [10, 34]. Detailed procedures are available for root tips, leaves, stems, plant embryos, and cell suspension cultures [10, 23, 25, 35–37].

Plastids (proplastids, etioplasts, chloroplasts, chromoplasts)

The pre-chilled tissue pieces (e.g. de-ribbed leaf fragments) are suspended in an appropriate grinding medium [38, 39] and homogenized within $3-5$ s in a blendor (e.g. *Starmix, Waring, Braun*) or a Polytron or Virtis homogenizer at high speed. For the recovery of intact plastids it is crucial to maintain low temperature and to disintegrate the tissue and separate the plastids from the cell debris, as quickly as possible, mostly by filtration through miracloth and subsequent centrifugation [38, 39]. The method is used with suitable grinding media for the isolation of chloroplasts from leaves of spinach, lettuce, peas and *Chenopodium quinoa* [38], of chromoplasts from daffodils [40], and of etioplasts from oat leaves [41]. Alternate disintegration procedures involve relatively long-term grinding of the plant material e.g. cultured tobacco cells [42] or pea roots [43] with mortar and pestle through a nylon net [4] to facilitate the rapid separation of the plastids from the cell debris. Intact plastids may also be isolated from intermediately formed protoplasts of leaves of grasses [26, 44, 45], spinach [22, 26], tobacco [44], chevril [26], and from castor bean endosperm [46] by gentle mechanical [22, 26, 44–46) or chemical [26] disruption. The use of protoplasts also permits in one step the disintegration of the cells and the separation of plastids from cytoplasmic fractions, if they are spun through a proper nylon net into a density gradient [45].

Mitochondria

General methods for the isolation of mitochondria have been published [6, 47, 48]. The tissue disintegration procedures range from gentle hand chopping [49] to high speed power homogenization (e.g. [50]). Leaf tissues are preferrentially homogenized for very short times not exceeding 2 – 3 s in a blender since longer grinding appears to destroy mitochondrial oxidative and phosphorylation capacities [48, 50]. Cotyledons or other parts of seedlings and fruits are chopped with razor blades and subsequently ground with mortar and pestle with or without abrasives [47 – 49, 51]. Massive storage tissues, such a potato tubers or cauliflower floral heads, are gently disintegrated with a domestic hand or powerdriven vegetable grater [6, 52] that is flooded with grinding medium. Tissue disruption within seconds is also achieved in a household vegetable juicer [6, 53]. The homogenization basket is lined with nylon cloth to filter the cell debris from the homogenate tossed out of the basket by the centrifugational power. Great advantages are the high speed disruption of the tissue and the virtually instantaneous removal of the cell organelles from the site of agitation. Mitochondria were also isolated from intermediately formed protoplasts [46]. The crude mitochondrial fraction obtained by differential centrifugation of the cell homogenates should be purified at least in discontinuous density gradients if high respiratory control ratios are desired [48, 54].

Microbodies (glyoxysomes, peroxisomes)

In fat-degrading tissues such as castor bean endosperm or fatty cotyledons (e.g. sunflower, water melon) numerous glyoxysomes are present, especially during the first days of germination [55, 56]. Leaf peroxisomes are found in green photosynthetic tissues, while non-green plant cells possess a few "unspecialized" microbodies [55]. Tissue disintegration for microbody isolation has to be very gentle because of the organelles' fragility. They are very easily damaged by mechanical forces, osmotic shock, and acid conditions. In general, the tissue is chopped with razor blades and then gently ground with mortar and pestle in a suitably buffered (pH 7 – 8) sucrose solution, 0.3 – 0.8 mol/l, with the protective reagents outlined in chapter 1.2.5.2 [55 – 59]. For leaf tissues short (5 – 15 s) disintegration times in a blender may also be acceptable [57]. The homogenate is filtered through cheese cloth and cell debris removed by centrifugation (200 – 500 g, 10 min). The supernatant solution is either layered directly onto a sucrose density gradient ranging from 338 g/l to 772 g/l or subjected to differential centrifugation [58, 59]. For large-scale isolates the tissue may be disintegrated by repeated 1 – 10 s bursts in a large blender [58] or in a special tissue slicer [60] that facilitates the quick homogenization of 200 g lots of tissues.

The Golgi apparatus and related cell components

The isolation of dictyosomes requires specific techniques. The reason is the unique architecture of the *Golgi* apparatus and its close association with other cell components of the endomembrane system [9, 12]. Favourable tissues are etiolated stems or non-green (storage) tissues [12]. All isolation steps should be conducted expeditiously and completed within 2 h. Low-shear disintegration of the tissue is achieved with a power-driven razor-blade chopper [12, 64] or with speedy hand-chopping with a razor blade for 5 – 10 min [12, 65]. For large scale isolation, short homogenization (2 min) of the tissue in a rotary knife homogenizer (e.g. Polytron, up to 4000 rpm) is also possible [12]. Short grinding with mortar and pestle of soybean suspension culture cells [66] or roots after pre-incubation in cellulolytic enzymes [67] has also been employed. After filtration of the homogenate through miracloth the dictyosomes are isolated by sucrose density gradient centrifugation [12]. Preparation of the homogenization medium and the sucrose gradients in cleared coconut milk and addition of dextran helps to preserve the typical morphology of the isolated dictyosomes. Furthermore, for stabilization of the dictyosome stacks, $1-2$ g glutardialdehyde per litre may be added to the freshly prepared homogenate [12, 65].

Vacuoles

The central vacuole of a mature higher plant cell is involved in turgor generation, storage of metabolic stand-by or inhibitory compounds and in the lysis of cytoplasmatic components [1, 22]. For the isolation of intact vacuoles in reasonable quantities tissue homogenization procedures have to avoid drastic mechanical forces to prevent rupture of the large size vacuoles. This may be achieved by the formation of intermediate protoplasts from various plant tissues [22, 27, 29, 68, 69] or cell suspension cultures [70] and subsequent osmotic, chemical, or gentle mechanical disruption. Also, cutting large amounts of strongly plasmolized, bulky tissues such as beet roots in a specially designed tissue slicer where the vacuoles are released into an undisturbed osmotic solution [60, 68] yielded reproducible results. Finally, gentle razor blade chopping or grinding with mortar and pestle of soft, plasmolized tissue (e.g. root tips) were also used for vacuole isolation [71]. Small vacuoles are also present in the latex of many plant species. They may be recovered from the collected latex by simple centrifugation [72].

Plant membranes

For the isolation of intracellular membranes, cells will have to be disintegrated as gently as for the separation of cell organelles in order to minimize cross-contamination of the desired membrane fraction by membrane fragments liberated from broken cell organelles [9].

The nuclear envelope may be obtained by subfractionation of a nuclear fraction after extraction of nucleoproteins in KCl solution, 2 mol/l [10]. Tonoplasts are best prepared from isolated and purified vacuoles after osmotic lysis [68] and sucrose density gradient centrifugation. Plasmalemma is usually isolated by sucrose density gradient centrifugation of homogenates prepared by razor blade chopping and gentle grinding with mortar and pestle of tissues [73, 74] or plant protoplasts [25, 75]. Membranes of the smooth or rough endoplasmic reticulum are obtained from such homogenates if the extraction medium contains $MgCl_2$, 3 mmol/l, which preserves the binding of ribosomes to the ER membranes without causing artificial aggregation of membranous structures [8, 9, 23, 37, 46]. By this means fractions of the smooth and rough ER are subsequently separated when the homogenate is centrifuged into density gradients [8, 76].

Plant storage compartments

In certain tissues plants store large amounts of proteins in protein bodies (protein vacuoles, aleurone grains), lipids in lipid bodies (spherosomes, oleosomes), and starch in starch granules. For the isolation of starch granules from e.g. cereal caryopses, the seeds should be imbibed in an aqueous solution of mercuric chloride (10 mmol/l) which inhibits amylase action and thus helps to preserve the intactness of the starch granules. The swollen caryopses or other starchy tissues, e.g. potato tubers, are chopped and a starch granule suspension is obtained by sieving the slurry through vibrational screens with 150−75 μm pore size [77]. Alternatively, grinding in a vegetable juicer was also reported to yield intact starch granules [6].

Protein bodies were isolated by centrifugation of homogenates mostly prepared from seed storage tissues e.g. mung bean cotyledons, by gentle homogenization with mortar and pestle [78], razor blade chopping [79], mechanical homogenizers [17, 80, 81], or from intermediately formed protoplasts [82]. For better preservation of the protein body integrity tissues may also be disintegrated in non-aqueous media containing hexane [81], or glycerol [17, 80].

Lipid bodies that consist mainly of neutral lipids (triglycerides) and small amounts of phospholipids and proteins are isolated from conventionally prepared homogenates after high speed centrifugation, floating as a fatty layer on top of the supernatant fluid [83]. For purification they should be floated several times to liberate membranous material and organelles trapped between the lipid bodies. Other types of lipid bodies which contain higher amounts of proteins and phospholipids may sediment with a buoyant density of 1.18 kg/l in sucrose density gradients [84].

References

[1] *Ph. Matile,* The Lytic Compartment of Plant Cells, Cell Biology Monographs, Vol. *1,* Springer-Verlag, Wien 1975.

[2] *P. L. Altman, D. S. Dittmer,* Metabolism, Biological Handbooks, Fed. Am. Soc. Exptl. Biol., Bethesda, 1968, pp. 9−47.

[3] *J. Wolf,* Der Diurnale Säurerhythmus, in: *W. Ruhland* (ed.), Encyclopedia of Plant Physiology, Vol. *XII/2,* Springer-Verlag, Berlin 1960, pp. 809−889.

[4] *P. S. Nobel,* Rapid Isolation Techniques for Chloroplasts, in: *S. P. Colowick, N. O. Kaplan* (eds.), Methods in Enzymology, Vol. *XXXI,* Academic Press, New York 1974, pp. 600−606.

[5] *C. W. Mehard,* Mitochondrien, in: *G. Jacobi* (ed.), Biochemische Cytologie der Pflanzenzelle, Thieme Verlag, Stuttgart 1974, pp. 109−126.

[6] *G. G. Laties,* Isolation of Mitochondria from Plant Material, in: *S. P. Colowick, N. O. Kaplan* (eds.), Methods in Enzymology, Vol. *XXXI,* Academic Press, New York 1974, pp. 589−600.

[7] *N. E. Good, G. D. Winget, W. Winter, T. N. Connolly, S. Izawa, R. M. M. Singh,* Hydrogen Ion Buffers for Biological Research, Biochem. *5,* 467−477 (1966).

[8] *J. M. Lord, T. Kagawa, T. S. Moore, H. Beevers,* Endoplasmic Reticulum as the Site of Lecithin Formation in Castor Bean Endosperm, J. Cell. Biol. *57,* 659−667 (1973).

[9] *P. H. Quail,* Plant Cell Fractionation, Annu. Rev. Plant Physiol. *30,* 425−484 (1979).

[10] *W. W. Franke,* Zellkerne und Kernbestandteile, in: *G. Jacobi* (ed.), Biochemische Cytologie der Pflanzenzelle, Thieme Verlag, Stuttgart 1974, pp. 14−40.

[11] *S. I. Honda,* Fractionation of Green Tissue, in: *S. P. Colowick, N. O. Kaplan* (eds.), Methods in Enzymology, Vol. *XXXI,* Academic Press, New York 1974, pp. 544−553.

[12] *D. J. Morré, T. J. Buckhout,* Isolation of Golgi Apparatus, in: *E. Reid* (ed.), Methodological Surveys (B), Vol. *9*, Plant Organelles, Ellis Horwood, Chichester 1979, pp. 117 – 134.

[13] *W. D. Loomis,* Overcoming Problems of Phenolics and Quinones in the Isolation of Plant Enzymes and Organelles, in: *S. P. Colowick, N. O. Kaplan* (eds.), Methods in Enzymology, Vol. *XXXI*, Academic Press, New York 1974, pp. 528 – 544.

[14] *T. Galliard,* Techniques for Overcoming Problems of Lipolytic Enzymes and Lipooxygenases in the Preparation of Plant Organelles, in: *S. P. Colowick, N. O. Kaplan* (eds.), Methods in Enzymology, Vol. *XXXI*, Academic Press, New York 1974, pp. 520 – 528.

[15] *H. Stern,* Isolation and Purification of Plant Nucleic Acids from Whole Tissues and from Isolated Nuclei, in: *S. P. Colowick, N. O. Kaplan* (eds.), Methods in Enzymology, Vol. *XIIB*, Academic Press, New York 1968, pp. 100 – 112.

[16] *C. R. Stocking,* Chloroplasts: Nonaqueous, in: *S. P. Colowick, N. O. Kaplan* (eds.), Methods in Enzymology, Vol. *XXIIIA*, Academic Press, New York 1971, pp. 221 – 228.

[17] *L. Y. Yatsu, T. J. Jacks,* Association of Lysosomal Activity with Aleurone Grains in Plant Seeds, Arch. Biochem. Biophys. *124*, 466 – 471 (1968).

[18] *A. Ruesink,* Protoplasts of Plant Cells, in: *S. P. Colowick, N. O. Kaplan* (eds.), Methods in Enzymology, Vol. *69*, Academic Press, New York 1980, pp. 69 – 84.

[19] *E. C. Cocking,* The Isolation of Plant Protoplasts, in: *S. P. Colowick, N. O. Kaplan* (eds.), Methods in Enzymology, Vol. *XXXI*, Academic Press, New York 1974, pp. 578 – 583.

[20] *D. N. Kuhn, P. K. Stumpf,* Preparation and Use of Protoplasts for Studies of Lipid Metabolism, in: *S. P. Colowick, N. O. Kaplan* (eds.), Methods in Enzymology, Vol. *72*, Academic Press, New York 1981, pp. 774 – 783.

[21] *E. Galun,* Plant Protoplasts as Physiological Tools, Annu. Rev. Plant Physiol. *32*, 237 – 266 (1981).

[22] *T. Boller, H. Kende,* Hydrolytic Enzymes in the Central Vacuole of Plant Cells, Plant Physiol. *63*, 1123 – 1132 (1979).

[23] *M. Nishimura, D. Graham, T. Akazawa,* Isolation of Intact Chloroplasts and other Cell Organelles from Spinach Leaf Protoplasts, Plant Physiol. *58*, 309 – 314 (1976).

[24] *K. Ohyama, L. E. Pelcher, D. Horu,* A Rapid, Simple Method for Nuclei Isolation from Plant Protoplasts, Plant Physiol. *60*, 179 – 181 (1977).

[25] *D. S. Perlin, R. M. Spanswick,* Labeling and Isolation of Plasma Membranes from Corn Leaf Protoplasts, Plant Physiol. *65*, 1053 – 1057 (1980).

[26] *J. P. Mascarenhas, M. Berman-Kurtz, R. R. Kulikowski,* Isolation of Plant Nuclei, in: *S. P. Colowick, N. O. Kaplan* (eds.), Methods in Enzymology, Vol. *XXXI*, Academic Press, New York 1974, pp. 558 – 565.

[27] *G. J. Wagner, H. W. Siegelman,* Large Scale Isolation of Intact Vacuoles and Isolation of Chloroplasts from Protoplasts of Mature Plant Tissues, Science *190*, 1298 – 1299 (1975).

[28] *R. Haas, E. Heinz, G. Popovici, G. Weissenböck,* Protoplasts from Oat Primary Leaves as Tools for Experiments on the Compartmentation in Lipid and Flavanoid Metabolism, Z. Naturforsch. *34c*, 854 – 864 (1979).

[29] *C. Buser, Ph. Matile,* Malic Acid in Vacuoles Isolated from Bryophyllum Leaf Cells, Z. Pflanzenphysiol. *82*, 462 – 466 (1977).

[30] *J. E. Lamb, H. Riezman, W. M. Becker, C. J. Leaver,* Regulation of Glyoxysomal Enzymes During Germination of Cucumber. 2. Isolation and Immunological Detection of Isocitrate Lyase and Catalase, Plant Physiol. *62*, 754 – 760 (1978).

[31] *H. J. Hohorst, F. H. Kreutz, Th. Bücher,* Über Metabolitgehalte und Metabolit-Konzentrationen in der Leber der Ratte, Biochem. Z. *332*, 18 – 46 (1959).

[32] *N. S. Radin,* Preparation of Lipid Extracts, in: *S. P. Colowick, N. O. Kaplan* (eds.), Methods in Enzymology, Vol. *XIV*, Academic Press, New York 1969, pp. 245 – 254.

[33] *N. S. Radin,* Extraction of Tissue Lipids with a Solvent of Low Toxicity, in: *S. P. Colowick, N. O. Kaplan* (eds.), Methods in Enzymology, Vol. *72*, Academic Press, New York 1981, pp. 5 – 7.

[34] *J. McLeish,* Quantitative Relationship between Deoxyribonucleic and Ribonucleic Acid in Isolated Plant Nuclei, Proc. Roy. Soc. *B158*, 261 – 278 (1963).

[35] *C. A. Price,* A Note on Isolation of Plant Nuclei, in: *E. Reid* (ed.), Methodological Surveys (B), Vol. *9*, Plant Organelles, Ellis Horwood, Chichester 1979, pp. 200 – 206.

[36] *Y. M. Chen, C. Y. Lin, H. Chang, T. J. Guilfoyle, J. L. Key,* Isolation and Properties of Nuclei from Control and Auxin-treated Soybean Hypocotyl, Plant Physiol. *56*, 78 – 82 (1975).

[37] *E. I. Philipp, W. W. Franke, T. W. Keenan, J. Stadler, E. D. Jarasch,* Characterization of Nuclear Membranes and Endoplasmic Reticulum Isolated from Plant Tissue, J. Cell. Biol. *68*, 11 – 29 (1976).

[38] *S. G. Reeves, D. O. Hall,* Higher Plant Chloroplasts and Grana. General Preparative Procedure. (Excluding High Carbon Dioxide Fixation Ability Chloroplasts), in: *S. P. Colowick, N. O. Kaplan* (eds.), Methods in Enzymology, Vol. *69*, Academic Press, New York 1980, pp. 85 – 94.

[39] *D. A. Walker,* Preparation of Higher Plant Chloroplasts, in: *S. P. Colowick, N. O. Kaplan* (eds.), Methods in Enzymology, Vol. *69*, Academic Press, New York 1980, pp. 94 – 104.

[40] *B. Liedvogel, P. Sitte, H. Falk,* Chromoplasts in the Daffodil: Finestructure and Chemistry, Cytobiol. *12*, 155 – 174 (1976).

[41] *A. R. Wellburn, F. A. M. Wellburn,* A New Method for the Isolation of Etioplasts with Intact Envelopes, J. exp. Bot. *22*, 972 – 979 (1971).

[42] *I. Washitani, S. Sato,* Studies on the Function of Proplastids in the Metabolism of in Vitro Cultured Tobacco Cells. I. Localization of Nitrite Reductase and NADP-dependent Glutamate Dehydrogenase, Plant Cell Physiol. *18*, 117 – 125 (1977).

[43] *B. J. Milflin, H. Beevers,* Isolation of Intact Plastids from a Range of Plant Tissues, Plant Physiol. *53*, 870 – 874 (1974).

[44] *C. K. M. Rathnam, G. E. Edwards,* Protoplasts as a Tool for Isolating Photosynthetically Active Chloroplasts from Grass Leaves, Plant Cell Physiol. *17*, 177 – 186 (1976).

[45] *S. P. Robinson, D. A. Walker,* Rapid Separation of the Chloroplast and Cytoplasmic Fractions from Intact Leaf Protoplasts, Arch. Biochem. Biophys. *196*, 319 – 323 (1979).

[46] *M. Nishimura, H. Beevers,* Isolation of Intact Plastids from Protoplasts from Castor Bean Endosperm, Plant Physiol. *62*, 40 – 43 (1978).

[47] *W. D. Bonner, Jr.,* A General Method for the Preparation of Plant Mitochondria, in: *S. P. Colowick, N. O. Kaplan* (eds.), Methods in Enzymology, Vol. *X*, Academic Press, New York 1967, pp. 126 – 133.

[48] *C. Jackson, A. L. Moore,* Isolation of Intact Higher-plant Mitochondria, in: *E. Reid* (ed.), Methodological Surveys (B), Vol. *9*, Plant Organelles, Ellis Horwood, Chichester 1979, pp. 1 – 12.

[49] *H. S. Ku, H. K. Pratt, A. R. Spurr, W. M. Harris,* Isolation of Active Mitochondria from Tomato Fruit, Plant Physiol. *43*, 883 – 887 (1968).

[50] *R. Douce, A. L. Moore, M. Neuburger,* Isolation and Oxidative Properties of Intact Mitochondria Isolated from Spinach Leaves, Plant Physiol. *60*, 625 – 628 (1977).

[51] *D. A. Day, J. B. Hanson,* On Methods for the Isolation of Mitochondria from Etiolated Cornshoots, Plant Sci. Lett. *11*, 99 – 104 (1977).

[52] *R. Thebud, K. A. Santarius,* Effects of Freezing on Isolated Plant Mitochondria, Planta *152*, 242 – 247 (1981).

[53] *D. A. Day, J. T. Wiskich,* Isolation and Properties of the Outer Membrane of Plant Mitochondria, Arch. Biochem. Biophys. *171*, 117 – 123 (1975).

[54] *J. Nedergaard, B. Cannon,* Overview-Preparation and Properties of Mitochondria from Different Sources, in: *S. P. Colowick, N. O. Kaplan* (eds.), Methods in Enzymology, Vol. *LV*, Academic Press, New York 1979, pp. 3 – 28.

[55] *H. Beevers,* Microbodies in Higher Plants, Annu. Rev. Plant Physiol. *30*, 159 – 193 (1979).

[56] *E. L. Vigil, G. Wanner, R. R. Theimer,* Isolation of Plant Microbodies, in: *R. Reid* (ed.), Methodological Surveys (B), Vol. *9*, Plant Organelles, Ellis Horwood, Chichester 1979, pp. 89 – 102.

[57] *N. E. Tolbert,* Isolation of Leaf Peroxisomes, in: *S. P. Colowick, N. O. Kaplan* (eds.), Methods in Enzymology, Vol. *XXIII*, Academic Press, New York 1971, pp. 665 – 682.

[58] *H. Beevers, R. W. Breidenbach,* Glyoxysomes, in: *S. P. Colowick, N. O. Kaplan* (eds.), Methods in Enzymology, Vol. *XXXI*, Academic Press, New York 1974, pp. 565 – 571.

[59] *H. Beevers, R. R. Theimer, J. Feierabend,* Microbodies (Glyoxysomen, Peroxisomen), in: *G. Jacobi* (ed.), Biochemische Cytologie der Pflanzenzelle, Thieme Verlag, Stuttgart 1974, pp. 127 – 146.

[60] *R. A. Leigh, D. Branton,* Isolation of Vacuoles from Root Storage Tissue of Beta Vulgaris L., Plant Physiol. *58*, 656 – 662 (1976).

[61] *R. R. Theimer, H. Beevers,* Uricase and Allantoinase in Glyoxysomes, Plant Physiol. *47*, 246 – 251 (1971).

[62] *R. R. Theimer, G. Anding, P. Matzner,* Kinetin Action on the Development of Microbody Enzymes in Sunflower Cotyledons in the Dark, Planta *128,* 41 – 47 (1976).

[63] *A. H. C. Huang, H. Beevers,* Isolation of Microbodies from Plant Tissues, Plant Physiol. *48,* 637 – 641 (1971).

[64] *D. J. Mooré,* Isolation of Golgi Apparatus, in: *S. P. Colowick, N. O. Kaplan* (eds.), Methods in Enzymology, Vol. *XXII,* Academic Press, New York 1971, pp. 130 – 148.

[65] *P. M. Ray, W. R. Eisinger, D. G. Robinson,* Organelles Involved in Cell Wall Polysaccharide Formation and Transport in Pea Cells, Ber. Dt. Bot. Ges. *89,* 121 – 146 (1976).

[66] *T. S. Moore, H. Beevers,* Isolation and Characterization of Organelles from Soybean Suspension Cultures, Plant Physiol. *53,* 261 – 265 (1974).

[67] *C. T. Brett, D. H. Northcote,* The Formation of Oligoglucans Linked to Lipid During Synthesis of β-Glucan by Characterized Membrane Fractions Isolated from Peas, Biochem. J. *148,* 107 – 117 (1975).

[68] *R. A. Leigh, D. Branton, F. Marty,* Methods for the Isolation of Intact Vacuoles and Fragments of Tonoplasts, in: *E. Reid* (ed.), Methodological Surveys (B), Vol. *9,* Plant Organelles, Ellis Horwood, Chichester 1979, pp. 69 – 80.

[69] *R. Kringstad, W. H. Kenyon, C. C. Black, Jr.,* The Rapid Isolation of Vacuoles from Leaves of Crassulacean Acid Metabolism Plants, Plant Physiol. *66,* 379 – 382 (1980).

[70] *F. Sasse, D. Backs-Hasemann, W. Barz,* Isolation and Characterization of Vacuoles from Cell Suspension Cultures of Daucus Carota, Z. Naturf. *34c,* 848 – 853 (1979).

[71] *Ph. Matile, A. Wiemken,* Vacuoles and Spherosomes, in: *S. P. Colowick, N. O. Kaplan* (eds.), Methods in Enzymology, Vol. *XXXI,* Academic Press, New York 1974, pp. 572 – 578.

[72] *Ph. Matile, B. Jans, R. Rickenbacher,* Vacuoles of Chelidonium Latex: Lysosomal Property and Accumulation of Alcaloids, Biochem. Physiol. Pflanzen *161,* 447 – 458 (1970).

[73] *J. L. Hall, A. R. D. Taylor,* Isolation of the Plasma Membrane from Higher Plant Cells, in: *E. Reid* (ed.), Methodological Surveys (B), Vol. *9,* Plant Organelles, Ellis Horwood, Chichester 1979, pp. 103 – 111.

[74] *T. K. Hodges, R. T. Leonard,* Purification of a Plasma Membrane-bound Adenosine Triphosphatase from Plant Roots, in: *S. P. Colowick, N. O. Kaplan* (eds.), Methods in Enzymology, Vol. *XXXII,* Academic Press, New York 1974, pp. 392 – 406.

[75] *W. E. Boss, A. W. Ruesink,* Isolation and Characterization of Concanavaline-A-labeled Plasma Membranes of Carrot Protoplasts, Plant Physiol. *64,* 1005 – 1011 (1979).

[76] *R. L. Jones,* The Isolation of Endoplasmic Reticulum from Barley Aleuron Layers, Planta *150,* 58 – 69 (1980).

[77] *W. Banks, D. D. Muir,* Structure and Chemistry of the Starch Granule, in: *P. K. Stumpf, E. E. Conn* (eds.), The Biochemistry of Plants, Vol. *3,* Academic Press, New York 1980, pp. 321 – 369.

[78] *N. R. Gilkes, M. J. Chrispeels,* Endoplasmic Reticulum of Mung Bean Cotyledons. Accumulation During Seed Maturation and Catabolism During Seedling Growth, Plant Physiol. *65,* 600 – 604 (1980).

[79] *L. Beevers, R. M. Mense,* Glycoprotein Biosynthesis in Cotyledons of Pisum Sativum L. Involvement of Lipid-linked Intermediates, Plant Physiol. *60,* 703 – 708 (1977).

[80] *C. L. Jelsema, M. Ruddat, D. J. Morré, F. A. Williamson,* Specific Binding of Gibberellin A_1 to Aleuron Grain Fractions from Wheat Endosperm, Plant Cell Physiol. *18,* 1009 – 1019 (1977).

[81] *R. J. Youle, A. H. C. Huang,* Protein Bodies from the Endosperm of Castor Bean. Subfractionation. Protein Components, Lectins, and Changes During Germination, Plant Physiol. *58,* 703 – 709 (1976).

[82] *W. Van der Wilden, E. M. Herman, M. J. Chrispeels,* Protein Bodies of Mung Bean Cotyledons as Autophagic Organelles, Proc. Natl. Acad. Sci. *77,* 428, 432 (1980).

[83] *L. Y. Yatsu, T. J. Jacks, T. P. Hensarling,* Isolation of Spherosomes (Oleosomes) from Onion, Cabbage, and Cotton Seed Tissues, Plant Physiol. *48,* 675 – 682 (1971).

[84] *C. L. Jelsema, D. J. Mooré, M. Ruddat, C. Turner,* Isolation and Characterization of the Lipid Reserve Bodies, Spherosomes, from Aleurone Layers of Wheat, Bot. Gaz. *138,* 138 – 149 (1977).

1.3   Methods for Protein Determination

Georg-Burkhard Kresze

The determination of protein concentration is important to provide a reference point for other measurements such as metabolite concentration or enzyme activity, as well as in clinical chemistry for diagnostic purposes [1]. Several different methods for protein quantitation are available. Proteins obviously have a number of basic properties in common. However, different proteins vary greatly in their amino acid composition and sequence as well as in their size and shape. Since the different assay methods depend on different properties of the proteins, consistency between the methods may be low. Furthermore, the various methods can hardly be expected to give equal responses with different proteins. Thus, no single method is optimal for every application.

When selecting a method for the estimation of protein, several criteria have to be considered:

− detectability;
− interference by other substances present;
− variability of response with different proteins;
− facility of performance.

The methods presented below will be described with special regard to these criteria. A more detailed review on the techniques of protein determination can be found in [2].

1.3.1   Protein Standards

Most of the methods for protein determination require standardization. Ideally, the assay should be calibrated with the protein itself whose concentration is to be measured. Since this will not be possible in most cases, the use of crystalline bovine serum albumin as the standard has been generally adopted. Since proteins may contain up to 0.1 g of water per g of protein [2], the concentration of the standard should be checked spectrophotometrically. Protein quantities are then obtained as "serum albumin units" rather than as absolute amounts. If this deviation is expected to be marked (e.g. in dye-binding assays), another protein which more closely matches the behaviour of the sample under study should be employed as the standard.

1.3.2   Ultraviolet Absorption Methods

Most proteins exhibit ultraviolet-light absorption spectra with absorption maxima around 275 – 280 nm due to their content of tyrosine and tryptophan. At shorter wavelengths, other amino acids contribute to the absorption, and below 220 nm absorption of the peptide bond itself is important.

If a pure protein is studied, the absorbance at 280 nm can be used for quantitation. $A_{280}^{1\%}$ values* of a large number of proteins have been compiled by *Kirschenbaum* [3]; many of them are about 10 (i.e., a solution containing 1 mg of protein per ml has an absorption of 1.0 at 280 nm) but may vary by more than an order of magnitude (from <1 to about 30) due to the different contents of aromatic amino acids. Thus, A_{280} measurements of mixtures or impure proteins can only give approximate results which nevertheless may be sufficient in some cases, e.g. for detecting proteins in the effluent of chromatographic columns.

A major problem with all UV-spectroscopic methods for protein determination is interference by other materials (such as nucleic acids or buffer substances) which absorb in this range. To correct for nucleic acid content, the method of *Warburg & Christian* [4] can be used as follows.

Assay

Assay conditions: wavelengths 260 nm and 280 nm; light path 10 mm; assay volume 3.00 ml; room temperature.

Measurement

Pipette into quartz cuvette:	
water	2.80 ml
read A_1 at 260 nm and 280 nm against air,	
sample solution (1 to 5 mg protein/ml)	0.20 ml
mix, read A_2 at 260 nm and 280 nm.	

Calculation [5]

$$\Delta A = 1.55 \times (A_2 - A_1)_{280} - 0.76 \times (A_2 - A_1)_{260}$$

---

* $A_x^{1\%}$, absorbance at x nm of a solution containing 10 mg substance per ml (light path, 10 mm).

Protein concentration in sample solution is

$$\rho = 15 \times \Delta A \ g/l$$

The detectability of the assay can be enhanced to about 100 μg of protein per assay by using semimicro cuvettes and a total assay volume of 1 ml.

Comments

The UV-spectroscopic method is non-destructive and can be easily and rapidly performed. However, variability with different proteins is large (the formula given above is strictly valid only for proteins with $A_{280}^{1\%} = 8.9$ and $A_{280}/A_{260} = 1.75$ [2]), and other substances can interfere strongly. This especially applies when the composition of the sample solution is not precisely known, so that an appropriate buffer blank cannot be used.

To increase detectability, measurement at shorter wavelength pairs has been used (230/260 [6], 215/225 [7, 8], 205/280 [9], 205/210 [10], 224/233.3 [11], 228.5/234.5 [12], 235/280 [13], or 191 – 194 nm [14]). Although as little as 10 μg of protein may be sufficient for assay, interference by UV-absorbing materials (e.g. NaOH, acetate, citrate) is even more serious in this wavelength region.

1.3.3   The Biuret Method

The biuret method is based on the reaction of peptide bonds with copper ions in alkali leading to the formation of a violet-coloured complex which can be quantitated photometrically. Standard biuret procedures have been described by *Gornall et al.* [15] and *Beisenherz et al.* [16]. The following assay is a modification of the procedure given in [16].

Assay

Preparation of solutions

1. Sodium hydroxide (0.2 mol/l):

   dissolve 8 g NaOH in water to give 1000 ml.

2. Biuret reagent:

dissolve 9 g sodium potassium tartrate, $NaKC_4H_4O_6 \cdot 4 H_2O$ p.a. with 400 ml NaOH (1), add 3 g pulverized $CuSO_4 \cdot 5 H_2O$ and finally 5 g KI. Each salt must be dissolved completely before the next addition. The mixture is made up to 1000 ml with NaOH (1) and can be stored in a polyethylene bottle protected from light for at least several weeks at room temperature. Biuret reagent (formulated as in [15]) can also be obtained commercially (*Sigma Chem. Co.,* St. Louis, Mo., U.S.A.).

3. Trichloroacetic acid (3 mol/l):

dissolve 490 g TCA in water to give 1000 ml.

4. Bovine serum albumin standard:

dissolve crystalline serum albumin in water to give a concentration of 2.5 mg/ml; the exact concentration should be checked spectrophotometrically ($A_{280}^{1\%} = 6.60$ [2]). Prepare several dilutions containing from 0.2 to 2.5 mg/ml.

Sample solution: protein solutions containing 0.2 to 2.5 mg/ml can be used without dilution. More concentrated solutions should be diluted appropriately with water.

Assay conditions: wavelength Hg 546 nm; light path 10 mm; assay volume 1.2 ml; room temperature.

Measurement

Pipette into centrifuge tubes:		sample	blank
sample or standard (0.2 – 2.5 mg of protein)	(4)	1.00 ml	–
trichloroacetic acid	(3)	0.20 ml	–
mix, centrifuge for 2 to 5 min; carefully discard supernatant, drain tubes by placing them upside down onto filter paper,			
water		0.20 ml	0.20 ml
biuret reagent	(2)	1.00 ml	1.00 ml
mix to dissolve precipitate; incubate for 30 min at room temperature, read absorbance of sample and blank. $\Delta A = A_{sample} - A_{blank}$.			

Calculation: the biuret factor used in calculation may vary with different reagent batches as well as with the age of the reagent. Therefore it is advisable to calibrate the assay with known amounts of bovine serum albumin. The calibration curve is linear up to at least 2.5 mg of protein per assay. From this curve, the biuret factor f = $(\Delta A/\text{mg of protein})^{-1}$ can be calculated. From the data given in [16], f = 4.1 is obtained; similar values are usually found. Protein concentration in the sample solutions is then calculated according to

$$\rho = f \times \Delta A \ \ \text{g/l}$$

Comments

The biuret assay depends on the concentration of peptide bonds rather than individual amino acid residues. Therefore, variability with different proteins is smaller than in most other methods (calibration factors for a number of proteins are given in [16]). Relatively few substances interfere. However, since there is interference by common compounds such as ammonium sulphate, Tris, glycerol, or sucrose, it is generally advisable to precipitate the protein with trichloroacetic acid before the assay. This also abolishes problems with coloured proteins unless the chromophore is bound covalently to the protein (in that case, the absorbance of the protein can be corrected for by a suitable blank). Lipids also interfere with the biuret assay because they cause turbidity of the solution. To cope with this problem, the solution can either be extracted with ether [5] or treated with sodium deoxycholate (30 mg/ml) [2] before the assay, or the solution is decolorized after reading the absorbance with KCN (25 mg/ml) and $A_{546\text{nm}}$ is read again [17]. ΔA is then calculated according to

$$\Delta A = (A_{\text{sample}} - A_{\text{blank}})_{\text{before KCN}} - (A_{\text{sample}} - A_{\text{blank}})_{\text{after KCN}} \cdot$$

The sensitivity of the biuret method is rather low. Several micro-biuret procedures have been described which make use of the UV absorption of the biuret complex in the range of 260 to 330 nm [2]. However, interferences and blank absorbances are much more disturbing in this region. Separation of protein-bound copper by gel filtration or ion-exchange chromatography, followed by quantitation by reaction with chloramine T [18], as the copper-diethyldithiocarbamate complex [19], or by atomic absorption spectrometry [20] has also been used. These procedures have a very good detectability (down to 20 ng of protein per assay) but are rather complex to perform and subject of strong interferences. It appears that another procedure such as the *Lowry* assay is more appropriate when better detectability is required. On the other hand, if sufficient material is available, the biuret assay may well stand out as the method of choice due to its high precision and ready performance. The biuret assay has been recommended as a reference method for determination of total serum protein [21, 22].

1.3.4   The Lowry Method

The method described by *Lowry et al.* [23] combines the use of the biuret reaction of proteins with copper ions in alkali with the reduction of the *Folin-Ciocalteu* phenol reagent (phosphomolybdic-phosphotungstic acid) by tyrosine and tryptophan residues. The latter reaction, in a complicated way, is intensified by the copper-protein complex. A detailed review of the *Lowry* method is available [24].

The following procedure is a modification of the original method of *Lowry et al.* [23] based on the results of *Bensadoun & Weinstein* [25] and *Hess et al.* [26].

Assay

Preparation of solutions

1. Sodium deoxycholate (10 mg/ml):

   dissolve 1 g sodium deoxycholate, $NaC_{24}H_{39}O_4$ p.a., in water to give 100 ml.

2. Trichloroacetic acid (240 mg/ml):

   dissolve 24 g TCA in water to give 100 ml.

3. Sodium hydroxide (0.8 mol/l):

   dissolve 32 g NaOH in water to give 1000 ml.

4. Reagent A:

   dissolve 0.2 g disodium tartrate $Na_2C_4H_4O_6 \cdot 2\ H_2O$, p.a., and 10 g Na_2CO_3 in 69 ml NaOH (3), dilute to 100 ml with water.

5. Reagent B:

   dissolve 2 g disodium tartrate and 1 g $CuSO_4 \cdot 5\ H_2O$, p.a. with 12.5 ml NaOH (3) and dilute to 100 ml with water. Store protected from light.

6. Reagent mixture:

   mix 25 volumes NaOH solution (3), 18 volumes reagent A (4), and 2 volumes reagent B (5). Prepare immediately before use.

7. Reagent C:

   dilute 1 volume of commercial Folin-Ciocalteu reagent with 2 volumes of water. This mixture should be prepared daily just before use.

8. Bovine serum albumin standard:

   dissolve crystalline bovine serum albumin with water to give a concentration of 1.0 mg/ml. The exact concentration should be checked spectrophotometrically ($A_{280\,nm}^{1\%} = 6.60$ [2]). Prepare several dilutions containing from 0.01 to 0.10 mg/ml.

Sample solution: protein solutions containing 0.01 to 0.10 mg/ml can be used without dilution. More concentrated solutions must be diluted adequately with water before the assay. At least three different sample dilutions should be used. Samples containing less than 0.01 mg protein/ml can be concentrated by repeated precipitation with trichloroacetic acid in the same tube as described below.

Assay conditions: wavelength 750 nm; light path 10 mm; assay volume 1.05 ml; room temperature.

Measurement

Pipette into centrifuge tubes or 1.5-ml reaction vials:		sample	blank
sample or standard (5 to 60 µg of protein)	(8)	0.60 ml	–
water		–	0.60 ml
sodium deoxycholate	(1)	0.01 ml	0.01 ml
mix, incubate for 15 min			
trichloroacetic acid	(2)	0.20 ml	0.20 ml
mix, centrifuge for 30 min at 3300 min^{-1} in a swinging-bucket rotor or (reaction vials) for 7 min in a microcentrifuge (*Eppendorf* 5414 or equivalent), remove supernatant carefully by aspirating with a *Pasteur* pipette connected to a water-suction device. The last part of the solution is removed by tilting the tube and maintaining the tip of the suction pipette off the bottom and in contact with the side of the tube.			
reagent mixture	(6)	0.45 ml	0.45 ml
vortex to dissolve precipitate			
reagent C	(7)	0.30 ml	0.30 ml
mix immediately and vigorously			
reagent C	(7)	0.30 ml	0.30 ml
mix vigorously; with reaction vials, centrifuge briefly to bring the solution completely to the bottom of the vial; incubate for 30 to 60 min protected from light, read absorbance.			

Calculation: the colour obtained is not strictly proportional to protein concentration although the procedure described has been optimized to improve linearity [26]. Therefore, it is necessary to prepare a calibration curve with different amounts of serum albumin each time the assay is performed. The standard curve is drawn onto graph paper and used for determination of the amount of protein present in the sample.

Comments

Owing to its simplicity, precision, and sensitivity, the method of *Lowry et al.* [23] has become the most widely used procedure for protein determination. For enhanced detectability, micro-adaptations have been described which can be used with as little as 0.2 µg of protein [23, 24] whereas the original procedure [23] is suitable for 25 to 300 µg of protein. Thus, the *Lowry* method is about 10 to 20 times more sensitive than UV absorption measurement and 50 to 100 times more sensitive than the standard biuret assay. If no spectrophotometer capable of providing readings at 750 nm is available, other wavelengths in the range of 500 to 750 nm (e.g. Hg 578 nm) can be used, but of course with reduced detectability.

The *Lowry* method is known to give a non-linear calibration curve [23]. Thus, for better precision it is advisable to assay at least three different dilutions of the sample. The non-linear behaviour appears to be inherent in the reaction mechanism [24]. Nevertheless, a number of attempts have been made to improve linearity, either by variation of the reaction conditions [2, 24, 26] or by the application of curve-fitting techniques [24].

The colour given by a protein in the *Lowry* assay depends on its content of tyrosine and tryptophan residues. Therefore, variation of response with various proteins can be quite significant and is certainly greater than in the biuret assay, although generally smaller than in UV absorption measurement [23, 24].

A large number of substances interfere with protein quantitation by the *Lowry* assay [2, 23 – 25] (Table 1).

Many modifications of the original procedure have been proposed to cope with interfering substances; details and references can be found in [24]. However, unless one is dealing with solutions which either are devoid of interfering substances or are so precisely defined that the same amount of the interfering substance(s) can be incorporated into the reagent blank and into the standard samples as is present in the unknown sample, it is highly advisable to separate the protein from interfering substances by precipitation with trichloroacetic acid. This also serves to concentrate proteins from very dilute solutions. Trichloroacetic acid alone does not precipitate proteins reliably and quantitatively at low levels (1 to 25 µg) of protein. This difficulty can be overcome by the combined use of trichloroacetic acid and deoxycholate [25] or trichloroacetic acid and yeast soluble ribonucleic acid (0.125 mg/ml) [27]. The latter method can also be applied to precipitate proteins quantitatively in the presence of detergents such as dodecylsulphate, digitonin, or sulphobetaines [27].

Special conditions for determination of membrane and lipoprotein samples have been given in [28].

Table 1. Substances interfering with the *Lowry* assay. Numbers in parentheses give the tolerable concentration limit as given in the literature [2, 24, 25]

Buffers

Tris (100 μmol/l), phosphate (100 mmol/l), Tricine (100 μmol/l), MOPS (10 μmol/l), Hepes (1 μmol/l), Bicine (5 μmol/l), citrate (1 mmol/l), glycylglycine (100 μmol/l)

Amino acids

Histidine (150 nmol/l), tyrosine (150 nmol/l), tryptophan (150 nmol/l), cysteine (40 μmol/l), cystine (40 μmol/l), glycine (1 mmol/l), glutamate (no value in literature).

Carbohydrates

Sucrose (4 mmol/l), monosaccharides (glucose, fructose, mannose, sorbose, xylose, rhamnose) (150 μmol/l), glucosamine (5 mg/l), glycerol (100 ml/l), ethylene glycol (1 g/l), Ficoll, Metrizamide, polyvinyl pyrrolidone

Reductants

Phenols (except nitrophenols), dithiothreitol (20 μmol/l), glutathione (40 μmol/l), 2-mercaptoethanol (700 μmol/l), hydrazine (5 mg/l), ascorbic acid (100 μg/l)

Salts, metals, etc.

$(NH_4)_2SO_4$ (1.5 g/l), K^+ (12 mmol/l), Hg^{2+} (15 μmol/l), Mn^{2+} (15 μmol/l), Co^{2+} (1 μmol/l), NaCl (0.7 mol/l), NaI (15 mmol/l)

Miscellaneous

EDTA (70 μmol/l ?), haematin (15 μg/l), Triton X-100 (10 mg/l), oxidized lipids (10 – 200 μmol/l), fatty acids (1 mmol/l, Ampholines, salicylate (20 μg/l), penicillin (0.5 units/ml), acetylacetone (2 μmol/l)

1.3.5   Protein-dye Binding Methods

A number of dyes have been used to stain proteins after electrophoretic separation in gels. These staining procedures are very sensitive, so attempts have been made to devise methods for protein quantitation with dyes such as Xylene Brilliant Cyanine G [29, 30], Amido Black [31 – 33], bromosulphalein [34, 35], or Coomassie Brilliant Blue R-250 [36, 37]. Although these procedures possess high sensitivity, they require separation of the protein-dye complex from unbound dye by filtration or centrifugation. However, very simple methods have been reported which use Coomassie Brilliant Blue G-250 [38, 39] or bromophenol blue [40]. The following procedure was given by *Bradford* [38]; it is based on the shift of the absorption maximum of the dye from 465 to 595 nm which occurs upon binding of the dye to protein.

Assay

Preparation of solutions

1. Dye reagent:
   dissolve 0.10 g Coomassie Brilliant Blue G-250 (C.I. 42 655, Xylene Brilliant

Cyanine G, *Serva* Blue G) in 50 ml ethanol ($\varphi = 0.95$). To this solution, add 100 ml phosphoric acid (H_3PO_4 p.a. $w = 0.85$). Dilute the resulting solution to 1000 ml with water, and filter. The final reagent is stored at room temperature and is stable for at least two weeks [41]. Five-fold concentrated dye reagent can be obtained commercially (*Bio-Rad Laboratories,* Richmond, Cal., U.S.A.).

2. Protein standard solution:

dissolve the standard protein with water or NaCl (0.15 mol/l) to give a concentration of 1.4 mg/ml. Check the exact concentration using the appropriate absorption coefficient [3]. Prepare several dilutions containing 0.2 to 1.4 mg/ml for the standard assay or 0.01 to 0.1 mg/ml for the micro-assay.

Sample solution: protein solutions containing 0.2 to 1.4 mg/ml (for the standard assay) or 0.01 to 0.1 mg/ml (for the micro-assay) can be used without dilution. More concentrated protein solutions must be diluted with water or NaCl (0.15 mol/l).

Assay conditions: wavelength 595 nm; light path 10 mm; assay volume 5.1 ml (standard assay) or 1.1 ml (micro-assay); room temperature.

Measurement

Standard assay

Pipette into 12 × 100-mm test tube:		sample	blank
sample or standard	(2)	0.10 ml	–
(20 to 140 µg of protein)			
water or appropriate buffer		–	0.10 ml
dye reagent	(1)	5.00 ml	5.00 ml
mix by inversion or vortexing; after 5 min and before one hour, read absorbance against blank in glass or plastic cuvettes.			

Microassay

Pipette into 12 × 100-mm test tube:		sample	blank
sample or standard	(2)	0.10 ml	–
(1 to 20 µg of protein)			
water or appropriate buffer		–	0.10 ml
dye reagent	(1)	1.00 ml	1.00 ml
mix by inversion or vortexing; after 5 min and before one hour, read absorbance against blank in glass or plastic cuvettes.			

Calculation: prepare a calibration curve with appropriate amounts of a standard protein. Plot the weight of protein against the corresponding absorbance on graph paper. Use this standard curve to determine the protein in the unknown sample.

Comments

Although the *Bradford* method stands out for its simplicity and high detectability, it possesses some important drawbacks which limit its application.

The standard curves are non-linear for many proteins, especially with more than 60 µg of protein [38, 39, 41 – 43]. This inherent non-linearity is caused by the reagent itself since there is an overlap in the spectrum of the two different colour forms of the dye so that the background value for the reagent is continually decreasing as more dye is bound to protein [38]. However, linear calibration curves have been reported for some proteins [42, 43]. Linearity can be improved by plotting the results on a log-log scale [44], by measuring the ratio A_{595}/A_{465} rather than A_{595} alone [45], or by using different volumes of the dye reagent [41]. More simply, satisfactory results are obtained by running the assay with a set of standards and using the calibration curve obtained in this manner rather than *Beer*'s Law to determine the amount of protein in the unknown sample. Absorbance may also vary with the age of the dye reagent [41].

A very serious problem with all assays based on dye binding is variation of response with different proteins which can be severe [42 – 44, 46]. When studied with 23 different proteins, the standard deviation in estimates of protein concentration by the *Bradford* method was twice the value obtained by the *Lowry* method [43]. Therefore it is important to use a standard protein which gives a similar colour yield as the protein or mixture under study. Unfortunately, bovine serum albumin is a poor standard in the *Bradford* method since it gives a much higher colour yield than most other proteins [43]. Bovine gamma globulin was recommended as a better standard which gives a more normal response and a linear calibration curve [43].

Several common laboratory substances such as Tris, acetic acid, 2-mercaptoethanol, sucrose, glycerol, and EDTA have small but detectable effects when present in the Coomassie Blue-binding assay. To correct for this interference, the blank and standard samples should contain the appropriate amounts of the interfering substance providing its concentration is known. Serious interferences are caused by phenol (>1 mmol/l), urea (>2 mol/l) [39], guanidine hydrochloride (6 mol/l) [45], ampholytes [41], alkaline buffers [38, 45], and detergents such as Triton X-100 or sodium dodecylsulphate (>1 mg/ml) [38, 43, 45]. If the protein concentration is high enough, the sample solution can be diluted in order to lower the concentration of the interfering substance to an innocuous value. Alkaline solutions must be neutralized before the assay. To remove the interference by dodecylsulphate, precipitation of the detergent with potassium phosphate was described [47]. The *Bradford* assay has also been used for special purposes (see also [43]) such as determination of the protein of cells immobilized in gels [48] or of biliproteins which cause problems due to their absorbance in the range of 550 – 600 nm [49].

The Coomassie Blue-binding assay should be performed in glass or plastic cuvettes since the protein-dye complex tends to bind to quartz cuvettes. Blue cuvettes can be cleaned either by rinsing with concentrated glassware detergent, followed by water and acetone, or by soaking in HCl (0.1 mol/l) [38].

It appears that the method of *Bradford* [38], while providing a rapid test for the presence of proteins (e.g. in chromatographic fractions), should be viewed with some caution. Reliable application is presently limited to measurement of proteins or protein mixtures which have been carefully standardized with other methods, or to situations where only very relative information is required [24].

1.3.6   Other Methods

Methods of high detectability have been described which use either fluorescent techniques such as binding of fluorescamine [50, 51], o-phthalaldehyde [52, 53], or cyclo-heptaamylose-dansyl chloride [54], radioactive labelling with [³H]-dansyl chloride [55] or [³H]-fluorodinitrobenzene [56], or enzymatic assay of glutamate derived from hydrolyzed protein [52]. Although some of these assays (as with the modifications of the biuret assay mentioned on p. 86) allow quantitation of less than 20 ng of protein, most are considerably complex to perform and require special instrumentation, so that their use will be restricted to applications where extreme sensitivity is required. In performing these high-sensitivity assays, special care should be given to cleaning of all glass- and plastic ware since a single fingerprint may contain up to 0.5 µg of protein [36].

Other methods which do not appear to offer appreciable advantages compared to the *Lowry* or dye-binding assays include estimation of protein with ninhydrin after alkaline hydrolysis [57], with trinitrobenzene sulphonic acid [58], by competitive binding [59], by turbidimetry [5, 60] or nephelometry [61, 62]. Turbidimetric and nephelometric methods are simple to perform but lack specificity. Estimation of protein by nephelometry has been optimized for the assay of protein in cerebrospinal fluid [61].

The nitrogen content of proteins is relatively invariable at 0.16 g per g of protein. Thus, nitrogen determination by the methods of *Kjeldahl* or *Dumas* [63 – 66] is an accurate but rather complicated procedure to determine protein concentration. Finally, mention should be made of determination of the dry weight of proteins [67] which is important for investigations of protein structure.

Appendix

Determination of Ammonium Sulphate

Samples subjected to enzyme isolation and purification procedures contain ammonium sulphate up to a concentration of ca. 3.5 mol/l. This may lead to inhibition of the catalytic activities of the enzymes and may affect the optical measurements, if the assay system contains additional phosphate and magnesium ions. The determination of the ammonium sulphate concentration in aqueous extracts from animal tissues is simple.

The determination is based on the titration of sulphate ions with barium chloride solution [68]; barium ions in excess produce an orange-coloured complex with the indicator alizarin S. This procedure has advantages in routine enzyme isolation and is feasible because animal tissues contain only very small amounts of sulphate.

Phosphate interferes with the determination; if the phosphate concentration is known, the result can be corrected for this value. In most cases the phosphate concentration can be neglected in comparison to the amount of ammonium sulphate present.

Preparations of solutions

1. Indicator solution:

   dissolve 100 mg alizarin S in water and make up to 50 ml.
   Add 450 ml ethanol and 3 ml hydrochloric acid, 2 mol/l.

2. Barium chloride solution (25 mmol/l):

   dissolve 6.1 g $BaCl_2 \cdot 2H_2O$ in water and make up to 1 000 ml.

Assay system

Pipette into a 50 ml conical flask:
5 ml water, 0.1 ml sample solution, and 5 ml indicator solution (1).

Titrate with barium chloride solution (2) until the colour changes from yellow to reddish-brown.

Calculation

The ammonium sulphate concentration is

$$c = V/4 \quad mol/l$$

V   is the volume of $BaCl_2$ solution required (ml).

References

[1] *L. M. Killingsworth,* Clinical Applications of Protein Determinations in Biological Fluids Other Than Blood, Clin. Chem. *28*, 1093–1102 (1982).
[2] *C. J. R. Thorne,* Techniques for Determining Protein Concentration, in: *H. L. Kornberg* (ed.), Techniques in Protein and Enzyme Biochemistry, Vol. B 104, Elsevier/North Holland, Amsterdam 1978, pp. 1–18.

[3] *D. M. Kirschenbaum,* Molar Absorptivity and $A_{1cm}^{1\%}$ values for proteins at selected wavelengths of the ultraviolet and visible regions, in: *G. D. Fasman* (ed.), Handbook of Biochemistry and Molecular Biology, 3rd edn., Vol. 2, CRC Press, Cleveland, Ohio, 1976, pp. 383 – 545; Anal. Biochem. *68,* 465 – 484 (1975), *80,* 193 – 211 (1977), *81,* 220 – 246 (1977), *82,* 83 – 100 (1977), *87,* 223 – 242 (1978), *90,* 309 – 330 (1978); Int. J. Pept. Prot. Res. *13,* 479 – 492 (1979); Int. J. Biochem. *11,* 487 – 500 (1980), *13,* 621 – 636 (1981).

[4] *O. Warburg, W. Christian,* Isolierung und Kristallisation des Gärungsferments Enolase, Biochem. Z. *310,* 384 – 421 (1941).

[5] *E. Layne,* Spectrophotometric and Turbidimetric Methods for Measuring Proteins, in: *S. P. Colowick, N. O. Kaplan* (eds.), Methods in Enzymology, Vol. *III,* Academic Press, New York 1957, pp. 447 – 454.

[6] *V. F. Kalb, R. W. Bernlohr,* A New Spectrophotometric Assay for Protein in Cell Extracts, Anal. Biochem. *82,* 362 – 371 (1977).

[7] *W. J. Waddell,* A simple ultraviolet spectrophotometric method for the determination of protein, J. Lab. Clin. Med. *48,* 311 – 314.

[8] *J. B. Murphy, M. W. Kies,* Note on Spectrophotometric Determination of Proteins in Dilute Solutions, Biochem. Biophys. Acta *45,* 382 – 384 (1960).

[9] *R. K. Scopes,* Measurement of Protein by Spectrophotometry at 205 nm, Anal. Biochem. *59,* 277 – 282 (1974).

[10] *A. R. Goldfarb, L. J. Saidel, E. Mosovich,* The Ultraviolet Absorption Spectra of Proteins, J. Biol. Chem. *193,* 397 – 404 (1951).

[11] *W. E. Groves, F. C. Davis, B. H. Sells,* Spectrophotometric Determination of Microgram Quantities of Protein without Nucleic Acid Interference, Anal. Biochem. *22,* 195 – 210 (1968).

[12] *B. Ehresmann, P. Imbault, J. H. Weil,* Spectrophotometric Determination of Protein Concentration in Cell Extracts Containing tRNA's and rRNA's, Anal. Biochem. *54,* 454 – 463 (1973).

[13] *J. R. Whitaker, P. E. Granum,* An Absolute Method for Protein Determination Based on Difference in Absorbance at 235 and 280 nm, Anal. Biochem. *109,* 156 – 159 (1980).

[14] *M. M. Mayer, J. A. Miller,* Photometric Analysis of Proteins and Peptides at 191 – 194 mµ, Anal. Biochem. *36,* 91 – 100 (1970).

[15] *A. G. Gornall, C. J. Bardawill, M. M. David,* Determination of Serum Proteins by Means of the Biuret Reaction, J. Biol. Chem. *177,* 751 – 766 (1949).

[16] *G. Beisenherz, H. J. Boltze, T. Bucher, R. Czok, K. H. Garbade, E. Meyer-Arendt, G. Pfleiderer,* Diphosphofructose-Aldolase, Phosphoglyceraldehyd-Dehydrogenase, Milchsäure-Dehydrogenase, Glycerophosphat-Dehydrogenase and Pyruvat-Kinase aus Kaninchenmuskulatur in einem Arbeitsgang, Z. Naturf. *8b,* 555 – 577 (1953).

[17] *J. W. Keyser, J. Vaughn,* Turbidities in the Estimation of Serum Proteins by the Biuret Method, Biochem. J. *44,* xxii (1949).

[18] *M. L. Goldberg,* Quantitative Assay for Submicrogram Amounts of Protein, Anal. Biochem. *51,* 240 – 246 (1973).

[19] *M. K. Johnson,* Variable Sensitivity in the Microbiuret Assay of Proteins, Anal. Biochem. *86,* 320 – 323 (1978).

[20] *D. Davies, E. S. Holdsworth,* A Method for the Estimation of Proteins in Colored or Turbid Solutions, Anal. Biochem. *100,* 92 – 94 (1979).

[21] *B. T. Doumas, D. D. Bayse, R. J. Carter, T. Peters, jr., R. Schaffer,* A Candidate Reference Method for Determination of Total Protein in Serum. I. Development and Validation, Clin. Chem. *27,* 1642 – 1650 (1981).

[22] *B. T. Doumas, D. D. Bayse, K. Borner, R. J. Carter, F. Elevitch, C. C. Garber, R. A. Graby, L. L. Hause, A. Mather, T. Peters, jr., R. N. Rand, D. J. Reeder, S. M. Russell, R. Schaffer, J. O. Westgard,* A Candidate Reference Method for Determination of Total Protein in Serum. II. Test for Transferability, Clin. Chem. *27,* 1651 – 1654.

[23] *O. H. Lowry, N. J. Rosebrough, A. L. Farr, R. J. Randall,* Protein Measurement with the *Folin* Phenol Reagent, J. Biol. Chem. *193,* 265 – 275 (1951).

[24] *G. L. Peterson,* Review of the *Folin* Phenol Protein Quantitation Method of *Lowry,* Rosebrough, Farr and Randall, Anal. Biochem. *100,* 201 – 220 (1979).

[25] *A. Bensadoun, D. Weinstein,* Assay of Proteins in the Presence of Interfering Materials, Anal. Biochem. *70,* 241 – 250 (1976).

[26] *H. H. Hess, M. B. Lees, J. E. Derr,* A Linear Lowry-Folin Assay for Both Water-Soluble and Sodium Dodecyl Sulfate-Solubilized Proteins, Anal. Biochem. *85,* 295 – 300 (1978).

[27] *I. Polacheck, E. Cabib,* A Simple Procedure for Protein Determination by the *Lowry* Method in Dilute Solutions and in the Presence of Interfering Substances, Anal. Biochem. *117,* 311 – 314 (1981).

[28] *M. A. K. Markwell, S. M. Haas, N. E. Tolbert, L. L. Bieber,* Protein Determination in Membrane and Lipoprotein Samples. Manual and Automated Procedures, in: *S. P. Colowick, N. O. Kaplan* (eds.), Methods in Enzymology, Vol. *72,* Academic Press, New York 1981, pp. 296 – 303.

[29] *S. Bramhall, N. Noack, M. Wu, J. R. Loewenberg,* A Simple Colorimetric Method for Determination of Protein, Anal. Biochem. *31,* 146 – 148 (1969).

[30] *A. T. Høstmark, L. Sørensen, R. Askevold,* Influence of Sodium Hydroxide on Protein Determination with the Xylene Brilliant Cyanin G Micromethod, Anal. Biochem. *83,* 782 – 784 (1977).

[31] *A. Heil, W. Zillig,* Reconstitution of Bacterial DNA-Dependent RNA-Polymerase from Isolated Subunits as a Tool for the Elucidation of the Role of the Subunits in Transcription, FEBS Lett. *11,* 165 – 168 (1970).

[32] *W. Schaffner, C. Weissmann,* A Rapid, Sensitive, and Specific Method for the Determination of Protein in Dilute Solutions, Anal. Biochem. *36,* 502 – 514 (1973).

[33] *V. Neuhoff, K. Philipp, H.-G. Zimmer, S. Mesecke,* A Simple, Versatile, Sensitive, and Volume-Independent Method for Quantitative Protein Determination which is Independent of Other External Influences, Hoppe-Seyler's Z. Physiol. Chem. *360,* 1657 – 1670 (1979).

[34] *J. McGuire, P. Taylor, L. A. Greene,* A Modified Bromosulfalein Assay for the Quantitative Estimation of Protein, Anal. Biochem. *83,* 75 – 81 (1977).

[35] *L. J. Wallace, L. M. Partlow,* A Sensitive Microassay for Protein in Cells Cultured on Collagen, Anal. Biochem. *87,* 1 – 10 (1978).

[36] *G. S. McKnight,* A Colorimetric Method for the Determination of Submicrogram Quantities of Protein, Anal. Biochem. *78,* 86 – 92 (1977).

[37] *A. Esen,* A Simple Method for Quantitative, Semiquantitative, and Qualitative Assay of Protein, Anal. Biochem. *89,* 264 – 273 (1978).

[38] *M. M. Bradford,* A Rapid and Sensitive Method for the Quantitation of Microgram Quantities of Protein Utilizing the Principle of Protein-Dye Binding, Anal. Biochem. *72,* 248 – 254 (1976).

[39] *J. J. Sedmak, S. E. Grossberg,* A Rapid, Sensitive, and Versatile Assay for Protein Using *Coomassie* Brilliant Blue G250, Anal. Biochem. *79,* 544 – 552 (1977).

[40] *R. Flores,* A Rapid and Reproducible Assay for Quantitative Estimation of Proteins Using Bromophenol Blue, Anal. Biochem. *88,* 605 – 611 (1978).

[41] *T. Spector,* Refinement of the *Coomassie* Blue Method of Protein Quantitation, Anal. Biochem. *86,* 142 – 146 (1978).

[42] *J. Pierce, C. H. Suelter,* An Evaluation of the *Coomassie* Brilliant Blue G-250 Dye-Binding Method for Quantitative Protein Determination, Anal. Biochem. *81,* 478 – 480 (1977).

[43] *Bio-Rad* Protein Assay, *Bio-Rad* Bulletin 1069 EG, *Bio-Rad Laboratories,* Richmond, Cal., USA (1979).

[44] *F. Chiappelli, A. Vasil, D. F. Haggerty,* The Protein Concentration of Crude Cell and Tissue Extracts as Estimated by the Method of Dye Binding: Comparison with the *Lowry* Method, Anal. Biochem. *94,* 160 – 165 (1979).

[45] *J. C. Bearden, jr.,* Quantitation of Submicrogram Quantities of Protein by an Improved Protein-Dye Binding Assay, Biochim. Biophys. Acta *533,* 525 – 529 (1978).

[46] *H. Van Kley, S. M. Hale,* Assay for Protein by Dye Binding, Anal. Biochem. *81,* 485 – 487 (1977).

[47] *Z. Zaman, R. L. Verwilghen,* Quantitation of Proteins Solubilized in Sodium Dodecyl Sulfate-Mercaptoethanol-Tris Electrophoresis Buffers, Anal. Biochem. *100,* 64 – 69 (1979).

[48] *A. Freeman, T. Blank, Y. Aharonowitz,* Protein Determination of Cells Immobilized in Cross-Linked Synthetic Gels, Eur. J. Appl. Microbiol. Biotechnol. *14,* 13 – 15 (1982).

[49] *R. Almog, D. S. Berns,* A Sensitive Assay for Proteins and Biliproteins, Anal. Biochem. *114,* 336 – 341 (1981).

[50] *P. Böhlen, S. Stein, W. Dairman, S. Udenfriend,* Fluorometric Assay of Proteins in the Nanogram Range. Arch. Biochem. Biophys. *155,* 213 – 220 (1973).

[51] *J. V. Castell, M. Cervera, R. Marco,* A Convenient Micromethod for the Assay of Primary Amines and Proteins with Fluorescamine. A Reexamination of the Conditions of Reaction, Anal. Biochem. *99,* 379 – 391 (1979).

[52] *E. C. Butcher, O. H. Lowry,* Measurement of Nanogram Quantities of Protein by Hydrolysis Followed by Reaction with Orthophthalaldehyde or Determination of Glutamate, Anal. Biochem. *76,* 502 – 523 (1976).

[53] *S. A. Robrish, C. Kemp, W. H. Bowen,* The Use of the o-Phthalaldehyde Reaction as a Semiquantitative Assay for Protein and to Determine Protein in Bacterial Cells and Dental Plaque, Anal. Biochem. *84,* 196 – 204 (1978).

[54] *T. Kinoshita, F. Iinuma, A. Tsui,* Microassay of Proteins on Membrane Filter in the Nanogram Range Using Cycloheptaamylose-Dansyl Chloride Complex, Anal. Biochem. *66,* 104 – 109 (1975).

[55] *R. M. Schultz, P. M. Wassarman,* [^{3}H]Dansyl Chloride. A Useful Reagent for the Quantitation and Molecular Weight Determination of Nanogram Amounts of Protein, Anal. Biochem. *77,* 25 – 32 (1977).

[56] *R. M. Schultz, J. D. Bleil, P. M. Wassarman,* Quantitation of Nanogram Amounts of Protein Using [^{3}H]Dinitrofluorobenzene, Anal. Biochem. *91,* 354 – 356 (1978).

[57] *R. McGrath,* Protein Measurement by Ninhydrin Determination of Amino Acids Released by Alkaline Hydrolysis, Anal. Biochem. *49,* 95 – 102 (1972).

[58] *L. C. Mokrash,* Use of 2,4,6-Trinitrobenzene Sulfonic Acid for the Coestimation of Amines, Amino Acids, and Proteins in Mixtures, Anal. Biochem. *18,* 64 – 71 (1967).

[59] *R. Best, E. Howell, A. Decillis, K. J. Schray,* A Novel Method for Protein Quantitation Using a Competitive Binding Assay, Anal. Biochem. *119,* 299 – 303 (1982).

[60] *S. G. Jackson, E. L. McCandless,* Simple, Rapid Turbidometric Determination of Inorganic Sulfate and/or Protein, Anal. Biochem. *90,* 802 – 808 (1978).

[61] *H. Reiber,* Eine schnelle und einfache nephelometrische Bestimmungsmethode für Protein im *liquor cerebrospinalis,* J. Clin. Chem. Clin. Biochem. *18,* 123 – 127 (1980).

[62] *H. Reiber,* Ein neues physikalisch-chemisches Prinzip zur Bestimmung von Protein-Konzentrationen in biologischen Proben, Fresenius Z. Analyt. Chem. *311,* 374 – 375 (1982).

[63] *C. A. Lang,* Simple Microdetermination of Kjeldahl Nitrogen in Biological Materials, Anal. Chem. *30,* 1692 – 1694 (1958).

[64] *O. Minari, D. B. Zilversmit,* Use of KCN for Stabilization of Color in Direct Nesslerization of Kjeldahl Digests, Anal. Biochem. *6,* 320 – 327 (1963).

[65] *M. Nube, C. P. M. van den Aarsen, J. P. Gilliams, W. Th. J. M. Hekkens,* The Determination of Ammonium in *Kjeldahl* Digests Using the Gas-Sensing Ammonia Electrode. Comparison of the Direct Method with the Known-Addition Method. Clin. Chim. Acta *100,* 239 – 244 (1980).

[66] *M. Kreisner,* Vereinfachte Form der potentiometrischen Eiweißbestimmung im biologischen Material, Fresenius Z. Analyt. Chem. *307,* 285 (1981).

[67] *D. W. Kupke, T. E. Dornier,* Protein Concentration Measurement: The Dry Weight, in: *S. P. Colowick, N. O. Kaplan* (eds.), Methods in Enzymology, Vol. *XVLIII,* Academic Press, New York 1978, pp. 155 – 162.

[68] *H. U. Bergmeyer, G. Holz, E. M. Kauder, H. Möllering, O. Wieland,* Kristallisierte Glycerokinase aus *Candida Mykoderma,* Biochem. Z. *333,* 471 – 480 (1961).

2 Reagents for Enzymatic Analysis

2.1   Handling of Reagents

Hans Ulrich Bergmeyer and Marianne Graßl

Numerous enzymes of high purity are available commercially. Their stability is generally good. It has also become possible to stabilize an increasing number of very sensitive enzymes. As a result, solutions of coenzymes or metabolites are often more labile than solutions or suspensions of the enzymes themselves. It should also be remembered that other analytical reagents (e. g. very dilute NaOH, thiosulphate solutions, etc., required for titrations) also decompose easily but, like biochemical reagents, are absolutely reliable when handled correctly. Knowledge of the essential properties of these compounds is a prerequisite for their correct handling.

Analysts in general are not nearly as familiar with the biochemicals involved in enzymatic analysis as they are with inorganic reagents, for instance. Familiarization is not helped by the number and diversity of the names of substances, abbreviations, and definitions of quality; therefore, these concepts are in need of further elucidation.

2.1.1   Nomenclature, Abbreviations, and Consequences of Standardization

The nomenclature, and particularly the abbreviations, for biochemical substances still have a historical basis, though considerable progress has been made internationally in recent years towards the establishment of uniform and systematic names.

2.1.1.1   Nomenclature and Abbreviations

Recommendations for nomenclature and corresponding abbreviations for complex biochemical substances have been made by the Commission on Biochemical Nomenclature, International Union of Pure and Applied Chemistry (IUPAC) and the International Union of Biochemistry (IUB) [1]. A revised edition was published in 1978 [2]. A systematic arrangement and an internationally recognized nomenclature for enzymes were particularly essential. Such a nomenclature was first published in 1961, was greatly extended in 1972 and 1975, and was further completed and revised in 1978 (cf. Vol. I, chapter 1.3, also for references). Abbreviated enzyme names are thus made unambiguous by the addition of the Enzyme Commission's classification number (EC No.).

The abbreviations for phosphorylated biochemical substances are less straightforward. Adenosine 5'-triphosphoric acid, for example, contains four dissociable hydrogen atoms. The salts are therefore represented by $ATP-NaH_3$ to $ATP-Na_4$. In formulae, the state of dissociation may be given, e. g. ATP^{4-}, but the abbreviation ATP is more generally used. This denotes either adenosine 5'-triphosphoric acid or triphosphate, the degree of dissociation of the salt being ignored. The abbreviations NAD^+, NADH, $NADP^+$ and NADPH have become accepted for the pyridine coenzymes. In these abbreviations, H indicates that the pyridine ring is hydrogenated in position 4, and does not refer to the state of dissociation of the pyrophosphate residue in the molecule. For example, the disodium salt of the dibasic acid NADH is represented by the abbreviation $NADH-Na_2$.

Arbitrary abbreviations have been created for many substances used in biochemistry, as well as for naturally occurring and model substrates, many of which have been generally adopted (cf. Appendix 2).

2.1.1.2 Consequences of Standardization

Standardization efforts have been of major importance in achieving a proper definition of biochemical substances (cf. Vol. I, chapter 1.3). They have extended in recent years to the unification of all parameters according to the international system of units (cf. Vol. I, chapter 1.1.2), which is essentially characterized by the fundamental units metre, kilogram, second, and mole. The consistent application of these units leads to the volume units m^3 and litre, in chemistry to mol/l, mol/kg, kg/l, kg/kg, l/l, and, for the unit of time, to seconds (s) instead of minutes (min).

Consequently, a new unit had to be defined for rates of substrate conversion by enzymes: μmol was replaced by mol and minutes (min) by seconds (s). IUPAC and IUB now [3] differentiate between "reaction rate" in $mol \times l^{-1} \times s^{-1}$ and "rate of substance conversion" in $mol \times s^{-1}$. Thus, the catalytic activity of an enzyme is measured by the rate of conversion of its substrate (mol/s). The international Unit of catalytic activity U (μmol/min) is replaced by the Katal, kat (mol/s). 1 U = 16.67 nkat. It remains to be seen how quickly this new unit will become accepted. Both new and old units are used in the present edition of this book. The reaction rate, defined in physical chemistry ($-dc/dt$) as the change in substrate concentration per time unit ($mol \times l^{-1} \times s^{-1}$), corresponds to the catalytic activity concentration of the catalyzing enzyme (kat/l; formerly U/l). So far, however, there is no great change concerning the characterization of enzymes used for analytical purposes.

The standardization of coenzymes and metabolites, on the other hand, is comparatively simple. The most important point in this respect is the correct and comprehensive description of contents and purity.

2.1.2   Quality Requirements

Enzymatic analysis is mainly carried out on biological material, the substances to be analyzed generally being natural products. The samples contain a large number of chemically similar substances. The reagents used for the analysis must therefore ensure that the analysis is specific and that the other substances present in the sample do not interfere. The quality requirements for the reagents used must consequently be high.

Reagents for enzymatic analysis are mainly buffers, inorganic cations and anions, and natural products, especially enzymes, coenzymes, and metabolites. Synthetic model substrates are also employed for the determination of enzyme activities (cf. Vol. I, chapter 2.3.2.3).

2.1.2.1   The Concept of Quality [4]

Difficulties are occasionally encountered in defining the purity even of such well-characterized substances as buffers and organic substrates. For example, 4-nitrophenyl phosphate, a model substrate of alkaline phosphatase, should be essentially free from 4-nitrophenol. Because of the high and specific absorbance of this contaminant acceptable tolerances can easily be established. It is much more difficult, however, to define and adhere to purity criteria for enzymes and coenzymes isolated from natural sources in which many unknown by-products may be present.

The quality requirements for the reagents used in enzymatic analysis comprise

- purity   - type of preparation
- activity   - packaging.
- stability

The problems start with the definition of purity. It would generally be sufficient to adhere to the principle "as pure as necessary, not as pure as possible". However, this principle cannot be used, since a method, together with the reagents that it involves, must be capable of being applied to a wide range of test materials. These may differ considerably in their composition, and therefore in the nature of the interfering reactions which may take place with contaminants present in the reagents. Purity is thus a relative concept.

The activity, particularly the catalytic activity of enzymes, should remain constant over a reasonable period, and should be high enough for enzymatic reactions to proceed in an acceptable time.

The stability of the reagent should refer both to the pure substance used and to the solution of the substance. In both cases, the substance should not change during storage.

The type of preparation should be application-orientated. It should not cause diffi-culties in the assay, e. g. because of very viscous solutions, hygroscopic lyophilisates, or reagent tablets that dissolve too slowly. Chemical reaction steps (e. g. the conver-sion of an insoluble barium salt into the sodium salt) should not be necessary.

The packaging material usually receives too little attention. Bottles, stoppers, and foil wrappers may release substances that inactivate their contents, or they may be made in such a way that the contents are damaged by access of atmospheric moisture or atmospheric oxygen.

None of the five points mentioned can be considered in isolation. For example, the problem of moisture-permeable stoppers in reagent bottles influences both stability and purity. Absorption of water by NADH, for example, leads to inhibitor forma-tion.

The quality requirements for the reagents form a complex system of inter-dependent parameters. The number of factors that influence quality can be limited if (expressed mathematically) the parameters "type of preparation" and "packaging" are kept con-stant. One can then say that quality is a function of purity, activity, and stability:

$$\text{quality} = f(\text{purity, activity, and stability})$$

However, these three parameters again form a trilateral function among themselves:

i. e. purity depends on stability and activity, stability depends on purity and activity, and activity depends on purity and stability. These three parameters must be consider-ed together.

2.1.2.2   The Individual Quality Requirements

Type of preparation

The fundamental requirements for an application-orientated type of preparation are maximum simplicity of use of the reagent and a guarantee of the highest possible qual-ity.

Simplicity of use: soluble alkali metal salts are obviously easier to use than sparing-ly soluble barium salts. Formerly it was usual to prepare sugar phosphates, for ex-ample, as barium salts, which had to be converted into the desired alkali salt before use. A crystalline substance can be weighed out more easily than a hygroscopic lyo-philisate. Pre-weighed reagent tablets (e.g. 4-nitrophenyl phosphate as the substrate for alkaline phosphatase) save weighing out the substrate, but they must dissolve quickly and completely.

Manipulation of the substance before use, e. g. dialysis of enzyme suspensions, should be avoided. This dialysis is necessary when ammonium ions present in an enzyme suspension interfere with the determination. Example: in the determination of the catalytic activities of the aminotransferase in serum, if the serum contained glutamate dehydrogenase, this would react with the oxoglutarate used and the ammonium ions of the enzyme suspension. In this case the indicator enzymes malate dehydrogenase and lactate dehydrogenase are used as a solution in 50% glycerol or as lyophilized material.

Guarantee of Quality: this parameter is again closely connected with the simplicity of use, and also with stability and activity. An enzyme solution is easier to use than a suspension. Solutions containing preservatives save daily preparation of fresh solutions, and guarantee constant quality for periods of days or weeks. However, preservatives are permissible only if they have no effect on the catalytic activity of either the dissolved enzyme or the enzyme to be determined. Example: enzyme solutions in 50% glycerol guarantee not only satisfactory determination of the catalytic activities of the aminotransferases but also lasting quality of the indicator enzymes.

Similarly, the use of NAD and CoA preparations in the form of the lithium salts instead of the free acids offers greater stability and hence satisfactory quality for a longer time.

Packaging

The best of reagents will deteriorate in time if poorly packaged. Glass that gives off heavy metals leads to inhibition of enzymes. Plastics that release plasticizers are no better. Films and stoppers that are permeable to water vapour allow lyophilisates to adsorb water and deliquesce. Example: one of the causes of inhibitor formation in NADH is moisture, cf. Fig. 1.

A satisfactory packaging material is particularly important for mixtures of reagents which are used in clinical chemistry for routine determinations. They usually contain a number of different components. It is therefore necessary to rule out a variety of external influences.

The main criterion must be that no component of the packaging material has any effect on its contents; the package itself must provide effective protection against harmful external influences. Conscientious manufacturers of reagents for enzymatic analysis overcome these problems.

It is quite possible to meet the requirements with regard to type of preparation and packaging to such a degree that the reagents are suitable for universal use, or at least for the foreseeable range of applications. Among the various factors that influence quality, therefore, these two can be kept constant. This leaves the parameters purity, stability, and activity.

Purity, activity, stability

Enzymes, coenzymes, and metabolites are typical of the reagents used in enzymatic analysis.

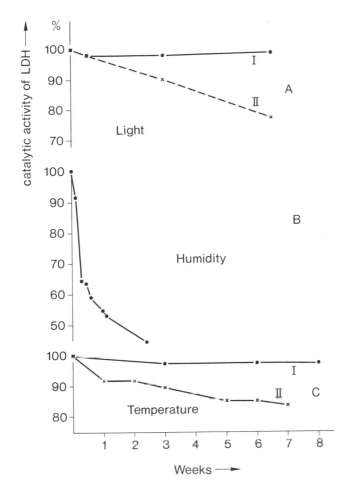

Fig. 1. Stability of NADH preparations (formation of inhibitors as measured by LDH activity).
A. Effect of light: I dark, II diffuse daylight. Room temperature (ca. 21 °C), exclusion of moisture by use of stoppered containers; brown glass (not special glass).
B. Effect of moisture: NADH stored in a thin layer at 22 °C in ca. 80% relatively humidity. Light excluded.
C. Effect of temperature: I stored at 4 °C (refrigerator), II at 33 °C. Moisture and light excluded.

Enzymes

The best criterion of purity for an enzyme to be used in enzymatic analysis is the specific catalytic activity, with accompanying information on the catalytic activities of contaminating enzymes. These "contaminating activities" are expressed as percentages of the specific catalytic activity of the reagent enzyme if the definitions of units are on the same basis.

The indication of "side activities" * serves to characterize the enzyme, but provides no further information on its purity. Crystallization is no criterion of purity.

The following examples show how seriously contaminating activities can interfere. In the determination of the catalytic activity of pyruvate kinase (PK)** in serum, a few thousandths of a percent of pyruvate kinase in the indicator enzyme lactate dehydrogenase (LDH), which is used in excess, lead to errors of more than 5% at normal levels of PK activity. A similar situation occurs in the determination of the catalytic activities of the aspartate aminotransferase (AST) and alanine aminotransferase (ALT) (cf. Table 1).

Table 1. Errors due to contaminating activities in indicator enzymes.

Assay	Value measured U/l (nkat/l)	Indicator enzyme and contamination	Error in the value found
PK	12 (200)	LDH (0.006% PK)	5%
AST	7 (117)	MDH (0.03% AST)	6%
ALT	5 (83)	LDH (0.03% ALT)	6%

allowed: 0.001% PK   in LDH
         0.01% ALT   in LDH
         0.005% AST   in MDH

The number of contaminating activities that must be checked for depends on the expected range of application. A selection is necessary in view of the large number of enzymes known. If intended for a specific purpose, the auxiliary and indicator enzymes must be tested for contaminating activities in the complete assay systems in which they are to be used. Only in this way are the effective figures for interfering catalytic activities obtained. For example, the indicator enzyme, LDH, in the assay of the catalytic activity of alanine aminotransferase, ALT, must be tested for contaminating activities in the optimized assay system for ALT which includes pyridoxal phosphate [5, 6]. This is because apo-ALT in the LDH preparation is detectable only in the presence of pyridoxal phosphate, which is not ordinarily part of the assay system for LDH. Corresponding remarks apply to creatine kinase[†] present as an inactive impurity in the auxi-

---

* Definition: an enzyme may react with compounds other than its true substrate. Note: such side activities cannot be removed by purification procedures, in contrast to contaminating activities.
** Enzymes:
  PK   Pyruvate kinase, ATP: pyruvate 2-O-phosphotransferase, EC 2.7.1.40
  LDH   Lactate dehydrogenase, L-lactate: NAD oxidoreductase, EC 1.1.1.27
  AST   Aspartate aminotransferase (glutamate oxaloacetate transaminase, GOT), L-aspartate: 2-oxoglutarate aminotransferase, EC 2.6.1.1
  ALT   Alanine aminotransferase (glutamate pyruvate transaminase, GPT), L-alanine: 2-oxoglutarate aminotransferase, EC 2.6.1.2
  MDH   Malate dehydrogenase, L-malate: NAD oxidoreductase, EC 1.1.1.37
  [†] ATP: creatine N-phosphotransferase, EC 2.7.3.2.

liary and indicator enzymes used for determination of this activity: sulphydryl compounds present in the reagents will re-activate the contaminant creatine kinase, falsely elevating the measured activity.

Freedom from proteases is also important, since these have a decisive influence on the stability of enzyme preparations.

The catalytic activity of enzymes is expressed in international units (units, U, or katals, kat, cf. chapter 3.1.2.1). The catalytic activity content of an enzyme preparation, formerly "specific catalytic activity" is referred to the mass of protein (U/mg, µkat/mg). Only in exceptional cases (e. g. for some hydrolases) is this definition impracticable. In some cases it is necessary to specify fixed measuring conditions (e. g. the nature and concentration of the substrate); where necessary the definition of other units or the corresponding experimental conditions have to be stated.

The measuring temperature also forms part of a completely defined specific activity of an enzyme, e. g. 400 U/mg (30 °C), or 6.7 µkat/mg (30 °C).

The stability of enzymes is generally better than that of coenzymes. Instability is often due to the growth of micro-organisms that produce proteases. Preservatives may be added to prevent this. However, it is necessary to check whether these inhibit the enzyme and hence interfere with the determination. A stabilizing effect can often be achieved by addition of the substrate or a homologue of it (e. g. the stabilization of glycerol kinase* with 1% (v/v) of ethylene glycol; this is the only way in which stability for well over 1 year can be achieved).

Coenzymes and Metabolites

The concept of "activity" must be understood here as a supplementary criterion of purity. For example, a coenzyme may be chemically almost 100% pure, but it may still inhibit the enzyme-catalyzed system because of practically undetectable traces of heavy metals.

The preparation of natural substances in a pure form, and hence the characterization of their purity for use as reagents in enzymatic analysis, is also made difficult by the fact that it is often not possible to crystallize them. The use of the melting point as a criterion of purity almost always fails. The isolation methods available at present for preparation of these compounds from biological material frequently yield compounds containing impurities of similar structure or function as well as degradation products.

The purity of coenzymes and substrates is best determined by enzymatic analysis. In practice, only the fraction active in the enzymatic assay is of importance. All other analytical data have only supplementary value.

Examples

It is of little value to determine the purity of a fructose 1,6-bisphosphate preparation by its optical rotation, because contamination by other fructose-phosphate esters causes interference. A colorimetric determination of fructose (e. g. with the

---

* ATP: glycerol 3-phosphotransferase, EC 2.7.1.30.

resorcinol-HCl reaction) is also inappropriate, because the main contaminants contribute to the colour development.

A necessary addition to the results of the enzymatic analysis are details concerning the cation content of the preparation, especially in the case of non-crystalline salts of polybasic acids.

Example
Four sodium salts of fructose 1,6-bisphosphoric acid are possible. It is difficult to maintain the conditions of preparation such that only one type of salt is produced. In this case, the data on the degree of purity should most certainly contain the sodium content as well as the percentage of fructose $1,6-P_2$ (free acid) and water content.

The second necessary supplement to the results of the enzymatic analysis is specification of the water content of the preparation. If, apart from the enzymatically active compound and cations, a preparation contains only water, it can be termed "pure". We do not designate a compound containing 90% active material + 10% water as "90% pure"; the substance is pure. On the other hand, the designation "pure" for a non-crystalline preparation is false if its analysis adds up to 100% by including a correction for a fictitious (calculated) water content.

It is misleading to state a definite water content for non-crystalline substances, such as NAD(P), NAD(P)H, alkali salts of some phosphate esters, etc., because this depends on the conditions under which the substance was dried. Even substances which crystallize with water of crystallization, such as adenosine, can lose water on storage, while other compounds are hygroscopic. We do not include the water content in the formulae of non-crystalline preparations (e.g. NAD $\cdot$ $4H_2O$), because such a formula suggests a degree of purity and content of the active substance which is not warranted. The purity of a non-crystalline coenzyme or substrate is therefore given by the sum of the enzymatically active substance, the relevant cation or anion, and water.

Substances that are often used, such as ATP, should also be examined for their ability to function in an assay system that is particularly complicated and hence relatively susceptible to interference (cf. Table 2). Function tests are essential for ready-to-use reagent kits containing many individual components. Purity and activity are best verified here by measurements on standardized control sera with values in the normal and high ranges.

During the past few years more sophisticated chromatographic methods such as high performance liquid chromatography (HPLC) have been introduced for the quality control of nucleotides, nucleosides and several substrates. This procedure is very sensitive and allows the detection of extremely low amounts of impurities (e.g. 0.001% (w/w) of GTP in ATP). For the characterization of various substrates such as maltoheptaose and its derivatives (for the measurement of the catalytic activity of α-amylase) this is nowadays the only method to be employed for this purpose.

Other important parameters to be measured in widely used cofactors are earth alkali and heavy metals. One of the most reliable methods to detect and determine

Table 2. Specification for adenosine 5'-triphosphate (ATP), cryst. disodium salt, special quality.

Formula:	$C_{10}H_{14}N_5O_{13}P_3Na_2 \cdot 3\,H_2O$
Molecular weight:	ATP 507.2
	ATP-Na$_2$H$_2$ $\cdot$ 3 H$_2$O 605.2
Appearance:	colourless crystalline substance
Solubility:	clearly soluble in water (c = 50 mg/ml)
ATP-Na$_2$H$_2$ $\cdot$ 3 H$_2$O:	100%
ATP (enzymatically):	84%
ATP (absorbance at 260 nm):	84%
P$_i$	$\leq 0.05\%$
AMP (enzymatically): ⎫	
ADP (enzymatically): ⎭	$\leq 0.05\%$
GTP (HPLC):	$\leq 0.01\%$
Fe (atomic absorption):	≤ 10 ppm
Mg (atomic absorption):	≤ 10 ppm
Ca (atomic absorption):	≤ 5 ppm
Zn (atomic absorption):	≤ 5 ppm
V (atomic absorption):	≤ 1 ppm
A_{250}/A_{260}	0.79 ± 0.02
A_{280}/A_{260}	0.15 ± 0.01
A_{290}/A_{260}	< 0.01
Stability:	stable at $+4\,°C$, stored dry; no significant decomposition within 12 months.

Suitability for CK assays checked by performance tests.

very small amounts of those impurities is atomic absorption. In case of ATP (cf. Table 2) this is a valuable help in the characterization of a commerical product.

Sometimes very unexpected by-products are present in substrates and interfere with various enzyme assay systems. As was demonstrated by *Lowenstein* and his coworkers [7, 8], creatine phosphate can be contaminated by traces of oxalate and/or pyrophosphate which impair the activity of creatine kinase, pyruvate kinase and adenylate deaminase*, respectively.

Summary

If care is taken to ensure that no external influences (due to the packaging) adversely affect the quality of biochemical reagents, and that the type of preparation has no influence on the quality while at the same time guaranteeing maximum simplicity of use of the preparation, the quality of reagents for enzymatic analysis can be defined by the parameters purity, activity, and stability.

This trilateral function stability/activity/purity is, e.g., very clearly evident in the case of NADH. As soon as a hygroscopic NADH lyophilisate adsorbs water (possibly deliquesces and turns yellow), not only the purity but also the activity in the enzymatic system is impaired; the stability has been lost.

---

* AMP deaminase, AMP aminohydrolase, EC 3.5.4.7.

There is a another parameter that facilitates routine work and minimizes errors, but in principle is of no relevance to the quality. This is the consistency of the quality of different production lots. Well-known manufacturers take steps to ensure such consistency. If any doubt exists, it is necessary for the investigator himself to check the quality of the reagents in the laboratory before they are used for analysis.

2.1.3   Storage, Stability, and Control of Substances and Solutions

2.1.3.1   Substances

Enzymes in stabilized aqueous glycerol solution, as suspensions in ammonium sulphate solution, or as lyophilisates should be stored at $0\,°C$ to $+4\,°C$, unless other directions are given. When stored under these conditions, the loss of activity is generally minimal, even over some months; freezing of crystalline suspensions can frequently lead to a considerable loss of activity. On the other hand, enzyme solutions are generally more stable in the frozen state, particularly if the product is thawed infrequently. Any access of moisture to lyophilized enzymes must be prevented; for example, cold vials of lyophilisate must first be warmed up to room temperature before being opened.

Coenzymes such as NAD(P), NAD(P)H, CoA, and FAD must be stored dry at $0\,°C$ to $+4\,°C$ and protected against light. Incorrect storage of NADH, for example, results in the formation of inhibitors of dehydrogenases. This phenomenon is already apparent before any decrease in the NADH content of the preparation can be measured.

Exposure to light, moisture, and – to a lesser extent – elevated temperature are the main factors which cause destruction of NADH. From systematic studies it appears that exposure to atmospheric moisture and oxygen are the most important factors in the formation of inhibitors in the solid substance (Fig. 1, p. 107). The measured values were referred to a standard LDH preparation, the activity of which had been measured with an NADH preparation which could not be purified further (100% activity). NADH is stable if the following conditions are maintained: absolute exclusion of light, oxygen and moisture, low temperature (warming to room temperature, up to $33\,°C$, for short periods causes virtually no damage). CoA is oxidized rapidly by atmospheric oxygen; moreover, the pyrophosphate bond is easily hydrolyzed. Most coenzymes decompose by hydrolysis; therefore moisture must be excluded. Lyophilized preparations take up water particularly easily, so that they must be stored in a desiccator. In many cases storage at $-20\,°C$ is recommended.

2.1.3.2   Solutions

Stability of solutions is a relative term. One should differentiate between stability of stock solutions, mainly stored in the refrigerator, and working solutions for daily use. Whereas stock solutions in general should be stable for weeks or months, working solutions (which are typically kept on the bench during daily use and stored at night in the refrigerator) should keep their quality for one or more days, desirably for one working week.

Freshly distilled water should be used for the preparation of solutions of biochemical reagents. The term "re-distilled water" no longer needs to be stressed. In most cases, it is sufficient to distill de-ionized water through a glass still in order to remove micro-organisms and substances introduced by the ion-exchange resin. Purification of tap water by commercially available systems is practicable and economical. There are a number of such units on the market which can be easily adapted to the tap- or de-ionized water supply (e.g. *Millipore* System, "Organic pure" from *Barnstead*, "Ultra pure water" from the *Elga Group*, "Zinser Ionfilter", Aquademat from *Hecht*). These units mostly consist of a charcoal filter to adsorb organic material, a mixed bead de-ionizer and a membrane filter to remove micro-organisms. It is important that the water is not more than one day old. Special care is required with commercial so-called "sterile aqua dest."; it often contains reducing compounds.

For various analyses (e.g., in pH-metry) CO_2-free water is necessary. For this purpose the re-purified water has to be boiled for approx. $5 - 10$ min, cooled under exclusion of CO_2 (air) and used within a few hours.

In fluorimetry the purification of water needs special care. The water applied for the preparation of the various solutions must be freshly distilled from a ca. 20% solution of $KMnO_4$ to remove any contaminating fluorescing compounds by oxidation. To avoid any traces of fine particles which disturb the measurement the water should be filtered carefully through a glass-sintered filter. Most kinds of filter paper – except some special chromatography papers – contain very small amounts of fluorescing material, and should not be used to remove solid impurities (cf. Vol. I, chapter 3.4.4.1).

Compounds leached from stoppers (usually rubber stoppers) can cause interference in some enzyme reactions. Bottles containing solutions of substrates, coenzymes, or enzymes should therefore be stoppered only with reliable rubber stoppers, or preferably with stoppers made of polyethylene or silicone rubber.

Diluted solutions of coenzymes and substrates should always be stored in a refrigerator, not only because of their chemical instability, but also because of the rapid growth of micro-organisms at higher temperatures. The shelf life of these solutions is stated in the methods desribed in the following volumes. These data should be considered when reagents are prepared for the assays. Buffer solutions, especially phosphate, acetate, amino acid, and sugar solutions, should be stored in thoroughly cleaned and sterilized dark bottles with tapered glass stoppers. The daily requirement should be poured out rather than pipetted. Coenzyme solutions, with the exception of NADH and NADPH, are best frozen in small portions so as to avoid excessive freezing and thawing. Solutions of NADH and NADPH are acid-labile; the pH of the solutions should not fall below 7.5. On the other hand NAD and NADP are alkali-labile. CoA is most stable around pH 4; ATP around pH 9.

Sedimented crystals of enzyme suspensions in ammonium sulphate solution should not be resuspended by shaking, but by careful swirling of the vessels. The high catalytic activity of highly purified enzymes means that only a fraction of a milligram is required to provide excess of enzyme for substrate and coenzyme assays. Hence the volumes of enzyme suspensions are small. To avoid losses due to dilution by water, precipitation of ammonium sulphate crystals, or contamination by micro-organisms, the containers should always be kept stoppered.

Diluted enzyme solutions should be made up only with ice-cold "re-distilled" water or buffer solution. Since enzymes are unstable in highly diluted solutions, only the volume required for 1 – 2 hours should be made up (see the corresponding chapters). In many cases the addition of bovine serum albumin stabilizes the diluted enzyme or prevents its binding to the glass surface.

It is recommended that all the necessary solutions be stored in an ice-bath on the bench during the working period (Fig. 2). Insulated plastic containers with metal racks that take tubes of various sizes are available commercially *.

Fig. 2. Ice-bath for biochemical reagents.

All reagents required for assays, including stock buffer solutions and "re-distilled water", should be kept covered; otherwise additional contamination may occur by airborne bacteria or chemical compounds. This can result in interferences such as displacement of the pH of buffer solutions by absorbed carbon dioxide, the inactivation of enzymes in dilute solution by proteases from micro-organisms, or the falsification of the results of ethanol determinations by traces of alcohol in the laboratory atmosphere.

---

* e.g. from *Fritz Kniese,* D-3550 Marburg-Marbach, GFR.

2.1.3.3   Control of Reagents

It is recommended that the activities of the enzymes used and the concentrations of coenzyme and substrate solutions be checked before starting a series of measurements (cf. p. 112). The functioning of a complete assay system for an end-point determination can be roughly tested by adding a trace of the pure substance being measured to the cuvette after the reaction has been completed. A renewed reaction should occur immediately.

Similarly, if there is no reaction in a determination of enzyme activity, a small amount of the pure enzyme should be added to test whether the assay system is in order.

For a true quality control of an assay system, including the performance of the reagents cf. chapter 3.2.

2.1.4   Reagent Kits

The definition of what constitutes a reagent kit has been given by IFCC in 1979 [9]: "Two or more different clinical or general laboratory materials (excluding reconstituting materials), with or without other components packaged together and designed for the performance of a procedure for which directions are supplied with the package. Custom-made and investigational materials are not included". A discussion group at the Center for Desease Control in Atlanta, Georgia, had earlier (in 1969) stated that "a kit, reagent set or diagnostic aid is a collection or assembly of reagents, devices, or equipment, or a combination thereof, offered for sale or distribution and containing all of the major components and the written instructions necessary to perform one or more designated diagnostic tests or procedures". This broad definition includes the large variety of products in clinical chemistry ranging from single tests to the big multi-channel laboratory machines. Reagent kits designed for use in food analysis, in other fields of application, and in research work are also now available.

2.1.4.1   Reasons for the Use of Reagent Kits

If an analytical method is used frequently, it is convenient to pre-mix the reagent solutions in order to reduce the number of pipettings in the assay (e.g. cf. Vol. I, p. 49). Complete reagent kits meet the need for economy and simplicity which is especially necessary for routine laboratories because

a) they enable the user to employ the small amounts of reagents that are needed for single assays,

b) they eliminate the uneconomic use of reagents and time involved in weighing and mixing reagents by the user himself.

Considering especially the explosive growth in clinical laboratory sciences (cf. e.g. [10]), one recognizes that

a) the number of determinations has approximately doubled within 5 years. In the laboratory of a 2000-bed hospital the daily workload reaches approximately 6000 determinations. This only can be achieved by application of reliable reagent kits including well proved instructions combined with well established quality-control procedures, and by the application of mechanized or automated instrumentation.
   Resources of personnel and time do not permit preparation and quality control of reagent solutions within the clinical laboratory, and laboratory-made reagents are not suitable for some automated analysis systems.

b) The number of methods of clinical relevance has increased considerably during the last decade. Routine methods for the determination of, e.g., hormones, tumour markers, etc., require a range of expertise which cannot be enlarged within a single laboratory to such a level that it would be possible for the user himself to prepare all the reagent solutions required for daily use.
   The benefit of using reagent kits prepared by highly qualified experts is that they are convenient and economic, and that they guarantee high quality, since reputable manufacturers use carefully controlled chemicals, prove the reliability of the entire kit by function tests and give information about limits of application, interferences by drugs and pathological metabolites. Such data generally also include documentation of the external evaluations of the kit which are increasingly commissioned by professional bodies and by manufacturers themselves.

c) Considering the costs of health care all over the world it is absolutely necessary to save time by using reagent kits, since time is the most important cost factor in a laboratory (cf. Vol. I, chapter 1.4). Quality of laboratory data is improved by better accuracy and precision and this is an essential factor in improved medical care.

These advantages of reagent kits were quickly recognized and their use has now become widespread. Reagent kits are the decisive tools (besides mechanization and automation of methods) which make available not only complicated methods, but also the huge variety of different methods now required by clinical laboratories.
   The trend towards greater convenience in the use of prepared reagents continues: chemistries are already offered for some closed systems in which even the preparation of solutions is no longer necessary. Chemicals are bound on the surface of test tubes or cuvettes, or are fixed to solid phases, while reactions may take place in multi-layer films.

2.1.4.2   Developments in Handling of Reagents

Enzymatic analysis has always been developed in response to the needs of the user in the various fields of application. Reagent kits reflect these needs, and developments in handling of reagents cannot be demonstrated more clearly than in the course of continuous improvement of kits. Kits also reflect the latest state of the art in general, but particularly with regard to feasibility. A new generation of reagent kits has been created in recent years.

General requirements of reagent kits

Improvement in handling of biochemical reagents consists of reduction of pipetting steps in the assay procedure, thus minimizing errors introduced by pipetting, reducing the time required for analysis, and diminishing the work of preparation and processing of samples.

The simplification of all the steps of an analysis is of most interest for the user, since time is an important cost factor (cf. Vol. I, chapter 1.4). However, simplification must not impair the quality of results.

The optimization process of industrially manufactured reagent kits is designed to bring into balance the costs, the convenience and the quality of laboratory data according to the following scheme [11].

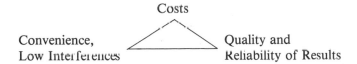

This balanced triangular relationship is influenced by the actual state of the art in chemical techniques, by medical needs with respect to the quality of laboratory results and by the financial resources available.

The possibility of creating test kits according to these requirements has been pointed out in Vol. I, chapter 1.2, with regard to convenience and low interference (choice of method and method design). The achievement of goals with regard to the quality and reliability of results depends basically on the stability of the reagents and their solutions. The efforts demanded in stabilizing reagents in turn are related to the requirements and conditions under which they are intended to be used.

The stability of a reagent kit from the point of view of the user should take into account the following considerations:

– in general, long term stability under storage in the refrigerator is preferred to shorter stability at room temperature,

– stability from the date of arrival at the user should be at least half a year (refrigerator),

– stability of working solutions should be assured for one week if possible, and at least for one working day.

Conventional kits

The simplest design of a reagent combination is to supply each single component or solution of it in a separate bottle. However, the requirements of the user call for special formulations in most cases. Preservatives must be added to inhibit growth of micro-organisms in working solutions, but they must not impair the function of the assay. Single or several dry chemicals can be bottled as powders or lyophilisates, or offered as tablets with defined weight. For better solubility, dry reagents can be dispensed in the form of granulates. Glycerol or ethylene glycol can be added to stabilize enzyme solutions.

The next stage of simplification is to combine those components which do not interact with each other, thus preparing stable reagent mixes.

Since in a multi-step analysis such as creatine kinase determination (cf. Vol. III, chapter 7.4, p. 508), too many pipetting steps and too many reagent solutions are involved for convenient working with minimum error, the goal was to create one-vial reagent mixes. However, this cannot be realized in many cases because of the instability of the various substances when brought together in a dissolved or lyophilized form.

Enzymes can be protected by entrapping, e.g. in Sephadex [12]. Labile or expensive enzymes can be used for series of analyses by immobilization [13] (cf. Vol. I, chapter 3.6.6).

An early and still useful approach in stabilizing reagents is to adhere them to paper. The well-known test strips for use in urine and blood, capable of detecting up to nine constituents, need only be mentioned here. They can be used for quantitative determinations by means of reflectance photometry (cf. Vol. I, chapter 3.3). Also, for analysis of urinary sediment, test strips can be used to restrict microscopic examination to those samples which are positive to at least one test strip indicating erythrocytes, leukocytes or other urine constituents, thus rationalizing urine analysis while maintaining diagnostic accuracy (cf. [14, 15]).

Kits of the new generation

Immobilization of reagents simply by fixation on paper, or by covalent bonding to vessel surfaces or magnetizable particles, can be applied beneficially for their stabilization in reagent kits, and at the same time for facilitating separation of bound and free phases in enzyme-immunoassays.

A special case of carrier-fixed reagents are the so-called coated tubes. These are mainly applied in enzyme-immunoassays, where antigens or antibodies are bound covalently or by other means to the inner surfaces of test tubes. This enables rapid bound/free-separation in immunological assay procedures (cf. Vol. I, chapter 2.7). The advantage of this procedure is that all interfering components of the sample and of the assay mixture are poured off after the binding reaction, so that only the pure antibody-antigen complex remains at the surface of the tube.

Tubes coated on their inner walls with oxidizing agents provide another example of the value of reagents bound to the surfaces of vessels. Pre-incubation of urine samples in such tubes allows measurements of urinary constituents to be made which would otherwise be subject to interference by reducing substances such as ascorbic acid.

A special application of the paper technology is the development of test kits which contain a buffer and a strip on which the other reagents are adsorbed on different zones (Peridochrom®, [16, 17]). In this case the various components necessary for the assay are separated and well stabilized, because each substance is fixed on the paper under its optimal conditions. Since they are not in contact with each other possible interference is avoided. In practice, the reagent is made ready for use by dipping the strip into a given aliquot of the buffer. This type of kit is available for the determination of, e.g., uric acid and cholesterol [16, 17].

For the determination of ascorbic acid in biological material, e.g. fruit juices in food chemistry, papers with attached ascorbic acid oxidase are used to destroy ascorbic acid in mixtures which contain this substance together with various other reducing substances, thus making the determination of the sum of reducing constituents in the sample specific for ascorbic acid [18, 19].

Another technology is to print the components of an analytical system on paper strips. After dipping these strips into the sample, e.g. into urine, diffusion of the reagents to this spot yields dye development. A test strip exists for the detection of catalase in urine or milk according to this principle [20].

A further step in this direction is the development of reagent carriers for the measurement of the catalytic activity of enzymes in plasma [21]. Whole blood is used, blood cells are removed on the strip by the use of glass fibre paper, the enzyme reaction with plasma occurs on the strip, and the result is measured by reflectance photometry. In this procedure, the pre-analytical steps such as separation of plasma by centrifugation, transfer of plasma into the cuvette etc. have been automated and no longer require any laboratory time.

The functioning of this most advanced procedure is demonstrated by the Reflo-quant® system of *Boehringer Mannheim* (Fig. 3 and Table 3).

Accuracy and precision of this procedure are as good as with conventional procedures in the photometer cuvette [21].

Closed systems

A closed system is a combination of instrument and reagents in which no reagents can be used other than the specially designed reagent combination for a particular method applied to one sample. The reagents are combined as solutions, tablets or powders.

One example is the ACA system of *Du Pont-de-Nemours* in Wilmington, Delaware, USA (cf. Vol. I, chapter 3.7). The test kit, which is a transparent bag with several compartments, serves as a reaction vessel and as cuvette for the photometric measurement. A single test pack exists for each determination which is labelled with

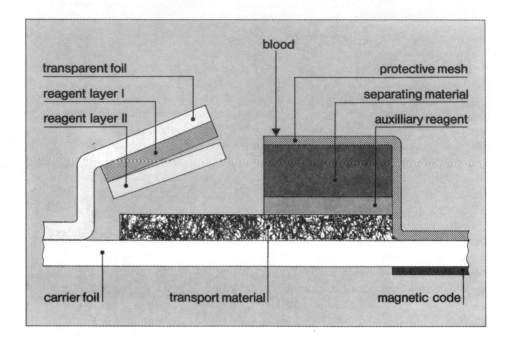

Fig. 3. Construction of the Refloquant® strip.

Plasma is separated from blood placed on the separation area by diffusion through the separating material within 15 to 30 s. The plasma then enters a second glass fibre layer for transport to further reagent layers.

After 30 s the chemical reaction is initiated. The reagent layers bearing all necessary reagents are brought into contact with the second glass fibre paper by the instrument.

Reaction temperature is controlled by the instrument.

After a defined time of about 2 min the reflectance is measured automatically by a small bench-top instrument, Reflotron®.

Table 3. Operation Mode of the Reflotron®-System

Operator	Instrument
– collects about 30 μl of whole blood from fingertip, ear-lobe or vein in a pipette or other device	– reads parameter characteristics from the magnetic code on the test strip
– applies blood on marked area of test strip	– brings test strip up to 37 °C and starts reaction
– inserts strip into adapter of the instrument	– measures reflectance of colour after predetermined time
– presses starting knob	– calculates concentration
– reads result after about 2 min	– presents result as numeric display or printout

the method name and a binary code for programming the instrument. This occurs by inserting the test pack behind the sample vessel.

The analyzer automatically adds the necessary amounts of sample and buffer into the corresponding test pack. In the breaker-mixer system, the reagent compartments are opened and the optimal amounts of reagents mixed with the diluted sample. After the incubation period a cuvette of precise dimensions is formed in the photometer and the absorbance is measured through the transparent test pack membranes.

All functions of the instrument are controlled by a specially developed micro-computer, the measurement data are calculated, and the results printed in concentration or activity units, respectively, as desired. For each patient a report is printed which contains patient identification, the name of the method performed and the analysis results.

Another interesting development in this direction is the *Kodak* Ektachem System [22]. The instrument is fully mechanized, consists of various modules and can be used for series of determinations as well as for single measurements. All procedures and functions are controlled by microcomputers yielding finally a report containing all data relevant to the patient.

The essential part of the system, however, is the multi-layer slide (about the size of a postage stamp) which consists of all reagents needed for the performance of the particular method. The kind and the number of the various layers are dependent on the nature of the special method. In the case of the determination of urea there is the following sequence (from top to bottom):

- spreading layer to allow an even distribution of the sample over the whole area and an isotropic diffusion to the next layer
- reagent layer containing a hydrophilic matrix with urease to form ammonia and carbon dioxide
- semi-permeable membrane which is highly selective for ammonia
- indicator layer in which a dyestuff is formed from ammonia and merocyanin
- supporting layer, the film base, made from transparent polyester.

The slides are inserted into the instrument according to the corresponding program for the analysis sequence. After the measured addition of the sample and the method-dependent inccubation the colour intensity is measured by a reflectance photometer. This system cannot use blood for analysis, serum is needed; thus, the most time-consuming step of the pre-treatment of patient specimens, i.e. centrifugation, is left to the experimenter.

2.1.4.3 Quality Requirements for Test Kits

The use of reagent kits in daily practice of routine laboratories can only result in the potential technical and practical advantages if certain requirements as to their quality can be guaranteed. It is self-evident that the manufacturer is responsible for quality.

However, the user, e.g. the head of laboratory, must be in a position to select the appropriate kit or system from the various manufacturers according to the individual needs of his laboratory.

Reliability in quality of reagent kits can be assured by

a) maintaining Good Manufacturing Practice (GMP), including broad evaluation of the products, and

b) comprehensive labelling.

In clinical laboratory sciences, especially, individual researchers [23 – 25] as well as international and national authorities such as WHO [26, 27], IFCC [9, 28], ECCLS [29], NCCLS [30], DIN [31], VDGH [32], have given guidelines for the description of quality requirements of reagent kits.

Good Manufacturing Practice and evaluation

For an efficient quality control of test kits for diagnostic and scientific purposes or for food chemistry some fundamental prerequisites are imperative:

– suitable methods must be available with the aid of which the degree of purity of the raw materials, of the intermediate products, and of the end products can be clearly fixed,

– an infallible control system must be available which immediately brings to light any irregularities, starting from delivery of the raw materials and continuing up to dispatch of the finished goods.

Because of the importance of this subject the American Food and Drug Administration is pushing forward the application of the Good Manufacturing Practices for diagnostics (cf. e.g. [33]).

These regulations concern

– organization and personnel (training, responsibility)
– buildings and facilities
– equipment
– control of components, containers, and in-process materials
– production and process controls
– packaging and labelling control
– holding and distribution
– laboratory controls
– records and reports

To ensure a smooth manufacture of those reagents and to comply with the official rules the following analytical controls have to be built into the operation sequence:

a) Raw material control
   All chemicals or biochemicals which are used for the manufacture have to be analyzed according to well worked out testing procedures (specifications) including physical parameters, identity test, state of purity (by titration, enzymatic and/or chromatographic or photometric methods, atomic absorption, flame photometry), content of impurities by suitable analytical procedures and often function tests.

b) Packing material control
   Packing material which is in contact with the reagents has to be inspected for physical parameters (appearance, cleanliness, size, strains). In addition, tests have to be performed to ensure that no component which might contaminate the reagent will be released from the material.

c) In-process control
   During the formulation of the test kits in-process controls have to be carried out to assure that all ingredients are added in the amount required and that the manufacturing operations comply exactly with the instructions.

d) Final quality control
   The final quality control of a test kit comprises the following steps:
   — representative sampling of the various vials belonging to one test kit
   — physical inspection of the test kit, the vials and the labelling
   — check of the appearance of the content of the various bottles
   — identity check of all ingredients
   — quantitation of all critical components (e.g. enzymes, coenzymes, substrates)
   — performance test using various control sera or patients' sera, checking linearity and recovery
   — stability tests

The *external evaluation* of reagent kits is one of the most important steps during the development of a test kit. The quality of a kit would be questionable if the evaluation had not been performed under defined conditions. Guidelines for the evaluation of kits in the clinical chemistry laboratories have been worked out (cf. e.g. [25, 28]).

Labelling

Labelling means [9] all written, printed, graphic of other matter on or accompanying clinical laboratory materials. Labelling includes both labels and package inserts. A package insert or brochure must not be permanently attached to the package.

   What the label and what the insert have to contain has been clearly defined (cf. e.g. [9]). More recently, the German Diagnostics Group [34] has suggested that the package insert should be split. A working instruction should be attached to each package, while a documentation about the scientific background of the method, its limitation in application and all results of the external evaluation should be available to the user on request.

This documentation with a qualifying functional description is felt to be necessary since the user, who is responsible for the quality of his laboratory results, must be able to decide if a certain kit meets the requirements of his special medical field of application.

Manufacturer's kit documentation should cover, e.g., statements about

- precision and accuracy of the method,
- limits of application under various diseases
- applicability to various biological materials,
- interferences by drugs and pathological metabolites.

It is the aim of all analytical work to produce "true results". For comparison of the results from a new method with those of a reference method, it is necessary that a reference method exists. The status of the efforts in this respect is reviewed by *Tietz* [35]. Furthermore, criteria have to be worked out on which the acceptability of a kit will be based and which take account of the clinical decisions that will be based on the results of the analyses (cf. e.g. [23, 24]).

References

[1] IUPAC-IUB Combined Commission on Biochemical Nomenclature. Abbreviations and Symbols for Chemical Names of Special Interest in Biological Chemistry. Revised Tentative Rules (1965), J. Biol. Chem. *241*, 527 – 533 (1966). Biochemistry *5*, 1445 – 1453 (1966). Arch. Biochem. Biophys. *115*, 1 – 12 (1966).

[2] International Union of Biochemistry: Biochemical Nomenclature and Related Documents, The Biochemical Society, 7 Warwick Court, London 1978.

[3] IUPAC Physical Chemistry Division, Subcommittee on Chemical Kinetics. Symbolism and Terminology in Chemical Kinetics, Pure Appl. Chem. *53*, 753 – 771 (1981).

[4] *H. U. Bergmeyer,* Quality Requirements for Reagents Used in Enzymatic Analyses in Biological Materials, 8th Int. Congr. Clin. Chemistry, Copenhagen, June 1972. Abstracts of papers 12.2; – Chem. Rundschau *25*, Nr. 28, (1972).

[5] *H. U. Bergmeyer, P. Scheibe, A. W. Wahlefeld,* Optimization of Methods for Aspartate Aminotransferase and Alanine Aminotransferase, Clin. Chem. *24*, 58 – 73 (1978).

[6] Expert Panel on Enzymes; Committee on Standards (IFCC): Provisional Recommendations on IFCC Methods for the Measurement of Catalytic Concentrations of Enzymes. Part 3. IFCC Method for Alanine Aminotransferase, Clin. Chim. Acta *105*, 147F – 154F (1980). – J. Clin. Chem. Clin. Biochem. *18*, 521 – 534 (1980).

[7] *K. Tornheim, J. M. Lowenstein,* Creatine Phosphate Inhibition of Heart Lactate Dehydrogenase and Muscle Pyruvate Kinase Is Due to a Contaminant, J. Biol. Chem. *254*, 10586 – 10587 (1979).

[8] *Th. J. Wheeler, J. M. Lowenstein,* Creatine Phosphate Inhibition of Adenylate Deaminase Is Mainly Due to Pyrophosphate, J. Biol. Chem. *254*, 1484 – 1486 (1978).

[9] Expert Group on Diagnostic Kits and Reagents, International Federation of Clinical Chemistry (IFCC), Provisional Recommendation (1978) on Evaluation of Diagnostic Kits, Part 1. Recommendations for Specifications on Labelling of Clinical Laboratory Materials, J. Clin. Chem. Clin. Biochem. *17*, 657 – 664 (1979); – Clin. Chim. Acta *93*, 153F – 188F (1979).

[10] *R. Kattermann,* 5-Jahres-Bericht 1975 – 1979, Klin. Chem. Institut, Klinikum der Stadt Mannheim, 1980.

[11] *F. Stähler,* Kits: Their Place, and Cost Evaluation, Lecture X. Intern. Congr. Clin. Chem. March 3 – 5, 1976, Mexico City.

[12] *Z. Schneider, H. C. Friedmann,* Stabilization of Enzymes in Sephadex — A Long-Term Experiment, J. Appl. Biochem. *3,* 135 – 146 (1981).

[13] *B. Mattiasson,* Reversible Immobilization of Enzymes with Special Reference to Analytical Applications, J. Appl. Biochem. *3,* 183 – 194 (1981).

[14] *D. Kutter,* Teststreifen zur Rationalisierung der mikroskopischen Harnuntersuchung, Dtsch. med. Wschr. *105,* 1246 – 1249 (1980).

[15] *C. Bonard, E. Weber, P. U. Koller, K. D. Willamowski, F. Bachmann,* Rationalisierung im Urinlaboratorium ohne Verzicht auf diagnostische Sicherheit, Dtsch. med. Wochenschr. *107,* 249 – 251 (1982).

[16] Boehringer Mannheim GmbH, Reagent Bands for Analytical Purposes, European Patent Office (Munich), EP-A 0054679, published 1982.

[17] Boehringer Mannheim GmbH, Device and Process for the Determination of H_2O_2, German Patent P 3135667, applied 1981.

[18] *H. O. Beutler, G. Beinstingl, G. Michal,* Bestimmung von Ascorbinsäure in Frucht- und Gemüsesäften, International Federation of Fruit Juice Producers, Scientific-Technical Commission, Reports p. 325 – 331, Bled 1980.

[19] *G. Henniger,* Enzymatische Bestimmung von L-Ascorbinsäure in Lebensmitteln, Pharmazeutika und biologischen Flüssigkeiten, Alimenta *20,* 12 – 14 (1981).

[20] General Electric, Printed Reagent Strips, Selected Business Ventures *6* (1982).

[21] *F. Stähler,* Real Time Clinical Chemistry using Dry Chemistries, 2nd Asian & Pacific Congress of Clinical Biochemistry, Singapore, 1982. *W. Werner,* New Development in Dry Chemistry, Annal. Biol. Clin. *42,* 384 (1982).

[22] *H. G. Curme* et al., Multilayer Film Elements for Clinical Analysis: General Concepts, Clin. Chem. *24,* 1335 – 1342 (1978).

[23] *R. N. Barnett,* Medical Significance of Laboratory Results. Am. J. Clin. Pathol. *50,* 671 – 676 (1968).

[24] *J. O. Westgard, R. N. Carey, S. Wold,* Criteria for Judging Precision and Accuracy in Method Development and Evaluation. Clin. Chem. *20,* 825 – 833 (1974).

[25] *I. W. Percy-Robb, P. M. G. Broughton, R. D. Jennings, J. J. McCormack, D. W. Neill, R. A. Saunders, Mary Warner,* A Recommended Scheme for the Evaluation of Kits in the Clinical Chemistry Laboratory, Ann. Clin. Biochem. *17,* 217 – 226 (1980).

[26] Guiding Principles and Recommendations on Labelling of Clinical Laboratory Materials — a WHO Memorandum; Bulletin of the WHO *56,* 881 (1978).

[27] World Health Organization, WHO Expert Committee on Biological Standardization, 31st Report: Requirements for Immunoassay Kits, WHO Technical Report Series 1981, No. 658.

[28] Expert Group on Diagnostic Kits and Reagents, International Federation of Clinical Chemistry (IFCC), Provisional Recommendations on the Evaluation of Clinical Chemistry Kit Methods. Part 2. Guidelines for the Evaluation of Diagnostic Kits, Clin. Chim. Acta *95,* 163 F – 168 F (1979).

[29] European Committee for Clinical Laboratory Standards, ECCLS, Standard for the Labelling of Clinical Laboratory Materials, 2nd Draft, ECCLS Document Vol. 2, Nr. 3, February 1982.

[30] National Committee for Clinical Laboratory Standards (NCCLS), Standard for Procedure Manuals, PSL 2, June 1980.

[31] German Institute for Standardization (Deutsches Institut für Normung, DIN) 58937 Blatt 4, DK 616-07 : 061.6 : 65.012 (083), Anforderungen an die Beschreibung von Methoden. Januar 1975 und 1. Vorlage Dezember 1980.

[32] Association of Diagnostics and Diagnostic Instruments Manufacturers (Verband der Diagnostica- und Diagnosticageräte-Hersteller, VDGH), Guidelines for the Printed Material supplied with Clinicochemical Diagnostic Products, Oct. 2, 1981.

[33] Department of Health, Education and Welfare, Food and Drug Administration: Manufacture, Packing, Storage and Installation of Medical Devices, Regulations Establishing Good Manufacturing Practices, Federal Register, Vol. 43, No. 141, July 21, 1978, p. 31508 – 31532.

[34] German Diagnostics Group (Deutsche Diagnostica Gruppe, DDG): Provisional Recommendations, Part 1, Draft 1, Principles of Standardization in Clinical Chemistry, March 6, 1981.

[35] *N. W. Tietz,* A Model for a Comprehensive Measurement System in Clinical Chemistry, Clin. Chem. *25,* 833 – 839 (1979).

2.2   Biochemical Reagents for General Use

The biochemical reagents described in chapters 2.2.1 and 2.2.2 are all commercial preparations and have been taken from the suppliers' catalogues current in Spring 1983. We do not feel it necessary to indicate manufacturers and dealers since these are now well known, in contrast to former times. However, we want to draw attention to the Directory of Biologicals, 1983, published by the Editorial Offices of Nature, London.

2.2.1   Enzymes

Hans Ulrich Bergmeyer, Marianne Graßl and Hans-Elmar Walter

In this chapter commercially available enzyme preparations are listed which are frequently used in analytical methods; however, enzymes used only once or twice in the methods described in the following volumes may not be included. If necessary the specific data for those omitted from this chapter will be presented in the relevant methodological chapters. For those preparations which are employed in distinct methods but are not supplied on the market, a summarized isolation procedure will be given in an appendix to the chapter concerned.

The characterization of an enzyme includes the indication of its source, the reaction it catalyzes, the principal effectors, the basic physical and physicochemical data, as well as the trivial and official systematic names.

These data are presented for the enzymes described on the following pages.

The section "assay" comprises

- principle (omitted if identical with the catalyzed reaction)
- measurement and incubation conditions, if necessary including e.g. wavelength, light path, volume of assay (in certain cases incubation mixture, temperature)
- composition of the assay and/or incubation mixture
- concentration of the various constituents in the mixture

These brief indications are generally sufficient for checking the quality of the enzyme preparation. The assay conditions are based essentially on procedures developed in our laboratories.

Although 30 °C is recommended as the measurement temperature by international committees, we state the temperature at which the methods have been established.

The description of proteolytic enzymes has been omitted because they are mainly used for the elucidation of protein structure. On the other hand the determination of a number of peptidases and proteases is described in detail in Vol. V.

References

The data are mainly based on the following surveys and on papers published in journals listed below (up to 1982):

P. D. Boyer, H. Lardy, K. Myrbäck, 2nd edn., The Enzymes, Academic Press, New York 1960–1963.
P. D. Boyer, The Enzymes, 3rd edn., Academic Press, New York 1970–1973.
H. M. Rauen, Biochemisches Taschenbuch, 2nd edn., Springer-Verlag Berlin, Göttingen, Heidelberg 1964.
S. P. Colowick, N. O. Kaplan, Methods in Enzymology, Academic Press, New York.
Hoppe-Seyler Thierfelder, Handbuch der physiol. und path.-chem. Analyse, Vol. VIa 1964, Vol. VIb, VIc, Springer, Berlin 1966.
Th. E. Barman, Enzyme Handbook, Springer, Berlin 1969.
Arch. Biochem. Biophys. – Biochim. Biophys. Acta – Biochemistry – Biochem. J. – FEBS Letters – J. Biol. Chem. – Eur. J. Biochem. – Proc. Natl. Acad. Sci. USA.

Acetate Kinase

from *Escherichia coli*

ATP: acetate phosphotransferase, EC 2.7.2.1

$$\text{Acetate} + \text{ATP} \xrightarrow{\text{AK}} \text{acetyl-P} + \text{ADP}$$

Equilibrium constant (Tris buffer, pH 7.4; 29 °C)

$$K = \frac{[\text{acetyl-P}] \times [\text{ADP}]}{[\text{acetate}] \times [\text{ATP}]} = 8 \times 10^{-3}$$

Michaelis constants (Tris buffer, pH 7.4; 29 °C)

		Relative rate [1]
ATP	2×10^{-3} mol/l (7×10^{-3} mol/l)	–
Acetate	3×10^{-1} mol/l (1×10^{-1} mol/l)	1.0
Propionate	4.7×10^{-1} mol/l	0.1
Mg^{2+}	5×10^{-3} mol/l	–
ADP	1.5×10^{-3} mol/l	–
Acetyl-P	5×10^{-3} mol/l	–

Specificity: the enzyme reacts with acetate or propionate, but is inactive with formate or butyrate. Phosphate donors are ATP (1.00), GTP (0.83) or ITP (0.68). ADP or IDP is inactive in the reverse direction.

Assay

$$\text{Acetate} + \text{ATP} \xrightarrow{\text{AK}} \text{acetyl-P} + \text{ADP}$$

$$\text{ADP} + \text{PEP} \xrightarrow{\text{PK}} \text{ATP} + \text{pyruvate}$$

$$\text{Pyruvate} + \text{NADH} + \text{H}^+ \xrightarrow{\text{LDH}} \text{L-lactate} + \text{NAD}^+.$$

Wavelength 339, Hg 334 or Hg 365 nm; light path 10 mm; final volume 2.99 ml; 25 °C.

2.00 ml	triethanolamine buffer (0.1 mol/l; pH 7.6)	67 mmol/l
0.50 ml	acetate, Na salt, (2 mol/l)	333 mmol/l
0.20 ml	ATP, Na salt, (50 mg/ml)	5.4 mmol/l
0.10 ml	PEP, CHA salt, (15 mg/ml)	1.1 mmol/l
0.08 ml	NADH, Na salt, (10 mg/ml)	0.32 mmol/l
0.04 ml	MgCl$_2$ (0.1 mol/l)	1.33 mmol/l
0.02 ml	PK (2 mg/ml)	2.7 kU/l
0.01 ml	LDH (5 mg/ml)	8.3 kU/l
0.02 ml	MK (2 mg/ml)	4.8 kU/l
0.02 ml	enzyme in buffer	

Stability: in suspensions in ammonium sulphate solution, 3.2 mol/l; pH ca. 6, a decrease in activity may occur within 6 months at 4°C.

Purity required: specific activity $\geq$ 170 U/mg protein (25 °C). Contaminants (related to the activity of AK) $\leq$ 0.01% ALAT (GPT), LDH, MK and "NADH-oxidase", 0.02% ASAT (GOT).

Reference

[1] *J. A. Rose et al.,* J. Biol. Chem. *211,* 737 (1954).

Acetylcholinesterase

from *Electrophorus electricus*

Acetylcholine acetylhydrolase, EC 3.1.1.7

$$\text{Acetylcholine} + \text{H}_2\text{O} \xrightarrow{\text{acetylcholinesterase}} \text{acetate} + \text{choline}$$

Michaelis constants [1, 2]

Acetylcholine	9.5 $\times$ 10^{-5} mol/l
Acetylhomocholine	1.3 $\times$ 10^{-3} mol/l

Dimethylaminoethyl acetate	6.5×10^{-4} mol/l
Methylaminoethyl acetate	7.8×10^{-3} mol/l
Aminoethyl acetate	1.56×10^{-2} mol/l

Effectors: natural amino acids and a large number of amides, dipeptides and esters inhibit [3]. Further inhibitors are organophosphorus compounds [4]. Alkyl phosphates such as tetraethyl pyrophosphate and diisopropyl fluorophosphate are irreversible inhibitors [5, 6].

Molecular weight [7]: 230000.

Specificity [5]: the enzyme is not specific for acetylcholine and hydrolyzes various esters of acetic acid.

pH Optimum: 7.

Assay

$$\text{Acetylthiocholine} + H_2O \xrightarrow[\text{esterase}]{\text{acetylcholine}} \text{acetate} + \text{thiocholine}$$

$$\text{Thiocholine} + \text{DTNB} \longrightarrow \text{thiocholine thionitrobenzoate disulphide} +$$
$$\text{thionitrobenzoate.}$$

Wavelength Hg 405 nm; light path 10 mm; final volume 3.15 ml; 25 °C; $\varepsilon = 1.33$ $1 \times$ mmol$^{-1} \times$ mm^{-1}.

3.00 ml	buffer (sodium phosphate 0.05 mol/l; pH 7.2; DTNB, 5,5'-dithio-bis-(2-nitrobenzoic acid) 0.1 mg/ml)	phosphate	48 mmol/l
		DTNB	0.24 mmol/l
0.10 ml	acetylthiocholine iodide (0.156 mol/l)		5 mmol/l
0.05 ml	enzyme in sodium phosphate buffer, 0.05 mol/l; pH 7.2		

Stability: lyophilized enzymes are stable for several months at 4 °C.

Purity required: activity $\geqslant$ 1000 U/mg substance.

References

[1] *I. B. Wilson, C. Quan*, Arch. Biochem. Biophys. *73*, 131 (1958).
[2] *I. B. Wilson, E. Cabib*, J. Am. Chem. Soc. *78*, 202 (1956).
[3] *F. Bergmann et al.*, J. Biol. Chem. *186*, 693 (1950).
[4] *W. B. Neely et al.*, Biochemistry *3*, 1477 (1964).
[5] *I. B. Wilson*, in: *P. D. Boyer* (ed.), The Enzymes, Vol. *V*, Academic Press, New York 1971, p. 87.
[6] *R. Zech*, Hoppe-Seyler's Z. physiol. Chem. *350*, 1415 (1969).
[7] *L. T. Kremzner, J. B. Wilson*, Biochemistry *3*, 1902 (1964).

β-N-Acetyl-D-glucosaminidase

from bovine kidney

2-Acetamido-2-deoxy-β-D-glucoside acetamidodeoxyglucohydrolase, EC 3.2.1.30

$$\text{β-Phenyl-2-acetamido-2-deoxy-D-glucoside} + H_2O \xrightarrow{\text{N-acetylglucosaminidase}}$$
$$\text{phenol} + 2\,\text{acetamido-2-deoxy-D-glucose}$$

Michaelis constants [1]

Source	Substrate	K_m	Conditions
Diplococcus pneumoniae	4-NP-GNAc*	2.2×10^{-4} mol/l	P_i, citrate buffer, pH 5.3; 37 °C
Rat epididymis	phenyl-N-acetyl-β-glucosaminide	7.1×10^{-4} mol/l	citrate buffer, pH 4.3; 38 °C
	4-NP-GNAc	4×10^{-4} mol/l	
Limpet	phenyl-N-acetyl-β-glucosaminide	6.4×10^{-3} mol/l	citrate buffer, pH 4.3; 38 °C
	4-NP-GNAc*	3.8×10^{-3} mol/l	
Rat kidney	phenyl-N-acetyl-β-glucosaminide	1.78×10^{-3} mol/l	citrate buffer, pH 4.3; 37 °C

Effectors [1]: with respect to phenyl-N-acetyl-β-glucosaminide, acetamide ($K_I = 1.02 \times 10^{-2}$ mol/l), acetate ($K_I = 1.7 \times 10^{-2}$) and N-acetylglucosamine competitively inhibit the enzyme from rat kidney.

Molecular weight [1]: 150 – 160000 (rat kidney).

Specificity [1]: the enzyme attacks both N-acetyl-β-glucosaminides and N-acetyl-β-galactosaminides. It hydrolyzes the terminal glucosaminidic bonds of odd-numbered oligosaccharides to yield N-acetylglucosamine and the next lower even-numbered oligosaccharide.

pH Optimum: 4 – 5.

Assay

$$\text{4-NP-N-Ac-β-D-glucosaminide} + H_2O \xrightarrow{\text{N-acetylglucosaminidase}} \text{4-NP} +$$
$$\text{N-Ac-D-glucosamine.}$$

---

* Abbreviation: 4-nitrophenyl 2-acetamido-2-deoxy-β-D-glucopyranoside.

Wavelength Hg 405 nm; light path 10 mm; incubation volume 1.10 ml; final volume 3.10 ml; 25 °C; $\varepsilon = 1.85\ l \times mmol^{-1} \times mm^{-1}$.

0.50 ml	citrate buffer (80.1 mol/l; pH 4.5)	45 mmol/l
0.50 ml	4-NP-N-Ac-β-D-glucosaminide (1 mg/ml)	0.45 g/l
0.10 ml	enzyme in buffer	

after exactly 5 min add

2.00 ml	borate buffer (0.2 mol/l; pH 9.8)	0.13 mol/l

Stability: suspensions in ammonium sulphate solution, 3.2 mol/l; pH ca. 6, are stable for several months.

Purity required: specific activity $\geqslant 4$ U/mg (25 °C). Contaminants (related to the activity of N-acetylglucosaminidase) $\leqslant 0.1\%$ α-mannosidase, $\leqslant 2\%$ α-fucosidase (4-NP-glycosides as substrates).

Reference

[1] *Th. E. Barman,* Enzyme Handbook, Vol. *II*, Springer, Berlin 1969, p. 588.

Acyl-CoA Oxidase

from micro-organisms

Acyl-CoA: oxygen oxidoreductase, EC 1.3.3.-

$$\text{Acyl-CoA} + 1/2\,O_2 \xrightarrow{\text{acyl-CoA oxidase}} \text{enoyl-CoA} + H_2O_2$$

Michaelis constants

		source
Palmitoyl-CoA	3.3×10^{-5} mol/l	*Candida* species [1]
Palmitoyl-CoA	1.2×10^{-4} mol/l	*Arthrobacter* species

Effectors: heavy metal ions (Hg^{2+}, Ag^+) inhibit the enzyme from *Candida*.

Molecular weight: approx. 600000 (*Candida* species)
600000 210000 (by gel filtration; *Arthrobacter* species)

Specificity

	Data for the enzyme from		
Arthrobacter species		*Candida* species	
Acyl-CoA (carbon chain)	Relative activity (%)	Acyl-CoA (carbon chain)	Relative activity (%)
Hexanoyl-CoA (6:0)	11	Butyryl-CoA (4:0)	7
Octanoyl-CoA (8:0)	60	Hexanoyl-CoA (6:0)	13
Decanoyl-CoA (10:0)	73	Octanoyl-CoA (8:0)	18
Dodecanoyl-CoA (12:0)	87	Decanoyl-CoA (10:0)	78
Tetradecanoyl-CoA (14:0)	99	Lauroyl-CoA (12:0)	100
Hexadecanoyl-CoA (16:0)	53	Myristoyl-CoA (14:0)	69
Octadecanoyl-CoA (18:0)	16	Palmitoyl-CoA (16:0)	28
Isosanoyl-CoA (20:0)	7	Stearoyl-CoA (18:0)	12
9-Tetradecenoyl-CoA (14:1)	100	Oleoyl-CoA (18:1)	36
9-Hexadecenoyl-CoA (16:1)	65		
9,12-Hexadecadienoyl-CoA (16:2)	55		
cis-9-Octadecenoyl-CoA (18:1)	45		
trans-9-Octadecenoyl-CoA (18:1)	31		
d-12-Hydroxy-*trans*-9-octa-decenoyl-CoA (18:1)	7		
cis-9,*cis*-12-Octadecenoyl-CoA (18:2)	31		
cis-6,*cis*-9,*cis*-12-Octadecenoyl-CoA (18:3)	95		
15-Tetracosenoyl-CoA (24:1)	7		

pH Optimum: 7.5 (serum acyl-CoA)
                   8.5 (Palmitoyl-CoA) $\Big\}$ *Arthrobacter* species
                   7.5 – 8   *Candida* species

Assay

$$\text{Palmitoyl-CoA} + 1/2\,O_2 \xrightarrow{\text{acyl-CoA oxidase}} \text{2-hexadecenoyl-CoA} + H_2O_2$$

$$H_2O_2 + CH_3OH \xrightarrow{\text{catalase}} HCHO + 2\,H_2O$$

$$HCHO + \text{AHMT*} + OH^- \xrightarrow{KIO_4} \text{red-purple tetrazine dye}.$$

Wavelength 550 nm; light path 10 mm; incubation volume 0.5 ml; final volume 2.0 ml; 37 °C; $\varepsilon = 3.67\,l \times mmol^{-1} \times mm^{-1}$.

0.05 ml	palmitoyl-CoA (0.8 mmol/l)	80 µmol/l
0.15 ml	water	

0.25 ml   catalase/methanol (1.6 kU/ml;   catalase   800 kU/l
          potassium phosphate buffer;   methanol   40 ml/l
          0.1 mol/l, pH 7.5; 80 µl methanol/ml)
0.05 ml   enzyme in potassium phosphate buffer,
          50 mmol/l; pH 7.5

incubate for exactly 10 min at 37 °C, add

0.50 ml   KOH (2 mol/l)
0.50 ml   AHMT* (0.6% w/v in KOH, 0.5 mol/l)

incubate for exactly 5 min at 37 °C, add

0.50 ml   KIO$_4$ (0.75% w/v in KOH, 0.2 mol/l)

mix, keep for about 10 min at room temperature, measure absorbance at 550 nm against water. Run a blank in which the enzyme has been added after addition of KOH.

Stability: the lyophilized, stabilized enzyme is stable for several months at 4 °C.

Purity: activity $\geqslant$ 2.0 U/mg substance (37 °C). Contaminants (related to the activity of acyl-CoA oxidase) $\leqslant$ 100% catalase, $\leqslant$ 0.1% lipase.

Reference

[1] S. Shimizu et al., Biochem. Biophys. Res. Commun. 91, 108 (1979).

Acyl-CoA Synthetase

from *Pseudomonas* species

Acid : CoA ligase (AMP-forming), EC 6.2.1.3

$$R\text{-COOH} + ATP + CoA\text{-SH} \xrightarrow{\text{ACS}} AMP + PP_i + RCO-S-CoA$$

Michaelis constants

Palmitate 6.3×10^{-5} mol/l
CoA 3.4×10^{-4} mol/l

---

* 4-amino-3-hydrazino-5-mercapto-1,2,4-triazole.

Effectors: heavy metals and ionic detergents inhibit.

Specificity [1]: the enzyme acts on acids from C-6 to C-20.

Carbon chain of free fatty acid	Relative rate
6:0	9
8:0	74
10:0	99
12:0	84
14:0	76
16:0	100
18:0	76
18:1	90

pH Optimum: 8.5 – 8.8.

Assay

$$\text{ATP} + \text{palmitate} + \text{CoA} \xrightarrow{\text{acyl-CoA synthetase}} \text{palmitoyl-CoA} + \text{AMP} + \text{PP}_i$$

$$\text{Palmitoyl-CoA} + O_2 \xrightarrow{\text{acyl-CoA oxidase}} \text{2-hexadecenoyl-CoA} + H_2O_2$$

$$H_2O_2 + CH_3OH \xrightarrow{\text{catalase}} HCHO + 2\,H_2O$$

$$HCHO + \text{AHMT}* + OH^- \xrightarrow{\text{KIO}_4} \text{red-purple tetrazine dye}.$$

Wavelength 550 nm; light path 10 mm; incubation volume 0.55 ml; final volume 2.05 ml; 37 °C; $\varepsilon = 3.67\ l \times \text{mmol}^{-1} \times \text{mm}^{-1}$.

0.05 ml	ATP-Mg (82.6 mmol/l; Mg, 0.1 mol/l)	ATP	7.5 mmol/l
		Mg^{2+}	0.1 mmol/l
0.05 ml	CoA, Na salt, (6.7 mmol/l)	CoA	0.6 mmol/l
0.05 ml	acyl-CoA oxidase (2.0 U/ml)		180 U/l
0.25 ml	catalase/methanol (1.6 kU/ml; potassium phosphate buffer, 0.1 mol/l; pH 7.5; 80 µl methanol/ml)	catalase methanol	750 kU/l 36 ml/l
0.05 ml	enzyme in potassium phosphate buffer, 50 mmol/l; pH 7.5		
0.10 ml	palmitate (1.0 mmol/l Triton X-100, 1% w/v)	palmitate	0.18 mmol/l

incubate for exactly 10 min at 37 °C

0.50 ml   KOH (2 mol/l)
0.50 ml   AHMT (0.6% w/v in KOH, 0.5 mol/l)

---

* 4-amino-3-hydrazino-5-mercapto-1,2,4-triazole.

incubate for exactly 5 min at 37 °C

0.50 ml KIO_4 (0.75% w/v in KOH, 0.2 mol/l)

mix, keep for about 10 min at room temperature, measure absorbance at 550 nm against water. Run a blank without palmitate.

Stability: a decrease in the activity of the lyophilized enzyme, stabilized with ATP and sugars, of approx. 20% may occur within 6 months at -4 °C.

Purity: activity $\geqslant$ 0.25 U/mg substance (37 °C). Contaminants (related to the activity of acyl-CoA synthetase) $\leqslant$ 0.5% "NADH oxidase", ATPase, 0.01% catalase.

Reference

[1] *S. Shimizu et al.*, Anal. Biochem. *98*, 341 (1979).

Adenosine Deaminase

from calf intestine

Adenosine aminohydrolase, EC 3.5.4.4

$$\text{Adenosine} + H_2O \xrightarrow{\text{ADA}} \text{inosine} + NH_3$$

Michaelis constants (phosphate buffer, pH 7.4; 20 °C) [1]

		Relative rate
Adenosine	3.53×10^{-5} mol/l	1.00
2'-Deoxyadenosine	2.30×10^{-5} mol/l	0.96
3'-Deoxyadenosine	2.25×10^{-5} mol/l	0.65
2,6-Diaminopurine riboside	3.17×10^{-5} mol/l	0.29
AMP, A-3-MP, A-2-MP	$-$	0

Inhibitor constants (phosphate buffer, pH 7.0; 25 °C) [2]

Purine 9'-ribonucleotide	8.8×10^{-6} mol/l
Various 6-substituted (9-hydroxyalkyl) purines	$3.4 - 38 \times 10^{-5}$ mol/l

For inhibitor properties of various adenosine analogues, see [2].

Effectors: Ag^+, Hg^{2+}, Cu^{2+}, sulphydryl reagents (4-chloromercuribenzoate and methylmercuric nitrate), tosylphenyl-chloro-ketone and diisopropyl-fluorophosphate are strongly inhibiting [3].

Molecular weight: 33 500 [3].

Assay

Wavelength 265 nm; light path 10 mm; final volume 3.12 ml; 25 °C; $\varepsilon = 0.81$ l $\times$ mmol^{-1} $\times$ mm^{-1}.

3.00 ml	phosphate buffer (50 mmol/l; pH 7.4)	49.5 mmol/l
0.10 ml	adenosine (0.36 mg/ml buffer)	0.04 mmol/l
0.02 ml	enzyme in buffer	

Stability: suspensions in ammonium sulphate solution, 3.2 mol/l, or solutions in 50% glycerol + phosphate, 10 mmol/l; pH ca. 6, are stable for several months at 4 °C.

Purity required: specific activity $\geqq$ 200 U/mg protein (25 °C). Contaminants (related to the specific activity of ADA) $\leqq$ 0.01% AP, A-5-MP-deaminase, guanase and NP.

References

[1] *A. Coddington,* Biochim. Biophys. Acta *99,* 442 (1965).
[2] *J. G. Cory, R. J. Suhadolnik,* Biochemistry *4,* 1729 (1965).
[3] *N. Pfrogner,* Arch. Biochem. Biophys. *119,* 141 (1967).

Alanine Aminotransferase

Glutamate-Pyruvate Transaminase

from pig heart

L-Alanine: 2-oxoglutarate aminotransferase, EC 2.6.1.2

$$2\text{-Oxoglutarate} + \text{L-alanine} \underset{}{\overset{\text{ALAT}}{\rightleftharpoons}} \text{L-glutamate} + \text{pyruvate}$$

Equilibrium constant glycylglycine buffer, pH 8.0; 25 °C)

$$K = \frac{[\text{L-glutamate}] \times [\text{pyruvate}]}{[2\text{-oxoglutarate}] \times [\text{L-alanine}]} = 2.2$$

Michaelis constants (Tris buffer, pH 8.1; 25 °C) [1]

L-Alanine	2.8×10^{-2} mol/l
2-Oxoglutarate	4×10^{-4} mol/l
Pyruvate	3×10^{-4} mol/l
L-Glutamate	2.5×10^{-2} mol/l

Inhibitor constants (Tris buffer, pH 8.1; 37 °C) [1]

L-Glutamate	2.5×10^{-2} mol/l
L-Alanine	2.8×10^{-2} mol/l
L-Norleucine	3.0×10^{-2} mol/l
L-Norvaline	7.0×10^{-2} mol/l
L-2-Aminobutyrate	9.0×10^{-2} mol/l
L-2-Aminoadipate	9.0×10^{-2} mol/l
Glycine	5.0×10^{-1} mol/l
L-Aspartate	5.5×10^{-1} mol/l

Effectors: the transfer of the amino group from glutamate to either 2-oxoglutarate or pyruvate is inhibited by formate [1]. The enzyme exhibits high sensitivity to sulphydryl and carboxyl reagents and is particularly sensitive to cycloserine, amino oxyacetate and cysteine [2].

Molecular weight (pH 6.6) [1]: 115000.

Specificity: the enzyme is specific for L-alanine and L-glutamate. Less than 2% of the rate is obtained with: 2-oxoglutarate and L-2-aminobutyrate, L-norleucine, L-ornithine or L-aspartate in the forward reaction, and with pyruvate and L-2-aminobutyrate, L-norvaline, L-ornithine, L-aspartate, DL-3-hydroxyglutamate or L-2-aminoadipate in the reverse reaction. The following do not react: glycine, D-alanine, L-serine, DL-O-methylserine, L-leucine, L-asparagine, L-glutamine, DL-O-phosphoserine, L-kynurenine, D-glutamate, DL-2-methylglutamate and 4-aminobutyrate [1].

Assay

$$\text{2-Oxoglutarate + L-alanine} \xrightarrow{\text{ALAT}} \text{L-glutamate + pyruvate}$$

$$\text{Pyruvate + NADH + H}^+ \xrightarrow{\text{LDH}} \text{L-lactate + NAD}^+.$$

Wavelength 339, Hg 334 or Hg 365 nm; light path 10 mm; final volume 3.23 ml; 25 °C.

3.00 ml	potassium phosphate buffer (50 mmol/l; pH 7.5; 97 mg L-alanine/ml)	phosphate alanine	46.2 mmol/l 1.0 mol/l
0.15 ml	2-oxoglutarate, Na salt, (39 mg/ml; pH ca. 7)		9.6 mmol/l
0.05 ml	NADH, Na salt, (10 mg/ml)		0.2 mmol/l
0.01 ml	LDH (5 mg/ml)		9.1 kU/l
0.02 ml	enzyme in buffer		

Stability: suspensions in ammonium sulphate solution, 3.2 mol/l; pH ca. 6, are stable for several months at 4 °C.

Purity required: specific activity $\geq$ 80 U/mg protein (25 °C). Contaminants (related to the specific activity of ALAT) $\leq$ 0.01% GlDH, ASAT, LDH and MDH.

References

[1] *M. H. Saier, W. T. Jenkins,* J. Biol. Chem. *242*, 91, 101 (1967).
[2] *S. Hoppe, H. L. Segal,* J. Biol. Chem. *237*, 3189 (1962).

L-Alanine Dehydrogenase

from *Bacillus subtilis*

L-Alanine: NAD oxidoreductase (deaminating), EC 1.4.1.1

$$\text{L-Alanine} + \text{NAD}^+ + \text{H}_2\text{O} \xrightleftharpoons{\text{alanine dehydrogenase}} \text{pyruvate} + \text{NADH} + \text{NH}_4^+$$

Equilibrium constant (Tris buffer; 25 °C) [1]: $K = 3.1 \times 10^{-14}$

Michaelis constants
(deamination reaction: carbonate buffer, pH 10; 25 °C)

L-Alanine	1.7×10^{-3} mol/l
NAD	1.8×10^{-4} mol/l
L-α-Aminobutyrate	1.8×10^{-1} mol/l
L-Valine	1.5×10^{-1} mol/l
L-Isoleucine	2.5×10^{-1} mol/l
L-Serine	5.7×10^{-2} mol/l

(amination reaction: Tris buffer, pH 8.0; 25 °C)

Pyruvate	5.4×10^{-4} mol/l
NADH	2.3×10^{-5} mol/l
NH$_4^+$	3.8×10^{-2} mol/l
2-Oxobutyrate	2.3×10^{-2} mol/l
Glyoxylate	1.6×10^{-2} mol/l
Hydroxypyruvate	2.5×10^{-3} mol/l

Inhibitor constants (carbonate buffer, pH 10; 25 °C) [1]

D-Alanine	2×10^{-2} mol/l
Glycine	3.8×10^{-2} mol/l
D-Cysteine	6.4×10^{-4} mol/l
D-Homocysteine	7.8×10^{-3} mol/l
Sarcosine	4.3×10^{-3} mol/l

Effectors: alanine dehydrogenase is inhibited by heavy metal ions and SH-blocking reagents (e.g. 4-chloromercuribenzoate). The inhibition by 4-chloromercuribenzoate is completely reversed by addition of cysteine [1, 2]. Numerous D-amino acids as well as alanine analogues inhibit competitively [1].

Molecular weight [1]: approx. 228000.

Specificity: alanine dehydrogenase reacts specifically with L(+)-alanine. L-Alanine analogues are deaminated (L-α-aminobutyrate, L-valine, L-isoleucine, L-serine and L-norvaline) or aminated (2-oxoglutarate, glyoxalate, hydroxypyruvate, 2-oxovalerate, 2-oxoisovalerate, 2-oxoisocaproate and 2-oxocaproate) with decreasing velocity [1].

pH Optimum [1]: amination reaction pH 8.8 – 9.0, deamination reaction pH 10.0 – 10.5.

Assay

Wavelength 339, Hg 334 or Hg 365 nm; light path 10 mm; final volume 3.17 ml; 25 °C.

3.00 ml	carbonate buffer (50 mmol/l; pH 10.0)	47.3 mmol/l
0.10 ml	L-alanine, 0.50 mol/l, (44 mg L-alanine/ml)	15.7 mmol/l
0.05 ml	NAD, approx. 27 mmol/l, (20 mg NAD/ml)	0.5 mmol/l
0.02 ml	enzyme in water	

Stability: suspensions in ammonium sulphate solution, 2.4 mol/l; pH ca. 7, are stable for several months at 4 °C.

Purity required: specific activity $\geq$ 30 U/mg protein (25 °C; L-alanine as substrate). Contaminants (related to the specific activity of alanine-DH) $\leq$ 0.01% MDH and LDH (L-malate and L-lactate as substrates, respectively).

References

[1] *A. Yoshida, E. Freese,* Biochim. Biophys. Acta *92,* 33 (1964); *96,* 248 (1965).
[2] *D. M. Greenberg,* in: *D. M. Greenberg,* Metabolic Pathways, Vol. *3,* Academic Press, New York 1969, p. 97.

Alcohol Dehydrogenase

from yeast

Alcohol: NAD oxidoreductase, EC 1.1.1.1

$$\text{Ethanol} + \text{NAD}^+ \xrightleftharpoons{\text{ADH}} \text{acetaldehyde} + \text{NADH} + \text{H}^+$$

Equilibrium constant (phosphate or glycine buffer; 20 °C) [3]

$$K = \frac{\text{acetaldehyde} \times [\text{NADH}] \times [\text{H}^+]}{\text{ethanol} \times [\text{NAD}^+]} = 8.0 \times 10^{-12}\,\text{mol/l}$$

Michaelis constants (phosphate buffer, pH 7.15; 25 °C) [1]

Ethanol	1.3×10^{-2} mol/l
NAD	7.4×10^{-5} mol/l
Acetaldehyde	7.8×10^{-4} mol/l
NADH	1.08×10^{-5} mol/l

Inhibitor constants (phosphate buffer, pH 7.15; 25 °C) [1]

Ethanol	4.3×10^{-2} mol/l
NAD	6.1×10^{-4} mol/l
Acetaldehyde	6.7×10^{-4} mol/l
NADH	1.8×10^{-5} mol/l

Molecular weight (pH 7.0): 148 000 [3], 141 000 [4].

Specificity: the enzyme oxidizes primary alcohols. Isopropanol and sec. butanol are slowly oxidized, while higher secondary and tertiary alcohols do not react. Numerous aldehydes are reduced in the reverse reaction (cf. also [2]). The enzyme does not react with NADP.

Assay

Wavelength 339, Hg 334 or Hg 365 nm; light path 10 mm; final volume 2.93 ml; 25 °C.

2.50 ml	sodium pyrophosphate buffer	85.5 mmol/l
	(0.1 mol/l; pH 9.0; containing glycine,	
	1.67 mg/ml)	19.1 mmol/l
0.10 ml	semicarbazide hydrochloride	
	(250 mg/ml; pH ca. 6.5)	6.2 mmol/l
0.10 ml	ethanol (96%, v/v)	0.6 mol/l
0.20 ml	NAD (20 mg/ml)	1.8 mmol/l
0.01 ml	GSH (90 mg/ml)	1.0 mmol/l
0.02 ml	enzyme in 0.1% albumin solution	

Stability: the lyophilized enzyme stabilized with sucrose is stable for several months at 4 °C; the enzyme dissolved in water is stable for about 1 week at 4 °C. In crystalline suspensions in ammonium sulphate solution 3.2 mol/l; pH approx. 6, a decrease in activity of approx. 40% may occur within 6 months at 4 °C.

Purity required: specific activity $\geq$ 300 U/mg protein (25 °C). Contaminants (related to the specific activity of ADH) $\leq$ 0.01% LDH.

References

[1] *C. C. Wratten, W. W. Cleland,* Biochemistry *2,* 935 (1963).
[2] *H. Sund, H. Theorell,* in: *P. D. Boyer, H. Lardy, K. Myrbäck* (eds.), The Enzymes, Vol. 7, Academic Press, New York 1963, p. 25.
[3] *K. I. Backlin,* Acta Chem. Scand. *12,* 1279 (1958).
[4] *M. Buehner, H. Sund,* Eur. J. Biochem. *11,* 73 (1969).

Alcohol Dehydrogenase

from *Leuconostoc mesenteroides*

Alcohol: NAD(P) oxidoreductase, EC 1.1.1.2

$$\text{Alcohol} + \text{NADP}^+ \underset{}{\overset{\text{ADH}}{\rightleftharpoons}} \text{aldehyde} + \text{NADPH} + \text{H}^+$$

Equilibrium constant [1]: the reaction is reversible.

Michaelis constants

NADP[1] 1.6×10^{-4} [mol/l] (Tris buffer, pH 7.5)

Native enzyme [2]:

NADP	5.9×10^{-5} mol/l
Ethanol	7.1×10^{-2} mol/l

Subunit S_2 [2]:

NADP	6.1×10^{-5} mol/l
Ethanol	8.3×10^{-3} mol/l

Effectors: Ag^+, Hg^{2+}, Cu^{2+}, 4-chloromercuribenzoate and 8-hydroxyquinoline are strong inhibitors. Pyrazole, iodoacetate and o-phenanthroline inactivate the enzyme.

Molecular weight [2]: the enzyme is a dimeric protein containing two different subunits.

Native enzyme	240000
Subunit S_1	80000 (NAD specific, labile)
Subunit S_2	160000 (NADP specific, stable)

Specificity: in the forward direction the following alcohols are oxidized in the presence of NADP

	Relative rate
Ethanol	100
n-Propanol	23
n-Butanol	10
n-Pentanol	3
Methanol	2
Ethane-1.2-diol	1

In the reverse reaction the aldehydes listed below are reduced in the presence of NADPH

	Relative rate
Acetaldehyde	100
n-Butyraldehyde	32
Glyoxal	16
Glycolaldehyde	14
Formaldehyde	8

pH Optimum: 7.5 – 9.2.

Assay

Wavelength 339, Hg 334 or Hg 365 nm; light path 10 mm; final volume 2.96 ml; 25 °C.

2.50 ml	sodium pyrophosphate buffer (75 mmol/l; pH 9.0	63 mmol/l
	containing 1.67 mg glycine/ml)	19 mmol/l
0.10 ml	semicarbazide (2.2 mol/l)	74 mmol/l
0.10 ml	ethanol (96%; undenatured)	0.7 mol/l
0.20 ml	NADP, Na salt, (25 mg/ml)	2.1 mmol/l
0.01 ml	GSH (0.3 mol/l)	1.0 mmol/l
0.05 ml	enzyme in 0.1% albumin solution	

Stability: the lyophilized enzyme is stable for several months at 4 °C.

Purity required: activity $\geqslant 20$ U/mg substance (25 °C; $\underline{\Delta}$ ca. 70 U/mg enzyme protein). Contaminants (related to the specific activity of ADH) $\leqslant 0.01\%$ "NADPH oxidase" and aldehyde dehydrogenase.

References

[1] *R. D. De Moss,* in: *S. P. Colowick, N. O. Kaplan* (eds.), Methods in Enzymology, Vol. *I*, Academic Press, New York 1955, p. 504.
[2] *A. Hatanaka et al.,* Agric. Biol. Chem. *38*, 1819 (1974).

Alcohol Oxidase

from *Candida boidinii*

Alcohol: oxygen oxidoreductase, EC 1.1.3.13

$$\text{Alcohol (primary)} + O_2 \xrightleftharpoons{\text{alcohol-OD}} \text{aldehyde} + H_2O_2$$

Michaelis constants [1]

Methanol	$2 \ \times 10^{-3}$ mol/l
Ethanol	4.5×10^{-3} mol/l

Effectors: 4-chloromercuribenzoate, Cu^{2+} and CN^- inhibit at concentrations of 10^{-3} mol/l.

Molecular weight [1]: ca. 600000.

Specificity

	Relative rate
Methanol	100
Ethanol	75
n-Propanol	25
i-Propanol	5
n-Butanol	15
n-Amyl alcohol	5
Allyl alcohol	65

The enzyme does not react with i-butanol, isoamyl alcohol, n-hexanol, glycerol, glycine and L-serine.

pH Optimum: 7.5 – 8.

Assay

$$\text{Methanol} + O_2 \xrightarrow{\text{alcohol-OD}} \text{formaldehyde} + H_2O_2$$

$$H_2O_2 + DH_2 \xrightarrow{\text{POD}} 2\,H_2O + D$$

DH_2 = leuco dyestuff; D = dyestuff.

Wavelength Hg 405 nm; light path 10 mm; $\varepsilon = 3.68$ l $\times$ mmol^{-1} $\times$ mm^{-1}; final volume 2.97 ml; 25 °C.

2.80 ml	phosphate buffer/ABTS (0.1 mol/l; pH 7.5;	94 mmol/l
	ABTS 2 mmol/l)	1.9 mmol/l
0.05 ml	enzyme in buffer	
0.01 ml	POD (10 mg/ml)	8.4 kU/l
0.01 ml	H_2O_2 (0.1 ml 30% H_2O_2/l)	3 µmol/l
0.10 ml	methanol (0.01 ml/ml)	10.5 mmol/l

Stability: in the lyophilized state a decrease in activity of approx. 10% may occur within 6 months at 4°C.

Purity required: specific activity $\geqslant$ 5 U/mg enzyme protein (25°C).

Reference

[1] *H. Sahm, F. Wagner,* Eur. J. Biochem. *36,* 250 (1973).

Aldehyde Dehydrogenase

from yeast

Aldehyde: NAD(P) oxidoreductase, EC 1.2.1.5

$$\text{Aldehyde} + \text{NADP}^+ + H_2O \xrightarrow{\text{AlDH}} \text{acid} + \text{NAD(P)H} + H^+$$

Equilibrium constant [1]: the reaction is irreversible.

Michaelis constants

Substrate

NAD	1.3×10^{-4} mol/l	acetaldehyde (Tris buffer, pH 8.7) [2]
NAD	0.3×10^{-4} mol/l	acetaldehyde (Tris buffer, pH 7.7) [2]
NAD	2×10^{-5} mol/l	benzaldehyde (Tris buffer, pH 8) [1]
NADP	5×10^{-5} mol/l	benzaldehyde (Tris buffer, pH 8)
Acetaldehyde	9×10^{-6} mol/l	
Formaldehyde	7×10^{-4} mol/l	NAD (Tris buffer, pH 8) [1]
Propionaldehyde	3×10^{-6} mol/l	
Benzaldehyde	7×10^{-6} mol/l	

Effectors: the yeast enzyme needs K^+ ions [2] for activity and is inhibited by SH-reagents such as N-ethylmaleimide, 2-iodobenzoic acid, iodoacetic acid and 4-chloromercuribenzoic acid. NAD, NADP and acetaldehyde protect the enzyme from inhibition by the SH-reagents [3].

Heavy metals, especially Cu^{2+}, acetaldehyde (at high concentrations) inhibit [4]. EDTA, histidine, K^+, NH_4^+ and Rb^{2+} are activators [2, 3].

Molecular weight: ca. 200 000 [5, 6].

Specificity: the enzyme oxidizes a number of aliphatic and aromatic aldehydes. Acetyl-GSH is not hydrolyzed [1, 2, 7].

pH Optimum: pH 8.75 for several buffers [2], pH 9 for potassium pyrophosphate buffer, 0.1 mol/l [8].

Assay [2]

Wavelength 339, Hg 334 or Hg 365 nm; light path 10 mm; final volume 2.91 ml; 25 °C.

2.50 ml	potassium pyrophosphate buffer (0.1 mol/l; pH 9.0, adjusted with citrate, 1 mol/l)	86 mmol/l
0.20 ml	NAD (20 mg/ml)	2.1 mmol/l
0.10 ml	pyrazole (6.8 mg/ml)	3.4 mmol/l
0.10 ml	enzyme in 1% albumin solution	
0.01 ml	acetaldehyde (3 µl/ml)	approx. 0.23 mmol/l

Stability: in the lyophilized state (stabilized with potassium phosphate) a decrease in activity of approx. 10% may occur within 6 months at 4 °C.

Purity required: specific activity $\geqslant$ 15 U/mg enzyme protein (25 °C). Contaminants (related to the specific activity of AlDH): $\leqslant 0.01\%$ "NADH oxidase", ADH, LDH.

References

[1] *C. R. Steinmann, W. B. Jakoby,* J. Biol. Chem. *243*, 730 (1968).
[2] *S. Black,* in: *S. P. Colowick, N. O. Kaplan* (eds.), Methods in Enzymology, Vol. *I*, Academic Press, New York 1955, p. 508.
[3] *A. O. M. Stoppani, C. Milstein,* Biochem. J. *67*, 406 (1957).
[4] *M. K. Jacobsen, C. Bernofsky,* Biochim. Biophys. Acta *350*, 277 (1974).
[5] *C. R. Steinmann, W. B. Jakoby,* J. Biol. Chem. *242*, 5019 (1967).
[6] *J. F. Clark, W. B. Jakoby,* J. Biol. Chem. *245*, 6065 (1970).
[7] *S. L. Bradbury, W. B. Jakoby,* J. Biol. Chem. *246*, 1834 (1971).
[8] *H. Möllering, Boehringer Mannheim,* unpublished.

Aldolase

from rabbit muscle

D-Fructose-1,6-bisphosphate D-glyceraldehyde 3-phosphate-lyase, EC 4.1.2.13

$$\text{Fructose-1,6-P}_2 \xrightleftharpoons{\text{Aldolase}} \text{GAP + DAP}$$

Equilibrium constant (30 °C) [1]

$$K = \frac{[\text{GAP}] \times [\text{DAP}]}{[\text{fructose-1,6-P}_2]} = 8.1 \times 10^{-5}\,\text{mol/l}$$

Michaelis constants (glycylglycine buffer, pH 7.1; 25 °C) [2]

Fructose-1,6-P$_2$	1.4×10^{-5} mol/l
Fructose-1-P	7.0×10^{-3} mol/l

Effectors: Cu^{2+}, Ag^+, Zn^{2+} and 2-phenanthroline are inhibiting [3]. The enzyme is inactivated by N-bromoacetyl ethanolamine phosphate [4] and pyridoxal phosphate [5].

Molecular weight [6]: 161 000.

Specificity: in the cleavage reaction the enzyme reacts with fructose-1,6-P$_2$ (1.00) and fructose-1-P (0.02) [7]. In the condensation reaction the enzyme is specific for dihydroxyacetone, while glyceraldehyde phosphate cannot be replaced by other aldehydes [8].

Assay

$$\text{Fructose-1,6-P}_2 + H_2O \xrightarrow{\text{Aldolase}} \text{GAP + DAP}$$

$$\text{GAP} \xrightarrow{\text{TIM}} \text{DAP}$$

$$2\,\text{DAP} + 2\,\text{NADH} + 2\,H^+ \xrightarrow{\text{GDH}} 2\,\text{glycerol-3-P} + 2\,\text{NAD}^+.$$

Wavelength 339, Hg 334 or Hg 365 nm; light path 10 mm; final volume 3.19 ml; 25 °C.

3.00 ml	triethanolamine buffer (0.1 mol/l; pH 7.6)	94.0 mmol/l
0.05 ml	NADH, Na salt, (10 mg/ml)	0.2 mmol/l
0.10 ml	fructose-1,6-P$_2$, tricyclohexylammonium salt, (30 mg/ml)	1.1 mmol/l
0.02 ml	GDH/TIM (2 mg/ml)	GDH 1 kU/l TIM 6.7 kU/l
0.01 ml	enzyme in buffer	

Stability: crystalline suspensions in ammonium sulphate solution, 3.2 mol/l; pH ca. 6, are stable for several months at 4 °C.

Purity required: specific activity $\geq$ 9 U/mg protein (25 °C). Contaminants (related to the specific activity of aldolase) $\leq$ 0.01% GAPDH, 0.03% TIM and 0.05% GDH.

References

[1] *W. J. Rutter,* in: *P. D. Boyer, H. Lardy, K. Myrbäck* (eds.), The Enzymes, Vol. *5,* Academic Press, New York 1961, p. 341.
[2] *P. D. Spolters, R. C. Adelman, S. Weinhouse,* J. Biol. Chem. *240,* 1327 (1965).
[3] *K. Kobashi, B. L. Horecker,* Arch. Biochem. Biophys. *121,* 178 (1967).
[4] *F. C. Hartmann et al.,* J. Biol. Chem. *248,* 8233 (1973).
[5] *L. C. Davis et al.,* Proc. Natl. Acad. Sci. USA, *68,* 416 (1971).
[6] *S. Szuchet, D. A. Yphantis,* Biochem. *12,* 5115 (1973).
[7] *W. J. Rutter, O. C. Richards, B. M. Woodfin,* J. Biol. Chem. *236,* 3193 (1961).
[8] *O. Meyerhof, K. Lohmann, P. Schuster,* Biochem. Z. *286,* 301, 319 (1936).

Aldose Mutarotase

from *Penicillium notatum* and hog kidney, resp.

Aldose 1-epimerase, EC 5.1.3.3

$$\alpha\text{-}D\text{-Glucose} \xrightleftharpoons{\text{mutarotase}} \beta\text{-}D\text{-glucose}$$

Equilibrium constant [1]: the reaction is reversible.

Effectors [2]: several sugars and sugar alcohols are competitive inhibitors of the enzyme.

Molecular weight [1]: ca. 70000.

Specificity [2]

Substrate	Relative rate
α-D-Glucose	1.00
α-D-Galactose	1.05
β-Cellobiose	0.19
β-L-Arabinose	1.19
α-D-Xylose	0.36

pH Optimum [1]: 5.7.

Assay

Wavelength 546 nm; light path 100 mm; final volume 10 ml; 25 °C; measurement of the optical rotation with a precision polarimeter.

10 ml   EDTA buffer (5 mmol/l; pH 7.4)
500 mg   glucose
0.1 ml   enzyme in EDTA buffer

the spontaneous mutarotation has to be accounted for by a blank reaction.

Stability: suspensions in ammonium sulphate solution, 3.2 mol/l; pH ca. 6, are stable for many months at 4 °C.

Purity required: specific activity $\geqslant$ 5000 units/mg protein (25 °C).

Reference

[1] *R. Bentley, D. S. Bathe,* J. Biol. Chem. *235*, 1219 and 1225 (1960).

D-Amino Acid Oxidase

from pig kidney

D-Amino acid: oxygen oxidoreductase (deaminating), EC 1.4.3.3

$$\text{D-Amino acid} + H_2O + O_2 \xrightarrow{\text{D-AOD}} \text{2-oxo acid} + NH_3 + H_2O_2$$

Michaelis constants

			Relative rate
D-Alanine	$1.8 \times 10^{-3}\,\text{mol/l}$	⎫	1.00
D-2-Aminobutyrate	$4.6 \times 10^{-4}\,\text{mol/l}$		0.29
D-Norvaline	$6.3 \times 10^{-5}\,\text{mol/l}$		0.20
D-Norleucine	$4.0 \times 10^{-4}\,\text{mol/l*}$	*	0.97
D-Valine	$1.4 \times 10^{-3}\,\text{mol/l}$		0.89
D-Methionine	$8.3 \times 10^{-4}\,\text{mol/l}$		2.75
D-Proline	$1.7 \times 10^{-3}\,\text{mol/l}$	⎭	3.02
O₂	$1.8 \times 10^{-4}\,\text{mol/l}$	⎫	1.00
2,6-Dichlorophenol indophenol	$1.5 \times 10^{-5}\,\text{mol/l}$	**	0.007
Methylene blue	$6.7 \times 10^{-5}\,\text{mol/l}$	⎭	0.003
FAD	$4.5 \times 10^{-7}\,\text{mol/l***}$		–

---
　　* Pyrophosphate buffer, pH 8.5; air; 25 °C.
　** Pyrophosphate buffer, pH 8.5; 25 °C; donor: D-alanine.
*** Phosphate buffer, pH 7.5; 20 °C.

Effectors: the enzyme is competitively inhibited by AMP, dCMP, UMP and to a slight extent by IMP, GMP, dGMP, CMP, dUMP and dTMP and by straight chain fatty acids, 2-oxo acids and 2-hydroxy acids [1, 2].

Molecular weight: 125 000 [3], 90 000 [4], monomeric form: 38 000 − 39 000 [5, 6].

Specificity: L-amino acids do not react with the enzyme. O_2 can be replaced by dyes (except phenazine methosulphate). Riboflavin, FMN, quinone, menadione, cytochrome c and ferricyanide are inactive [4].

Assay

Wavelength 339, Hg 334 or Hg 365 nm; light path 10 mm; final volume 3.09 ml; 25 °C.

2.50 ml	Tris buffer (0.2 mol/l; pH 8.3; O_2 saturated)	161.5 mmol/l
0.50 ml	D-alanine (20 mg/ml)	37.5 mmol/l
0.05 ml	NADH, Na salt, (10 mg/ml)	0.20 mmol/l
0.01 ml	catalase (0.1 mg/ml)	21 kU/l
0.01 ml	LDH (5 mg/ml)	9.1 kU/l
0.02 ml	enzyme in water	

Stability: crystalline suspensions in ammonium sulphate solution, 3.2 mol/l; pH ca. 6.5, are stable for several months at 4 °C.

Purity required: specific activity $\geq$ 15 U/mg protein (25 °C). Contaminants (related to the specific activity of D-AOD) $\leq$ 0.1 unit proteases (*Kunitz* [7])/mg protein.

References

[1] *D. B. McCormick, B. M. Chassy, J. C. M. Tsibris,* Biochim. Biophys. Acta *89,* 447 (1964).
[2] *M. Dixon, K. Kleppe,* Biochim. Biophys. Acta *96,* 357, 368, 383 (1965).
[3] *Y. Miyake et al.,* Biochim. Biophys. Acta *105,* 86 (1965).
[4] *E. Antonini et al.,* J. Biol. Chem. *241,* 2358 (1966).
[5] *B. Curti et al.,* Biochim. Biophys. Acta *327,* 266 (1973).
[6] *S. C. Tu, S. J. Edelstein, D. B. McCormick,* Arch. Biochem. Biophys. *159,* 889 (1973).
[7] *M. Kunitz,* J. Gen. Physiol. *30,* 291 (1947).

L-Amino Acid Oxidase

from *Crotalus durissus*

L-Amino acid: oxygen oxidoreductase (deaminating), EC 1.4.3.2

$$\text{L-Amino acid} + H_2O + O_2 \xrightarrow{\text{L-AOD}} \text{2-oxo acid} + NH_3 + H_2O_2$$

Effectors *: various amino acids (e.g. L-leucine) stabilize the enzyme [1], while at high concentrations they exhibit a significant inhibitory action (e.g. L-leucine, L-methionine and L-valine) [2 – 5]. Various aliphatic and aromatic sulphonic acids, sulphonamides and benzoic, salicylic and mandelic acid, hydroxylamine and semicarbazide exert a competitive inhibition [6]. Riboflavin and analogous compounds (isoriboflavin, atebrin and alloxazine) do not inhibit competitively [7]. The inhibition is due to the formation of a ternary enzyme-inhibitor complex [5].

Molecular weight * [1, 2]: approx. 140000. (L-Amino acid oxidase is a glycoprotein containing 2 moles of FAD per mol L-AOD. The enzyme has been separated into three isoenzymes which differ in their amino acid composition [8]).

Specificity *: in the presence of O_2 the enzyme converts L-amino acids into the respective 2-oxo acids [5]. It is specific for the L-isomers of leucine, methionine, phenylalanine, norvaline, norleucine, cysteine, tyrosine, tryptophan and citrulline. Histidine, arginine and ornithine are oxidized slowly and glycine not at all. The L-amino acid must possess a free carboxyl group and an unsubstituted α-amino-group. L-Amino acids with a secondary amino group (e.g. N-methyl leucine and proline) are not attacked. D-Amino acids do not react [6, 9].

pH Optimum [10] *: 7 – 7.5.

Assay

$$\text{L-Leucine} + O_2 \xrightarrow{\text{L-AOD}} \text{oxo-isocapronate} + H_2O_2$$

$$H_2O_2 + \text{o-dianisidine} \xrightarrow{\text{POD}} \text{dyestuff.}$$

Wavelength 436 nm; light path 10 mm; final volume 3.03 ml; 25 °C;
$\varepsilon = 0.83 \, 1 \times \text{mmol}^{-1} \times \text{mm}^{-1}$.

3.00 ml	buffer/substrate (TEA 0.2 mol/l; pH 7.6; leucine, 1 mg/ml; o-dianisidine-HCl, 65 µg/ml)	TEA leucine o-dianisidine	0.2 mol/l 7.6 mmol/l 64 mg/ml
0.01 ml	POD (10 mg/ml)		8.2 kU/l
0.02 ml	enzyme in water		

Stability: suspensions in ammonium sulphate solution, 3.2 mol/l; pH ca. 6, are stable for several months.

Purity: specific activity ⩾ 7 U/mg (25 °C). Contaminant ⩽ 0.01 protease units/mg acc. to *Kunitz.*

---

* These data apply to the enzyme from *Crotalus adamanteus.*

References

[1] *W. K. Paik, S. Kim*, Biochim. Biophys. Acta *139*, 49 (1967).
[2] *A. Meister, D. Wellner*, in: *P. D. Boyer, H. Lardy, K. Myrbäck* (eds.), The Enzymes, Vol. 7, Academic Press, New York 1963, p. 609.
[3] *D. Wellner, A. Meister*, J. Biol. Chem. *236*, 2357 (1961).
[4] *A. Meister*, Biochemistry of the Amino Acids, Vol. *1*, Academic Press, New York 1965, p. 304.
[5] *V. Massey, B. Curti*, J. Biol. Chem. *242*, 1259 (1967).
[6] *E. A. Zeller, A. Maritz*, Helv. Chim. Acta *27*, 1888 (1944); *28*, 365 und 1615 (1945).
[7] *T. B. Singer, E. B. Kearney*, Arch. Biochem. *27*, 348 (1950).
[8] *D. Wellner, B. Hayes*, Ann. N.Y. Acad. Sci. *151*, 118 (1968).
[9] *E. A. Zeller*, Advances in Enzymology *8*, 476 (1948).
[10] *D. Wellner, A. Meister*, J. Biol. Chem. *235*, 2013 (1960).

α-Amylase

from porcine pancreas

1,4-α-D-Glucan glucanohydrolase, EC 3.2.1.1

The enzyme catalyzes the hydrolysis of α-1,4-glucan linkages.

Michaelis constant: starch 6×10^{-4} g/ml

Effectors: the enzyme is activated by Cl^-. α-Urea and other amide reagents are inhibiting [1].

Molecular weight [2]: 50000.

Specificity: α-amylase catalyzes the hydrolysis of internal α-1,4-glucan bonds in polysaccharides (e.g. starch and glycogen), which consist of 3 or more α-1,4-linked D-glucose units, resulting in the formation of dextrins and a mixture of reducing sugars.

Assay

Wavelength Hg 546 nm; light path 10 mm; incubation volume 1.0 ml; final volume 12.00 ml; 25 °C.

0.50 ml	starch solution (10 mg *Zulkowski* starch/ml phosphate buffer, 20 mmol/l; pH 7.0; NaCl, 6 mmol/l)	starch	5 mg/ml
		phosphate	10 mmol/l
		NaCl	3 mmol/l
0.48 ml	water		

warm to 25 °C in a water-bath

0.02 ml   enzyme in water

incubate for exactly 5 min at 25 °C.

1.00 ml   colour reagent (1 g 3,5-dinitrosalicylic acid in 20 ml NaOH, 2 mol/l, add 30 g K-Na tartrate and dilute to 100 ml with water)

heat for 5 min in a boiling water-bath, cool, add 10.00 ml water, read absorbance against a blank and obtain the maltose content as reducing equivalents from a standard curve.

Stability: crystalline suspensions in ammonium sulphate solution 3.2 mol/l; pH ca. 6, are stable for several months at 4°C.

Purity required: specific activity $\geq$ 1000 U/mg protein (25 °C).

References

[1] *G. C. Toralballa, M. Eitington,* Arch. Biochem. Biophys. *119, 519 (1967).
[2] *M. Granger et al.,* FEBS letters *5b,* 189 (1975).

β-Amylase

from sweet potato (*Ipomoea batatas*)

1,4-α-D-Glucan maltohydrolase, EC 3.2.1.2

Hydrolysis of α-1,4-glucan linkages in polysaccharides with cleavage of maltose from the non-reducing end of the chain.

Michaelis constant (acetate buffer, pH 4.8; 35 °C)

Polysaccharide with 44 glucose units [1] 7.3×10^{-5} mol/l

Inhibitor constants (acetate buffer, pH 4.8)

Cyclohexa-amylose [2]	1.7×10^{-4} mol/l
Methyl-α-glucoside	4.0×10^{-2} mol/l

Effectors: the enzyme is sensitive to SH reagents and is inhibited by heavy metal ions, 4-chloromercuribenzoate, iodoacetamide and urea. Ascorbate inhibits by reduction of copper ions and formation of an inactive enzyme-copper complex [3].

Molecular weight [4]: 206000.

Specificity: the enzyme hydrolyzes α-1,4-glucan linkages in polysaccharides, in which the maltose units are cleaved from the non-reducing end of the chain. α-1,6-Glucan linkages are not hydrolyzed and act as a barrier. The smallest molecule attacked is maltotetraose.

Maltotriose, methyl-α-maltotriose and maltose are hydrolyzed at very low rates. Cyclohexa-amylose does not react.

Assay

Wavelength Hg 546 nm; light path 10 mm; incubation volume 1.01 ml; final volume 12.00 ml; 25 °C.

1.00 ml   starch solution (5 mg/ml acetate buffer, 10 mmol/l; pH 4.8; NaCl 3 mmol/l)

incubate for several min at 25 °C

0.01 ml   enzyme in water

incubate for exactly 5 min at 25 °C

1.00 ml   colour reagent (1 g 3,5-dinitrosalicylic acid in 20 ml NaOH, 2 mol/l, add 30 g
            K-Na tartrate and dilute to 100 ml with water)

heat for 5 min in a boiling water-bath, cool, add 10 ml water, read absorbance against blank and obtain content of reducing equivalents as maltose from standard curve.

Stability: suspensions in ammonium sulphate solution, 3.2 mol/l; pH 3.5, are stable for several months at 4°C.

Purity required: specific activity ≧ 500 U/mg protein (25 °C).

References

[1] *J. M. Bailey, D. French,* J. Biol. Chem. *226,* 1 (1957).
[2] *J. A. Toma et al.,* Biochem. Biophys. Res. Commun. *12,* 184 (1963).
[3] *A. W. Rowe, C. E. Weill,* Biochim. Biophys. Acta *65,* 245 (1962).
[4] *P. M. Colman, B. W. Matthews,* J. Mol. Biol. *60,* 163 (1971).

Amyloglucosidase

from *Aspergillus niger*

1,4-α-D-Glucan glucohydrolase, EC 3.2.1.3

Hydrolysis of α-1,4- and α-1,6-glucan bonds in polysaccharides.

Molecular weight [1]: 97000.

Specificity: the enzyme hydrolyzes α-1,4- and α-1,6-glucan linkages in polysaccharides, such as starch, amylose, amylopectin, amylodextrin [2], glycogen [3] and a number of glucosyl oligosaccharides, cleaving glucose units from the non-reducing end of the chain. The rates of hydrolysis at the pH-optimum [2] of pH 4.8 and 37 °C relative to maltose (100) are as follows [4].

α-1,4-Oligosaccharides: maltose (100), maltotriose (360), maltotetraose (770), maltopentaose (1000), 6-α-maltosylglucose (260), 3-α-maltosylglucose (140), 3-α-maltotriosylglucose (520), 4^G-α-glucosylsucrose (97).

α-1,6-Oligosaccharides: isomaltose (1.3), isomaltotriose (8.0), isomaltotetraose (15), isomaltopentaose (23), panose (4-α-isomaltosylglucose) (73).

α-1,3-Oligosaccharides: nigerose (1.1), 4-α-nigerosylglucose (3.7), 4-α-nigero-triosylglucose (2.6).

Amyloglucosidase is a glycoprotein containing 10 – 35% carbohydrate [5, 6]. The optimum temperature for hydrolysis is 50 °C.

Assay

$$\text{Glycogen} + (n - 1)\,H_2O \xrightarrow{\text{amyloglucosidase}} n\,\text{glucose}.$$

Wavelength 339, Hg 334 or Hg 365 nm; light path 10 mm; incubation volume 1.6 ml; assay volume 2.92 ml; 25 °C.

Incubation mixture

1.00 ml	acetate buffer (0.1 mol/l; pH 4.0)	62.5 mmol/l
0.50 ml	glycogen solution (8 mg/ml)	2.5 mg/ml
0.10 ml	enzyme in water	

incubate for exactly 5 min at 24 °C

0.40 ml	Tris solution (0.3 mol/l)

Assay mixture

2.50 ml	triethanolamine buffer (0.1 mol/l; pH 7.6)	86 mmol/l
0.10 ml	$MgCl_2$ (0.2 mol/l)	6.7 mmol/l
0.10 ml	ATP, Na salt, (10 mg/ml)	0.5 mmol/l
0.10 ml	NADP, Na salt, (10 mg/ml)	0.39 mmol/l
0.10 ml	incubation mixture	–
0.01 ml	HK (10 mg/ml)	4.7 kU/l
0.01 ml	G6P-DH (5 mg/ml)	2.4 kU/l

Stability: suspensions in ammonium sulphate solution, 3.2 mol/l; pH ca. 6, are stable for several months at 4°C.

Purity required: specific activity $\geq$ 14 U/mg protein (25 °C).

References

[1] *J. H. Pazur, K. Kleppe,* J. Biol. Chem. *237,* 1002 (1962).
[2] *J. H. Pazur, T. Ando,* J. Biol. Chem. *235,* 297 (1960).
[3] *J. A. Johnson, R. M. Fusaro,* Anal. Biochem. *15,* 140 (1966).
[4] *M. Abdullah, I. D. Fleming, P. M. Taylor, W. J. Whelan,* Biochem. J. *89,* 35 P (1963).
[5] *J. A. Pazur, K. Kleppe, E. M. Ball,* Arch. Biochem. Biophys. *103,* 515 (1963).
[6] *I. D. Fleming, B. A. Stone,* Biochem. J. *97,* 13 P (1965).

Arginase

from mammalian liver

L-Arginine amidinohydrolase, EC 3.5.3.1

$$\text{L-arginine} + H_2O \xrightarrow{\text{L-arginase}} \text{L-ornithine} + \text{urea}$$

Michaelis constant [1]: L-arginine 1.16×10^{-2} mol/l for the enzyme from bovine liver.

Inhibitor constants [1]

L-Ornithine	7.7×10^{-3} mol/l	⎫
L-Lysine	4.8×10^{-3} mol/l	⎭ for the enzyme from bovine liver

Effectors: Mn^{2+}, Ni^{2+} and Co^{2+} are activators [2]. L-ornithine, L-lysine and other amino acids [1], citrate and borate as well as Hg^{2+}, Ag^+ and Zn^{2+}, inhibit [3].

Molecular weight [4]: ca. 115 – 120000 (bovine liver arginase).

Specificity [3]: arginase hydrolyzes arginine to ornithine and urea. Both the free guanidino group and the carboxyl group in the arginine molecule are required for arginase activity. The free amino group can be substituted, or replaced by a hydroxyl group. A carbon atom in the carbon chain can be replaced by oxygen; e.g. canavanine is hydrolyzed.

pH Optimum [3]: 10 (Mn^{2+} activated).

Assay

$$Arginine + H_2O \xrightarrow{\text{L-arginase}} \text{L-ornithine} + urea$$

$$Urea + 2\,H_2O \xrightarrow{\text{urease}} 2\,NH_4^+ + CO_3^{--}$$

$$2\,NH_4^+ + 2\text{-oxoglutarate} + 2\,NADH \xrightarrow{\text{GlDH}} 2\,\text{L-glutamate} +$$
$$2\,NAD^+ + 2\,H_2O\,.$$

Wavelength 339, Hg 334 or Hg 365 nm; light path 10 mm; final volume 3.00 ml; 25 °C.

2.50 ml	Tris-HCl buffer (0.2 mol/l; pH 8.6)	0.17 mol/l
0.15 ml	2-oxoglutarate (0.5 mol/l)	25 mmol/l
0.05 ml	NADH, Na salt, (10 mg/ml)	0.2 mmol/l
0.20 ml	L-arginine (0.2 mol/l)	13.3 mmol/l
0.04 ml	GlDH (10 mg/ml 50% glycerol)	16 kU/l
0.01 ml	urease (30 mg/ml)	10 kU/l
0.05 ml	enzyme in manganese maleate buffer (10 mmol/l; pH 7.5)	

Stability: the lyophilized enzyme stabilized with manganese sulphate and sodium maleate is stable for many months at 4 °C.

Purity required: activity $\geqslant$ 70 U/mg substance (25 °C). Contaminant (related to the activity of arginase) $\leqslant$ 0.02% trypsin (25 °C; BAEE as substrate).

References

[1] *A. Hunter, C. E. Downs,* J. Biol. Chem. *157*, 427 (1945).
[2] *D. M. Greenberg et al.,* Arch. Biochem. Biophys. *62*, 446 (1956).
[3] *D. M. Greenberg,* in: *P. D. Boyer, H. Lardy, K. Myrbäck* (eds.), The Enzymes, Vol. *4*, Academic Press, New York 1960, p. 257.
[4] *D. Harell, M. Sokolovsky,* Eur. J. Biochem. *25*, 102 (1972).

Arylsulphatase

from snails (*Helix pomatia*)

Arylsulphate sulphohydrolase, EC 3.1.6.1

$$\text{Phenolsulphate} + \text{H}_2\text{O} \xrightarrow{\text{arylsulphatase}} \text{phenol} + \text{sulphate}$$

Specificity: the enzyme specifically hydrolyzes sulphuric acid esters of phenols.

Assay

$$\text{4-Nitrophenyl sulphate} + \text{H}_2\text{O} \xrightarrow{\text{arylsulphatase}} \text{4-nitrophenol} + \text{sulphate}.$$

Wavelength Hg 405 nm; light path 10 mm; incubation volume 1.03 ml; 25°C; $\varepsilon = 1.85 \; \text{l} \times \text{mmol}^{-1} \times \text{mm}^{-1}$.

0.97 ml	acetate buffer (0.1 mol/l; pH 6.2)	94 mmol/l
0.03 ml	4-nitrophenyl sulphate, Na salt, (50 mmol/l)	1.5 mmol/l
0.03 ml	enzyme in water	

incubate for 20 min at 25°C, then add 2 ml NaOH, 0.5 mol/l, and read absorbance against blank.

Stability: aqueous stabilized solutions and suspensions in ammonium sulphate solution, 3.2 mol/l; pH approx. 6, are stable for several months at 4°C.

Purity required: activity $\geq$ 2.5 U/ml (25°C).

Reference

[1] *K. S. Dodgson, G. M. Powell,* Biochem. J. *73*, 666 (1959).

Ascorbate Oxidase

from *Cucurbita* species

L-Ascorbate: oxygen oxidoreductase, EC 1.10.3.3

$$\text{2 L-Ascorbate} + \text{O}_2 \xrightarrow{\text{ascorbate-OD}} \text{2 dehydroascorbate} + \text{2 H}_2\text{O}$$

Michaelis constant [1]: 2×10^{-4} mol/l (citrate/phosphate buffer, pH 5.6, 25°C).

Inhibitor constant [2]: azide 2.1×10^{-4} mol/l.

Effectors [3]: 4-chloromercuribenzoate, CN^-, Na_2S, diethyl-dithiocarbamate, 8-hydroxyquinoline and potassium ethylxanthate inhibit.

Molecular weight [4]: 140000.

Specificity [3]: the enzyme oxidizes several analogues of ascorbate.

pH Optimum [2]: 5.5.

Assay

Wavelength 245 nm; light path 10 mm; incubation volume 1.10 ml; final volume 4.10 ml; 25 °C; $\varepsilon = 1.0 \text{ l} \times \text{mmol}^{-1} \times \text{mm}^{-1}$.

1.00 ml	buffer/ascorbate (0.1 mol/l; pH 5.6;	phosphate	91 mmol/l
	EDTA, 0.5 mmol/l; ascorbate 0.5 mmol/l)	EDTA	0.45 mmol/l
		ascorbate	0.45 mmol/l
0.1 ml	enzyme in phosphate/bovine serum albumin solution		
	(phosphate, 4 mmol/l; bovine serum albumin 0.05%)		

incubate for 5 min, add

3.00 ml   HCl (0.2 mol/l)

read against a blank in which 0.1 ml enzyme solution was added after acidification with HCl.

Stability: in the lyophilized state a decrease in activity of approx. 20% may occur within 6 months at 4 °C.

Purity required: activity $\geqslant$ 170 U/mg substance (25 °C); $\underline{\Delta}$ ca. 1700 U/mg enzyme protein.

References

[1] *C. R. Dawson, W. B. Tarpley,* in: *J. P. Summer, K. Myrbäck* (eds.), The Enzymes *II*, Academic Press, New York 1951, p. 454.
[2] *T. Nakamura,* J. Biochem. *64*, 189 (1968).
[3] *G. R. Stark, C. R. Dawson,* in: *P. D. Boyer, H. Lardy, K. Myrbäck* (eds.), The Enzymes *8*, Academic Press, New York 1963, p. 297.
[4] *K. Tokuyama et al.,* Biochem. *4*, 1362 (1965).

L-Asparaginase

from *Escherichia coli*

L-Asparagine amidohydrolase, EC 3.5.1.1

$$\text{L-Asparagine} + H_2O \xrightarrow{\text{L-asparaginase}} \text{L-aspartate} + NH_3$$

Michaelis constant [1]: asparagine 6×10^{-6} mol/l.

Molecular weight [2]: 141 000.

Specificity: the isoenzyme EC_2 reacts with L-glutamine at ca. 2% of the rate with L-asparagine [3].

Assay

$$\text{Asparagine} + H_2O \xrightarrow{\text{Asparaginase}} \text{aspartate} + NH_4^+$$

$$\text{2-Oxoglutarate} + NADH + NH_4^+ \xrightarrow{\text{GlDH}} \text{L-glutamate} + MAD^+ + H_2O.$$

Wavelength 339, Hg 334 or Hg 365 nm; light path 10 mm; final volume 2.77 ml; 25 °C.

2.50 ml	Tris buffer (0.05 mol/l; pH 8.0)	45 mmol/l
0.10 ml	asparagine (0.2 mol/l; pH 8.0)	7.2 mmol/l
0.01 ml	2-oxoglutarate, Na salt, (36.5 mg/ml)	0.9 mmol/l
0.05 ml	NADH, Na salt, (10 mg/ml)	0.2 mmol/l
0.01 ml	GlDG (10 mg/ml glycerol, 50%)	4.3 U/ml
0.02 ml	enzyme in water	

Stability: solutions in 50% glycerol, pH ca. 6.5, are stable for several months at 4 °C.

Purity required: specific activity $\geq$ 80 U/mg protein (25 °C). Conversion with L-glutamine (side reaction of the enzyme) approx. 2% in relation to the activity of L-asparaginase.

References

[1] *A. Arens et al.,* Hoppe-Seyler's Z. Physiol. Chem. *351,* 197 (1970).
[2] *T. Maita, L. Morokuma, G, Matsuda,* J. Biochem. *76,* 1351 (1974).
[3] *H. A. Campbell et al.,* Biochemistry *6,* 721 (1967).

Aspartate Aminotransferase

Glutamate-Oxaloacetate Transaminase

from pig heart (mitochondrial)

L-Aspartate: 2-oxoglutarate aminotransferase, EC 2.6.1.1

$$\text{2-Oxoglutarate + L-aspartate} \xrightleftharpoons{\text{ASAT}} \text{L-glutamate + oxaloacetate}$$

Equilibrium constant (arsenate buffer, pH 7.4; 37 °C) [1]

$$K = \frac{[\text{L-glutamate}] \times [\text{oxaloacetate}]}{[\text{2-oxoglutarate}] \times [\text{L-aspartate}]} = 0.16 - 0.17$$

Michaelis constants (arsenate buffer, pH 7.4; 37 °C) [1]

L-Aspartate	3.9×10^{-3} mol/l	
2-Oxoglutarate	4.3×10^{-4} mol/l	(mitochondrial ASAT)
Oxaloacetate	8.8×10^{-5} mol/l	
L-Glutamate	8.9×10^{-3} mol/l	

Inhibitor constants (arsenate buffer, pH 7.4)

L-Aspartate	3.48×10^{-3} mol/l	
2-Oxoglutarate	$7.1 \ \times 10^{-4}$ mol/l	37 °C [1]
Oxaloacetate	$5.0 \ \times 10^{-5}$ mol/l	(mitochondrial ASAT)
L-Glutamate	$8.4 \ \times 10^{-3}$ mol/l	
Succinate	$1.8 \ \times 10^{-2}$ mol/l	
Maleate	$2.3 \ \times 10^{-3}$ mol/l	
Glutarate	$3.0 \ \times 10^{-3}$ mol/l	25 °C [2]
Adipate	$1.2 \ \times 10^{-2}$ mol/l	
Fumarate	very large	

Effectors: pyridoxal or pyridoxamine activate the enzyme [3]. 4-chloromercuri-benzoate [4], L-α-methylaspartic acid [5], 2-amino-3-butenic acid [5], maleate [3], succinate [3], glutarate [3], and adipate [3] inhibit.

Molecular weights: 91 000 (mitochondrial ASAT) [6]; 94 000 (cytoplasmatic ASAT) [7].

Specificity: the enzyme can also react with cysteate, homocysteate, sulphinate and mesoxalate.

Assay

$$2\text{-Oxoglutarate} + \text{L-aspartate} \xrightarrow{\text{ASAT}} \text{L-glutamate} + \text{oxaloacetate}$$

$$\text{Oxaloacetate} + \text{NADH} + \text{H}^+ \xrightarrow{\text{MDH}} \text{L-malate} + \text{NAD}^+.$$

Wavelength 339, Hg 334 or Hg 365 nm; light path 10 mm; final volume 3.18; 25°C.

3.00 ml	phosphate buffer (0.1 mol/l; pH 7.4;	phosphate 94.5 mmol/l
5 mg	L-aspartate/ml)	aspartate 35.5 mmol/l
0.10 ml	2-oxoglutarate (30 mg/ml; pH ca. 7)	6.8 mmol/l
0.05 ml	NADH, Na salt, (10 mg/ml)	0.2 mmol/l
0.01 ml	MDH (2 mg/ml)	7.3 kU/l
0.02 ml	enzyme in phosphate buffer (0.1 mol/l; pH 7.4)	

Stability: suspensions in ammonium sulphate solutions, 3.2 mol/l; pH ca. 6, containing 2-oxoglutarate, 2.5 mmol/l and maleate, 50 mmol/l, are stable for several months at 4°C.

Purity required: specific activity $\leq$ 200 U/mg protein (25°C). Contaminants (related to the specific activity of ASAT) $\leq$ 0.01%, GlDH, GPT, LDH, MDH and oxaloacetate decarboxylase.

References

[1] *C. P. Henson, W. W. Cleland,* Biochemistry *3*, 338 (1964).
[2] *S. F. Velick, J. Vavra,* in: *P. D. Boyer, H. Lardy, K. Myrbäck* (eds.), The Enzymes, Vol. *4*, Academic Press, New York 1962, p. 219.
[3] *W. T. Jenkins,* J. Biol. Chem. *234*, 51 (1959).
[4] *M. Stankewics et al.,* Biochem. *10*, 2877 (1971).
[5] *R. R. Rando,* Biochem. *13*, 3859 (1974).
[6] *N. Feliss, M. Martinez-Carrion,* Biochem. Biophys. Res. Commun. *40*, 932 (1970).
[7] *O. L. Polyanowsky et al.,* FEBS Lett. *23*, 262 (1972).

Carboxypeptidase Y

from yeast

Glycine carboxypeptidase, EC 3.4.17.4

$$\text{Peptidyl-L-amino acid} + \text{H}_2\text{O} \xrightarrow{\text{carboxypeptidase Y}} \text{peptide} + \text{L-amino acid}$$

Michaelis constants [1]

measured by the hydrolysis of N-substituted L-dipeptides of the type Cbz-X-Leu.

Substrate	K_m mol/l
Cbz-Gly-Leu	2.95×10^{-3}
Cbz-Phe-Leu	0.24×10^{-3}
Cbz-Ala-Leu	1.28×10^{-3}
Cbz-Val-Leu	0.10×10^{-3}
Cbz-Leu-Leu	0.12×10^{-3}
Cbz-Ile-Leu	0.10×10^{-3}
Cbz-Ser-Leu	$1.7 \ \times 10^{-3}$
Cbz-His-Leu	$1.8 \ \times 10^{-3}$
Cbz-Pro-Leu	$1.8 \ \times 10^{-3}$
Cbz-Glu-Leu	$8.3 \ \times 10^{-3}$
Cbz-Nle-Leu	0.03×10^{-3}

Measured by the hydrolysis of N-substituted L-dipeptides of the type Cbz-Gly-X,

Substrate	K_m mol/l
Cbz-Gly-Leu	2.95×10^{-3}
Cbz-Gly-Phe	$1.2 \ \times 10^{-3}$
Cbz-Gly-Arg	16.3×10^{-3}
Cbz-Gly-Glu	24.1×10^{-3}

Effectors: carboxypeptidase Y is a DFP-sensitive enzyme [2] and is also stoichiometrically and irreversibly inhibited by PMSF [3]. Site specific reagents, i.e. the chloromethylketone derivative of benzyloxycarbonyl-L-phenylalanine inactivate both the peptidase and esterase activities of the exopeptidase [4]. The enzyme is inhibited by 4-hydroxymercuribenzoate and is sensitive to metal ions. Cu^{2+}, Ag^+ or Hg^{2+} result in a complete loss of activity, while Cu^+, Mg^{2+}, Ca^{2+}, Ba^{2+}, Cr^{2+}, Mn^{2+}, Fe^{2+}, Fe^{3+} or Ni^{2+} cause a partial loss of activity. EDTA and 2-phenanthroline have no effect on enzymatic activity. Trypsin inhibitors from soybean and the lima bean do not inhibit the activity. The inhibitions mentioned above are seen with both peptidase and esterase activities. Product- and substrate analogues act as reversible inhibitors: L-amino acids and NH_2-blocked L-amino acids show the competitive type of inhibition. Some phenylalanine analogues, e.g. β-phenylpropionate and trans-cinnamate, also inhibit the enzyme reversibly [5].

Molecular weight [6]: 61 000.

Specificity: the enzyme removes most amino acid residues, including that of proline, from the COOH-termini of proteins and peptides at pH 5.5 to 6.5. Catalysis is most effective when the penultimate and/or terminal residues have aromatic or aliphatic side-chains. The release of terminal amino acid is slow when glycine is located in the penultimate position. The release of terminal histidine, arginine or lysine is also relatively slow. Dipeptides are completely resistant to hydrolysis, and cleavage of tripeptides is minimal. The enzyme also hydrolyzes ester and amide substrates of chymotrypsin. For further details on specificity cf. references [2, 3].

pH Optimum [1]: pH 6 peptidase activity, pH 7.5 esterase activity.

Assay

$$\text{Cbz-Phe-Ala} + H_2O \xrightarrow{\text{carboxypeptidase Y}} \text{Z-Phe} + \text{alanine}$$

Wavelength 578 nm; light path 10 mm; incubation volume 2.52 ml; incubation temperature 37 °C; final volume 15.02 ml; measurement temperature 20 – 25 °C; $\varepsilon = 1.82$ $l \times mmol^{-1} \times mm^{-1}$.

2.50 ml	substrate acetate buffer (acetate 96 mmol/l; pH 6.0; carbobenzoxyphenylalanylalanine, 3.3 mmol/l; ethanol, 4% v/v)
0.02 ml	enzyme in water

incubate for exactly 15 min at 37 °C

2.50 ml	ninhydrin (400 mg ninhydrin, 60 mg hydrindantin dissolved in 15 ml dimethylsulphoxide, 5 ml lithium acetate buffer, 4.0 mol/l; pH 5.2, added)

heat for exactly 15 min in a boiling water-bath, cool in an ice-bath (= reaction mixture)

1.00 ml	reaction mixture
9.00 ml	ethanol

read the absorbance against blank in which the sample has been added after the addition of ninhydrin.

Stability: in the stabilized lyophilized state a decrease in activity of approx. 10% may occur within 6 months at 4 °C.

Purity required: activity ≥ 20 U/mg substance (37 °C).

References

[1] *R. W. Kuhn et al.,* Biochemistry *13*, 3871 (1974).
[2] *R. Hayashi et al.,* J. Biol. Chem. *248*, 8366 (1973).
[3] *R. Hayashi et al.,* J. Biochem. *77*, 1318 (1975).

[4] *R. Hayashi et al.*, J. Biol. Chem. *250*, 5221 (1975).
[5] *R. Hayashi* in: *S. P. Colowick, N. O. Kaplan*, Methods in Enzymology, Vol. *XLVII*, Academic Press, New York 1977, p. 84.
[6] *S. Aibara et al.*, Agric. Biol. Chem. *35*, 658 (1971).

Carnitine Acetyltransferase

from pigeon breast muscle

Acetyl-CoA: carnitine O-acetyltransferase, EC 2.3.1.7

$$\text{Acetyl-CoA} + \text{L-carnitine} \xrightleftharpoons{\text{carnitine acetyltransferase}} \text{acetyl-L-carnitine} + \text{CoA}$$

Equilibrium constant (Tris buffer, pH 7.0; 35 °C) [1]

$$K = \frac{[\text{CoA}] \times [\text{acetylcarnitine}]}{[\text{acetyl-CoA}] \times [\text{carnitine}]} = 1.67$$

Molecular weight (pH 7.2) [2]: 55000.

Specificity: the D-isomers of carnitine and acetylcarnitine are inactive.

Assay

Wavelength 233 nm; light path 10 mm; final volume 3.00 ml; 25 °C;
$\varepsilon = 4.5\,l \times \text{mmol}^{-1} \times \text{mm}^{-1}$.

2.73 ml	Tris buffer (0.1 mol/l; pH 8.0)	91 mmol/l
0.05 ml	CoA (10 mg/ml)	0.19 mmol/l
0.20 ml	acetyl-DL-carnitine-HCl (20 mg/ml)	5.5 mmol/l
0.02 ml	enzyme in buffer	

Stability: crystalline suspensions in ammonium sulphate solution 3.2 mol/l; pH ca. 6, are stable for several months at 4 °C.

Purity required: specific activity $\geq$ 80 U/mg protein (25 °C). Contaminant (related to the activity of carnitine acetyltransferase) $\leq$ 0.01% acetyl-CoA deacylase.

References

[1] *I. B. Fritz, S. K. Schultz, P. A. Srere*, J. Biol. Chem. *238*, 2509 (1963).
[2] *J. F. A. Chase, D. J. Pearson, P. K. Tubbs*, Biochim. Biophys. Acta *96*, 162 (1965).

Catalase

from beef liver

Hydrogen-peroxide: hydrogen-peroxide oxidoreductase, EC 1.11.1.6

$$2\ H_2O_2 \xrightarrow{\text{catalase}} 2\ H_2O + O_2$$

Molecular weights: 244000 [1, 2], 250000 [3,4] with 4 identical subunits of 60000 – 65000 [5].

Specificity and kinetic properties: a simplified scheme for the catalase reaction is given by the following equations [1, 6]:

$$Cat(OH)_4 + H_2O_2 \underset{k_2}{\overset{k_1}{\rightleftharpoons}} Cat(OH)_3OOH + H_2O$$

$$Cat(OH)_3OOH + Donor \xrightarrow{k_4} Cat(OH)_4 + H_2O + \text{oxidized donor}$$

$Cat(OH)_4$ = native catalase, $Cat(OH)_3OOH$ = complex I of *B. Chance* [1].

Catalase can also react with alkylhydrogen peroxides instead of H_2O_2. K_m = 1.1 mol/l at pH 7; 30 °C [6].

	k_1 (mol $\times$ l^{-1} $\times$ s^{-1}).
H_2O_2	$6\ \times 10^6$
Methylperoxide	8.5×10^5
Ethylperoxide	$2\ \times 10^4$

Many different compounds, e.g. methanol, ethanol, formate, SH-compounds and nitrite, can replace the second H_2O_2 molecule as the hydrogen donor [6].

	k_4 (mol $\times$ l^{-1} $\times$ s^{-1}) pH 7; 20 – 25 °C
H_2O_2	1.5×10^7
Methanol	$1\ \times 10^3$
Ethanol	$1\ \times 10^3$
Propanol	1.7×10
Formate	$9\ \times 10^5$
Nitrite	1.4×10^7

Assay

Wavelength 240 nm; light path 10 mm; final volume 3.02 ml; 25 °C; ε = 0.004 l $\times$ mmol^{-1} $\times$ mm^{-1}.

3.00 ml   phosphate buffer (50 mmol/l; pH 6.8;   phosphate   50 mmol/l
           1.2 μl 30% H_2O_2/ml; A_{240} must be
           0.500 ± 0.010 H_2O_2 10.5 mmol/l
0.02 ml   enzyme in phosphate buffer

Stability: crystalline suspensions in water (saturated with thymol, pH ca. 6) and solutions in 30% glycerol (with 10% ethanol) are stable for several months at 4 °C.

Purity required: specific activity ≥ 50000 U/mg protein (25 °C).

References

[1] *B. Chance,* Nature *161*, 914 (1948).
[2] *T. Samejima, M. Kamata, K. Shibata,* J. Biochem. *51*, 181 (1962).
[3] *R. C. Valentine,* Nature *204*, 1262 (1964).
[4] *N. A. Kiseler, C. L. Shpitzberg, B. K. Vainshtein,* J. Mol. Biol. *25*, 433 (1967).
[5] *H. Sund, K. Weber, E. Molbert,* Eur. J. Biochem. *1*, 400 (1967).
[6] *P. Nicholls, G. R. Schonbaum,* in: *P. D. Boyer, H. Lardy, K. Myrbäck* (eds.), The Enzymes, Vol. *8*, Academic Press, New York 1963, p. 147.

Cellulase

from *Trichoderma viride*

1,4(1,3; 1,4)-β-D-Glucan 4-glucanohydrolase, EC 3.2.1.4

Hydrolyzes β-1,4-glucan links in cellulose

Michaelis constants [1]

Cellobiose 1.9×10^{-2} mol/l
Cellotriose 3.1×10^{-3} mol/l
Cellotetraose 2.8×10^{-3} mol/l
Cellopentaose 7.0×10^{-4} mol/l
Cellohexaose 1.0×10^{-4} mol/l

Effectors [2]: the *Trichoderma viride* multi-enzyme complex is reported to be remarkably resistant to inhibitors, except carbohydrates, particularly cellobiose and excess cellulose.

Molecular weight: the enzyme complex contains at least three enzyme components both physically and enzymatically distinct.

C_1 [3]   ca. 42 000
Endoglucanase [1]   ca. 52 000
Exoglucanase [1]   ca. 76 000.

Specificity: the enzyme hydrolyzes 1,4-β-glucosidic linkages in cellulose, lichenin and cereal β-D-glucans.

pH Optimum [3]: 4.8.

Assay

$$\text{Carboxymethylcellulose} \xrightarrow[\text{H}_2\text{O}]{\text{cellulase}} \text{carboxymethyl oligosaccharides}$$

$$\text{Carboxymethyl oligosaccharides} + 3,5 \text{ dinitrosalicyclic acid} \longrightarrow \text{red colour}.$$

Wavelength 546 nm; light path 10 mm; incubation volume 1.03 ml; incubation temperature 37 °C; final volume 12.0 ml; temperature for colour reaction 100 °C.

1.00 ml   acetate buffer/carboxymethylcellulose (0.1 mol/l; pH 4.5;   97 mmol/l
           0.7% carboxymethylcellulose, w/v)
0.03 ml   enzyme in water

incubate for exactly 15 min at 37 °C

1.00 ml   colour reagent (1 g 3,5-dinitrosalicylic acid in 20 ml NaOH,
           2 mol/l, add 30 g K-Na tartrate and dilute to 100 ml with water)

heat for 10 min in a boiling water-bath, read absorbance against a blank in which the sample was added after the addition of the colour reagent and obtain the glucose content as reducing equivalents from a glucose standard curve.

Stability: the lyophilized enzyme is stable for many months at 4 °C.

Purity required: activity $\geqslant$ 0.5 U/mg substance (37 °C).

References

[1] *L. Li, H. Flora, R. M. King,* Arch. Biochem. Biophys. *111*, 439 (1965).
[2] *J. A. Howell, J. D. Stuck,* Biotechnol. Bioeng. *17*, 873 (1975).
[3] *L. E. R. Berghem et al.,* Eur. J. Biochem. *53*, 55 (1975).

Cholesterol Esterase

from micro-organisms

Sterol-ester acylhydrolase, EC 3.1.1.13

$$\text{A cholesterol ester} + H_2O \xrightarrow{\text{cholesterol esterase}} \text{cholesterol} + \text{a fatty acid anion}$$

Michaelis constants [1]

Cholesterol linoleate	3.86×10^{-5} mol/l	(isoenzyme I)
Cholesterol linoleate	1.43×10^{-5} mol/l	(isoenzyme II)

Effectors [1]: the enzyme is inhibited by heavy metals such as Ag^+, Hg^{2+} and Cu^{2+} but is stable against 4-chloromercuribenzoate. No inhibition is observed with other cations, metal chelating agents and diisopropyl fluorophosphate. Cholesterol esterase activity is low in presence of a suspension of cholesterol linoleate but the activity is considerably enhanced if the non-ionic detergent Triton X-100 is used to prepare the micellar aqueous substrate. Other surface-active agents such as polyvinyl alcohol, sodium dodecyl sulphate and the Tween series exhibit no stimulating effect on the enzyme activity. Optimum stimulation of the reaction by Triton X-100 is obtained at a final concentration of 0.3% (v/v); a further increase in its concentration results in reduced activity. Some bile salts such as cholic acid and glycocholic acid at final concentrations of 0.1 mmol/l enhance the cholesterol esterase activity to about 80% of the Triton X-100 substrate system. There is no significant difference between isoenzyme I and II of the cholesterol esterase in their requirement for these effectors.

Molecular weight [1]: ca. 129000 for each isoenzyme.

Specificity [1]: the enzyme splits preferentially long-chain fatty acid esters of cholesterol. It is more active on unsaturated fatty acid esters than on saturated fatty acids in the C_{18}-series. Of various cholesterol esters, linoleate is hydrolyzed most rapidly. The isoenzyme forms I and II exhibit similar patterns in fatty acid specificity.

Cholesterol ester		Relative rate (%)	
Name	Number of C-atoms to number of double bonds	Isoenzyme I	Isoenzyme II
Linoleate	(18:2)	100	100
Elaidate	(18:1)	78	81
Palmitate	(16:0)	65	63
Oleate	(18:1)	55	50
Stearate	(18:0)	22	19
Caprinate	(10:0)	12	12
Caproate	(6:0)	10	13
Propionate	(3:0)	7	7
Acetate	(2:0)	2	4

pH Optimum [1]: pH 7.3 for cholesterol esterase I and II.

Assay*

$$\text{Cholesterol oleate} + H_2O \xrightarrow{\text{cholesterol esterase}} \text{cholesterol} + \text{oleate}$$

$$\text{Cholesterol} + O_2 \xrightarrow{\text{cholesterol oxidase}} \text{cholestenone} + H_2O_2 .$$

Wavelength 240 nm; light path 10 mm; final volume 3.21 ml; 25 °C; $\varepsilon = 1.55$ $l \times mmol^{-1} \times mm^{-1}$.

3.00 ml	potassium phosphate buffer (0.7 mol/l; pH 7.5;	0.65 mol/l
	0.4% Thesit®)	3.7 g/l
0.10 ml	cholesterol oleate (4 mg/ml dioxan/Thesit®;	0.19 mmol/l
	dioxan : Thesit = 1 : 1)	
0.02 ml	H_2O_2 (0.48 mol/l)	3 mmol/l
0.02 ml	catalase (0.1 mg/ml buffer)	40 kU/l
0.02 ml	cholesterol oxidase (1 mg/ml)	0.16 kU/l
0.05 ml	enzyme in buffer	

Stability*: the stabilized lyophilized enzyme is stable for several months at 4 °C.

Purity required*: activity $\geqslant$ 25 U/mg substance (25 °C; $\underline{\Delta}$ ca. 100 U/mg enzyme protein). Contaminants (related to the specific activity of cholesterol esterase) $\leqslant$ 0.005% ATPase, HK, "NADH oxidase" and uricase, $\leqslant$ 0.001% GK and GOD.

Reference

[1] *T. Uwajiama, O. Terada,* Agric. Biol. Chem. *40,* 1957 (1976).

Cholesterol Oxidase

from *Nocardia erythropolis*

Cholesterol : oxygen oxidoreductase, EC 1.1.3.6

$$\text{Cholesterol} + O_2 \xrightarrow[\text{oxidase}]{\text{cholesterol}} \Delta^4\text{-cholestenone} + H_2O_2$$

Michaelis constant (phosphate buffer, pH 7.0): cholesterol approx. 1×10^{-6} mol/l.

---

* For the enzyme from *Pseudomonas fluorescens.*

pH Optimum: pH 5.0 – 8.0.

Assay

$$2\,H_2O_2 \xrightarrow{\text{catalase}} 2\,H_2O + O_2$$

$$\text{Cholesterol} + O_2 \xrightarrow{\text{ChOD}} \Delta^4\text{cholestenone} + H_2O_2.$$

Wavelength 240 nm; light path 10 mm; final volume 3.15 ml; 25°C; $\varepsilon = 1.55$ $l \times mmol^{-1} \times mm^{-1}$.

3.00 ml	K-phosphate buffer		
	(0.5 mol/l; pH 7.5, containing 0.4%	phosphate	0.48 mol/l
	Thesit*, O_2 saturated)	Thesit	3.8 g/l
0.10 ml	cholesterol (4 mg/ml 10% Thesit)		0.13 mg/ml
0.02 ml	H_2O_2 (0.05 ml 30% H_2O_2/ml)		3 mmol/l
0.01 ml	catalase (0.1 mg/ml)		21 kU/l
0.02 ml	enzyme in water		

Stability: suspensions in ammonium sulphate solution, 1 mol/l; pH 6, or solutions in NaCl solution, 3 mol/l, are stable for several months at 4°C.

Purity required: specific activity $\geq$ 25 U/mg protein (25°C; cholesterol as substrate). Contaminants (related to the activity of ChOD) $\leq$ 0.005% cholesterol esterase, $\leq$0.01%, uricase, GOD and "NADH oxidase".

Choline Kinase

from yeast

ATP:choline phosphotransferase, EC 2.7.1.32

$$\text{ATP} + \text{choline} \xrightarrow{\text{choline kinase}} \text{ADP} + o\text{-phosphocholine}$$

Michaelis constants [1]

Choline	1.5×10^{-5} mol/l
ATP	1.4×10^{-4} mol/l

---

* *Desitin-Werk, Karl Linke,* Hamburg, FRG.

Choline	2×10^{-5} mol/l	for the enzyme
2-Dimethylaminoethanol	1×10^{-4} mol/l	from brewer's
2-Diethylaminoethanol	3.3×10^{-4} mol/l	yeast [2].
2-Methylaminoethanol	6×10^{-4} mol/l	Conditions:
2-Ethylaminoethanol	8×10^{-4} mol/l	glycylglycine buffer,
Ethanolamine	1×10^{-2} mol/l	pH 8.5; 26°C

Effectors [1]: SH-reagents such as N-ethylmaleimide, 5.5-dithio-bis (2-nitrobenzoate) inhibit.
Mg^{2+} ions are essential for maximal activity. Cysteine enhances the reaction rate.

Molecular weight [1]: 67000.

Specificity: complete phosphorylation is obtained with choline as substrate. Ethanolamine and its methyl and ethyl derivatives are phosphorylated to some degree.

pH Optimum [2]: 8.5.

Assay

$$\text{Choline} + \text{ATP} \xrightarrow{\text{choline kinase}} \text{phosphocholine} + \text{ADP}$$

$$\text{ADP} + \text{PEP} \xrightarrow{\text{PK}} \text{ATP} + \text{pyruvate}$$

$$\text{pyruvate} + \text{NADH} + \text{H}^+ \xrightarrow{\text{LDH}} \text{L-lactate} + \text{NAD}^+.$$

Wavelength 339, Hg 334 or Hg 365 nm; light path 10 mm; final volume 3.07 ml; 25°C.

2.60 ml	glycylglycine buffer (0.2 mol/l; pH 8.6; GSH, 20 mmol/l; MgCl$_2$, 20 mmol/l)	glycylglycine GSH MgCl$_2$	0.17 mol/l 17 mmol/l 17 mmol/l
0.10 ml	PEP, cyclohexylammonium salt, (15 mg/ml)	PEP	1 mmol/l
0.10 ml	ATP, Na salt, (50 mg/ml)	ATP	2.7 mmol/l
0.10 ml	NADH, Na salt, (10 mg/ml)	NADH	0.46 mmol/l
0.05 ml	enzyme in buffer		
0.02 ml	PK/LDH (4 mg/ml)	PK LDH	3 kU/l 3 kU/l
0.10 ml	choline chloride (40 mg/ml)		9.3 mmol/l

Stability: in the lyophilized state a decrease in activity of approx. 10% may occur within 12 months at 4°C.

Purity required: activity $\geqslant$ 0.04 U/mg substance (25°C); $\underline{\Delta}$ ca. 0.5 U/mg enzyme protein.

References

[1] *M. A. Brostrom, E. T. Browring,* J. Biol. Chem. *248,* 2364 (1973).
[2] *J. Wittenberg, A. Kornberg,* J. Biol. Chem. *202,* 431 (1953).

Choline Oxidase

from *Arthrobacter globiformis*

Choline: oxygen 1-oxidoreductase, EC 1.1.3.17

$$\text{Choline} + O_2 \xrightleftharpoons{\text{choline oxidase}} \text{betaine aldehyde} + H_2O_2$$

Michaelis constants

Choline	1.2×10^{-3} mol/l
Betaine aldehyde	8.7×10^{-3} mol/l'

Effectors: no activation by divalent cations or inhibition by complexing agents such as EDTA. Detergents, e.g. Triton X-100 or sodium dodecylsulphate, are tolerated up to relatively high concentrations.

Molecular weight [1]: 71 000 – 83 000.

Specificity [1, 2]: choline oxidase is specific for choline. Ethanolamine is virtually not attacked.

	Relative rate
Choline	100
N,N-Dimethylaminoethanol	69.1
N-Methylaminoethanol	11.7
N,N-Diethylaminoethanol	15.7
N-Ethylaminoethanol	6.5
Ethanolamine	0.2

Methanol, ethanol, propanol, formaldehyde, acetaldehyde and propionic aldehyde do not react. Triethanolamine and diethanolamine are converted at 2.6% or 0.8% respectively of the reaction rate with choline.

pH Optimum: 7.5.

Assay

$$\text{Choline} + O_2 \xrightarrow{\text{choline oxidase}} \text{betainaldehyde} + H_2O_2$$

$$\text{Betainaldehyde} + O_2 + H_2O \xrightarrow{\text{choline oxidase}} \text{betain} + H_2O_2$$

$$2\,H_2O_2 + \text{4-aminoantipyrine} + \text{phenol} \xrightarrow{\text{POD}}$$
$$4\,H_2O + \text{4-(p-benzoquinonemonoimino)-phenazone}.$$

Wavelength 500 nm; light path 10 mm; final volume 3.05 ml; 37 °C;
$\varepsilon = 0.6\,l \times \text{mmol}^{-1} \times \text{mm}^{-1}$.

3.00 ml	buffer/substrate (Tris 0.1 mol/l; pH 8.0; aminoantipyrine, 0.1 mg/ml, phenol 0.2 mg/ml; choline chloride, 21 mg/ml; POD, 3 U/ml)	Tris	98 mmol/l
		aminoantipyrine	0.1 g/l
		phenol	0.2 g/l
		choline chloride	21 g/l
		POD	3 kU/l
0.05 ml	enzyme in buffer		

Stability: solutions in NaCl, 4 mol/l; EDTA, 10 mmol/l, are stable for several months at 4 °C.

Purity required: specific activity $\geqslant$ 10 U/mg (37 °C). Contaminant (related to the specific activity of choline oxidase) $\leqslant$ 0.02% GOD.

References

[1] *S. Ikuta et al.*, J. Biochem. *82*, 1741 (1977).
[2] *M. Takayama et al.*, Clin. Chim. Acta *79*, 93 (1977).

Citrate Lyase

from *Aerobacter aerogenes*

Citrate oxaloacetate-lyase, EC 4.1.3.6

$$\text{Citrate} \xrightleftharpoons{\text{citrate lyase}} \text{oxaloacetate} + \text{acetate}$$

Equilibrium constant (pH 8.4; 25 °C; Mg^{2+}, 0.5 to 10 mmol/l) [1]

$$K = \frac{[\text{acetate}^-] \times [\text{oxaloacetate}^{2-}]}{[\text{citrate}^{3-}]} = 0.325\,\text{mol/l}$$

Michaelis constants (Tris buffer, pH 7.4; 25 °C) [2]

Citrate 2.1×10^{-4} mol/l
Mg^{2+} 3.0×10^{-3} mol/l

Effectors: the enzyme is inhibited by oxaloacetate [1].

Molecular weight [3]: 550000.

Specificity: citrate lyase is specific for citrate. Isocitrate, aconitate, oxalate, succinate, fumarate, 2-oxoglutarate, L-glutamate, tartrate, lactate, malate, acetate, ascorbate, glucose, fructose or ethanol do not react [4].

Assay

$$Citrate \xrightarrow{\text{citrate lyase}} oxaloacetate + acetate$$

$$Oxaloacetate + NADH + H^+ \xrightarrow{\text{MDH}} \text{L-malate} + NAD^+$$

$$(Oxaloacetate \xrightarrow{\text{decarboxylase}} pyruvate + CO_2)$$

$$Pyruvate + NADH + H^+ \xrightarrow{\text{LDH}} \text{L-lactate} + NAD^+.$$

Wavelength 339, Hg 334 or Hg 365 nm; light path 10 mm; final volume 3.17 ml; 25 °C.

3.00 ml	triethanolamine buffer (0.1 mol/l; pH 7.6)	94.5 mmol/l
0.05 ml	$ZnCl_2$ (4 mg/ml)	0.5 mmol/l
0.05 ml	NADH, Na salt, (10 mg/ml)	0.2 mmol/l
0.02 ml	citrate, Na salt, (0.1 mol/l)	6.7 mmol/l
0.01 ml	LDH (2 mg/ml)*	3.5 kU/l
0.01 ml	MDH (2 mg/ml)	7.6 kU/l
0.03 ml	enzyme in triethanolamine buffer (10 mmol/l; pH 7.6; 0.04 mg $ZnCl_2$, 60 mg ammonium sulphate/ml)	

Stability: lyophilized enzymes, containing 24% citrate lyase, 24% albumin, 48% sucrose and 4% $MgSO_4 \cdot 7 H_2O$ (pH of solution ca. 7.0) are stable several months at 4 °C.

Purity required: specific activity $\geq$ 8 U/mg protein (25 °C). Contaminants (related to the activity of citrate lyase) < 0.05% ICDH, (NAD-specific), and "NADH oxidase".

---

* To react with the pyruvate formed from oxaloacetate, because certain citrate lyase preparations contain oxaloacetate decarboxylase.

References

[1] *S. S. Tata, S. P. Datta*, Biochem. J. *95*, 470 (1965).
[2] *H. H. Duron, I. C. Gunsalus*, in: *S. P. Colowick, N. O. Kaplan* (eds.), Methods in Enzymology, Vol. *V*, Academic Press, New York 1962, p. 622.
[3] *P. Dimroth, H. Eggerer*, Eur. J. Biochem. *53*, 227 (1975).
[4] *H. Möllering, W. Gruber*, Anal. Biochem. *17*, 369 (1966).

Citrate Synthase

("condensing enzyme") from pig heart

Citrate oxaloacetate-lyase, EC 4.1.3.7

$$\text{Citrate} + \text{CoA} \xrightleftharpoons{\text{CS}} \text{acetyl-CoA} + \text{oxaloacetate} + H_2O$$

Equilibrium constant (phosphate buffer; 22 °C)

$$K = \frac{[\text{acetyl CoA}] \times [\text{oxaloacetate}] \times [H_2O]}{[\text{citrate}] \times [\text{CoA}]} = 1.2 \times 10^{-4}\,\text{mol/l}$$

Michaelis constants (imidazole/acetate buffer, pH 6.1; 22 °C)

CoA $2.8 \times 10^{-5}\,\text{mol/l}$
Citrate $2.5 \times 10^{-4}\,\text{mol/l}$

For K_m for acetyl-CoA and oxaloacetate, cf. [1]; these depend on the pH and substrate concentration.

Inhibitor constants (triethanolamine buffer, pH 7.0; 23 °C)

Fluoroacetyl-CoA $1.3 \times 10^{-6}\,\text{mol/l}$

Inhibition by palmitoyl-CoA and citroyl-CoA, cf. [2].

Molecular weight (pH 8.2) [3]: 86000.

Specificity: the enzyme reacts with fluoroacetate, citroyl-CoA, S-malonyl-CoA and fluoroacetyl-CoA. Propionyl-CoA shows only 0.1% of the activity with acetyl-CoA. Butyryl-CoA is inactive.

Assay

Wavelength 232 nm; light path 10 mm; final volume 3.12 ml; 25 °C; $\varepsilon = 0.54\ 1 \times$ mmol^{-1} × mm^{-1}.

3.00 ml	Tris buffer (0.1 mol/l; pH 8.0)	96.2 mmol/l
0.05 ml	oxaloacetic acid (1.32 mg/ml)	0.17 mmol/l
0.05 ml	acetyl-CoA, Li salt, (10 mg/ml)	<0.2 mmol/l
0.01 ml	enzyme in buffer	

Stability: suspensions in ammonium sulphate solution, 3.2 mol/l; pH ca. 7, are stable for several months at 4 °C.

Purity required: specific activity $\geq$ 110 U/mg protein (25 °C). Contaminant (related to the specific activity of CS) $\leq$ 0.1% MDH.

References

[1] *G. W. Kosicki, P. A. Srere,* J. Biol. Chem. *236,* 2560 (1961).
[2] *P. A. Srere,* Biochim. Biophys. Acta *77,* 693 (1963); *106,* 445 (1965).
[3] *P. A. Srere,* J. Biol. Chem. *241,* 2157 (1966).

Creatine Kinase

from rabbit muscle

ATP : creatine N-phosphotransferase, EC 2.7.3.2

$$\text{Creatine} + \text{ATP} \xrightleftharpoons{\text{CK}} \text{creatine-P} + \text{ADP}$$

Equilibrium constant [1]

$$K = \frac{[\text{creatine-P}] \times [\text{ADP}]}{[\text{creatine}] \times [\text{ATP}]} = \begin{array}{l} 7.2 \times 10^{-9}\,(\text{pH}\,7.4;\,30\,°\text{C}) \\ 2.98 \times 10^{-9}\,(\text{pH}\,9.8;\,30\,°\text{C}) \end{array}$$

Michaelis constants [2]

ATP	$5\ \times 10^{-4}$ mol/l	
Creatine	1.6×10^{-2} mol/l	glycine buffer, pH 8.8; 30 °C
Mg^{2+}	$6\ \times 10^{-4}$ mol/l	

ADP	8×10^{-4} mol/l	
Creatine phosphate	5×10^{-3} mol/l	glycylglycine buffer, pH 7.0; 30 °C
Mg^{2+}	2×10^{-3} mol/l	

Inhibitor constant (glycine buffer, pH 8.8; 30 °C) [2]

ADP $\qquad$ 2×10^{-3} mol/l non-competitive (creatine)

3×10^{-4} mol/l competitive (ATP)

Effectors: various adenosine phosphate derivatives, L-thyroxine, L-tri-iodothyronine, malonate and various inorganic ions (e.g. Ca^{2+}) are competitors.

Molecular weight [3]: 81 000 (2 subunits).

Specificity: the enzyme is specific for creatine, ATP and ADP. N-Ethylglycocyamine and glycocyamine react at considerably slower rates, while creatinine, D- or L-arginine, histidine and taurocyamine do not react. The enzyme has slight ATPase activity.

Assay

$$\text{Creatine-P + ADP} \xrightarrow{\text{CK}} \text{creatine + ATP}$$

$$\text{Glucose + ATP} \xrightarrow{\text{HK}} \text{glucose-6-P + ADP}$$

$$\text{Glucose-6-P + NADP}^+ \xrightarrow{\text{G6P-DH}} \text{gluconate-6-P + NADPH + H}^+.$$

Wavelength 339, Hg 334 or Hg 365 nm; light path 10 mm; final volume 2.94 ml; 25 °C.

2.50 ml	substrate/buffer solution (0.756 g creatine-P, Na salt, in 50 ml imidazole buffer, 0.1 mol/l; pH 6.6)	imidazole	85 mmol/l
0.05 ml	glucose (240 mg glucose · H_2O/ml imidazole buffer, 0.1 mol/l; pH 6.6)		21 mmol/l
0.05 ml	magnesium acetate (130 mg magnesium acetate · 4 H_2O/ml)		10 mmol/l
0.10 ml	EDTA (34 mg EDTA-Na_2H_2 · 2 H_2O/ml imidazole buffer, 0.1 mol/l, pH 6.6)		3.1 mmol/l
0.05 ml	ADP, free acid, (30 mg/ml imidazole buffer, 0.1 mol/l; pH 6.6)		1.2 mmol/l
0.05 ml	NADP, Na salt, (30 mg/ml imidazole buffer, 0.1 mol/l; pH 6.6)		0.42 mmol/l
0.10 ml	GSH (dissolve 90 mg GSH in 0.9 ml water and adjust to pH 5.5 with 0.1 ml NaOH, 2 mol/l)		9.8 mmol/l
0.01 ml	HK (2 mg/ml)		1 kU/l
0.01 ml	G6P-DH (1 mg/ml)		0.5 kU/l
0.02 ml	enzyme in buffer		

Stability: in lyophilized state a decrease in activity of approx. 15% may occur within 6 months at 4°C. CK solutions can be dialyzed against ammonium citrate solution, 0.05 mol/l; pH 9.0, at 4°C without appreciable loss of activity.

Purity required: specific activity $\geq$ 25 U/mg protein (25°C). Contaminants (related to the specific activity of CK) $\leq$ 0.01% ATPase, HK and MK.

References

[1] *S. A. Kuby, E. A. Noltmann,* in: *P. D. Boyer, H. Lardy, K. Myrbäck* (eds.), The Enzymes, Vol. *6*, Academic Press, New York 1962, p. 515.
[2] *S. A. Kuby, L. Noda, H. A. Lardy,* J. Biol. Chem. *210*, 65 (1954).
[3] *D. C. Watts,* in: *P. D. Boyer* (ed.), The Enzymes, Vol. *VIII*, Academic Press, New York 1973, p. 838.

Creatininase

from *Alkaligenes* species [1]

Creatinine amidohydrolase, EC 3.5.2.10

$$\text{Creatinine} + H_2O \xrightarrow{\text{creatininase}} \text{creatine}$$

Equilibrium constant (glycylglycine buffer, pH 8.0; 37°C) [1]

$$K_{H_2O} = \frac{[\text{creatine}]}{[\text{creatinine}]} = 1.27$$

Michaelis constant (glycylglycine buffer, pH 8.0; 37°C) [1]: creatinine 3.3×10^{-2} mol/l.

Specificity: the enzyme is specific for creatinine [2 – 4].

Assay [1]

$$\text{Creatinine} + H_2O \xrightarrow{\text{creatininase}} \text{creatine}$$

$$\text{Creatine} + \text{ATP} \xrightarrow{\text{CK}} \text{creatine-P} + \text{ADP}$$

$$\text{ADP} + \text{PEP} \xrightarrow{\text{PK}} \text{ATP} + \text{pyruvate}$$

$$\text{Pyruvate} + \text{NADH} + H^+ \xrightarrow{\text{LDH}} \text{L-lactate} + \text{NAD}^+.$$

Wavelength 339, Hg 334 or Hg 365 nm; light path 10 mm; final volume 3.10 ml; 25 °C.

2.30 ml	glycylglycine buffer (0.1 mol/l; pH 8.0)	74 mmol/l
0.10 ml	NADH, Na salt, (10 mg/ml)	0.4 mmol/l
0.50 ml	ATP, Na salt, (10 mg/ml)	1 mmol/l
0.10 ml	PEP, tricyclohexylammonium salt, (10 mg/ml)	0.6 mmol/l
0.20 ml	creatinine (50 mg/ml)	28.5 mmol/l
0.10 ml	MgCl$_2$ (0.1 mol/l)	3.2 mmol/l
0.01 ml	LDH (10 mg/ml)	18 kU/l
0.01 ml	PK (10 mg/ml)	6 kU/l
0.05 ml	CK (60 mg/ml 0.5% NaHCO$_3$ solution)	24 kU/l
0.03 ml	enzyme in buffer	

Stability: solutions in 50% glycerol, pH ca. 8, are stable for several months at 4 °C. Creatininase can be dialyzed without any decrease in activity against phosphate buffer, 0.05 mol/l; pH 8.0; storage of the dialyzed solution at −20 °C.

Purity required: specific activity ≥ 70 U/mg protein (25 °C). Contaminants (related to the activity of creatininase) ≤ 0.01% ATPase and hexokinase, 0.1% creatinase.

References

[1] *Boehringer Mannheim GmbH,* German Patents P 2122255.9, P 2122294.6 & P 2122298.0 (1971).
[2] *J. Roche, G. Lacombe, H. Girard,* Biochim. Biophys. Acta 6, 210 (1950).
[3] *S. Akamatsu, R. Miyashita,* Enzymologia *15*, 122, 158 and 173 (1951).
[4] *B. F. Miller, R. Dubos,* J. Biol. Chem. *121*, 429 and 457 (1937).

Diaphorase

from micro-organisms

NAD(P)H: dye oxidoreductase, EC 1.6.99

$$\text{NADH} + \text{H}^+ + \text{hydrogen acceptor} \xrightarrow{\text{diaphorase}} \text{NAD}^+ + \text{reduced acceptor}$$

Michaelis constant [1]: NADH 9×10^{-4} mol/l.

Effectors [1]: N-ethylmaleimide inactivates at concentrations of less than 5 mmol/l. NADH, NADPH and FMN protect the enzyme against denaturation by urea and guanidine.

Molecular weight [1]: ca. 24000.

Specificity: the enzyme oxidizes both NADH and NADPH. With NADPH as reductant, however, the reaction rate is less than 5% of that with NADH. The enzyme from *Clostridium kluyveri* does not react with oxygen or cytochrome.

pH Optimum: 8.5.

Assay

$$\text{L-Lactate} + \text{NAD}^+ \xrightarrow{\text{LDH}} \text{pyruvate} + \text{NADH} + \text{H}^+$$

$$\text{NADH} + \text{INT} \xrightarrow{\text{diaphorase}} \text{formazan} + \text{NAD}^+.$$

Wavelength 492 nm; light path 10 mm; final volume 3.01 ml; 25 °C;
$\varepsilon = 1.94\,1 \times \text{mmol}^{-1} \times \text{mm}^{-1}$.

2.65 ml	Tris buffer/lactate (Tris, 0.1 mol/l; pH 8.8;	Tris	8.8 mmol/l
	D,L-lactate 0.175 mol/l; 1.5 µl Triton X-100/ml)	lactate	0.154 mol/l
0.01 ml	LDH (from rabbit muscle, 1 mg/ml)		1.6 kU/l
0.10 ml	NAD (10 mg/ml phosphate buffer, 0.1 mol/l; pH 7.4)		0.45 mmol/l
0.10 ml	INT (1.87 mg iodonitrotetrazolium chloride/ml)		0.12 mmol/l
0.10 ml	BSA (30 mg bovine serum albumin/ml water)		1 g/l
0.05 ml	enzyme in phosphate buffer (20 mmol/l;		
	ammonium sulphate, 0.5 mol/l, FAD, 10 µmol/l; pH 6.5)		

Stability: in the lyophilized state a decrease in activity of approx. 20% may occur within 12 months at 4 °C.

Purity required: activity $\geqslant$ 15 U/mg substance (25 °C).

Reference

[1] *F. Kaplan et al.,* Arch. Biochem. Biophys. *132*, 91 (1969).

Diaphorase (Lipoamide Dehydrogenase)

from pig heart

NADH: lipoamide oxidoreductase, EC 1.6.4.3

$$\text{Lipoamide} + \text{NADH} + \text{H}^+ \underset{}{\overset{\text{diaphorase}}{\rightleftarrows}} \text{dihydrolipoamide} + \text{NAD}^+$$

Equilibrium constant: the reaction is reversible. The back reaction is favoured.

Michaelis constants [1]

1. Substrate	2. Substrate		Relative rate
Lipoamide	NADH	5×10^{-3} mol/l	1.00
Dihydrolipoamide	NAD	1.4×10^{-4} mol/l	0.12
Ferricyanide	NADH	2.7×10^{-4} mol/l	0.10
Lipoate	NADH	2×10^{-3} mol/l	0.01

Molecular weight [2]: 100000.

Specificity: the enzyme reacts with numerous electron donors in both directions. Electron acceptors are oxidized lipoyl derivatives, NAD and several pyridine analogues, ferricyanide, methylene blue and many other dyes and O_2.

Assay

$$6,8\text{-Dithiooctate} + NADH + H^+ \xrightarrow{\text{diaphorase}} 6,8\text{-dihydrothiooctate} + NAD^+.$$

Wavelength 339, Hg 334 or Hg 365 nm; light path 10 mm; final volume 2.90 ml; 25 °C.

2.00 ml	phosphate buffer (0.1 mol/l; pH 5.9)	69 mmol/l
0.10 ml	EDTA (10 mg/ml)	0.90 mmol/l
0.50 ml	DL-dithio-n-octanoic acid (lipoic acid, 15 mg/ml; pH ca. 5.9)	12.5 mmol/l
0.20 ml	serum albumin (10 mg/ml)	0.67 mg/ml
0.05 ml	NADH, Na salt, (10 mg/ml)	0.20 mmol/l
0.03 ml	NAD (10 mg/ml)	0.13 mmol/l
0.02 ml	enzyme in buffer	

Stability: suspensions in ammonium sulphate solution, 3.2 mol/l; pH ca. 6, are stable for several months at 4 °C.

Purity required: specific activity $\geq$ 210 U/mg protein (25 °C; measured with lipoamide as substrate) or $\geq$ 25 U/mg protein (25 °C; measured with lipoic acid). Turnover with O_2 as electron acceptor is ca. 5% of the specific activity measured with lipoic acid.

References

[1] *V. Massey,* Biochim. Biophys. Acta *37,* 314 (1960).
[2] *V. Massey, T. Hofmann, G. Palmer,* J. Biol. Chem. *237,* 3820 (1960).

Enolase

from rabbit muscle

2-Phospho-D-glycerate hydro-lyase, EC 4.2.1.11

$$\text{Glycerate-2-P} \; \underset{}{\overset{\text{enolase}}{\rightleftarrows}} \; \text{PEP} + H_2O$$

Equilibrium constant (25 °C) [1]

$$K = \frac{[\text{PEP}]}{[\text{glycerate-2-P}]} = 6.7$$

Michaelis constants (imidazole buffer, pH 6.8) [2]

		Relative rate
Glycerate-2-P	7×10^{-5} mol/l	1.0
Erythronate-3-P	3×10^{-4} mol/l	1.0
PEP	9.2×10^{-5} mol/l	–
Mg^{2+}	1.3×10^{-4} mol/l	–

Molecular weight (pH 7.1): 82000.

Specificity: the enzyme from rabbit muscle reacts with erythronate-3-P.

Assay

$$\text{Glycerate-2-P} \; \xrightarrow{\text{enolase}} \; \text{PEP} + H_2O$$

$$\text{PEP} + \text{ADP} \; \xrightarrow{\text{PK}} \; \text{pyruvate} + \text{ATP}$$

$$\text{Pyruvate} + \text{NADH} + H^+ \; \xrightarrow{\text{LDH}} \; \text{L-lactate} + NAD^+$$

wavelength 339, Hg 334 or Hg 365 nm; light path 10 mm; final volume 3.01 ml; 25 °C.

2.50 ml	triethanolamine buffer (0.1 mol/l; pH 7.6)	83 mmol/l
0.10 ml	$MgSO_4$ (0.1 mol/l)	3.3 mmol/l
0.05 ml	NADH, Na salt, (10 mg/ml)	0.2 mmol/l
0.10 ml	glycerate-2-P, Na salt, (10 mg/ml)	0.9 mmol/l
0.20 ml	ADP, Na salt, (10 mg/ml)	1.1 mmol/l
0.02 ml	LDH (5 mg/ml)	18.5 kU/l
0.02 ml	PK (2 mg/ml)	2.7 kU/l
0.02 ml	enzyme in triethanolamine buffer	

Stability: crystalline suspensions in ammonium sulphate solution, 3.2 mol/l; pH 6, are stable for several months at 4 °C.

Purity required: specific activity $\geq$ 40 U/mg protein (25 °C). Contaminants (related to the specific activity of enolase) $\leq$ 0.2% PGM and PK.

References

[1] *J. F. Bealing, R. Czok, L. Eckert, I. Jaeger,* unpublished; cf. *R. Czok* in: *Hoppe-Seyler/Thierfelder,* Handbuch der physiol. und path.-chem. Analyse, Vol. *VIc,* Springer, Berlin 1966, p. 657.
[2] *F. Wold, R. Barker,* Biochim. Biophys. Acta *85,* 475 (1964).

Formate Dehydrogenase

from yeast

Formate : NAD oxidoreductase, EC 1.2.1.2

$$\text{Formate} + \text{NAD}^+ \xrightarrow{\text{FDH}} CO_2 + \text{NADH} + H^+$$

Equilibrium constant: the forward reaction is strongly favoured [1, 2].

Michaelis constants

Formate	1.3×10^{-2} mol/l [1]
Formate	7.7×10^{-5} mol/l [2]
NAD	6.4×10^{-4} mol/l [2]

Effectors: Cu^{2+}, Hg^{2+} and 4-chloromercuribenzoate are strong inhibitors. Azide, cyanate, nitrite, rhodanide and nitrate inactivate.

Molecular weight [1]: 74000.

Specificity: the enzyme is highly specific for formate. There is no reaction with acetate, oxalate, lactate or succinate. NADP is not reduced.

pH Optimum [1]: 7.5 – 9.

Assay

Wavelength 339, Hg 334 or Hg 365 nm; light path 10 mm; final volume 3.20 ml; 25 °C.

2.50 ml	potassium phosphate buffer (0.1 mol/l; pH 7.6)	78 mmol/ml
0.10 ml	NAD-Li · 2 H_2O (40 mg/ml)	1.8 mmol/ml
0.50 ml	formate (68 mg sodium formate/ml)	156 mmol/ml
0.10 ml	enzyme in water	

Stability: the lyophilized enzyme is stable for several months at 4°C.

Purity required: activity $\geqslant$ 0.4 U/mg substance (25°C); $\underline{A}$ ca. 3 U/mg enzyme protein.

References

[1] *H. Schuette et al.,* Eur. J. Biochem. *62,* 151 (1976).
[2] *J. R. Quayle,* in: *S. P. Colowick, N. O. Kaplan* (eds.), Methods in Enzymology, Vol. *IX,* Academic Press, New York 1966, p. 360.

β-D-Fructofuranosidase

(Invertase, Saccharase, β-h-Fructosidase) from yeast

β-D-Fructofuranoside fructohydrolase, EC 3.2.1.26

$$\text{Sucrose} + \text{H}_2\text{O} \xrightarrow{\text{β-fructosidase}} \text{D-glucose} + \text{D-fructose}$$

Michaelis constant (maleate buffer, pH 4.6) [1]: sucrose 9.1×10^{-3} mol/l.

Molecular weight (pH 5.4) [2]: 270000. (The enzyme is a glycoprotein and contains about 50% carbohydrate).

Specificity: the yeast enzyme catalyzes the hydrolysis of sugars that have an unsubstituted end terminal β-D-fructofuranosyl residue. The following do not react: α-fructofuranosides, fructopyranosides, β-L-sorbofuranosides, β-D-xyloketofuranosides and sugars with substituents on the β-fructofuranosyl residue.

Assay

$$\text{Sucrose} + \text{H}_2\text{O} \xrightarrow{\text{β-fructosidase}} \text{D-glucose} + \text{D-fructose}$$

$$\text{D-Glucose} + \text{D-fructose} + 2\,\text{ATP} \xrightarrow{\text{HK}} \text{glucose-6-P} + \text{fructose-6-P} + 2\,\text{ADP}$$

$$\text{Fructose-6-P} \xrightarrow{\text{PGI}} \text{glucose-6-P}$$

$$2\,\text{Glucose-6-P} + 2\,\text{NADP}^+ \xrightarrow{\text{G6P-DH}} \text{2-gluconate-6-P} + 2\,\text{NADPH} + 2\,\text{H}^+.$$

Wavelength 339, Hg 334 or Hg 365 nm; light path 10 mm; incubation volume 1.60 ml; assay volume 2.48 ml; 25 °C.

Incubation mixture

1.00 ml	acetate buffer (0.1 mol/l; pH 4.65)	62.5 mmol/l
0.50 ml	sucrose (100 mg/ml)	91.5 mmol/l
0.10 ml	enzyme in water,	

incubate for exactly 5 min at 25 °C, add 0.4 ml Tris solution (0.3 mol/l)

Assay mixture

2.50 ml	triethanolamine buffer (0.1 mol/l; pH 7.6)	89 mmol/l
0.10 ml	MgCl$_2$ (0.2 mol/l)	7.1 mmol/l
0.10 ml	ATP, Na salt, (10 mg/ml)	0.57 mmol/l
0.10 ml	NADP, Na salt, (10 mg/ml)	0.4 mmol/l
0.01 ml	HK (10 mg/ml)	5 kU/l
0.01 ml	PGI (10 mg/ml)	12.5 kU/l
0.01 ml	incubation solution	
0.01 ml	G6P-DH (5 mg/ml)	2.5 kU/l

Stability: lyophilized enzymes are stable for several months at 4 °C. The aqueous enzyme solution can be used for several weeks when kept at 4 °C.

Purity required: activity ≥ 150 U/mg substance (25 °C). Contaminants (related to the specific activity of β-fructosidase) ≤ 0.01% α-glucosidase, β-glucosidase and β-galactosidase, ≤ 0.001% α-galactosidase.

References

[1] *M. V. Tracey,* Biochim. Biophys. Acta 77, 147 (1963).
[2] *N. P. Neumann, J. O. Lampen,* Biochemistry 6, 468 (1967).

Fructose-6-phosphate Kinase

from rabbit muscle

ATP : D-fructose-6-phosphate 1-phosphotransferase, EC 2.7.1.11

$$\text{Fructose-6-P} + \text{ATP} \xrightarrow{\text{F-6-PK}} \text{fructose-1,6-P}_2 + \text{ADP}$$

Molecular weight (pH 8) [1]: 360 000.

Specificity [2]: the muscle enzyme phosphorylates sedoheptulose 7-phosphate, but is inactive with: D-fructose-1-P, α-D-glucose-1-P, D-glucose-6-P, D-mannose-6-P, D-ribose-5-P, D-ribulose-5-P, L-sorbose-1-P and L-sorbose-6-P. For properties of F-6-PK from sheep heart and yeast, cf. [3, 4].

Assay

$$\text{Fructose-6-P} + \text{ATP} \xrightarrow{\text{F-6-PK}} \text{fructose-1,6-P}_2 + \text{ADP}$$

$$\text{ADP} + \text{PEP} \xrightarrow{\text{PK}} \text{ATP} + \text{pyruvate}$$

$$\text{Pyruvate} + \text{NADH} + \text{H}^+ \xrightarrow{\text{LDH}} \text{L-lactate} + \text{NAD}^+.$$

Wavelength 339, Hg 334 or Hg 365 nm; light path 10 mm; final volume 2.86 ml; 25 °C.

2.00 ml	Tris buffer (0.1 mol/l; pH 8.5)	Tris	70 mmol/l
0.10 ml	MgSO$_4 \cdot$ 7 H$_2$O (10 mg/ml;	MgSO$_4$	1.4 mmol/l
	KCl, 10 mg/ml)	KCl	4.5 mmol/l
0.10 ml	PEP, tricyclohexylammonium salt (10 mg/ml)	PEP	0.71 mmol/l
0.10 ml	fructose-1,6-P$_2$, Na salt, (10 mg/ml)		0.64 mmol/l
0.20 ml	fructose-6-P, Na salt, (10 mg/ml)		1.8 mmol/l
0.20 ml	ATP, Na salt, (10 mg/ml)		1.1 mmol/l
0.10 ml	NADH, Na salt, (10 mg/ml)		0.4 mmol/l
0.03 ml	PK (2 mg/ml)		4.2 kU/l
0.01 ml	LDH (5 mg/ml)		9.6 kU/l
0.02 ml	enzyme in ammonium sulphate, 2.0 mol/l; pH 7.5		

Stability: suspensions in ammonium sulphate solution, 3.2 mol/l, pH ca. 7.5; (containing phosphate, 10 mmol/l and adenosine phosphates, 1 mmol/l) are stable for several months at 4 °C.

Purity required: specific activity $\geq$ 60 U/mg protein (25 °C). Contaminants (related to the specific activity of F-6-PK) $\leq$ 0.01% aldolase (fructose-1,6-P$_2$-specific, fructose-1-P-specific) and PGI. Reaction with fructose and fructose-1-P (side reaction of the enzyme) approx. 0.15% and 1% resp.

References

[1] *K. H. Ling, F. Marcus, H. A. Lardy,* J. Biol. Chem. *240*, 1893 (1965).
[2] *H. A. Lardy,* in: *P. D. Boyer, H. Lardy, K. Myrbäck* (eds.), The Enzymes, Vol. *6*, Academic Press, New York 1962, p. 67.
[3] *Th. E. Barman,* Enzyme Handbook, Vol. *1*, Springer, Berlin 1969, p. 386.
[4] *T. J. Lindell, E. Stellwagen,* J. Biol. Chem. *243*, 907 (1968).

L-Fucose Dehydrogenase

from mammalian liver

6-Deoxy-L-galactose: NAD$^+$ 1-oxidoreductase, EC 1.1.1.122

$$\text{L-Fucose + NAD} \xrightarrow{\text{fucose-DH}} \text{L-fucono-1,5-lactone + NADH + H}^+$$

Equilibrium constant [1]: the reaction is essentially irreversible as L-fucono-1,5-lactone hydrolyzes spontaneously to L-fuconate.

Michaelis constants [1]

Substrate	Enzyme preparation from	
	pig liver	sheep liver
L-Fucose	0.32×10^{-3} mol/l	0.074×10^{-3} mol/l
D-Arabinose	2.1×10^{-3} mol/l	0.40 mol/l
D-Lyxose	48×10^{-3} mol/l	
L-Xylose	44×10^{-3} mol/l	
3-Amino-D-arabinose	8×10^{-3} mol/l	
L-Galactose	8×10^{-3} mol/l	
NAD	0.02 mol/l	0.19 mol/l

Molecular weight [2]: ca. 96000 for the enzyme from sheep liver.

Specificity [2]: the enzyme from pig liver shows less or no reaction with D-arabitol, L-arabinose, D-fucose, D-mannose, D-galactose, L-rhamnose, D-fructose, lactose, melibiose, D-galactono-1,4-lactone, D-gulono-1,4-lactone, D-glucosamine, D-galactosamine, N-acetyl-D-glucosamine, N-acetyl-D-galactosamine and D-glucose.

For the enzyme from sheep liver, L-fucose can be replaced by 2-deoxy-D-ribose and D-ribose but not by D-glucose, D-galactose, D-xylose or L-arabinose [2].

NADP is inactive with the enzymes from pig and sheep liver.

pH Optimum: 8.7.

Assay

Wavelength 339, Hg 334 or Hg 365 nm; light path 10 mm; final volume 1.00 ml; 37 °C.

0.98 ml   buffer/substrate (Tris-HCl, 50 mmol/l; pH 8.0; L-fucose, 2.5 mmol/l; NAD, 0.5 mmol/l)
0.02 ml   enzyme in Tris buffer

Stability: the lyophilized enzyme is stable for several months at $-20\,°C$.

Purity required: activity $\geqslant 120\ U/g$ substance $(37\,°C)$.

References

[1] *H. Schachter et al.,* J. Biol. Chem. *244*, 4785 (1969).
[2] *Th. E. Barman,* Enzyme Handbook, Supplement I, Springer, Berlin 1974, p. 48.

α-L-Fucosidase

from mammalian liver

α-L-Fucoside fucohydrolase, EC 3.2.1.51

An α-L-fucoside + H_2O $\longrightarrow$ an alcohol + L-fucose

Michaelis constant [1]: 4-nitrophenyl α-L-fucopyranoside $1.9 \times 10^{-4}\,mol/l$

Inhibitor constant [1]: fucose $1.6 \times 10^{-3}\,mol/l$.

Effectors [1]: L-fucose is a potent physiological effector. Hg^{2+}, Cu^{2+} and Ni^{2+} inhibit, as also does 4-chloromercuriphenyl sulphonic acid at a concentration of $2 \times 10^{-6}\,mol/l$.

Molecular weight [1]: ca. 217000.

Specificity [2]: the enzyme hydrolyzes $1 \rightarrow 2$, $1 \rightarrow 3$ and $1 \rightarrow 4$ fucosyl linkages and is active on glycopeptides of rat preputial gland β-glucuronidase but not on native glycoproteins such as IgG, β-glucuronidase and bromelain.

pH Optimum [1]: $5.5 - 5.8$.

Assay

4-nitrophenyl-α-L-fucoside + H_2O $\xrightarrow{\text{fucosidase}}$ 4-nitrophenol + L-fucose.

Wavelength Hg 405 nm; light path 10 mm; incubation volume 1.02 ml; final volume 3.02 ml; $25\,°C$; $\varepsilon = 1.85\ l \times mmol^{-1} \times mm^{-1}$.

0.50 ml   acetate buffer (0.1 mol/l; pH 5)   49 mmol/l
0.50 ml   4-nitrophenyl-α-L-fucoside (1 mg/ml)   1.7 mmol/l
0.02 ml   enzyme in acetate buffer

incubate for exactly 5 min at 25 °C, add

2.00 ml   borate buffer (0.2 mol/l; pH 9.8)   0.13 mol/l

read absorbance; run a blank, in which enzyme is added after incubation.

Stability: suspensions in ammonium sulphate solution, 3.2 mol/l; pH ca. 6, are stable for several months at 4 °C.

Purity required: specific activity $\geqslant$ 2 U/mg (25 °C). Contaminants (related to the activity of α-L-fucosidase) $\leqslant$ 5% N-acetyl-β-D-glucosaminidase, $\leqslant$ 0.2% α-mannosidase.

References

[1] *D. J. Opheim, O. Touster,* in: *S. P. Colowick, N. O. Kaplan* (eds.), Methods in Enzymology, Vol. *V*, Academic Press, New York 1978, pp. 505 – 510.
[2] *D. I. Opheim, O. Touster,* J. Biol. Chem. *252*, 739 (1977).

Fumarase

from pig heart

L-Malate hydro-lyase, EC 4.2.1.2

$$\text{L-Malate} \xrightleftharpoons{\text{fumarase}} \text{fumarate} + H_2O$$

Equilibrium constant (pH 6.5; 25 °C) [1]

$$K = \frac{[\text{fumarate}]}{[\text{L-malate}]} = 0.23$$

Michaelis constants (Tris/acetate buffer, 10 mmol/l; 21 °C)

		K_0^*
L-Malate	3.79×10^{-6} mol/l	1130 [2]
Fumarate	1.74×10^{-6} mol/l	2520

* pH-independent.

Inhibitor constants (pH 6.35; 23 °C) [3]

Adipate	1.0×10^{-1} mol/l
Succinate	5.2×10^{-2} mol/l
Glutarate	4.6×10^{-2} mol/l
Malonate	4.0×10^{-2} mol/l
D-Tartrate	2.5×10^{-2} mol/l
Mesaconate	2.5×10^{-2} mol/l
L-2-Hydroxy-3-sulphopropionate	1.65×10^{-2} mol/l
Maleate	1.1×10^{-2} mol/l
D-Malate	6.3×10^{-3} mol/l
Citrate	3.5×10^{-3} mol/l
Trans-aconitate	6.3×10^{-4} mol/l

Molecular weight (pH 7.3) [2]: 194000.

Absorption coefficients

Fumarate A_{240} $\varepsilon = 0.244$ $1 \times \text{mmol}^{-1} \times \text{mm}^{-1}$; pH 7.4
$\quad\quad\quad A_{250}$ $\varepsilon = 0.14$ $1 \times \text{mmol}^{-1} \times \text{mm}^{-1}$; pH 4.5 [1]
$\quad\quad\quad A_{250}$ $\varepsilon = 0.145$ $1 \times \text{mmol}^{-1} \times \text{mm}^{-1}$; pH 7.3 [1]
$\quad\quad\quad A_{290}$ $\varepsilon = 0.011$ $1 \times \text{mmol}^{-1} \times \text{mm}^{-1}$; pH 7.3 [1]

Specificity: fumarase is absolutely specific for fumarate and L-malate. The following compounds do not react: D-malate, DL-thiomalate, maleate, *cis-* and *trans-*aconitate, citrate, mesaconate, citraconate, D-, L- and mesotartrate and the mono- and diesters of fumarate.

Assay

Wavelength 240 nm; light path 10 mm; final volume 3.02 ml; 25 °C; $\varepsilon = 0.244$ $1 \times \text{mmol}^{-1} \times \text{mm}^{-1}$.

3.00 ml	substrate/phosphate buffer	phosphate	100 mmol/l
	(0.1 mol/l; pH 7.6; L-malate, 6.7 mg/ml)	malate	50 mmol/l
0.02 ml	enzyme in 0.1% bovine serum albumin		

Stability: crystalline suspensions in ammonium sulphate solution, 3.2 mol/l; pH ca. 7.5, are stable for several months at 4 °C.

Purity required: specific activity ≥ 350 U/mg protein (25 °C). Contaminant (related to the specific activity of fumarase) $\leq 0.02\%$ MDH.

References

[1] *R. M. Bock, R. A. Alberty,* J. Am. Chem. Soc. *75,* 1921 (1953).
[2] *D. A. Brant, L. B. Barnett, R. A. Alberty,* J. Am. Chem. Soc. *85,* 2204 (1963).
[3] *V. Massey,* Biochem. J. *55,* 172 (1953).

GABAse

(enzyme mixture)

from *Pseudomonas fluorescens,* containing

4-Aminobutyrate: 2-oxoglutarate aminotransferase, EC 2.6.1.19

Succinate-semialdehyde: NAD(P) oxidoreductase, EC 1.2.1.16

$$\text{4-Aminobutyrate + 2-oxoglutarate} \xrightleftharpoons{\text{GAB-AT}}$$
$$\text{L-glutamate + succinate semialdehyde}$$

$$\text{Succinate semialdehyde + NAD(P)}^+ + H_2O \xrightarrow{\text{SS-DH}} \text{succinate}$$
$$+ \text{NADP(H)} + H^+$$

Equilibrium [1, 2]: the transamination reaction is reversible, the semialdehyde oxidation irreversible.

Michaelis constants [1, 2]

GAB-GT	
4-Aminobutyrate	3×10^{-3} mol/l
SS-DH	
Succinate semialdehyde	5.6×10^{-3} mol/l
NADP	2.8×10^{-6} mol/l

Inhibitor constants [2]

		K_1
SS-DH	Glyoxalate	3×10^{-4} mol/l
	Malonate semialdehyde	1×10^{-4} mol/l

Effectors [1, 2]: hydroxylamine, 4-chloromercuribenzoate, KCN and aminoacetate strongly inhibit the aminobutyrate aminotransferase. Glyoxalate and malonate semialdehyde inhibit the succinate-semialdehyde dehydrogenase.

Specificity [1, 2]: GAB-GT is highly specific for 4-aminobutyrate. β-Alanine, 5-aminovalerate, 6-aminocaproate, ω-aminocaprylate, ornithine or lysine do not react and only 3-hydroxy-4-aminobutyrate can replace the substrate to a limited extent.

SS-DH is specific for the oxidation of succinate semialdehyde and shows no reaction with a wide variety of aliphatic and aromatic aldehydes (including glyoxalate- and malonate semialdehyde).

NADP is 8.2 times as active as NAD.

pH Optimum [2]: GAB-GT   pH 7.4 – 8.8
              SS-DH   pH 8.5

Assay

$$\text{4-Aminobutyrate + 2-oxoglutarate} \xrightleftharpoons{\text{GAB-AT}}$$

$$\text{L-glutamate + succinate semialdehyde}$$

$$\text{Succinate semialdehyde + NADP}^+ + \text{H}_2\text{O} \xrightarrow{\text{SS-DH}}$$

$$\text{succinate + NADPH + H}^+.$$

Wavelength 339, Hg 334 or Hg 365 nm; light path 10 mm; final volume 2.95 ml; 25 °C.

2.30 ml	potassium diphosphate buffer (0.15 mol/l; pH 8.0)	117 mmol/l
0.10 ml	mercaptoethanol (0.10 mol/l)	3.4 mmol/l
0.20 ml	NADP, Na salt, (20 mg/ml)	1.7 mmol/l
0.20 ml	4-aminobutyric acid (873 mmol/l; pH adjusted to 8.0)	59 mmol/l
0.10 ml	2-oxoglutarate-Na$_2$ · 2 H$_2$O (38.4 mg/ml)	5.8 mmol/l
0.05 ml	enzyme in water	

Stability: in the lyophilized state a decrease in activity of approx. 10% may occur within 12 months at 4 °C.

Purity required: activity $\geqslant$ 0.5 U/mg substance (25 °C); $\triangleq$ ca. 4 U/mg enzyme protein.

References

[1] *W. B. Jakoby,* in: *S. P. Colowick, N. O. Kaplan* (eds.), Methods in Enzymology, Vol. *V*, Academic Press, New York 1962, p. 771.
[2] *E. M. Scott, W. B. Jakoby,* J. Biol. Chem. *234*, 932 – 940 (1959).

β-Galactose Dehydrogenase

from *Pseudomonas fluorescens*

D-Galactose: NAD 1-oxidoreductase, EC 1.1.1.48

$$\beta\text{-}D\text{-Galactose} + NAD^+ + H_2O \xrightleftharpoons{\text{galactose-DH}}$$

$$D\text{-galactono-}\gamma\text{-lactone} + NADH + H^+$$

Equilibrium constants (Tris buffer, pH 6.7; 30 °C) [1]

$$K = \frac{[D\text{-galactonolactone}] \times [NADH] \times [H^+]}{[D\text{-galactose}] \times [NAD^+]} = 570 \qquad \text{mol/l}$$

Michaelis constants (Tris buffer, pH 8.6; 30 °C) [1]

		Relative rate
D-Galactose	3.7×10^{-3} mol/l	3.0
L-Arabinose	1.0×10^{-2} mol/l	1.0
NAD	2.65×10^{-4} mol/l	

Molecular weight [2]: 100 000.

Specificity [1]: the enzyme also reacts with 6-*O*-methyl-D-galactose, D-fucose and 3-deoxy-D-galactose. 2-Deoxy-D-galactose, 2-deoxy-2-amino-D-galactose, 3,6-dideoxy-D-galactose (abequose) and L-arabinose react more slowly than D-galactose, but the reaction goes to completion. D-Glucose, L-glucose, D-mannose, D-ribose, D-xylose, α-D-mannoheptose, D-glucose-6-P, D-galactose-6-P, D-galacturonate or D-galactosamine do not react. The rate with NADP is ca. 2% of that with NAD.

Assay

Wavelength 339, Hg 334 or Hg 365 nm; light path 10 mm; final volume 3.23 ml; 25 °C.

3.00 ml	Tris buffer (0.1 mol/l; pH 8.6)	93 mmol/l
0.10 ml	NAD (50 mg/ml)	2.25 mmol/l
0.10 ml	D-galactose (100 mg/ml)	18.5 mmol/l
0.03 ml	enzyme in ammonium sulphate, 1 mol/l	

Stability: suspensions in ammonium sulphate solution, 2.2 mol/l; pH ca. 6, containing EDTA, 1 mmol/l, are stable for several months at 4 °C.

Purity required: specific activity $\geq$ 5 U/mg protein (25 °C). Contaminants (related to the activity of galactose-DH $\leq$ 0.01% ADH and β-galactosidase, $\leq$0.1% "NADH oxidase" and $\leq$0.5% LDH. Reaction with glucose (side reaction of the enzyme) $\leq$0.1%.

References

[1] *M. Doudoroff,* in: *S. P. Colowick, N. O. Kaplan* (eds.), Methods in Enzymology, Vol. *V,* Academic Press, New York 1962, p. 339.
[2] *K. Wallenfels, G. Kurz,* in: *S. P. Colowick, N. O. Kaplan* (eds.), Methods in Enzymology, Vol. *IX,* Academic Press, New York 1966, p. 112.

Galactose Oxidase

from *Polyporus circinatus*

D-Galactose: oxygen 6-oxidoreductase, EC 1.1.3.9

$$\text{D-Galactose} + O_2 \xrightarrow{\text{galactose oxidase}} \text{D-galacto-hexodialdose} + H_2O_2$$

Michaelis constants: (30 °C) [1]

		Relative rate
D-Galactose	2.4×10^{-1} mol/l	1.00
2-Deoxy-D-galactose	4.5×10^{-1} mol/l	0.32
Lactose	$>5.0 \times 10^{-1}$ mol/l	0.02
Melibiose	4.5×10^{-2} mol/l	0.80
Raffinose	2.5×10^{-2} mol/l	1.80
Stachyose	1.3×10^{-2} mol/l	6.10
Guaran	3.1×10^{-4} mol/l	1.80

Effectors: cyanide, diethyldithiocarbamate, azide, and hydroxylamine are inhibiting.

Molecular weight (pH 8.6) [3]: 68000.

Specificity: the enzyme from *P. circinatus* is inactive with D-glucose, L-galactose, L-arabinose, D-galacturonate, etc. [1].

Assay

$$\text{D-Galactose} + O_2 \xrightarrow{\text{galactose oxidase}} \text{galacto-hexodialdose} + H_2O_2$$

$$H_2O_2 + DH_2 \xrightarrow{\text{POD}} D + 2\,H_2O$$

DH_2 = leuco-dye, D = dye.

A unit is the catalytic activity which causes a change in absorbance of 1.0 per min under the conditions of the assay.

Wavelength 400 nm; light path 10 mm; final volume 4.00 ml; 25 °C.

1.90 ml	water		
0.10 ml	enzyme in water		
2.00 ml	reagent (phosphate buffer, 10 mmol/l;	phosphate	5 mmol/l
	pH 7.0; o-dianisidine, 0.2 mg/ml;	dianisidine	0.01 mg/ml
	galactose, 10 mg/ml; POD, 0.05 mg/ml,	galactose	28 mmol/l
	ca. 100 U/mg)	POD	2.5 kU/l

Stability: dry powder is stable for several months at 4 °C.

Purity required: activity $\geqq$ 5 units/mg substance (25 °C).

References

[1] G. Avigad et al., J. Biol. Chem. 237, 2736 (1962).
[2] D. Amaral et al., J. Biol. Chem. 238, 2281 (1963).
[3] D. J. Kosman et al., Arch. Biochem. Biophys. 165, 456 (1974).

α-Galactosidase

from green coffee beans

α-D-Galactosidase galactohydrolase, EC 3.2.1.22

$$\text{4-Nitrophenyl-}\alpha\text{-D-galactoside} + H_2O \xrightarrow{\;\;\alpha\text{-galactosidase}\;\;} \text{D-galactose} + \text{4-nitrophenol}$$

Michaelis constants

Phenyl-α-D-galactoside	1	$\times\ 10^{-3}$ mol/l	citrate buffer, pH 5.3 [1]
Phenyl-α-D-galactoside	3.8	$\times\ 10^{-3}$ mol/l	citrate buffer, pH 6.3; 37 °C [2][+]
4-Nitrophenyl-α-D-galactoside	3.3	$\times\ 10^{-4}$ mol/l	citrate buffer, pH 5.5; 30 °C [3]*
4-Nitrophenyl-α-D-galactoside	4.3	$\times\ 10^{-4}$ mol/l	acetate buffer, pH 5.8; 25 °C [4]**
4-Nitrophenyl-α-D-galactoside	6.57	$\times\ 10^{-4}$ mol/l	citrate buffer, pH 5.5; 30 °C [5]***

[+] For the enzyme from *Vicia sativa*.
* For the enzyme from *Cicer arietinum*.
** For the enzyme from *Mortierella vinacea*.
*** For the enzyme from *Phaseolus vulgaris*.

4-Nitrophenyl-α-D-galactoside	5.3×10^{-4} mol/l	acetate buffer; pH 5.5; 30°C [6][++]
Melibiose	3.9×10^{-4} mol/l	acetate buffer, pH 5.8; 25°C [3]*
Melibiose	2.24×10^{-3} mol/l	acetate buffer, pH 5.5; 30°C [6][++]
Raffinose	1.8×10^{-3} mol/l	acetate buffer, pH 4.4; 25°C [3]*
Raffinose	5.8×10^{-2} mol/l	citrate buffer, pH 6.3; 37°C [2][+]
Raffinose	1.25×10^{-2} mol/l	acetate buffer, pH 5.5; 30°C [6][++]

Effectors: the enzyme is inhibited by the D-galactose liberated, Hg^{2+}-, Ag^+-ions and myo-inositol [1, 3, 2].

Molecular weights: 30000 [2][+], 33000 [6][++], 26000[+++].

Specificity: α-galactosidase hydrolyzes a large number of α-D-galactopyranosidic linkages:

- glycosides of methanol, phenol and 4-nitrophenol
- disaccharides and derivatives: melibiose, epimelibiose, planteobiose, and melibionic acid
- trisaccharides: manninotriose, umbellipherose, raffinose, and planteose
- oligosaccharides, which contain several galactosylic residues, are split by hydrolyzing the non-substituted galactosylic residue from the non-reducing end of the chain. Stachiose is hydrolyzed first into galactose and raffinose, then raffinose into galactose and sucrose. For further details, cf. [1].

Assay

Wavelength Hg 405 nm; light path 10 mm; incubation volume 1.02 ml; 25°C; $\varepsilon = 1.85$ l × mmol^{-1} × mm^{-1}.

0.80 ml	phosphate buffer (0.1 mol/l; pH 6.5)	78.4 mmol/l
0.20 ml	4-nitrophenyl-α-D-galactoside (3 mg/ml)	2 mmol/l
0.02 ml	enzyme in buffer	

incubate for 5 min, stop the reaction by addition of

2.00 ml   borate buffer (0.2 mmol/l; pH 9.8)

read absorbance against blank.

[+] For the enzyme from *Vicia sativa*.   [+++] For the enzyme from coffee beans.
[++] For the enzyme from sweet almonds.   * For the enzyme from *Mortierella vinacea*.

Stability: suspensions in ammonium sulphate solution, 3.2 mmol/l; pH approx. 6, are stable for several months at 4 °C. Highly purified lyophilized enzymes (for EIA) stabilized with phosphate and sucrose are stable for many months at −20°C in the dry state under nitrogen.

Purity required: specific activity $\geq$ 10 U/mg protein (25 °C; 4-nitrophenyl-α-D-galactosidase as substrate). Contaminants (related to the specific activity of α-galactosidase) $\leq$ 0.05% N-acetyl-β-D-glucosaminidase, β-galactosidase, α-glucosidase, β-glucosidase and α-mannosidase (4-nitrophenylglycosides as substrate).

Purity required as enzyme label for enzyme-immunoassay: specific activity $\geq$ 600 U/mg protein (35 °C; 2-nitrophenyl-β-D-galactoside as substrate, mercaptoethanol as antioxidant; $\varepsilon_{2\text{-nitrophenol};\,405} = 0.35$ l × mmol^{-1} × mm^{-1}). There should be at least 12 moles of SH groups per mole enzyme available for derivatization (measured with *Ellman*'s Reagent).

References

[1] *J. E. Courtois, F. Petek,* in: *S. P. Colowick, N. O. Kaplan* (eds.), Methods in Enzymology, Vol. *VIII*, Academic Press, New York 1966, p. 565.
[2] *Ch. B. Sharma,* Biochem. Biophys. Res. Commun. *43*, 572 (1971).
[3] *H. Suzuki, S. C. Li, Y. T. Li,* J. Biol. Chem. *245*, 781 (1970).
[4] *K. M. L. Agrawal, O. P. Bahl,* J. Biol. Chem. *243*, 103 (1968).
[5] *F. Petek, E. Villarroya, J. E. Courtois,* Eur. J. Biochem. *8*, 395 (1968).
[6] *O. P. Mulhotra, P. M. Dey,* Biochem. J. *103*, 508 (1967).
[7] *P. M. Dey, J. B. Pridham,* Adv. Enzymology *36*, 91 (1972).

β-Galactosidase

from *Escherichia coli*

β-D-galactoside galactohydrolase, EC 3.2.1.23

$$\text{Lactose} + H_2O \xrightarrow{\text{β-galactosidase}} \text{β-D-galactose} + \text{D-glucose}$$

Michaelis constants (Tris buffer, pH 7.6; 20 °C)

		Relative rate
2-Nitrophenyl-β-galactoside	9.5×10^{-4} mol/l	1.00
Phenyl-β-D-galactoside	3.23×10^{-3} mol/l	0.05
Lactose	3.85×10^{-3} mol/l	0.06
4-Nitrophenyl-β-D-galactoside	4.45×10^{-4} mol/l	0.14

Effectors: Na ions inhibit the enzyme.

Molecular weight: 540000 (pH 7.5) [1]; 518000 (pH 7.6) [2].

Specificity: the enzyme hydrolyzes β-D-galactosides.

Assay

$$\text{Lactose} + H_2O \xrightarrow{\text{β-galactosidase}} \text{β-D-galactose} + \text{D-glucose}$$

$$\text{β-D-Galactose} + NAD^+ + H_2O \xrightarrow{\text{galactose-DH}}$$
$$\text{D-galactono-γ-lactone} + NADH + H^+.$$

Wavelength 339, Hg 334 or Hg 365 nm; light path 10 mm; final volume 3.16 ml; 25 °C.

2.50 ml	potassium phosphate buffer (0.1 mol/l; pH 7.0)	79 mmol/l
0.50 ml	lactose (70 mg/ml)	34 mmol/l
0.10 ml	NAD (50 mg/ml)	2.2 mmol/l
0.03 ml	MgSO$_4$ (0.1 mol/l)	1 mmol/l
0.01 ml	galactose-DH (5 mg/ml)	79 U/l
0.02 ml	enzyme in buffer	

Stability: crystalline suspensions in ammonium sulphate solution, 2.2 mol/l; pH ca. 6, are stable for several months at 4 °C.

Purity required: specific activity $\geq$ 30 U/mg protein (25 °C). Contaminants (related to the specific activity of β-galactosidase) $\leq$ 0.01% β-fructosidase, α-galactosidase, glucose-DH, α-glucosidase and "NADH oxidase".

References

[1] *G. R. Craven, E. Steers, C. B. Anfinsen,* J. Biol. Chem. *240,* 2468, 2478 (1965).
[2] *K. Weber, H. Sund, K. Wallenfels,* Biochem. Z. *339,* 498 (1964).

Gluconate Kinase

from *Escherichia coli*

ATP: D-gluconate 6-phosphotransferase, EC 2.7.1.12

$$\text{D-Gluconate} + ATP \xrightarrow{\text{gluconate kinase}} \text{D-gluconate-6-P} + ADP$$

Specificity: the enzyme is specific for D-gluconate; D-mannonate, 2-oxo-D-gluconate, D-altronate, D-glucuronate, D-arabinose, D-isoascorbate, D-ribonate, D-arabonate, D-ribose, D-galactonate, L-idonate, 5-oxo-D-gluconate and D-xylonate do not react [1].

Assay

$$\text{D-Gluconate} + \text{ATP} \xrightarrow{\text{gluconate kinase}} \text{D-gluconate-6-P} + \text{ADP}$$

$$\text{D-Gluconate-6-P} + \text{NADP}^+ \xrightarrow{\text{6-PGDH}}$$

$$\text{D-ribulose-5-P} + \text{NADPH} + \text{CO}_2 + \text{H}^+.$$

Wavelength 339, Hg 334 or Hg 365 nm; light path 10 mm; final volume 2.93 ml; 25 °C.

2.50 ml	glycylglycine buffer (0.1 mol/l; pH 8.0)	86 mmol/l
0.10 ml	MgCl$_2$ (0.1 mol/l)	3.3 mmol/l
0.10 ml	ATP, Na salt, (60 mg/ml; neutralized)	3.2 mmol/l
0.10 ml	NADP, Na salt, (10 mg/ml)	0.38 mmol/l
0.10 ml	gluconate, Na salt, (20 mg/ml)	0.3 mmol/l
0.01 ml	6-PGDH (10 mg/ml)	1.2 kU/l
0.02 ml	enzyme in water	

Stability: suspensions in ammonium sulphate solution, 3.2 mol/l, pH ca. 6, are stable for several months at 4 °C.

Purity required: specific activity $\geq$ 40 U/mg protein (25 °C). Contaminants (related to the specific activity of gluconate kinase) $\leq$ 0.05% G6P-DH, HK and "NADPH oxidase".

Reference

[1] *S. S. Cohen*, in: *S. P. Colowick, N. O. Kaplan* (eds.), Methods in Enzymology, Vol. *I*, Academic Press, New York 1955, p. 350.

Glucose Dehydrogenase

from *Bacillus megaterium*

β-D-Glucose: NAD(P)1-oxidoreductase, EC 1.1.1.47

$$\text{β-D-Glucose} + \text{NAD(P)}^+ \longrightarrow \text{D-glucono-δ-lactone} + \text{NAD(P)H} + \text{H}^+$$

Michaelis constants [1]

NAD	4.5×10^{-3} mol/l
Glucose	47.5×10^{-3} mol/l

Effectors [1]: the enzyme is insensitive to sulphydryl-group inhibitors, heavy metal ions and chelating agents. N-acetyl-2-deoxy-2-amino-D-glucose and myo-inositol inhibit. In contrast to the enzyme from mammalian liver [2], the *Bacillus* enzyme is unaffected by a great number of metabolic intermediates including fructose, glucose-6-P and glucose-1-P.

Molecular weight [1]: ca. 118000.

Specificity [1, 3]: D-Glucose dehydrogenase is highly specific for β-D-glucose and is capable of using either NAD or NADP.

Substrate	Relative rate (%)
2-Deoxyglucose	125
β-D-Glucose	100
D-Glucosamine	31
D-Xylose	15
D-Mannose	8
6-Amino-6-deoxyglucose	6
Cellobiose	1
D-Ribose	0.8
Lactose	0.7
2-Deoxyribose	0.1

pH Optimum [1]: pH 8 in Tris-HCl buffer.
pH 9 in acetate/borate buffer.

Assay

Wavelength 339, Hg 334 or Hg 365 nm; light path 10 mm; final volume 3.15 ml; 25 °C.

2.0 ml	phosphate buffer (0.12 mol/l; pH 7.6; NaCl, 0.15 mol/l)		phosphate NaCl	0.12 mol/l 0.15 mol/l
1.0 ml	glucose (0.1 g/ml buffer)			0.174 mol/l
0.1 ml	enzyme in buffer			
0.05 ml	NAD (80 mg/ml)			1.9 mmol/l

Stability: the lyophilized enzyme is stable for many months at 4 °C.

Purity required: specific activity $\geq$ 0.2 U/mg protein (25 °C). Contaminant (related to the specific activity of glucose dehydrogenase) $\leq$ 0.005% "NADH oxidase".

References

[1] *H. E. Pauly, G. Pfleiderer,* Hoppe-Seyler's Z. physiol. Chem. *356*, 1613 (1975).
[2] *R. P. Metzger et al.,* J. Biol. Chem. *239*, 1769 (1964).
[3] *D. Banauch et al.,* Z. Klin. Chem. Klin. Biochem. *13*, 101 (1975).

Glucose Oxidase

from moulds

β-D-Glucose: oxygen 1-oxidoreductase, EC 1.1.3.4

$$\beta\text{-D-Glucose} + H_2O + O_2 \xrightarrow{\text{GOD}} \text{D-glucono-}\delta\text{-lactone} + H_2O_2$$

Michaelis constant

GOD from *A. niger* (phosphate buffer, pH 5.6; 25 °C, air):

Glucose	3.3×10^{-2} mol/l	– [1]
Glucose	1.1×10^{-1} mol/l	1.00 [2]
2-Deoxyglucose	2.5×10^{-2} mol/l	0.06
O_2	2.0×10^{-4} mol/l	–

GOD from *P. notatum* (phosphate buffer, pH 5.6; 20 °C, air):

Glucose 9.6×10^{-3} mol/l [1]

Effectors: Ag^+, Hg^{2+}, Cu^{2+} show an inhibitory effect [3]. FAD binding is inhibited by several nucleotides [4].

Molecular weight GOD from *Aspergillus niger*: 160000 [5], from *Penicillium notatum*: 152000 [1].

Specificity: the enzyme is specific for β-D-glucose. GOD from *P. notatum* oxidizes the following sugars: β-D-glucose (1.00), 2-deoxy-D-glucose (0.25), 2-deoxy-6-fluoro-D-glucose (0.03), 6-methyl-D-glucose (0.02), 4,6-dimethyl-D-glucose (0.01), D-mannose (0.01), D-xylose, (0.01), α-D-glucose (0.006) and trehalose (0.003). O_2 can be replaced by hydrogen acceptors such as 2,6-dichlorophenol indophenol.

Assay

$$\text{D-Glucose} + H_2O + O_2 \xrightarrow{\text{GOD}} \text{D-glucono-}\delta\text{-lactone} + H_2O_2$$

$$H_2O_2 + DH_2 \xrightarrow{\text{POD}} 2\,H_2O + D$$

$$DH_2 = \text{leuko-dye};\ D = \text{dye}.$$

Wavelength Hg 436 nm; light path 10 mm; final volume 3.03 ml; 25 °C; $\varepsilon = 0.83\ l \times mmol^{-1} \times mm^{-1}$.

2.50 ml	phosphate buffer (0.1 mol/l; pH 7.0; 0.066 mg o-dianisidine-HCl/ml; O_2-saturated)	82.7 mmol/l 0.17 mmol/l
0.50 ml	D-glucose (100 mg/ml)	92.7 mmol/l
0.01 ml	POD (2 mg/ml)	1.2 kU/l
0.02 ml	enzyme in phosphate buffer	

Stability: the lyophilized enzyme is stable for several months at 4°C.

Purity required: activity $\geq$ 200 U/mg substance (25 °C). Contaminants (related to the specific activity of GOD) $\leq$ 0.01% amylase and saccharase, 10 U catalase/mg. For many purposes less pure preparations satisfy the requirements, e.g. with an activity of ca. 20 U/mg (25 °C) and 0.1% amylase and saccharase.

References

[1] *B. E. P. Swoboda, V. Massey*, J. Biol. Chem. *240*, 2209 (1965).
[2] *H. Q. Gibson, B. E. P. Swoboda, V. Massey*, J. Biol. Chem. *239*, 3927 (1964).
[3] *S. Nakamura, O. Ogura*, J. Biochem. *63*, 308 (1968).
[4] *B. E. P. Swoboda*, Biochim. Biophys. Acta *175*, 365 (1969).
[5] *H. Tsuge, O. Natsuaki, K. Ohashi*, J. Biochem. *78*, 835 (1975).

Glucose-6-phosphate Dehydrogenase

from yeast

D-Glucose 6-phosphate: NADP$^+$ 1-oxidoreductase, EC 1.1.1.49

$$\text{D-Glucose-6-P} + \text{NADP}^+ \xrightarrow{\text{G6P-DH}} \text{D-gluconate-6-P} + \text{NADPH} + \text{H}^+$$

Equilibrium constant: the oxidation reaction is favoured.

Michaelis constants

glucose-6-P	2×10^{-5} mol/l	Tris buffer, pH 8.0; 38 °C
NADP	2×10^{-6} mol/l	Mg^{2+}, 10 mmol/l [1]
glucose-6-P	2.7×10^{-4} mol/l	
NADP	5.6×10^{-5} mol/l	TEA buffer, pH 7.6, 25 °C
Galactose-6-P	1.4×10^{-2} mol/l	Mg^{2+}, 9.3 mmol/l [2]

Inhibitor constants

P_i	1×10^{-1} mol/l	Tris buffer, pH 8.0; 25 °C
D-Glucosamine-6-P	7.2×10^{-4} mol/l	MgCl$_2$, 10 mmol/l [3]
NADPH	2.7×10^{-5} mol/l	
Stearoyl-CoA	4.0×10^{-7} mol/l	TEA buffer, pH 7.6;
Palmitoyl-CoA	6.0×10^{-7} mol/l	25 °C [4]
Lauroyl-CoA	8.7×10^{-7} mol/l	

Effectors [2]: Mg^{2+} in a concentration of 5–10 mmol/l activates the enzyme, higher concentrations of Mg^{2+} and other divalent ions inhibit. Nucleoside monophosphates and several nucleoside di- and tri-phosphates such as ADP, ATP, GTP and UTP also inhibit.

Molecular weights: 128 000 [5], 102 000 (for the NADP-free-enzyme) [6], 104 000 [2].

Specificity: the enzyme is highly specific for glucose-6-P. The following react: glucose-6-P (100), galactose-6-P (0.25), 2-deoxyglucose-6-P (0.18) and glucosamine-6-P (0.02) [7].

Assay

Wavelength 339, Hg 334 or Hg 365 nm; light path 10 mm; final volume 2.92 ml; 25 °C.

2.50 ml	triethanolamine buffer (0.1 mol/l; pH 7.6)	86 mmol/l
0.20 ml	MgCl$_2$ (0.1 mol/l)	6.9 mmol/l
0.10 ml	glucose-6-P, Na salt, (10 mg/ml)	1.0 mmol/l
0.10 ml	NADP, Na salt, (10 mg/ml)	0.39 mmol/l
0.02 ml	enzyme in buffer	

Stability: crystalline suspensions in ammonium sulphate solution, 3.2 mol/l; pH ca. 6, are stable for several months at 4 °C. Solutions or suspensions can be dialyzed against phosphate buffer, 50 mmol/l, pH 7.6 at 4 °C without significant loss of activity.

Purity required: specific activity $\geq$ 140 U/mg protein (25 °C). Contaminants (related to the specific activity of G6P-DH) $\leq 0.01\%$ 6-PGDH and PGluM, $\leq 0.02\%$ HK and PGI, $\leq 0.2\%$ GR.

References

[1] O. H. Lowry et al., J. Biol. Chem. 236, 2746 (1961).
[2] G. F. Domagk et al., Hoppe-Seyler's Z. physiol. Chem. 350, 626 (1969).
[3] L. Glaser, D. H. Brown, J. Biol. Chem. 216, 67 (1955).
[4] T. J. Eger-Neufeldt et al., Biochem. Biophys. Res. Commun. 19, 43 (1965).
[5] P. Andrews, Biochem. J. 96, 595 (1965).
[6] R. H. Yue, E. A. Noltmann, S. A. Kuby, Biochemistry 6, 1174 (1967).
[7] H. Greiling, R. Kisters, Hoppe-Seyler's Z. physiol. Chem. 341, 172 (1965).

Glucose-6-phosphate Dehydrogenase

from *Leuconostoc mesenteroides*

D-Glucose 6-phosphate: NADP$^+$ 1-oxidoreductase, EC 1.1.1.49

$$\text{D-Glucose-6-P} + \text{NAD(P)}^+ \xrightarrow{\text{G6P-DH}}$$

$$\text{D-glucono-}\delta\text{-lactone-6-P} + \text{NAD(P)H} + \text{H}^+$$

Equilibrium constant: the forward reaction is strongly favoured.

Michaelis constants [1]

NAD	1.15×10^{-4} mol/l
NADP	7.4×10^{-6} mol/l
Glucose-6-P	3.6×10^{-5} mol/l (NADP as coenzyme)
Glucose-6-P	6.4×10^{-5} mol/l (NAD as coenzyme)

Effectors: Mg^{2+}, HCO_3^- up to concentrations of 0.3 mol/l are slightly activating. Phosphate inhibits at a concentration of 5×10^{-3} mol/l, NADPH is a competitive inhibitor in the NAD-dependent reaction [1, 2].

Unlike to the yeast enzyme, myristic acid, dehydroepiandrosterone and palmitoyl-CoA do not inhibit.

Molecular weight [3]: ca. 110000.

Specificity: the enzyme is highly specific for glucose-6-P and does not react with fructose-6-P, fructose-1.6-P$_2$, glucose-1-P or ribose-5-P. There is a slow unspecific reaction with D-glucose [1]. 2-Deoxyglucose-6-P is oxidized with NADP, but not with NAD [4]. The relative rate with NAD:NADP as coenzyme is 1.8:1 [1]. 3'-NADP is not reduced.

pH Optimum [5]: 7 – 8.5 with maximal activity at pH 7.8.

Assay

Wavelength 339, Hg 334 or Hg 365 nm; light path 10 mm; final volume: 3.05 ml: 25°C or 30°C.

2.60 ml	Tris buffer (0.1 mol/l; pH 7.8; MgCl$_2$ 3 mmol/l)	Tris	85 mmol/l
		Mg^{2+}	2.6 mmol/l
0.20 ml	NAD, free acid (70 mg/ml)		6.9 mmol/l
0.20 ml	glucose-6-P, Na salt, (0.150 mol/l)		9.8 mmol/l
0.05 ml	enzyme in water		

Stability: suspensions in ammonium sulphate solution, 3.2 mol/l; pH ca. 6, are stable for many months at 4°C.

Purity required: specific activity $\geqslant$ 550 U/mg protein (25°C) or $\geqslant$ 650 U/mg protein (30°C). Contaminants (related to the specific activity of G6P-DH) $\leqslant$ 0.001% CK, $\leqslant$ 0.01% GR, 6-PGDH and PGI, $\leqslant$ 0.02% "NADH oxidase", $\leqslant$ 0.05% HK.

References

[1] *Ch. Olive, H. R. Levy,* Biochemistry *6*, 730 (1967).
[2] *R. D. Moss,* in: *S. P. Colowick, N. O. Kaplan* (eds.), Methods in Enzymology, Vol. *I*, Academic Press, New York 1955, p. 328.
[3] *H. Möllering, Boehringer Mannheim GmbH,* unpublished.
[4] *H. Möllering, Boehringer Mannheim GmbH,* unpublished.
[5] *E. A. Noltmann, S. A. Kuby,* in: *P. D. Boyer, H. Lardy, K. Myrbäck* (eds.), The Enzymes, Vol. *7*, Academic Press, New York 1963, p. 223.

α-Glucosidase

(Maltase) from yeast

α-D-Glucoside glucohydrolase, EC 3.2.1.20

$$\alpha\text{-D-Glucoside} + H_2O \xrightleftharpoons{\alpha\text{-glucosidase}} \text{alcohol} + \text{D-glucose}$$

Equilibrium constant: the reversibility of the reaction has not been studied. The equilibrium lies to the right [1].

Molecular weight [1]: 68500.

Specificity: the enzyme specifically hydrolyzes α-glucosidic linkages, preferably α-1,4-linkages; α-1,6-linkages are only attacked slowly, while β-glucosidic linkages are not attacked [3]. Maltotriose is transformed at two thirds of the rate with maltose.

Assay

$$\text{Maltose} + H_2O \xrightarrow{\alpha\text{-glucosidase}} 2\ \text{D-glucose}$$

$$2\ \text{D-Glucose} + 2\ \text{ATP} \xrightarrow{HK} 2\ \text{glucose-6-P} + 2\ \text{ADP}$$

$$2\ \text{Glucose-6-P} + 2\ \text{NADP}^+ \xrightarrow{G6P\text{-}DH}$$
$$2\ \text{gluconate-6-P} + 2\ \text{NADPH} + 2\ H^+.$$

Wavelength 339, Hg 334 or Hg 365 nm; light path 10 mm; incubation volume 2.00 ml; final volume 3.00 ml; 25 °C.

Incubation mixture

1.49 ml	acetate buffer (0.1 mol/l; pH 6.0)	74.5 mmol/l
0.50 ml	maltose (200 mg/ml)	0.146 mol/l
0.01 ml	enzyme in water	

incubate for exactly 5 min at 25 °C, then place for 3 min in a boiling water-bath and centrifuge.

Assay mixture

2.56 ml	triethanolamine buffer (0.3 mol/l; pH 7.6)	0.256 mol/l
0.10 ml	MgCl$_2$ (0.1 mol/l)	3.3 mmol/l
0.10 ml	ATP, Na salt, (10 mg/ml)	0.55 mmol/l
0.10 ml	NADP, Na salt, (10 mg/ml)	0.38 mmol/l
0.10 ml	supernatant	
0.01 ml	G6P-DH (1 mg/ml)	0.47 kU/l
0.03 ml	HK (2 mg/ml)	2.8 kU/l

Stability: crystalline suspensions in ammonium sulphate solution, 3.2 mol/l; pH ca. 6, are stable for several months at 4 °C.

Purity required: specific activity $\geq$ 50 U/mg protein (25 °C). Contaminants (related to the specific activity of α-glucosidase) $\leq$ 0.01% α-galactosidase, β-galactosidase and β-glucosidase.

References

[1] *H. Halvorson,* in: *S. P. Colowick, N. O. Kaplan* (eds.), Methods in Enzymology, Vol. *III*, Academic Press, New York 1966, p. 559.
[2] *N. A. Khan, N. R. Eaton,* Biochim. Biophys. Acta *146*, 173 (1967).
[3] *W. Fischer,* unpublished.

β-Glucuronidase

from snails (*Helix pomatia*)

β-D-Glucuronide glucuronosohydrolase, EC 3.2.1.31

$$\text{A β-D-Glucuronide} + H_2O \xrightarrow{\text{β-glucuronidase}} \text{an alcohol} + \text{D-glucuronate}$$

Michaelis constants (pH 4.5)

Phenolphthalein-β-glucuronide 3.9×10^{-4} mol/l
Oestriol glucuronide 4.2×10^{-4} mol/l.

Effectors: deoxyribonucleic acid and numerous diamines are activators; D-glucuronate, D-galacturonate and D-saccharo-1,4-lactone are competitive inhibitors.

Molecular weight (β-glucuronidase from bovine liver [1]: 290000.

Specificity: β-glucuronidase hydrolyzes a large number of conjugated glucuronides, but does not react with α-glucuronides or β-glucosides. For properties of other β-glucuronidases, see [2, 3].

Assay

4-Nitrophenyl glucuronide + H_2O $\xrightarrow{\text{β-glucuronidase}}$ 4-nitrophenol + glucuronate.

Wavelength Hg 405 nm; light path 10 mm; incubation volume 1.06 ml; 25 °C; $\varepsilon =$ 1.85 l $\times$ mmol^{-1} $\times$ mm^{-1}.

1.00 ml	acetate buffer (0.1 mol/l; pH 4.5)	97 mmol/l
0.03 ml	4-nitrophenyl glucuronide (0.1 mol/l)	3 mmol/l
0.03 ml	enzyme in water	

incubate for 10 min at 25 °C.
Then add 2 ml NaOH, 0.5 mol/l, and read against blank after exactly 1 min.

Stability: stabilized aqueous solutions and suspensions in ammonium sulphate solution, 3.2 mol/l; pH approx. 6, are stable for several months at 4 °C.

Purity required: activity $\geqq$ 3 U/ml (25 °C).

References

[1] *M. Himeno et al.,* J. Biochem. 76, 1243 (1974).
[2] *Th. E. Barman,* Enzyme Handbook, Vol. 2, Springer-Verlag, Berlin 1969, p. 590.
[3] *G. A. Levy, C. A. Marsh,* in: *P. D. Boyer, H. Lardy, K. Myrbäck* (eds.), The Enzymes, Vol. 4, Academic Press, New York 1960, p. 397.

Glutamate Dehydrogenase

from bovine liver

L-Glutamate: NAD(P)$^+$ oxidoreductase (deaminating), EC 1.4.1.3

$$\text{L-Glutamate} + \text{NAD(P)}^+ + H_2O \underset{}{\overset{\text{GlDH}}{\rightleftharpoons}} \text{2-oxoglutarate} + \text{NAD(P)H} + NH_4^+$$

Michaelis constants (Tris buffer, pH 8.0; 23 °C) [1]:

L-Glutamate	1.8×10^{-3} mol/l
NADP	4.7×10^{-5} mol/l
2-Oxoglutarate	7.0×10^{-4} mol/l
NH$_4^+$	3.2×10^{-3} mol/l
NADPH	2.6×10^{-5} mol/l

K_m values for NAD or NADH are difficult to obtain due to their inhibitory action.

Molecular weight [2]: ca. 2200000 for the associated enzyme with 8 subunits, 280000 for one subunit.

Effectors: the oxidation of L-glutamate is stimulated by ADP and inhibited by GTP and diethylstilboestrol. In contrast, the oxidation of alanine, leucine, isoleucine, methionine, valine, norleucine, norvaline and 2-aminobutyrate is stimulated by GTP and diethylstilboestrol and inhibited by ADP [3].

Assay

Wavelength 339, Hg 334 or Hg 365 nm; light path 10 mm; final volume 2.93 ml; 25 °C.

2.50 ml	imidazole buffer (0.1 mol/l; pH 7.3)	85 mmol/l
0.20 ml	2-oxoglutarate (0.2 mol/l)	13.6 mmol/l
0.05 ml	ammonium acetate (12.8 mol/l)	0.22 mol/l
0.03 ml	NADH, Na salt, (10 mg/ml)	0.12 mmol/l
0.10 ml	EDTA, Na salt, (10 mg/ml)	0.9 mmol/l
0.03 ml	ADP, Na salt, (0.1 mol/l)	1 mmol/l
0.02 ml	enzyme in buffer	

Stability: suspensions in ammonium sulphate solution, 2.0 mol/l; pH ca. 7, and solutions in 50% glycerol are stable for many months at 4 °C. Enzyme suspensions or solutions can be dialyzed against phosphate buffer, 10 mmol/l, pH 7.5 for 24 h at 4 °C without loss of activity.

Purity required: specific activity $\geq$ 90 U/mg protein (25 °C). Contaminants (related to the specific activity of GlDH) $\leq$ 0.01% ADH, LDH and MDH.

References

[1] *L. A. Fahlen, B. O. Wiggert, P. P. Cohen,* J. Biol. Chem. *240,* 1083 (1965).
[2] *H. Sund, W. Burchard,* Eur. J. Biochem. *6,* 202 (1968).
[3] *G. M. Tomkins et al.,* J. Biol. Chem. *240,* 3793 (1965).

Glutamate-Oxaloacetate Transaminase,

cf. Aspartate Aminotransferase

Glutamate-Pyruvate Transaminase,

cf. Alanine Aminotransferase

Glutaminase

from *Escherichia coli*

L-Glutamine amidohydrolase, EC 3.5.1.2

$$\text{L-Glutamine} + H_2O \;\xrightleftharpoons{\text{glutaminase}}\; \text{L-glutamate} + NH_3$$

Equilibrium constant (pH 5.5; 20 °C) [1]

$$K = \frac{[NH_4^+] \times [\text{L-glutamate}]}{[\text{L-glutamine}] \times [H_2O]} = 320$$

Effectors: the enzyme is inhibited by heavy metals (e.g. Hg^{2+}, Ag^+, Pb^{2+}, Cu^{2+}). 6-Diazo-5-oxo-L-norleucine as analogue inhibitor causes irreversible inactivation.

Molecular weight [2]: 110 000.

Specificity: the enzyme from *E. coli* hydrolyzes L-glutamine (1.00) and α-methyl-DL-glutamine (0.25). L-Isoglutamine, D- or L-homoglutamine, α-methyl-DL-asparagine, L-isoasparagine, L-leucinamide, DL-prolinamide, DL-alaninamide, L-phenylalaninamide, L-tyrosinamide do not react.

Assay

Wavelength 480 nm; light path 10 mm; incubation volume 0.50 ml; final volume 8.5 ml; 37 °C.

Incubation mixture

0.25 ml	substrate/acetate buffer (0.1 mol/l; pH 4.9; L-glutamine, 80 mmol/l)	acetate	50 mmol/l
		glutamine	40 mmol/l
0.25 ml	enzyme in water		

incubate for 30 min at 37 °C add 0.50 ml 15% trichloroacetic acid and centrifuge off precipitate.

Assay mixture

0.5 ml   supernatant fluid
7.0 ml   water
1.0 ml   *Nessler*'s Reagent

read absorbance

Stability: lyophilized enzymes are stable for some months at -20 °C.

Purity required: activity $\geq$ 0.5 U/mg lyophilized material (37 °C).

References

[1] *T. Benzinger et al.,* Biochem. J. *71,* 400 (1959).
[2] *S. C. Hartmann,* J. Biol. Chem. *243,* 853 (1968).

Glutathione Reductase

from yeast

NAD(P)H: oxidized-glutathione oxidoreductase, EC 1.6.4.2

$$GSSG + NADPH + H^+ \xrightarrow{\text{GR}} 2\,GSH + NADP^+$$

Michaelis constants (phosphate buffer, pH 7.6; 25 °C) [1]

GSSG	6.1×10^{-5} mol/l
NADPH	7.6×10^{-6} mol/l

Turnover number (phosphate buffer, pH 7.6; 25 °C) [1]: oxidation of 1.3×10^4 mole NADPH per min per mole bound FAD.

Effectors: the reaction with NADH is inhibited by NaCl and KCl, and is stimulated by phosphate.

Molecular weights: 118 000 [2], 124 000 [1].

Specificity: the enzyme reacts with NADPH and NADH.

Assay

Wavelength 339, Hg 334 or Hg 365; light path 10 mm; final volume 2.87 ml; 25 °C.

2.50 ml	Tris buffer (0.1 mol/l; pH 8.0)	87.7 mmol/l
0.10 ml	EDTA, Na salt, (10 mg/ml)	0.94 mmol/l
0.20 ml	GSSG (50 mg/ml; pH ca. 7)	4.6 mmol/l
0.05 ml	NADPH, Na salt, (10 mg/ml)	0.16 mmol/l
0.02 ml	enzyme in buffer	

Stability: crystalline suspensions in ammonium sulphate solution, 3.2 mol/l; pH ca. 6, are stable for several months at 4 °C.

Purity required: specific activity $\geq$ 120 U/mg protein (25 °C). Contaminants (related to the specific activity of GR) $\leq$ 0.01% "NADPH oxidase" and 6-PGDH, $\leq$ 0.1% G6P-DH.

References

[1] R. D. Mavis, E. Stellwagen, J. Biol. Chem. 243, 809 (1968).
[2] R. F. Colman, S. Black, J. Biol. Chem. 240, 1796 (1965).

Glyceraldehyde-3-phosphate Dehydrogenase

from rabbit muscle

D-Glyceraldehyde 3-phosphate : NAD⁺ oxidoreductase(phosphorylating), EC 1.2.1.12

$$\text{GAP} + \text{NAD}^+ + \text{P}_i \xrightleftharpoons{\text{GAPDH}} \text{glycerate-1,3-P}_2 + \text{NADH} + \text{H}^+$$

Equilibrium constant (pyrophosphate buffer, pH 7.1) [1]

$$K = \frac{[\text{glycerate-1,3-P}_2] \times [\text{NADH}] \times [\text{H}^+]}{[\text{GAP}] \times [\text{P}_i] \times [\text{NAD}^+]} = 6.7 \times 10^{-8}$$

Michaelis constants

NADH	3.3×10^{-6} mol/l	imidazole buffer
Glycerate-1,3-P_2	8×10^{-7} mol/l	pH 7.4; 26°C

NAD	1.3×10^{-5} mol/l	
GAP	9.0×10^{-5} mol/l	$NaHCO_3$ buffer
P_i	2.9×10^{-4} mol/l	pH 8.6; 26°C
Arsenate	6.9×10^{-5} mol/l	

Inhibitor constants

	Type of inhibition		
NAD	comp. (NADH)	1.0×10^{-4} mol/l	imidazole buffer
GAP	comp. (glycerate-1,3-P_2)	6.0×10^{-8} mol/l	pH 7.4; 26°C [2]
NADH	comp. (NAD)	2.0×10^{-6} mol/l	$NaHCO_3$ buffer
Glycerate-1,3-P_2	comp. (GAP)	2.4×10^{-6} mol/l	pH 8.6; 26°C [2]
D-Threose-2,4-P_2	non comp. (GAP)	2.0×10^{-7} mol/l	pyrophosphate buffer, pH 8.5; 20–23°C [3]

Effectors: the enzyme is inhibited by reagents which react with the sulphydryl group at its catalytic site (e.g. heavy metal ions, 4-chloromercuribenzoate, iodoacetate, tetrathionate, 2-iodosobenzoate). Protective or activating effects are shown by reagents such as EDTA, cysteine and *Cleland's* reagent.

Molecular weight [4]: 144000.

Specificity: the turnover rate of the enzyme from rabbit muscle and yeast with D-glyceraldehyde is ca. 0.001% of the rate with GAP. Arsenate can replace P_i [5, 6]. For other properties of the yeast enzyme, cf. [7].

Assay

$$\text{Glycerate-3-P} + \text{ATP} \xrightarrow{\text{PGK}} \text{glycerate-1,3-}P_2 + \text{ADP}$$

$$\text{Glycerate-1,3-}P_2 + \text{NADH} + \text{H}^+ \xrightarrow{\text{GAPDH}} \text{GAP} + \text{NAD}^+ + P_i.$$

Wavelength 339, Hg 334 or Hg 365 nm; light path 10 mm; final volume 3.03 ml; 25°C.

2.50 ml	triethanolamine buffer (0.1 mol/l; pH 7.6)	82.5 mmol/l
0.20 ml	glycerate-3-P, dicyclohexylammonium salt, (50 mg/ml)	6 mmol/l
0.10 ml	ATP, Na salt, (20 mg/ml)	1.1 mmol/l
0.10 ml	EDTA (10 mg/ml in buffer)	0.9 mmol/l
0.05 ml	$MgSO_4$ (0.1 mol/l)	1.7 mmol/l

0.05 ml	NADH, Na salt, (10 mg/ml)	0.2 mmol/l
0.01 ml	PGK (10 mg/ml)	14.8 kU/l
0.02 ml	enzyme in buffer	

Stability: crystalline suspensions in ammonium sulphate solution, 3.2 mol/l, (containing, EDTA, 0.1 mmol/l); pH ca. 7.5; are stable for several months at 4°C.

Purity required: specific activity $\geq$ 80 U/mg protein (25°C. Contaminants (related to the specific activity of GAPDH) $\leq$ 0.01% GDH, $\leq$ 0.05% PGK, PGM and TIM.

References

[1] *C. F. Cori, S. F. Velick, G. T. Cori,* Biochim. Biophys. Acta *4*, 160 (1950).
[2] *C. S. Furfine, S. F. Velick,* J. Biol. Chem. *240*, 844 (1965).
[3] *A. V. Fluharty, C. E. Ballou,* J. Biol. Chem. *234*, 2517 (1959).
[4] *J. Bode, M. Blumenstein, M. A. Raftery,* Biochem. *14*, 1146 (1975).
[5] *O. Warburg, K. Gawehn, A. W. Geissler,* Z. Naturforschg. *12b*, 47 (1957).
[6] *S. F. Velick, C. Furfine,* in: *P. D. Boyer, H. A. Lardy, K. Myrbäck,* The Enzymes, Vol. 7, Academic Press, New York 1963, p. 243.
[7] *O. Warburg, W. Christian,* Biochem. Z. *303*, 40 (1939).

Glycerol Dehydrogenase

from *Enterobacter aerogenes*

Glycerol: NAD$^+$ 2-oxidoreductase, EC 1.1.1.6

$$\text{Glycerol} + \text{NAD}^+ \rightleftharpoons \text{dihydroxyacetone} + \text{NADH} + \text{H}^+$$

Equilibrium constant [1]

$$K = \frac{[\text{dihydroxyacetone}] \times [\text{NADH}] \times [\text{H}^+]}{[\text{glycerol}] \times [\text{NAD}^+]} = 5.1 \times 10^{-12} \, \text{mol/l}$$

Michaelis constants [1] (PP$_i$-buffer, 0.03 mol/l; pH 9; 25°C)

Glycerol	3.9×10^{-2} mol/l
1.2-Propane-2-diol	5.1×10^{-3} mol/l
2.3-Butane-diol	5.9×10^{-3} mol/l
NAD	1.5×10^{-4} mol/l
Dihydroxyacetone	1.3×10^{-3} mol/l
NADH	1.4×10^{-5} mol/l

Effectors: NH_4^+, K-, Th- and Rb ions activate in both directions [2, 3]. In the presence of higher salt concentrations dihydroxyacetone [4] causes product inhibition at a concentration of 4×10^{-4} mol/l which may be overcome by high concentrations of glycerol in the assay procedure. Zn^{2+}, Na^+, Li^+ and high ionic strength solutions inhibit [4], as well as 2,2-dipyridil, 8-quinolinol [2] and N-ethylmaleimide, iodoacetamide, dithioerythritol and 2-mercaptoethanol [3].

Molecular weight [5]: ca. 340000.

Specificity

Oxidizing activity [1, 2]	Relative rate
Glycerol	100
1.2-Propane-diol	100
2.3-Butane-diol	100
1.3-Butane-diol	37
Ethylene glycol	20
Inositol	18
1.4-Butane-diol	17
Isopropanol	17
Sorbitol	3
Ethanol	1

D-glucose, D-ribose, glycollate, ascorbate, mannitol and erythritol are not oxidized.

Reducing activity [1, 2]	Relative rate
Dihydroxyacetone	100
Methylglyoxal	56
Hydroxy-2-propanone acetate	27
Glycerolaldehyde	14

NAD cannot be substituted by NADP.

pH Optimum [2]: 9.

Assay

Wavelength 339, Hg 334 or Hg 365 nm; light path 10 mm; final volume 3.05 ml; 25°C.

2.40 ml	glycylglycine buffer (glycylglycine, 0.1 mol/l; ammonium sulphate, 0.04 mol/l; pH 9.0)	glycylglycine NH_4^+	79 mmol/l 32 mmol/l
0.30 ml	sucrose (500 mg/ml)		0.14 mol/l
0.20 ml	glycerol (7 mol/l; pH 9.0)		460 mmol/l
0.10 ml	NAD-Li · 2 H_2O (40 mg/ml)		1.9 mmol/l
0.05 ml	enzyme in sucrose (100 mg/ml)		

Stability: suspensions in ammonium sulphate, 3.2 mol/l; pH ca. 7.5, are stable for many months at 4°C.

Purity required: specific activity $\geqslant$ 25 U/mg (25°C). Contaminant $\leqslant$ 0.01% "NADH oxidase".

References

[1] *R. M. Burton,* in: *S. P. Colowick, N. O. Kaplan* (eds.), Methods in Enzymology, Vol. *I*, Academic Press, New York 1955, p. 397.
[2] *E. C. C. Lin, B. Magasanik,* J. Biol. Chem. *235*, 1820 (1960).
[3] *W. G. McGregor et al.,* J. Biol. Chem. *249*, 3132 (1974).
[4] *I. E. Strickland, O. N. Miller,* Biochim. Biophys. Acta *159*, 221 (1968).
[5] *M. I. Barret,* Ann Arbor, Microfilm *31* (8), 4476 (1971).

Glycerol-3-phosphate Dehydrogenase

from rabbit muscle

sn-**Glycerol 3-phosphate : NAD$^+$ 2-oxidoreductase, EC 1.1.1.8**

$$\text{L-Glycerol-3-P} + \text{NAD}^+ \underset{}{\overset{\text{GDH}}{\rightleftharpoons}} \text{DAP} + \text{NADH} + \text{H}^+$$

Equilibrium constant (Tris buffer; 25°C)

$$K = \frac{[\text{DAP}] \times [\text{NADH}] \times [\text{H}^+]}{[\text{L-glycerol-3-P}] \times [\text{NAD}^+]} = 1.0 \times 10^{-12}\,\text{mol/l}$$

Michaelis constants (phosphate buffer, pH 7.0; 23.3°C) [1]

Glycerol-3-P	$1.1 \times 10^{-4}\,\text{mol/l}$
NAD	$3.8 \times 10^{-4}\,\text{mol/l}$
DAP	$4.6 \times 10^{-4}\,\text{mol/l}$

Effectors: DAP and fructose-1,6-P$_2$ inhibit the enzyme.

Molecular weights: 78 000 (pH 6.28) [2], 173 000 (pH 5.6) [1].

Specificity: the enzyme only reacts with L-glycerol 3-phosphate and dihydroxyacetone phosphate. The rate with NADPH is 1/10 that with NADH.

Assay

$$\text{GAP} \xrightarrow{\text{TIM}} \text{DAP}$$

$$\text{DAP} + \text{NADH} + \text{H}^+ \xrightarrow{\text{GDH}} \text{glycerol-3-P} + \text{NAD}^+.$$

Wavelength 339, Hg 334 or Hg 365 nm; light path 10 mm; final volume 3.11 ml; 25°C.

2.80 ml	triethanolamine buffer (0.5 mol/l; pH 7.9;	45 mmol/l
	0.5% bovine serum albumin)	
0.20 ml	GAP (0.02 mol/l)	1.3 mmol/l
0.05 ml	NADH, Na salt, (10 mg/ml; Na_2CO_3, 1 mmol/l)	0.2 mmol/l
0.01 ml	TIM (2 mg/ml)	34 kU/l
0.02 ml	enzyme in buffer	

Stability: crystalline suspensions in ammonium sulphate solutions, 3.2 mol/l; pH ca. 6, are stable for several months at 4°C. Solutions and suspensions can be dialyzed against 200 volumes EDTA solutions, 1 mmol/l, for 24 h at 4°C without loss of activity.

Purity required: specific activity $\geq$ 170 U/mg protein (25°C). Contaminants (related to the specific activity of GDH) $\leq$ 0.001% aldolase and GAPDH, $\leq$ 0.01% LDH and TIM.

References

[1] *H. L. Young, N. Pace,* Arch. Biochem. Biophys. *75*, 125 (1958).
[2] *J. van Eys, B. J. Nuenke, M. K. Patterson,* J. Biol. Chem. *234*, 2308 (1959).

Glycerokinase

from *Candida mycoderma*

ATP : glycerol 3-phosphotransferase, EC 2.7.1.30

$$\text{Glycerol} + \text{ATP} \overset{\text{GK}}{\rightleftharpoons} \text{glycerol-3-P} + \text{ADP}$$

Michaelis constants (glycine buffer, pH 9.8; 25°C)

Glycerol	6×10^{-5} mol/l
ATP	9×10^{-5} mol/l [1]
ATP-Mg^{2-}	1.6×10^{-4} mol/l [2]

Molecular weight (pH 6.8): 251000.

Specificity: the enzyme reacts with GTP, ITP or UTP instead of ATP.

Assay

$$\text{Glycerol} + \text{ATP} \xrightarrow{\text{GK}} \text{glycerol-3-P} + \text{ADP}$$

$$\text{Glycerol-3-P} + \text{NAD}^+ \xrightarrow{\text{GDH}} \text{DAP} + \text{NADH} + \text{H}^+.$$

Wavelength 339, Hg 334 or Hg 365 nm; light path 10 mm; final volume 3.30 ml; 25°C.

3.00 ml	glycine buffer (0.2 mol/l; pH 9.8;	glycine	182 mmol/l
	208 mg 24% hydrazine hydrate/ml;	hydrazine	0.91 mmol/l
	MgCl$_2$ · 6 H$_2$O, 0.4 mg/ml)	MgCl$_2$	1.8 mmol/l
0.05 ml	ATP, Na salt, (50 mg/ml)		1.35 mmol/l
0.10 ml	NAD (10 mg/ml)		0.42 mmol/l
0.10 ml	glycerol (0.1 mol/l)		3.0 mmol/l
0.03 ml	GDH (10 mg/ml)		15 kU/l
0.02 ml	enzyme in glycerol, 10 mmol/l		

Stability: crystalline suspensions in ammonium sulphate solutions, 3.2 mol/l; pH ca. 6, containing 1% ethylene glycol (v/v) are stable for several months at 4°C.

Purity required: specific activity $\geq$ 85 U/mg protein (25°C). Contaminants (related to the specific activity of GK) $\leq$ 0.01% "NADH oxidase", $\leq$ 0.02% HK and MK.

References

[1] *H. U. Bergmeyer et al.,* Biochem. Z. *333*, 471 (1961).
[2] *N. Grummet, F. Lundquist,* Eur. J. Biochem. *3*, 78 (1967).

Glyoxalase I

from yeast

***S*-Lactoyl-glutathione methylglyoxal-lyase (isomerizing), EC 4.4.1.5**

$$\text{GSH} + \text{methylglyoxal} \xrightarrow{\text{Gl-I}} \text{S-lactoyl-GSH}$$

Equilibrium constant: the reaction is irreversible [1].

Michaelis constants

		Relative rate	
Methylglyoxal	$1 \quad \times 10^{-3}$ mol/l	1.0 ⎫	(phosphate
Phosphohydroxypyruvate aldehyde	$4 \quad \times 10^{-3}$ mol/l	1.0 ⎭	buffer,
			pH 6.6;
			21 °C) [1]
Glutathione	7.4×10^{-4} mol/l	1.0 ⎫	(NaHCO₃
Asparthione	3.7×10^{-3} mol/l	1.0 ⎬	solution,
Isoglutathione	1.23×10^{-2} mol/l	1.0 ⎭	0.2 mol/l;
			25 °C).

Inhibitor constant (phosphate buffer, pH 6.6; 20 °C) [2]:

Ophthalmate 9.5×10^{-4} mol/l

Effectors: the enzyme is inhibited by S-methylglutathione, γ-DL-glutamyl-DL-alanyl-glycine and S(N-ethylsuccinyl)-glutathione.

Molecular weight: 48 000.

Specificity: the enzyme also reacts with glyoxal, phosphohydroxypyruvate aldehyde and phenylglyoxal. Isoglutathione, asparthione, γ-D-Glu-L-Cys-Gly and S-acetylgluta-thione, but not thioglycollate, GSSG, thioneine, ascorbate, cysteine, cysteinylglycine, γ-glutamylcysteine or S-methylglutathione can replace glutathione.

Assay

Wavelength 235 nm; light path 10 mm; final volume 3.10 ml; 25 °C; $\varepsilon = 0.394$ l $\times$ mmol^{-1} $\times$ mm^{-1}.

3.00 ml	phosphate buffer (0.1 mol/l; pH 6.8)	97.4 mmol/l
0.05 ml	GSH (20 mg/ml; pH ca. 7)	1.1 mmol/l
0.02 ml	methylglyoxal (0.2 mg/ml)	0.02 mmol/l
0.02 ml	enzyme in buffer	

Stability: solutions in 30% glycerol; pH ca. 6.5, containing ammonium sulphate, 1 mol/l, are stable for months at 4 °C.

Purity required: specific activity ≧ 200 U/mg protein (25 °C). The preparation must be free from glyoxalase II.

References

[1] *R. H. Wearer, H. A. Lardy,* J. Biol. Chem. *236*, 313 (1961).
[2] *E. E. Cliffe, S. G. Waley,* Biochem. J. *79*, 475 (1961).
[3] *T. Jerzykowski, W. Matuszewski, R. Winter,* FEBS Letters *1*, 159 (1968).

Glyoxylate Reductase

from spinach

Glycollate: NAD$^+$ oxidoreductase, EC 1.1.1.26

$$\text{Glyoxylate} + \text{NADH} + \text{H}^+ \xrightleftharpoons{\text{Gly-R}} \text{glycollate} + \text{NAD}^+$$

Equilibrium constant (Tris buffer, 25°C)

$$K = \frac{[\text{glycollate}] \times [\text{NAD}^+]}{[\text{glyoxylate}] \times [\text{NADH}] \times [\text{H}^+]} = 6.06 \times 10^{14}\,\text{mol}^{-1} \times 1$$

Michaelis constants

Glyoxylate	1.54×10^{-2} mol/l	
Glyoxylate	$1.4 \ \times 10^{-2}$ mol/l [3]	phosphate buffer, pH 6.4 [2]
Glyoxylate	$9.1 \ \times 10^{-3}$ mol/l	phosphate buffer, pH 6.4;
NADH	$3.3 \ \times 10^{-5}$ mol/l	25°C [1]*

Effectors: cf. [2].

Specificity: the enzyme also catalyzes the reduction of hydroxypyruvate to D-glycerate at about 30% of the rate of the glyoxylate reduction [1]. The enzyme is inactive with pyruvate, 2-oxobutyrate, acetaldehyde, oxaloacetate and 2-oxoglutarate [2]. NADPH cannot replace NADH.

Assay

Wavelength 339, Hg 334 or Hg 365 nm; light path 10 mm; final volume 3.07 ml; 25°C.

2.50 ml	phosphate buffer (50 mmol/l; pH 6.4)	40.7 mmol/l
0.50 ml	glyoxylate, Na salt, (20 mg/ml; phosphate buffer, 50 mmol/l; pH 6.4)	25 mmol/l
0.05 ml	NADH, Na salt, (10 mg/ml)	0.2 mmol/l
0.02 ml	enzyme in water	

Stability: crystalline suspensions in ammonium sulphate solution, 3.2 mol/l; pH ca. 6, are stable for several months at 4°C.

Purity required: specific activity $\geq$ 55 U/mg protein (25°C). Contaminants (related to the specific activity of GlyR) $\leq$ 0.01% "NADH oxidase" and glycollate oxidase.

---

* For the enzyme from tobacco leaves.

References

[1] *I. Zelitch,* J. Biol. Chem. *216*, 553 (1955).
[2] *G. Laudahn,* Biochem. Z. *337*, 449 (1963).
[3] *H. Holzer, A. Holldorf,* Biochem. Z. *329*, 292 (1957).

Guanase

from rabbit liver

Guanine aminohydrolase, EC 3.5.4.3

$$\text{Guanine} + H_2O \xrightarrow{\text{guanase}} \text{xanthine} + NH_3$$

Michaelis constants (pyrophosphate/citrate buffer, 37 °C) [1]

		pH	Relative rate
Guanine	1.05×10^{-5} mol/l	7.7	1.00
8-Azaguanine	1.02×10^{-4} mol/l	5.9	–
1-Methylguanine	2.70×10^{-3} mol/l	7.1	–
Thioguanine	–	6.5	0.008
1-Methylthioguanine	–	6.5	0.004

Molecular weight: 170000.

Specificity: the enzyme is inactive with 2,6-diaminopurine, thioazaguanine disulphide and hydroxy-6-aminopyrazolo-(3,4d) pyrimidine.

Assay

Wavelength 293 nm; light path 10 mm; final volume 3.24 ml; 25 °C; $\varepsilon = 1.2$ l × $\text{mmol}^{-1} \times \text{mm}^{-1}$.

3.00 ml	Tris buffer (0.2 mol/l; pH 8.0; O_2 saturated)	185 mmol/l
0.02 ml	guanine (0.15 mg/ml)	0.07 mmol/l
0.02 ml	XOD (10 mg/ml)	25 U/l
0.02 ml	enzyme suspension	

Stability: suspensions in ammonium sulphate solution, 3.2 mol/l; pH ca. 6, are stable for several months at 4 °C. Enzyme solutions or suspensions can be dialyzed against potassium phosphate buffer, 20 mmol/l; pH 7.6, at 4 °C without significant loss of activity.

Purity required: specific activity $\geqq 0.06$ U/mg protein (25 °C).

Reference

[1] *R. Currie, F. Bergel, R. C. Bray,* Biochem. J. *104*, 634 (1967).

Guanosine-5′-monophosphate Kinase

from hog brain

ATP : (d)GMP phosphotransferase, EC 2.7.4.8

$$\text{ATP} + \text{GMP} \xrightleftharpoons{\text{GMPK}} \text{ADP} + \text{GDP}$$

Michaelis constants (Tris-acetate buffer, pH 7.5; 30 °C) [1]

		Relative rate
GMP	6×10^{-6} mol/l	1.00
dGMP	1.0×10^{-4} mol/l	0.47
8-Aza-GMP	1.6×10^{-4} mol/l	0.23
ATP	1.2×10^{-4} mol/l	–

Effectors [1]: the enzyme is inhibited by 6-thioguanosine 5′-phosphate, but not by IMP, XMP, CMP or AMP.

Molecular weight: 205000 (rat liver) [2], 88000 (*E. coli*) [3].

Specificity [1]: GMPK is specific for its substrates. IMP, XMP, CMP, AMP or UMP do not react.

Assay

$$\text{ATP} + \text{GMP} \xrightarrow{\text{GMPK}} \text{ADP} + \text{GDP}$$

$$\text{GDP} + \text{ADP} + 2\,\text{PEP} \xrightarrow{\text{PK}} \text{GTP} + \text{ATP} + 2\,\text{pyruvate}$$

$$2\,\text{Pyruvate} + 2\,\text{NADH} + 2\,\text{H}^+ \xrightarrow{\text{LDH}} 2\,\text{L-lactate} + 2\,\text{NAD}^+.$$

Wavelength 339, Hg 334 or Hg 365 nm; light path 10 mm; final volume 2.74 ml; 25 °C.

2.00 ml	triethanolamine buffer (0.1 mol/l; pH 7.6)	73 mmol/l
0.20 ml	PEP, tricyclohexylammonium salt, (5 mg/ml)	0.77 mmol/l
0.10 ml	$MgSO_4 \cdot 7\,H_2O$ (0.1 mol/l)	3.33 mmol/l

0.10 ml	KCl (2 mol/l)	73 mmol/l
0.20 ml	ATP, Na salt, (40 mg/ml; neutralized)	4.8 mmol/l
0.05 ml	GMP, Na salt, (10 mg/ml)	0.4 mmol/l
0.05 ml	NADH, Na salt, (10 mg/ml)	0.20 mmol/l
0.01 ml	PK (10 mg/ml)	7.3 kU/l
0.01 ml	LDH (5 mg/ml)	10 kU/l
0.02 ml	enzyme in water	

Stability: solutions in 50% glycerol, pH ca. 6, are stable for several months at 4°C.

Purity required: specific activity $\geq$ 10 U/mg protein (25°C). Contaminants (related to the specific activity of GMPK) $\leq$ 0.1% phosphatase, $\leq$ 1% myokinase.

References

[1] *R. R. Miech, R. E. Parks,* J. Biol. Chem. *240,* 351 (1965).
[2] *R. J. Buccino, J. S. Roth,* Arch. Biochem. Biophys. *132,* 49 (1969).
[3] *M. P. Oeschger, M. J. Bessman,* Eur. J. Biochem. *19,* 256 (1971).

Hexokinase

from yeast

ATP : D-hexose 6-phosphotransferase, EC 2.7.1.1

$$\text{D-Glucose} + \text{ATP} \xrightarrow{\text{HK}} \text{D-glucose-6-P} + \text{ADP}$$

Equilibrium constant (pH 6.0; 30°C)

$$K = \frac{[\text{D-glucose-6-P}] \times [\text{ADP}]}{[\text{D-glucose}] \times [\text{ATP}]}$$

$$= \frac{3.86 \times 10^2 \text{ (catalytic amounts of Mg}^{2+}) [1]}{1.55 \times 10^2 \text{ (excess Mg}^{2+}: 17 - 79 \text{ mmol/l)}}$$

Michaelis constants

		Relative rate	
D-Glucose	1.0×10^{-4} mol/l	1.0	
D-Fructose	7.0×10^{-4} mol/l	1.8	
D-Mannose	5.0×10^{-5} mol/l	0.8	phosphate buffer,
D-Galactose	$> 5.0 \times 10^{-2}$ mol/l	< 0.002	pH 7.4; 30°C [2]
D-Glucosamine	1.5×10^{-3} mol/l	0.7	
2-Deoxyglucose	3.0×10^{-4} mol/l	1.0	
ATP	2.0×10^{-4} mol/l	−	Tris buffer, pH 7.6; 28°C

Inhibitor constants

Glucose-6-P 9.1×10^{-3} mol/l (tetramethylammonium chloride buffer, pH 8.0; 25 °C)

Molecular weight [4]: 100 000.

Specificity: ATP can be partially replaced by other nucleotides [5].

Effectors: the enzyme is inhibited by compounds which react with SH-groups and by high Mg^{2+}-concentration. The enzyme is activated by catecholamines and related compounds [3] and requires Mg^{2+} for its catalytic activity.

Assay

$$\text{D-Glucose} + \text{ATP} \xrightarrow{\text{HK}} \text{D-glucose-6-P} + \text{ADP}$$

$$\text{D-Glucose-6-P} + \text{NADP}^+ \xrightarrow{\text{G6P-DH}} \text{D-gluconate-6-P} + \text{NADPH} + \text{H}^+.$$

Wavelength 339, Hg 334 or Hg 365 nm; light path 10 mm; final volume 2.53 ml; 25 °C.

1.00 ml	triethanolamine buffer (50 mmol/l; pH 7.6)	40 mmol/l
1.00 ml	glucose (100 mg/ml in buffer)	222 mmol/l
0.20 ml	$MgCl_2$ (0.1 mol/l)	8.0 mmol/l
0.20 ml	NADP, Na salt, (10 mg/ml)	0.91 mmol/l
0.10 ml	ATP, Na salt, (10 mg/ml)	0.64 mmol/l
0.01 ml	G6P-DH (1 mg/ml)	0.55 kU/l
0.02 ml	enzyme in buffer	

Specificity: crystalline suspensions in ammonium sulphate solution, 3.2 mol/l; pH ca. 6, are stable for several months at 4 °C.

Purity required: specific activity $\geq$ 140 U/mg protein (25 °C). Contaminants (related to the specific activity of HK) $\leq$ 0.01% G6P-DH, GR, MK and 6-PGDH, $\leq$ 0.002% PGI.

References

[1] *E. A. Robbins, P. D. Boyer,* J. Biol. Chem. *224,* 121 (1957).
[2] *A. Sols et al.,* Biochim. Biophys. Acta *30,* 92 (1958).
[3] *W. H. Harrison, R. M. Gray,* Arch. Biochem. Biophys. *151,* 357 (1972).
[4] *J. T. Schulze, S. P. Colowick,* J. Biol. Chem. *244,* 2306 (1969).
[5] *W. Gruber, H. Möllering, H. U. Bergmeyer,* Enzymol. Biol. Clin. *7,* 115 (1966).

Hyaluronoglucosaminidase (Hyaluronidase) testicular

Hyaluronate 4-glycanohydrolase, EC 3.2.1.35

Random hydrolysis of 1,4-linkages between 2-acetamido-2-deoxy-β-D-glucose and D-glucuronate residues in hyaluronate.

Effectors: high molecular weight polysaccharides such as heparin, chondroitin sulphate B, heparitin sulphate and polystyrene sulphonate are competitive inhibitors. Low molecular weight organic molecules with an acid function such as bile salts, dyes and sulphated detergents inhibit. Further inhibitors are heavy metal ions, such as Fe^{2+}, Fe^{3+}, Mn^{2+}, Cu^{2+}, Zn^{2+}, Hg^{2+}, and specific antibodies [1, 2]. NaCl stabilizes the enzyme [3].

Molecular weight [4]: ca. 55000.

Specificity: the enzyme hydrolyzes the endo-N-acetylhexosaminic bonds of hyaluronic acid and chondroitin. It also exhibits transglucosidase activity [5].

pH Optimum [6]: 4.5 – 6.

Assay [7]

$$\text{Hyaluronic acid} \xrightarrow[\text{H}_2\text{O}]{\text{hyaluronidase}} \text{oligosaccharides}$$

$$\text{Hyaluronic acid + albumin} \longrightarrow \text{insoluble complex.}$$

Wavelength 600 nm; light path 10 mm; incubation volume 0.50 ml; incubation temperature 38°C; final volume 3.00 ml; measurement temperature ca. 25°C.

0.25 ml	hyaluronic acid (4 mg/ml phosphate buffer, 0.3 mol/l, pH 5.30)	hyaluronic acid	2 mg/l
		phosphate	0.15 mol/l
0.25 ml	enzyme in phosphate buffer, 20 mmol/l; pH 6.9)		

incubate for 45 min at 38°C

2.50 ml	BSA (1 mg bovine serum albumin per 1 ml acetate buffer; pH 3.75, prepared from 4.56 ml glacial acetic acid, 3.26 g sodium acetate per 1000 ml, pH adjusted with HCl, 6 mol/l)	BSA	0.83 g/l

mix, read the absorbance at 600 nm against a blank. Obtain the activity from a standard curve established with an International Standard preparation.

Stability: the lyophilized enzyme is stable for many months at 4°C in the dry state; the enzyme dissolved in phosphate buffer, 20 mmol/l; pH 6.9, is stable for a few days at 4°C.

Purity required: activity $\geqslant$ 1000 units [7]/mg substance (38°C).

References

[1] *G. H. Warren et al.*, Nature *194*, 770 (1962).
[2] *M. B. Mathews*, in: *S. P. Colowick, N. O. Kaplan* (eds.), Methods in Enzymology, Vol. *VIII*, Academic Press, New York 1966, p. 654.
[3] *C. Yang, P. N. Srivastav*, J. Biol. Chem. *250*, 79 (1975).
[4] *A. Khorlin et al.*, FEBS Lett. *31*, 107 (1973).
[5] *J. Ludowieg, B. Vennesland*, J. Biol. Chem. *236*, 333 (1971).
[6] *M. De Salegui et al.*, Arch. Biochem. Biophys. *121*, 548 (1967).
[7] *A. Dorfman*, in: *S. P. Colowick, N. O. Kaplan* (eds.), Methods in Enzymology, Vol. *I*, Academic Press, New York 1955, p. 166.

3-Hydroxyacyl-CoA Dehydrogenase

from pig heart

L-3-Hydroxyacyl-CoA : NAD$^+$ oxidoreductase, EC 1.1.1.35

$$\text{L-3-Hydroxybutyryl-CoA} + \text{NAD}^+ \xrightleftharpoons{\quad\text{HOADH}\quad} \text{acetoacetyl-CoA} + \text{NADH} + \text{H}^+$$

Equilibrium constants (25°C)

$$K = \frac{[\text{acetoacetyl-CoA}] \times [\text{NADH}] \times [\text{H}^+]}{[\text{L-3-hydroxybutyryl-CoA}] \times [\text{NAD}^+]} = \begin{cases} 6.3 \ \times 10^{-11}\,\text{mol/l}\,[1] \\ 2.5 \ \times 10^{-11}\,\text{mol/l}\,[1] \\ 2.17 \times 10^{-10}\,\text{mol/l}\,[2] \\ 5.25 \times 10^{-9} \ \text{mol/l}\,[2] \end{cases}$$

Michaelis constants

S-Acetoacetyl-N-acetylcysteamine	1.0×10^{-2} mol/l
S-Acetoacetyl-pantetheine	8.0×10^{-6} mol/l
S-Acetoacetyl-dephospho-CoA	1.0×10^{-4} mol/l
S-Acetoacetyl-CoA	5.0×10^{-5} mol/l*

Effectors [4]: the enzyme is inhibited by 4-chloromercuribenzoate and sodium salyrgan. Glutathione and cysteine cause a partial reactivation. 2-Phenanthroline is a competitive inhibitor.

---

* Enzyme from sheep liver.

Specificity: the enzyme reacts with all normal L-3-hydroxyacyl-CoA derivatives from C_4 to C_{12} (enzyme from sheep liver) [3]. In addition, 3-hydroxyhexanoyl- and 3-hydroxy-4-hexenoyl-CoA derivatives are oxidized. NADP(H) reacts at 1/50 of the rate with NAD(H) [1].

Assay

Wavelength 339, Hg 334 or Hg 365 nm; light path 10 mm; final volume 3.16 ml; 25°C.

3.00 ml	pyrophosphate buffer (0.1 mol/l; pH 7.3)	95 mmol/l
0.05 ml	NADH, Na salt, (10 mg/ml)	0.2 mmol/l
0.10 ml	S-acetoacetyl-N-acetylcysteamine (31.4 mg/ml)	4.7 mmol/l
0.02 ml	enzyme in buffer	

Stability: suspensions in ammonium sulphate solution, 3.2 mol/l; pH ca. 6, are stable for months at 4°C.

Purity required: specific activity $\geq$ 125 U/mg protein (25°C; measured with acetoacetyl-CoA as substrate), or $\geq$ 25 U/mg protein (25°C; measured with S-acetoacetyl-N-acetylcysteamine as substrate).

References

[1] *S. J. Wakil,* in: *P. D. Boyer, H. Lardy, K. Myrbäck* (eds.), The Enzymes, Vol. 7, Academic Press, New York 1963, p. 97.
[2] *J. R. Stern,* Biochim. Biophys. Acta **26**, 448 (1957).
[3] *F. Lynen, O. Wieland,* in: *S. P. Colowick, N. O. Kaplan* (eds.), Methods in Enzymology, Vol. *I*, Academic Press, New York 1955, p. 566.
[4] *M. Grassl,* Dissertation, Munich, April 1957.

3-Hydroxybutyrate Dehydrogenase

from *Rhodopseudomonas spheroides*

D-3-Hydroxybutyrate: NAD$^+$ oxidoreductase, EC 1.1.1.30

$$\text{D-3-Hydroxybutyrate} + \text{NAD}^+ \xrightleftharpoons{\text{3-HBDH}} \text{acetoacetate} + \text{NADH} + \text{H}^+$$

Equilibrium constant (25°C) [1]

$$K = \frac{[\text{acetoacetate}] \times [\text{NADH}] \times [\text{H}^+]}{[\text{3-hydroxybutyrate}] \times [\text{NAD}^+]} = 1.42 \times 10^{-9} \text{ mol/l}$$

Michaelis constants [2]

D-3-Hydroxybutyrate	4.1×10^{-4} mol/l	Tris buffer, pH 8.5; 25 °C
NAD	8.0×10^{-5} mol/l	

Acetoacetate	2.8×10^{-4} mol/l	
3-Oxopentanoate	3.1×10^{-5} mol/l	phosphate buffer, pH 7.4; 25 °C
3-Oxohexanoate	5.4×10^{-4} mol/l	

NADH	5.4×10^{-5} mol/l	phosphate buffer, pH 7.0; 25 °C

Inhibitor constants (pH 7.4) [2]

DL-Lactate	7.0×10^{-4} mol/l	
DL-2-Hydroxybutyrate	1.7×10^{-3} mol/l	competitive
Laevulate	8.3×10^{-3} mol/l	
Succinate	–	complex type of
Malonate	–	inhibition

Inhibition of enzyme by thiol reagents, cf. [2].

Molecular weight (phosphate buffer, pH 7.4; 20 °C) [2]: 85 000.

Specificity: the enzyme reacts with D-3-hydroxybutyrate (1.00), 3-hydroxypentanoate (0.05) and 3-hydroxyhexanoate (0.04). L-3-Hydroxybutyrate and 3-hydroxypropionate do not react. Acetoacetate (1.00), 3-oxopentanoate (0.05) and 3-oxohexanoate (0.04) react in the reduction reaction. NADP(H) cannot replace NAD(H). 3-Acetylpyridine-adenine dinucleotide and thionicotinamide-adenine dinucleotide react at 1/10 of the rate with NAD.

Assay

Wavelength 339, Hg 334 or Hg 365 nm; light path 10 mm; final volume 2.72 ml; 25 °C.

2.00 ml	Tris buffer (0.2 mol/l; pH 8.0)	147 mmol/l
0.20 ml	NAD (20 mg/ml)	1.93 mmol/l
0.50 ml	DL-3-hydroxybutyrate, Na salt, (20 mg/ml)	29 mmol/l
0.02 ml	enzyme in buffer	

Stability: suspensions in ammonium sulphate solution, 3.2 mol/l; pH ca. 6, are stable for several months at 4 °C. Diluted enzyme solutions can be stabilized by the addition of NADH or Ca^{2+} [2].

Purity required: specific activity $\geq$ 3 U/mg protein (25 °C). Contaminants (related to the specific activity of 3-HBDH) $\leq$ 0.2% LDH, $\leq$ 5% MDH.

References

[1] *H. A. Krebs, J. Mellanby, D. H. Williamson,* Biochem. J. *82*, 96 (1962).
[2] *H. U. Bergmeyer et al.,* Biochem. J. *102*, 423 (1967).

3α,20β-Hydroxysteroid Dehydrogenase

from *Streptomyces hydrogenans*

17,20β,21-Trihydroxysteroid: NAD⁺ oxidoreductase, EC 1.1.1.53

$$20\text{-Dihydrocortisone} + NAD^+ \underset{\xrightarrow{\text{20-StDH}}}{\rightleftharpoons} \text{cortisone} + NADH + H^+$$

Michaelis constants

Reduction on C_{20}		Relative rate [2]
4-Pregnen-3,20-dione	2.2×10^{-5} mol/l [1]	1.00
4-Pregnen-17α-ol-3,20-dione	1.2×10^{-5} mol/l [1]	1.50
4-Pregnen-17α,21-diol-3,11,20-trione	5.1×10^{-5} mol/l	1.20
1,4-Pregnadien-17α,21-diol-3,11,20-trione	–	1.10
4-Pregnen-17α,21-diol-3,20-dione	6.3×10^{-6} mol/l	0.70
4-Pregnen-21-ol-3,20-dione	–	0.40
4-Pregnen-11β,17α,21-triol-3,20-dione	1.3×10^{-4} mol/l	0.15
1,4-Pregnadien-11β,17α,21-triol-3,20-dione	–	0.14
4-Pregnen-11β,21-diol-3,20-dione	2.4×10^{-4} mol/l	0.03

Reduction on C_3 [1]		V
5α-Androstan-3-one	3.3×10^{-5} mol/l	5.4
5α-Androstan-17β-ol-3-one	2.89×10^{-4} mol/l	19.6
5α-Androstan-17α-methyl-17β-ol-3-one	7.7×10^{-5} mol/l	1.0
5α-Androstan-3,16-dione	5.0×10^{-5} mol/l	167.2
5α-Adrostan-3,17-dione	3.06×10^{-4} mol/l	5.0

Oxidation on C_{20} [1]		V
4-Pregnen-20β-ol-3-one	2.7×10^{-5} mol/l	29

Oxidation on C_3 [1]		V
5α-Androstan-3α-ol	2.9×10^{-4} mol/l	20.3
5α(H)-Androstan-16-3α-ol	3.75×10^{-4} mol/l	26.1
5α-Androstan-3α-ol-17-one	2.5×10^{-4} mol/l	0.9
5α-Androstan-3α,17β-diol	3.5×10^{-5} mol/l	2.3
5α-Androstan-3α-ol-11,17- dione	4.2×10^{-4} mol/l	2.3
5α-Androstan-3α,11β-diol-17- one	2.5×10^{-4} mol/l	0.6
NADH (according to the substrate)	4.6×10^{-6} mol/l [1]	
to	1.2×10^{-5} mol/l	

Molecular weights: ca. 93 000 [3, 4], 118 400.

Specificity [1]: the enzyme exhibits both 20 β-hydroxysteroid : NAD oxidoreductase activity, which includes C_{21} steroids as well as the C_{20} position and the steric course in the β-position, and 3α-hydroxysteroid : NAD oxidoreductase activity with compounds with a 5 α-configuration of the androstane and pregnane series, 5 β-(A/B cis)-compounds and steroids with a double bond in the 4,5-position, 5,10-position or 1,2- and 4,5-position do not react. The rate of the enzyme activity is affected by the different substituents of the steroids.

Assay

Wavelength 339, Hg 334 or Hg 365 nm; light path 10 mm; final volume 3.17 ml; 25°C.

3.00 ml	phosphate buffer (0.1 mol/l; pH 6.5)	95 mmol/l
0.10 ml	cortisone (5 mg/ml in methanol)	0.44 mmol/l
0.05 ml	NADH, Na salt, (5 mg/ml)	0.1 mmol/l
0.02 ml	enzyme in water	

Stability: crystalline suspensions in ammonium sulphate solution, 3.2 mol/l; pH ca. 6, are stable for several months at 4°C.

Purity required: specific activity $\geq$ 10 U/mg protein (25°C). Contaminants (related to the specific activity of 20-StDH) $\leq$ 0.1% "NADH oxidase".

References

[1] *T. Pocklington, J. Jeffery,* Eur. J. Biochem. *7*, 63 (1968).
[2] *J. Schmidt-Thomé et al.,* Biochem. Z. *336*, 322 (1962).
[3] *H. J. Hübner,* in: *H. U. Bergmeyer,* Methoden der enzymatischen Analyse, Verlag Chemie, Weinheim 1962, p. 477.
[4] *G. Traexler,* in: *M. Gehatia,* Z. Natuforsch. *17b*, 432 (1962).

Isocitrate Dehydrogenase

from pig heart

threo-D_s-Isocitrate: NADP$^+$ oxidoreductase (decarboxylating), EC 1.1.1.42

$$\text{D-Isocitrate} + \text{NADP}^+ \xrightleftharpoons[\text{Mn}^{2+}(\text{Mg}^{2+})]{\text{ICDH}} \text{2-oxoglutarate} + CO_2 + \text{NADPH} + H^+$$

Equilibrium constant (glycylglycine buffer, pH 7.0; 22 °C)

$$K = \frac{[\text{2-oxoglutarate}] \times [CO_2] \times [\text{NADPH}] \times [H^+]}{[\text{D-isocitrate}] \times [\text{NADP}^+]} = 0.77 \,(\text{mol/l})^2 \,[1]$$

Michaelis constants [2, 3]

		Relative rate	
Isocitrate	2.6×10^{-6} mol/l	2.01	Tris buffer, pH 7.3; 24 °C
Oxalosuccinate	2.5×10^{-2} mol/l	3.22	citrate buffer, pH 5.6; 14 °C
Oxalosuccinate	5.6×10^{-4} mol/l	1.00	Tris buffer, pH 7.3; 24 °C
2-Oxoglutarate	1.3×10^{-4} mol/l	$-$	TEA buffer, pH 7.0; 37 °C
NADPH	9.2×10^{-6} mol/l		

Effectors: isocitrate is a competitive inhibitor of oxalosuccinate decarboxylation.

Molecular weight (pH 6.0) [4]: 61 000.

Specificity: the enzyme catalyzes the following reactions:
a) Isocitrate + NADP$^+$ $\rightleftharpoons$ 2-oxoglutarate + CO_2 + NADPH + H$^+$
b) Oxalosuccinate + NADPH + H$^+$ $\rightleftharpoons$ isocitrate + NADP$^+$
c) Oxalosuccinate $\rightleftharpoons$ 2-oxoglutarate + CO_2
NAD cannot replace NADP

Assay

Wavelength 339, Hg 334 or Hg 365 nm; light path 10 mm; final volume 2.85 ml; 25 °C.

2.50 ml	imidazole buffer (0.1 mol/l; pH 8.0)	88 mmol/l
0.10 ml	MgCl$_2$ (0.1 mol/l)	3.5 mmol/l
0.10 ml	NADP, Na salt, (10 mg/ml)	0.41 mmol/l
0.10 ml	isocitrate, Na salt, (4 mg/ml in buffer)	0.55 mmol/l
0.05 ml	enzyme in isocitrate/buffer	

Stability: solutions in 50% glycerol are stable for several months at 4°C.

Purity required: specific activity $\geq$ 2 U/mg protein (25°C). Contaminant (related to the specific activity of ICDH) $\leq$ 0.2% NAD-specific ICDH.

References

[1] *S. Ochoa,* in: *S. P. Colowick, N. O. Kaplan* (eds.), Methods in Enzymology, Vol. *I*, Academic Press, New York 1955, p. 699.
[2] *J. Moyle,* Biochem. J. *63,* 552 (1956).
[3] *Z. B. Rose,* J. Biol. Chem. *235,* 928 (1960).
[4] *G. Sichert et al.,* J. Biol. Chem. *226,* 965 (1957).

D-(−)-Lactate Dehydrogenase

from *Lactobacillus leichmannii*

D-Lactate: NAD$^+$ oxidoreductase, EC 1.1.1.28

$$\text{D-Lactate} + \text{NAD}^+ \xrightleftharpoons{\text{D-LDH}} \text{pyruvate} + \text{NADH} + \text{H}^+$$

Specificity: the enzyme is specific for D-(−)-lactate; L-(+)-lactate does not react.

Assay

a) Reduction reaction with pyruvate as substrate, see under L-(+)-LDH.
b) Oxidation reaction with D-lactate as substrate.

Wavelength 339, Hg 334 or Hg 365 nm; light path 10 mm; final volume 3.00 ml; 25°C.

2.28 ml	glycine/hydrazine buffer (glycine, 0.5 mol/l, hydrazine, 0.4 mol/l, pH 9.0)	glycine	380 mmol/l
		hydrazine	304 mmol/l
0.50 ml	D-(−)-lactate, Li salt, (0.5 mol/l)*		83 mmol/l
0.20 ml	NAD (30 mg/ml)		2.6 mmol/l
0.02 ml	enzyme in water		

read against blank containing water instead of substrate.

---

* The enzyme is not saturated at this substrate concentration.

Stability: suspensions in ammonium sulphate solution, 3.2 mol/l; pH ca. 6, are stable for several months at 4°C.

Purity required: specific activity $\geq$ 300 U/mg protein (25°C) in the reduction reaction with pyruvate as substrate. Specific activity $\geq$ 50 U/mg protein (25°C) in the oxidation reaction with D(−)-lactate as substrate. Contaminants (related to the specific activity of D-LDH) $\leq$ 0.01% ADH, GlDH and SDH, $\leq$ 0.1% MDH.

L-(+)-Lactate Dehydrogenase

from rabbit muscle*

L-Lactate: NAD$^+$ oxidoreductase, EC 1.1.1.27

$$\text{L-Lactate} + \text{NAD}^+ \xrightleftharpoons{\text{L-LDH}} \text{pyruvate} + \text{NADH} + \text{H}^+$$

Equilibrium constant (pH 7.0; 25°C)

$$K = \frac{[\text{pyruvate}] \times [\text{NADH}] \times [\text{H}^+]}{[\text{L-lactate}] \times [\text{NAD}^+]} = 2.76 \times 10^{-12} \text{mol/l [1]}$$

Michaelis constants

NAD	2.53×10^{-4} mol/l [2]
Lactate	6.7×10^{-3} mol/l [2]
NADH	1.07×10^{-5} mol/l [2]
Pyruvate	1.64×10^{-4} mol/l [2]
Glyoxylate	3.0×10^{-3} mol/l [3]

Inhibitor constants [2]

Lactate	2.1×10^{-1} mol/l
Pyruvate	2.0×10^{-3} mol/l

Molecular weight [4]: 140000.

Specificity: the enzyme is specific for L-(+)-lactate; D-(−)-lactate does not react. L-(+)-LDH oxidizes or reduces glyoxylate to oxalate or glycollate [3]. Glycerate, 3-amino- and 3-halogen-lactate are oxidized ca. 100 times more slowly than lactate [5].

---

* Predominantly isoenzymes 4 and 5.

Apart from pyruvate, a series of 2-oxo acids are reduced, with the rate decreasing with increasing chain length [6]. NADP(H) reacts at less than 1% of the rate with NAD(H) [6]. NAD analogues, in which the acid amide group in the 3-position of the pyridine ring is substituted by other groups, react at similar rates [7].

Assay

Wavelength 339, Hg 334 or Hg 365 nm; light path 10 mm; final volume 3.17 ml; 25°C.

3.00 ml	phosphate buffer (0.1 mol/l; pH 7.0)	94.5 mmol/l
0.10 ml	pyruvate, Na salt, (2.5 mg/ml)	0.77 mmol/l
0.05 ml	NADH, Na salt, (10 mg/ml)	0.2 mmol/l
0.02 ml	enzyme in buffer	

Stability: crystalline suspensions in ammonium sulphate solution, 3.2 mol/l; pH ca. 7, are stable for years at 4°C. Na_2SO_3 protects the lysine, cysteine and tyrosine residues of the enzyme [8].

Purity required: specific activity $\geq$ 550 U/mg protein (25°C). Contaminants (related to the specific activity of LDH) $\leq$ 0.001% aldolase and PK, $\leq$ 0.01% ALAT, MDH and myokinase.

References

[1] M. T. Hakala, A. J. Glaid, G. W. Schwert, J. Biol. Chem. 221, 191 (1956).
[2] V. Zewe, H. J. Fromm, J. Biol. Chem. 237, 1668 (1962).
[3] M. Romano, M. Cerra, Biochim. Biophys. Acta 177, 421 (1969).
[4] S. J. Lovell, D. J. Winzor, Biochemistry 13, 3527 (1974).
[5] W. Franke, E. Holz, Hoppe Seyler's Z. physiol. Chem. 314, 22 (1959).
[6] A. Meister, J. Biol. Chem. 184, 117 (1950).
[7] B. M. Anderson, N. O. Kaplan, J. Biol. Chem. 234, 1226 (1959).
[8] G. Pfleiderer, FEBS Lett. 1, 129 (1968).

Lactate Oxidase

from *Mycobacterium smegmatis*

L-Lactate: oxygen 2-oxidoreductase (decarboxylating), EC 1.13.12.4

$$\text{L-Lactate} + O_2 \xrightarrow{\text{lactate oxidase}} \text{acetate} + CO_2 + H_2O$$

Michaelis constant [1]: L-lactate 2.5×10^{-2} mol/l

Inhibitor constant [1]: pyruvate 14.5×10^{-3} mol/l

Effectors [1]: various α-hydroxy acids such as D-lactate (2.5 mmol/l), DL-glycerate (0.1 mmol/l), tartronate (30 mmol/l) and 4-chloromercuribenzoate are weakly inhibiting. Oxalate is a potent inhibitor (reversible inhibition in the dark and irreversible in the UV-range); $(NH_4)_2SO_4$ causes reversible enzyme inactivation.

Molecular weight [2]: 300000 – 400000.

Specificity [1]: the enzyme is specific for L-lactate and reacts very slowly with DL-α-hydroxybutyrate and DL-3-phenyllactate.

pH Optimum [1]: sharp optimum between pH 5.7 – 5.9.

Assay

Oxygen electrode; final volume 40.05 ml; 25 °C.

40.0 ml	L-lactate (0.1 mol/l; pH 6.0; saturated with air)	0.1 mol/l
0.05 ml	enzyme in water	

Stability: in the lyophilized state a decrease in activity of approx. 10% may occur within 12 months at 4 °C.

Purity required: activity $\geqslant$ 15 U/mg substance (25 °C); $\triangleq$ ca. 50 U/mg enzyme protein.

References

[1] *P. A. Sullivan,* Biochem. J. *110,* 363 (1968).
[2] *P. A. Sullivan et al.,* Biochem. J. *165,* 375 (1977).

Leucine Aminopeptidase

from hog kidney

α-Aminoacyl-peptide hydrolase (cytosol), EC 3.4.11.1

$$\text{L-Leucine peptide} + H_2O \xrightarrow{\text{LAP}} \text{L-leucine} + \text{peptide}$$

Michaelis constants (N-ethylmorpholine buffer, pH 8.4; °C)

Mg^{2+} activated			Relative rate
	L-Leucinamide	5.21×10^{-3} mol/l	1.00
	L-Leucylglycine	1.00×10^{-3} mol/l	0.70
	L-Leucylvaline	$5.1 \ \times 10^{-4}$ mol/l	0.73
	L-Leucylalanine	$7.9 \ \times 10^{-4}$ mol/l	0.66
	L-Leucine benzyl ester	1.56×10^{-3} mol/l	0.35

Mn^{2+} activated	L-Leucinamide	1.57×10^{-2} mol/l	4.45
	L-Leucylglycine	$8.1 \ \times 10^{-4}$ mol/l	2.02
	L-Leucylvaline	1.04×10^{-3} mol/l	3.08
	L-Leucylalanine	$6.5 \ \times 10^{-4}$ mol/l	2.02
	L-Leucine benzyl ester	$1.1 \ \times 10^{-3}$ mol/l	0.55

Effectors: Mg^{2+} or Mn^{2+} are essential for activity [1]. Cd^{2+}, Cu^{2+}, Hg^{2+}, Pb^{2+}, EDTA, alcohols and 4-chloromercuribenzoate inhibit the enzyme. 2-Phenanthroline, Na$_2$S and NaCN inactivate it [2].

Molecular weight (pH 8.5) [3]: 255 000.

Specificity: the enzyme hydrolyzes a large number of peptides and amino acid amides of the L-configuration. The enzyme has esterase activity [4]. It is inhibited by glycerol and n-butanol [5].

Assay

Wavelength 238 nm; light path 10 mm; final volume 3.00 ml; 25 °C; $\Delta \varepsilon$ is determined each time after completion of the reaction.

Activation mixture

2.7 ml	Tris buffer (50 mmol/l; pH 8.5)	45 mmol/l
0.2 ml	MnCl$_2$ solution (25 mmol/l)	1.7 mmol/l
0.1 ml	enzyme in buffer	

incubate for 2 – 3 h at 40 °C

Assay mixture

	Sample	Reference
L-leucine (60 mmol/l; pH 8.5)	–	2.5 ml
L-leucinamide (60 mmol/l; pH 8.5)	2.5 ml	–
MgCl$_2$ solution (62.5 mmol/l)	0.2 ml	0.2 ml
Tris buffer (0.5 mol/l; pH 8.5)	0.2 ml	0.2 ml

read absorbance against reference cuvette

enzyme solution (activated)	0.1 ml	

Stability: suspensions in ammonium sulphate solution, 3.2 mol/l; pH ca. 8, containing Tris, 20 mmol/l, and magnesium sulphate, 20 mmol/l, are stable for several months at 4 °C.

Purity required: specific activity $\geqq$ 100 U/mg protein (25 °C).

References

[1] *G. F. Bryce, B. R. Rabin,* Biochem. J. *90*, 513 (1964).
[2] *S. R. Himmelhoch,* Arch. Biochem. Biophys. *134*, 597 (1969).
[3] *P. Melius et al.,* Biochim. Biophys. Acta *221*, 62 (1970).
[4] *G. F. Bryce, B. R. Rabin,* Biochem. J. *90*, 509 (1964).
[5] *E. L. Smith, R. L. Hill,* in: *P. D. Boyer, H. Lardy, K. Myrbäck* (eds.), The Enzymes, Vol. *4*, Academic Press, New York 1960, p. 37.

Lipase

from *Rhizopus arrhizus*

Triacylglycerol acylhydrolase, EC 3.1.1.3

$$\text{Triacylglyceride} + H_2O \xrightarrow{\text{lipase}} \text{diacylglyceride} + \text{a fatty acid anion}$$

Effectors [1]: the enzyme is activated by Ca^{2+}, Mg^{2+} and albumin. EDTA, Zn^{2+}, Co^{2+}, Fe^{3+} are inhibiting.

Molecular weight [2]: ca. 40000.

Specificity [1, 3]: the enzyme with its high specific activity belongs to that group of esterases which hydrolyze emulsified substrates, especially the α and α'-esters of glycerol and long chain fatty acids.

pH Optimum: the enzyme is active in the range of pH $3.5 - 9$ and has two optima at pH 3.5 and pH 7.

Assay

titrator with 1 ml burette; recorder; 50 ml reaction vessel; magnetic stirrer; microprocessor-controlled dispenser; final volume approx. 26 ml; 37°C.

7.50 ml	substrate emulsion (25 ml Rhodoviol®, 25/140 M, 3%, 25 ml olive oil, purified, neutralized with NaOH, 0.1 mol/l, and 12.5 ml water mixed in a *Waring* blender and finally emulsified by ultrasonication)	Rhodoviol ca. 3.5 g/l olive oil ca. 115 ml/l
5.00 ml	$CaCl_2$ (0.1 mol/l)	19 mmol/l
1.00 ml	bovine serum albumin (20 mg/ml)	770 g/l

adjust mixture in reaction vessel to 37°C
and pH 7.5, add

0.1 ml   enzyme in $CaCl_2$, 1 mmol/l,

titrate at constant pH. Run a blank without sample.

Stability: suspensions in ammonium sulphate solution, 3.2 mol/l; pH ca. 6, are stable for several months at 4°C.

Purity required: specific activity $\geqslant$ 13000 U/mg protein (37°C). Contaminants (related to the specific activity of lipase) $\leqslant$ 0.001% ATPase, $\leqslant$ 0.00001% GK, $\leqslant$ 0.0001% alkaline phosphatase, $\leqslant$ 0.00005% acid phosphatase, $\leqslant$ 0.0001% HK.

References

[1] P. Laboureur, M. Labrousse, Bull. Soc. Chim. Biol. 48, 747 (1966).
[2] P. Laboureur, M. Labrousse, Bull. Soc. Chim. Biol. 50, 2179 (1968).
[3] P. Desnuelle et al., Bull. Soc. Chim. Biol. 49, 71 (1967).

Lipoprotein Lipase

from micro-organisms

Triacylglycero-protein acylhydrolase, EC 3.1.1.34

$$\text{Triacylglycerol} + H_2O \xrightarrow{\text{lipoprotein lipase}} \text{diacylglycerol} + \text{a fatty acid anion}$$

Effectors: Hg^{2+}, Ag^+, Cr^{2+}, Sn^{2+}, Cu^{2+}, and ionic detergents inhibit. Mg^{2+}, sodium cholate and bovine serum albumin stabilize the enzyme. 4-Chloromercuribenzoate (2 mmol/ml), monoiodoacetate (2 mmol/ml), NaF (20 mmol/ml), NaN_3 (20 mmol/ml), EDTA (5 mmol/ml) and 2-phenanthroline (2 mmol/ml) do not affect the enzyme activity while sodium dodecyl sulphate (0.1% w/v) is inactivating.

Molecular weight [1]: ca. 47000.

Specificity [1]: the enzyme has both lipolytic and sterol ester hydrolytic activities. It hydrolyzes triacylglycerols in chylomicrons, lipoproteins and diacylglycerols. With human plasma as substrate triglycerides are hydrolyzed more rapidly than cholesterol esters. The effects of pH and ionic strength on the enzymatic activity are somewhat different between the hydrolysis of triglyceride and of cholesterol ester depending on the different states of these substrates in the plasma or the transfer of the reaction products at the interface of substrates.

Substrate specificity (lipolytic activity)

Substrate	Number of C-atoms to number of double bonds	Relative rate
Olive oil		94
Triolein	(18:1)	100
Tripalmitin	(16:0)	2
Trimyristin	(14:0)	7
Trilaurin	(12:0)	4
Tricaprin	(10:0)	17
Tricaprylin	(8:0)	64
Tricaproin	(6:0)	2
Tributyrin	(4:0)	2
Tripropionin	(3:0)	2
Triacetin	(2:0)	1

pH Optimum [1]: 7.3.

Assay

$$\text{Triglyceride} + 3\,H_2O \xrightarrow{\text{lipoprotein lipase}} \text{glycerol} + 3 \text{ fatty acid anions}$$

$$\text{Glycerol} + \text{ATP} \xrightarrow{\text{GK}} \text{glycerol-3-P} + \text{ADP}$$

$$\text{ADP} + \text{PEP} \xrightarrow{\text{PK}} \text{ATP} + \text{pyruvate}$$

$$\text{Pyruvate} + \text{NADH} + H^+ \xrightarrow{\text{LDH}} \text{lactate} + \text{NAD}^+.$$

Wavelength 339, Hg 334, or Hg 365 nm; light path 10 mm; final volume 1.0 ml; 30°C.

0.96 ml	reagent (potassium phosphate, 0.1 mol/l; pH 7.0; ATP, 0.3 mmol/l; PEP 0.3 mmol/l; MgSO$_4$, 2 mmol/l; NADH, 0.24 mmol/l; PK, 4 U/ml; LDH, 1.5 U/ml; GK, 1 U/ml)	phosphate	96 mmol/l	
		ATP	0.29 mmol/l	
		PEP	0.29 mmol/l	
		MgSO$_4$	1.9 mmol/l	
		NADH	0.23 mmol/l	
		PK	3.8 kU/l	
		LDH	1.4 kU/l	
		GK	1 kU/l	
0.01 ml	triglyceride concentrate (supplied by *Miles*; 20 g/l)	triglyceride	0.2 g/l	
0.03 ml	enzyme in water			

Stability: the lyophilized enzyme stabilized with sodium cholate (70%) and Tris-HCl (23%) is stable for 2 years at 4°C and 2 months at 37°C in the dry state; the enzyme dissolved in water, 6 mg/ml, is stable for at least one week at 4°C.

Purity required: activity $\geqslant$ 100 units/mg substance (1 unit is that activity that catalyzes the formation of 1 nmole of glycerol from triglyceride concentration per min at 30°C). Contaminants: essentially free from proteases, GK, ATPase, "NADH oxidase", HK and MK.

Reference

[1] *M. Sigiuara et al.,* Chem. Pharm. Bull. *24*, 1202 (1976).

Lipoxygenase

from soybeans

Linoleate : oxygen oxidoreductase, EC 1.13.11.12

$$\text{Unsaturated fatty acid} + O_2 \xrightarrow{\text{lipoxygenase}}$$

$$\text{peroxide derivative of unsaturated fatty acid}$$

Michaelis constants (pH 9.0; 20°C) [1]

Linoleate	1×10^{-3} mol/l
O_2	3×10^{-4} mol/l

Effectors: *trans*-linoleate and 10,12-octadecadienoate are competitive inhibitors.

Molecular weight [2]: 102000.

Specificity: the enzyme catalyzes the incorporation of O_2 in poly-unsaturated fatty acids as follows [1, 3]:

$$CH_3-(CH_2)_4-\overset{\downarrow}{C}H=CH-CH_2-CH=CH-R$$

$$R''-CH=CH-CH_2-CH=\overset{\downarrow}{C}H-(CH_2)_7-COOH$$

Linoleate, arachidonate and linolenate are oxidized at the same rate.

Assay [4]

$$\text{Linoleate} + O_2 \xrightarrow{\text{lipoxygenase}} \text{13-hydroperoxyoctadeca-9,11-dienoate}$$

The increase in UV absorption at 238 nm which occurs during the lipoxygenase catalyzed oxidation is a measure of the activity. A unit is the enzyme activity which causes an increase in absorbance of 0.001/min under the assay conditions at 25°C.

Wavelength 238 nm; light path 10 mm; final volume 3.00 ml; 25°C.

2 ml	substrate/borate buffer (0.167 mol/l; pH 9.0; linoleate, 0.17 µl/ml, ethanol, 0.1 ml/ml)	borate	0.18 mol/l
		linoleate	0.18 mmol/l
1 ml	enzyme in borate buffer (0.2 mol/l; pH 9.0)		

Stability: lyophilized enzymes are stable for several months at 4°C.

Purity required: activity $\geq$ 50000 units/mg substance (25°C).

References

[1] *A. L. Tappel,* in: *S. P. Colowick, N. O. Kaplan* (eds.), Methods in Enzymology, Vol. *V*, Academic Press, New York 1962, p. 539.
[2] *H. Theorell, R. T. Holman, A. Akeson,* Acta Chem. Scand *1*, 571 (1947).
[3] *M. Hamberg, B. Samuelsson,* Biochem. Biophys. Res. Commun. *21*, 531 (1965).
[4] *H. Theorell, R. T. Holman, A. Akeson,* Arch. Biochem. Biophys. *14*, 250 (1947).

Luciferase

from *Photobacterium fischeri*

Alkanal, reduced-FMN-oxygen oxidoreductase (1-hydroxylating, luminescing), EC 1.14.14.3

$$\text{FMNH}_2 + \text{R-CHO} + \text{O}_2 \xrightarrow{\text{luciferase}} \text{FMN} + \text{RCOOH} + \text{H}_2\text{O} + \text{h} \times v_{495}$$

Equilibrium constant: the multi-step reaction is a mixed oxidation between FMNH_2 and an aliphatic aldehyde to yield FMN and a carboxylic acid accompanied by light emission according to the following reaction scheme:

Fast reaction

a)

$$\text{FMNH}_2 + \text{E} \rightleftharpoons \text{E-FMNH}_2 \xrightarrow{\text{O}_2} \text{E} \begin{cases} \text{FMNH} \cdot \\ \text{OOH} \end{cases}$$

Slow reaction

b)
$$E \Big\langle \begin{matrix} FMNH \\ OOH \end{matrix} \quad \xrightarrow[R-CHO]{R-CHO} \quad E \Big\langle \begin{matrix} FMNH \\ OOH \end{matrix} \quad \longrightarrow \quad E\text{-FMNHOH} + RCOOH$$

$$\Big\downarrow h \times \nu$$

$$E\text{-FMNHOH}$$

$$\Big\downarrow$$

$$E + FMN + H_2O$$

(For detailed information see also references [1, 2])

Michaelis constants [3]

FMNH$_2$		9.7×10^{-7} mol/l	
		Number of C-atoms	
Aldehyde	C-8	36	$\times 10^{-6}$ mol/l
	C-9	15	$\times 10^{-6}$ mol/l
	C-10	10	$\times 10^{-6}$ mol/l
	C-11	3	$\times 10^{-6}$ mol/l
	C-12	1.5	$\times 10^{-6}$ mol/l
	C-13	1.0	$\times 10^{-6}$ mol/l
	C-14	0.75	$\times 10^{-6}$ mol/l
	C-15	0.30	$\times 10^{-6}$ mol/l
	C-16	0.05	$\times 10^{-6}$ mol/l

Effectors: oxidase inhibitors such as 2-diethylaminoethyl-2.2-diphenylvalerate and 8-anilino-1-naphthalene sulphonate are strong inhibitors at concentrations of 10^{-5} mol/l. SH-reagents reacting with lysine or histidine, riboflavin, anaesthetic agents, cyanide, copper and other heavy metals inhibit [3].
For further information see references [4 – 6].

Molecular weight [4, 7]: 79000.

Specificity: highest light quantum efficiency results with FMNH$_2$, followed by flavin-isomers and flavin-analogues [7, 8].
The enzyme reacts only with aldehydes containing more than eight C-atoms [6, 9].

pH Optimum [6]: 6.8.

Assay

$$FMN + NADH + H^+ \quad \xrightarrow{\text{FMN reductase}} \quad FMNH_2 + NAD^+$$

$$FMNH_2 + \text{myristic aldehyde} + O_2 \quad \xrightarrow{\text{luciferase}} \quad FMN + \text{myristate} + H_2O$$

$$+ \text{light}$$

SA I Biometer 3000, calibrated with [^{14}C]-standard;
final volume: 2.00 ml; 25°C; sensitivity 6.8; zero ca. 5.32; delay 5 sec; assay 5 sec.

1.50 ml	potassium phosphate buffer (80 mmol/l; pH 7.0; glycerol, 20%; mercaptoethanol, 0.2 mmol/l)	phosphate glycerol mercapto-ethanol	60 mmol/l 15% 0.15 mmol/l
0.05 ml	FMN, Na salt, (0.5 mmol/l)	FMN	12.5 nmol/l
0.20 ml	myristic aldehyde (mix 100 mg myristic aldehyde, liquified by warming with 100 ml bovine serum albumin, 5%, containing Triton X-100, 1%)	myristic aldehyde BSA Triton X-100	0.47 mmol/l 5 g/l 1 g/l
0.20 ml	NADH, Na salt, (2.0 mmol/l)	NADH	0.2 mmol/l
0.01 ml	FMN reductase (10 U/ml)		50 U/l
0.04 ml	enzyme in phosphate buffer (see above)		

Count impulses and correct for the blank. Light emission (bioluminescence) is expressed by the impulses per sample volume per unit time ($C \times v^{-1} \times t^{-1}$). The catalytic concentration of the enzyme sample is

$$b = \frac{C}{t \times v \times 1.44 \times 10^8} \quad U/l$$

1.44×10^8 is an instrument specific factor.

Stability: in the lyophilized state a decrease in activity of approx. 20% may occur within 12 months at 4°C.

Purity required: activity $\geqslant$ 1.2 mU/mg substance (25°C); $\triangleq$ ca. 15 mU/mg enzyme protein. Contaminants $\leqslant$ 10 mU NADH : FMN oxidoreductase/mg protein, $\leqslant$ 10 mU myokinase/mg protein, $\leqslant$ 1 mU "NADH oxidase"/mg protein.

References

[1] *I. W. Hastings,* in: *S. P. Colowick, N. O. Kaplan* (eds.), Methods in Enzymology, Vol. *XVII*, Academic Press, New York 1978, p. 125.
[2] *I. W. Hastings, I. E. Becvar,* in: *S. P. Colowick, N. O. Kaplan* (eds.), Methods in Enzymology, Vol. *XVII*, Academic Press, New York 1978, p. 194.
[3] *I. W. Hastings, K. H. Nealson,* Annu. Rev. Microbiol. *31*, 549 (1977).
[4] *I. W. Hastings et al.,* J. Biol. Chem. *247*, 398 (1972).
[5] *S. C. Tu,* in: *S. P. Colowick, N. O. Kaplan* (eds.), Methods in Enzymology, Vol. *XVII*, Academic Press, New York 1978, p. 171.
[6] *I. W. Hastings et al.,* Biochem. *8*, 4681 (1969).
[7] *I. W. Hastings et al.,* in: *S. P. Colowick, N. O. Kaplan* (eds.), Methods in Enzymology, Vol. *XVII*, Academic Press, New York 1978, p. 135.
[8] *I. W. Hastings, G. W. Mitchell,* J. Biol. Chem. *244*, 2572 (1969).
[9] *S. Ulitzur, I. W. Hastings,* Proc. Natl. Acad. Sci. *76*, 265 (1979).

Luciferase

from *Photinus pyralis* (American Firefly)

Photinus luciferin : oxygen 4-oxidoreductase (decarboxylating, ATP-hydrolyzing), EC 1.13.12.7

$$\text{D-Luciferin} + \text{ATP} + O_2 \xrightarrow{\text{luciferase}} $$

$$\text{oxyluciferin} + CO_2 + H_2O + \text{AMP} + PP_i + h \times \nu_{562}$$

Michaelis constants:

ATP [1]	5.1×10^{-5} mol/l
ATP [2]	2.3×10^{-4} mol/l
D-Luciferin-AMP [3]	2.3×10^{-7} mol/l
D-Luciferin	2.4×10^{-6} mol/l

Effectors: cations such as Mg^{2+}, Mn^{2+}, Fe^{2+}, Co^{2+} and Zn^{2+} are necessary for light emission [1]. Pyrophosphate and adenyloxyluciferin are potent inhibitors of the luminescence reaction [3]. Dehydro-luciferin inhibits the luciferase [4]. Amines, copper, SH-reagents (4-chloromercuribenzoate) and high salt concentrations inhibit. Enzyme activity is not influenced by cyanide or azide. For inhibition by anions see reference 5. Temperatures above 35 °C inactivate the enzyme irreversibly [2].

Molecular weight [4, 6]: 100000.

Specificity: dATP [4], A-5'-tetraphosphate [7], diadenosine 5'-pentaphosphate [7] react with 5, 2.2 or 0.75 respectively, of the relative rate with ATP as substrate. ADP (in the absence of myokinase), AMP, ITP, CTP, UTP, creatine phosphate or other phosphorylated metabolites do not react [1]. Synthetic luciferyladenylate produces light emission with luciferase in the absence of ATP [8].

pH Optimum [4, 2]: 7.5 – 7.8.

Assay

$$\text{ATP} + \text{luciferin} \xrightarrow[Mg^{2+}]{\text{luciferase}} \text{luciferase} \cdot \text{luciferin} \cdot \text{AMP} + PP_i$$

$$\text{Luciferase} \cdot \text{luciferin} \cdot \text{AMP} \xrightarrow{O_2}$$

$$\text{luciferase} + \text{oxyluciferin} + \text{AMP} + CO_2 + \text{light}$$

$$PP_i + \text{UDPglucose} \xrightarrow[Mg^{2+}]{\text{UDPG-PP}} \text{UTP} + \text{glucose-1-P}$$

$$\text{Glucose-1-P} + \xrightarrow[\text{Glucose-1.2-P}_2]{\text{PGluM}} \text{glucose-6-P}$$

$$\text{Glucose-6-P} + \text{NADP}^+ \xrightarrow{\text{G6P-DH}} \text{gluconate-6-P} + \text{NADPH} + \text{H}^+$$

$$\text{NADPH} + \text{H}^+ + \text{NBT} \xrightarrow{\text{diaphorase}} \text{NADP}^+ + \text{formazan} .$$

Wavelength 546 nm; light path 10 mm; final volume 3.01 ml; 25 °C.

2.00 ml	Tris buffer (Tris, 0.1 mol/l; pH 7.75; EDTA, 3 mmol/l; bovine serum albumin, 20 mg/ml)	Tris	66 mmol/l
		EDTA	2 mmol/l
		BSA	13 g/l
0.10 ml	magnesium acetate (150 mmol/l)		5 mmol/l
0.10 ml	D-luciferin (ca. 3 mmol/l)		0.1 mmol/l
0.10 ml	ATP, Na salt, (20 mmol/l)		0.67 mmol/l
0.20 ml	UDPglucose, Na salt, (29 mmol/l)		1.9 mmol/l
0.10 ml	NADP, Na salt, (11 mmol/l)		0.36 mmol/l
0.05 ml	glucose-1.6-P_2, Na salt, (1 mmol/l)		0.017 mmol/l
0.20 ml	NBT (6 mmol/l)		0.4 mmol/l
0.02 ml	G6P-DH (2.5 mg/ml Tris buffer)		5.8 kU/l
0.02 ml	PGluM (2 mg/ml Tris buffer)		2.66 kU/l
0.05 ml	UDPG-PP (5 mg/ml glycerol, 50%)		8.3 kU/l
0.02 ml	diaphorase from micro-organisms (5 mg/ml Tris buffer)		0.5 kU/l
0.05 ml	enzyme in Tris buffer		

run a blank which contains Tris buffer instead of luciferase.

Stability: in the lyophilized state a decrease in activity of approx. 20% may occur within 12 months at 4 °C.

Purity required: specific activity $\geqslant 8$ mU/mg protein (25 °C). Contaminants $\leqslant 1$ mU ATPase/mg substance, $\leqslant 10$ mU nucleoside diphosphate kinase/mg substance.

References

[1] *B. L. Strehler* in: *H. U. Bergmeyer* (ed.), Methods of Enzymatic Analysis, 2nd English edition, Vol. *IV*, Verlag Chemie, Weinheim 1974, p. 2112.
[2] *A. A. Green, W. D. McElroy,* Biochim. Biophys. Acta *20*, 170 (1956).
[3] *W. C. Rhodes, W. D. McElroy,* J. Biol. Chem. *233*, 1528 (1958).
[4] *M. De Luca, W. D. McElroy,* in: *S. P. Colowick, N. O. Kaplan* (eds.), Methods in Enzymology, Vol. *LVII*, Academic Press, New York 1978, p. 3.
[5] *I. L. Denburg, W. D. McElroy,* Arch. Biochem. Biophys. *141*, 668 (1970).
[6] *J. Travis, W. D. McElroy,* Biochem. *5*, 2170 (1966).
[7] *G. Momsen,* Biochem. Biophys. Res. Commun. *84*, 816 (1978).
[8] *M. De Luca, W. D. McElroy,* Biochem. Biophys. Res. Commun. *18*, 836 (1965).

Lysine Decarboxylase

from *Bacterium cadaveris*

L-Lysine carboxy-lyase, EC 4.1.1.18

$$\text{L-Lysine} \xrightarrow{\text{lysine decarboxylase}} \text{cadaverine} + CO_2$$

Michaelis constants (phosphate buffer, pH 6; 30 – 37°C) [1]

Lysine	$3.7 \times 10^{-4}\,mol/l$
S-(2-aminoethyl)-L-cysteine	$4.5 \times 10^{-3}\,mol/l$

Molecular weight [2]: $1\,000\,000 \pm 50\,000$.

Specificity [2, 4]: the enzyme is specific for lysine and hydroxylysine. D-Lysine, L-arginine, L-glutamate, L-ornithine, L-histidine, DL-alanine, DL-valine, L-leucine, L-phenylalanine, L-tyrosine, L-tryptophan, DL-serine, L-proline, L-aspartate and the 2-acetyl, 2-methyl, 6-acetyl and 6-methyl derivates of lysine do no react.

Assay

Manometric determination; final volume 3.00 ml; 37°C.

Main compartment

0.5 ml	water	
1.0 ml	acetate buffer (0.5 mol/l; pH 6.0)	333 mmol/l
1.0 ml	enzyme in acetate buffer	

Side arm

0.5 ml	L-lysine (0.5 mol/l; pH 6.0)	83 mmol/l

Stability: lyophilized enzymes, stored dry at −20°C, are stable for several months.

Purity required: activity ≧ 0.1 U/mg substance (37°C).

References

[1] *K. Soda, M. Moriguchi,* Biochem. Biophys. Res. Commun. *34*, 34 (1969).
[2] *K. E. van Holde, R. L. Baldwin,* J. Phys. Chem. *62*, 734 (1958).
[3] *S. Linstedt,* Acta Chem. Scand. *5*, 486 (1951).
[4] *E. P. Gale, H. M. R. Epps,* Biochem. J. *38*, 232 (1944).

Malate Dehydrogenase

from pig heart (mitochondrial)

L-Malate: NAD$^+$ oxidoreductase, EC 1.1.1.37

$$\text{L-Malate} + \text{NAD}^+ \xrightleftharpoons{\text{MDH}} \text{oxaloacetate} + \text{NADH} + \text{H}^+$$

Equilibrium constant

$$K = \frac{[\text{oxaloacetate}] \times [\text{NADH}] \times [\text{H}^+]}{[\text{malate}] \times [\text{NAD}^+]} = \begin{cases} 2.33 \times 10^{-12}\,\text{mol/l}\,[1] \\ 1.2 \ \times 10^{-12}\,\text{mol/l}\,[1] \\ 7.5 \ \times 10^{-13}\,\text{mol/l}\,[1] \\ 6.4 \ \times 10^{-13}\,\text{mol/l}\,[2] \end{cases}$$

Michaelis constants [3]

		Relative rate
L-Malate	4.0×10^{-4} mol/l	100
L-Tartrate	9.0×10^{-3} mol/l	1.64
meso-Tartrate	1.2×10^{-3} mol/l	0.94
Oxaloacetate	3.3×10^{-5} mol/l	(phosphate buffer, 95 mmol/l; pH 8.3; 25°C) [4]

Effectors: iodinated compounds such as iodine cyanide, thyroxine, and molecular iodine inactivate the enzyme [5].

Molecular weight (pH 7.0) [6]: 70000.

Specificity [3]: the enzyme is specific for the L-configuration, D-malate and D-tartrate do not react. NAD can be replaced by its analogues, but not by NADP.

Assay

Wavelength 339, Hg 334 or Hg 365 nm; light path 10 mm; final volume 3.17 ml; 25°C.

3.00 ml	phosphate buffer (0.1 mol/l; pH 7.5)	94.6 mmol/l
0.10 ml	oxaloacetic acid (2 mg/ml buffer)	0.5 mmol/l
0.05 ml	NADH, Na salt, (10 mg/ml)	0.2 mmol/l
0.02 ml	enzyme solution in buffer	

Stability: crystalline suspensions in ammonium sulphate solution, 3.2 mol/l; pH ca. 6, or solutions in 50% glycerol; pH ca. 7, are stable for several months at 4°C.

Purity required: specific activity $\geq$ 1100 U/mg protein (25 °C). Contaminants (related to the specific activity of MDH) $\leq$ 0.005% ASAT, $\leq$ 0.01% fumarase and LDH.

References

[1] *F. B. Straub,* in: *Hoppe-Seyler/Thierfelder,* Handbuch der physiol. u. path.-chem. Analyse, Vol. *VIa,* Springer, Berlin 1964, p. 367.
[2] *A. Yoshida,* J. Biol. Chem. *240,* 1113, 1118 (1965).
[3] *C. J. R. Thorne,* Biochim. Biophys. Acta *59,* 624 (1962).
[4] *H. Möllering,* unpublished experiments.
[5] *S. Varrone et al.,* Eur. J. Biochem. *13,* 305 (1970).
[6] *C. J. R. Thorne, N. O. Kaplan,* J. Biol. Chem. *238,* 1861 (1963).

α-D-Mannosidase

from jack bean (*Canavalia ensiformis*)

α-D-Mannoside mannohydrolase, EC 3.2.1.24

$$\text{An α-D-mannoside} + H_2O \xrightleftharpoons{\text{α-mannosidase}} \text{an alcohol} + \text{D-mannose}$$

Michaelis constants

4-Nitrophenyl-α-D-mannoside	2.5×10^{-3} mol/l	⎫ (citrate buffer;
Benzyl-α-D-mannoside	3.1×10^{-2} mol/l	⎬ pH 4.5; 25 °C)
Methyl-α-D-mannoside	1.2×10^{-1} mol/l	⎭

Inhibitor constants [1]

Mannono-(1 → 4)-lactone	1×10^{-2} mol/l
Mannono-(1 → 5)-lactone	1.2×10^{-4} mol/l

Effectors: competitive inhibitors are mannono-(1 → 4)- and mannono-(1 → 5)-lactone [1]. The enzyme requires Zn^{2+} for catalytic activity [2, 3].

Specificity [1]: the enzyme hydrolyzes methyl-, benzyl- and 4-nitrophenyl-α-D-mannosides, α-(1 → 2)-, α-(1 → 3)- and α-(1 → 6)-oligomannosides. It cleaves the terminal mannose residues of ovalbumin, ovomucoid and orosomucoid. α-Mannosidase is thus a glycosyltransferase. β-Mannosides, α- or β-glucosides and α- or β-galactosides are not hydrolyzed.

pH Optimum [1]: 4 − 5.

Assay

$$\text{4-Nitrophenyl-}\alpha\text{-D-mannoside} + H_2O \xrightarrow{\alpha\text{-mannosidase}} \text{D-mannose} + \text{4-nitrophenol.}$$

Wavelength Hg 405 nm; light path 10 mm; incubation volume 1.02 ml; final volume 3.02 ml; 25°C; $\varepsilon = 1.85\ l \times mmol^{-1} \times mm^{-1}$.

0.50 ml	citrate buffer (0.1 mol/l; pH 8)	49 mmol/l
0.50 ml	4-nitrophenyl-α-D-mannoside (10 mmol/l)	4.9 mmol/l
0.02 ml	enzyme in water	

incubate for 5 min at 25°C, stop reaction by addition of

2.00 ml	borate buffer (0.2 mol/l, pH 9.8)	132 mmol/l

read absorbance against blank without sample.

Stability: suspensions in ammonium sulphate solution, 3.2 mol/l; pH ca. 6, are stable for several months at 4°C.

Purity required: specificity activity $\geqslant$ 10 U/mg protein (25°C). Contaminants (related to the specific activity of α-mannosidase) $\leqslant$ 0.01% α-galactosidase, β-galactosidase and α-glucosidase, $\leqslant$ 1% N-acetylglucosaminidase.

References

[1] *Y. T. Li,* J. Biol. Chem. *242,* 5474 (1967).
[2] *S. M. Snaith, G. A. Levy,* Nature *218,* 91 (1968).
[3] *S. M. Snaith, G. A. Levy,* Biochem. J. *110,* 663 (1968).

Myokinase, Adenylate Kinase

from rabbit muscle

ATP : AMP phosphotransferase, EC 2.7.4.3

$$\text{ATP} + \text{AMP} \xrightleftharpoons{MK} 2\,\text{ADP}$$

Equilibrium constant (pH 7.4; 25°C; Mg^{2+}, 10 mmol/l)

$$K = \frac{[\text{ADP}]^2}{[\text{ATP}] \times [\text{AMP}]} = 2.26$$

Michaelis constants (Tris buffer, pH 7.5; 30°C) [1]

ATP	3×10^{-4} mol/l
AMP	5×10^{-4} mol/l
ADP	1.58×10^{-3} mol/l

Inhibitor constants (Tris buffer, pH 7.5)

ATP	3.2×10^{-4} mol/l	
AMP	5×10^{-4} mol/l	
ADP	1.6×10^{-3} mol/l	30°C [1]
Adenosine monosulphate	1.86×10^{-5} mol/l	
ADP	3.3×10^{-4} mol/l	25°C [2]

Molecular weight [3]: 21 000.

Specificity: the enzyme is specific for AMP, dAMP, ADP, dADP, ATP and dATP [4, 5]. No other nucleoside mono-, di- and triphosphates react.

Assay

$$ATP + AMP \xrightarrow{MK} 2\,ADP$$

$$2\,ADP + 2\,PEP \xrightarrow{PK} 2\,ATP + 2\,pyruvate$$

$$2\,Pyruvate + 2\,NADH + 2\,H^+ \xrightarrow{LDH} 2\,L\text{-lactate} + 2\,NAD^+$$

Wavelength 339, Hg 334 or Hg 365 nm; light path 10 mm; final volume 2.80 ml; 25°C.

2.00 ml	triethanolamine buffer (0.1 mol/l; pH 7.6)	71.4 mmol/l
0.20 ml	AMP, Na salt, (10 mg/ml)	1.4 mmol/l
0.20 ml	ATP, Na salt, (10 mg/ml)	1.2 mmol/l
0.10 ml	PEP, tricyclohexylammonium salt, (5 mg/ml)	0.39 mmol/l
0.10 ml	$MgSO_4$ (32 mmol/l)	1.2 mmol/l
0.10 ml	KCl (4 mol/l)	142 mmol/l
0.05 ml	NADH, Na salt, (10 mg/ml)	0.2 mmol/l
0.01 ml	PK (10 mg/ml)	7.1 kU/l
0.02 ml	LDH (5 mg/ml)	19.6 kU/l
0.02 ml	enzyme in buffer	

Stability: suspensions in ammonium sulphate solution, 3.2 mol/l; pH ca. 6, are stable for several months at 4°C.

Purity required: specific activity $\geq$ 360 U/mg protein (25°C). Contaminant (related to the specific activity of MK) $\leq$ 0.01% ATPase.

References

[1] *O. H. Callaghan, G. Weber,* Biochem. J. *73*, 473 (1959).
[2] *L. Noda,* J. Biol. Chem. *232*, 237 (1958).
[3] *O. H. Callaghan,* Biochem. J. *67*, 651 (1957).
[4] *W. Gruber, H. Möllering, H. U. Bergmeyer,* Enzym. Biol. Clin. 7, 115 (1966).
[5] *K. Beaucamp,* unpublished experiments.

NADH-Peroxidase

from *Streptococcus faecalis*

NADH: hydrogen-peroxide oxidoreductase, EC 1.11.1.1

$$NADH + H^+ + H_2O_2 \xrightarrow{\text{NADH-POD}} NAD^+ + 2\,H_2O$$

Equilibrium constant: the reaction is irreversible.

Michaelis constants [1]

Substrate		Second Substrate
NADH	1.4×10^{-5} mol/l	H_2O_2
NADH	6.1×10^{-6} mol/l	ferricyanide
H_2O_2	2.0×10^{-4} mol/l	NADH
Ferricyanide	1.1×10^{-4} mol/l	NADH

(Na-acetate buffer, pH 5.6; 26°C)

Effectors [1, 2]: the enzyme contains FAD as prosthetic group and is inactivated by FAD-splitting or blocking substances such as KBr. Heavy metals are strong inhibitors; phosphate ions and higher substrate concentrations decrease the enzyme activity.
Various anions have activating effects, increasing in order of listing: chloride, sulphate, lactate, formate, oxalate, succinate, butyrate, propionate and acetate.

Molecular weight [1]: 120000.

Specificity [1]

	Relative rate (oxidant)
H_2O_2	100
Ferricyanide	30
1.4-Naphthoquinone	30
Menadione	20

No reaction was observed with glutathione (oxidized), cysteine, dehydroascorbate, cytochrome c, FAD, FMN, riboflavin, 2.6-dichlorophenol indophenol, methylene blue and cresol blue.

	Relative rate (prosthetic group)
NADH	100
APADH	84
Deamino-NADH	18
Pyridine-aldehyde	1.2

pH Optimum: 6.

Assay

Wavelength 339, Hg 334 or Hg 365 nm; light path 10 mm; final volume 3.30 ml; 25°C.

3.00 ml	Tris/acetate buffer (Tris, 0.2 mol/l; potassium acetate, 0.15 mol/l; pH 6.0)	Tris acetate	0.18 mol/l 0.14 mol/l
0.10 ml	NADH, Na salt, (12 mmol/l)		0.36 mmol/l
0.10 ml	enzyme in water		
0.10 ml	H_2O_2 (40 mmol/l)		1.2 mmol/l

Stability: suspensions in ammonium sulphate solution, 3.2 mol/l, are stable for several months at 4°C.

Purity required: specific activity ≥ 45 U/mg protein (25°C). Contaminant (related to the specific activity of NADH-POD) $\leq 2\%$ "NADH-oxidase".

References

[1] *M. I. Dolin,* J. Biol. Chem. *225,* 557 (1957).
[2] *M. I. Dolin,* J. Biol. Chem. *250,* 310 (1975).

NAD(P)H : FMN-Oxidoreductase

from *Photobacterium fischeri*

NAD(P)H : FMN oxidoreductase, EC 1.6.8.1

$$FMN + NAD(P)H + H^+ \xrightarrow{\text{FMN reductase}} FMNH_2 + NAD(P)^+$$

Michaelis constants [1]

NAD	8×10^{-5} mol/l	
NADPH	4×10^{-4} mol/l	(phosphate buffer, 1.3×10^{-2} mol/l; pH 7)
FMN	7.3×10^{-5} mol/l	EDTA, 0.1 mmol/l)
FAD	1.4×10^{-4} mol/l	

Effectors: electron transfer-blocking and oxidative phosphorylation-uncoupling agents are inhibitors [1].

Effector	concentration mol/l	% inhibition
Dicoumarol	10^{-7}	50
	10^{-3}	100
2.4-Dinitrophenol	10^{-3}	100
Rotenon	10^{-3}	100

Antimycin A, AMP and N-ethylmaleimide (both non-competitive) inhibit [2]. Amytal, 4×10^{-4} mol/l; does not inhibit the reaction or the coupled assay with luciferase. The enzyme is resistant towards proteolytic attack by trypsin [3].

Molecular weight [1, 4]: 45000.

Specificity: the enzyme is twice as active with NADH as with NADPH. Reduced lipoic acid is not a hydrogen donor [1]. FMN-isomers and analogues, FAD and riboflavin react at significantly slower rates [5]. Ferricyanide reacts at 15% of the rate with FMN, also 2.6-dichlorophenol indophenol and menadione in the presence of traces of FMN, 10^{-7} mol/l [1].

pH Optimum: $6-7$.

Assay

Wavelength 339, Hg 334 or Hg 365 nm; light path 10 mm; final volume 3.05 ml; 25°C.

2.75 ml	phosphate buffer (potassium phosphate 50 mmol/l; pH 7.0; EDTA, 0.1 mmol/l)	phosphate EDTA	45 mmol/l 90 µmol/l
0.10 ml	NADH, Na salt, (20 mmol/l)		0.65 mmol/l
0.10 ml	bovine serum albumin (30 mg/ml)		1 g/l
0.05 ml	FMN, Na salt, (Sigma F 2253, 7 mmol/l)		0.11 mmol/l
0.05 ml	enzyme in glycerol (glycerol, 40%; potassium phosphate, 50 mmol/l; pH 7.0; EDTA 0.1 mmol/l; DDT, 0.1 mmol/l)		

Stability: in the lyophilized state a decrease in activity of approx. 20% may occur within 12 months at 4°C.

Purity required: activity $\geqslant 7$ U/mg substance (25°C; $\triangleq$ ca. 100 U/mg enzyme protein).

References

[1] *W. Duane, I. W. Hastings,* Mol. Cell. Biochem. *6,* 53 (1975).
[2] *S. Tu, I. E. Becvar, I. W. Hastings,* Arch. Biochem. Biophys. *193,* 110 (1979).
[3] *K. Puget, A. M. Michelson,* Biochimie *54,* 1197 (1972).
[4] *E. Jablonsky, M. De Luca,* Biochemistry *16,* 2932 (1977).
[5] *G. Mitchell, I. E. Hastings,* J. Biol. Chem. *244,* 2572 (1969).

NAD-Pyrophosphorylase

from hog liver

ATP:NMN adenylyltransferase, EC 2.7.7.1

$$\text{NMN} + \text{ATP} \xrightleftharpoons{\text{NAD-pyrophosphorylase}} \text{NAD} + \text{PP}_i$$

Equilibrium constant (glycylglycine buffer, pH 7.4; 38°C) [1]

$$K = \frac{[\text{PP}_i] \times [\text{NAD}]}{[\text{ATP}] \times [\text{NMN}]} = 0.45$$

Michaelis constants (glycylglycine buffer, pH 7.4; 38°C)

		Relative rate [1]
ATP	4.6×10^{-4} mol/l	1.00
NMN	1.5×10^{-4} mol/l	1.00
PP$_i$	1.9×10^{-4} mol/l	0.48
NAD	8.3×10^{-5} mol/l	0.48
Mg^{2+}	2×10^{-4} mol/l	–

Specificity: nicotinic acid mononucleotide is the natural substrate under physiological conditions. The oxidized ribonucleotides can be replaced by the reduced forms. The following are inactive: nicotinamide riboside, ADP, AMP, NADP or FAD. PP$_i$ cannot be replaced by P$_i$ or metaphosphate.

Assay

$$\text{NMN} + \text{ATP} \xrightarrow{\text{NAD-pyrophosphorylase}} \text{NAD}^+ + \text{PP}_i$$

$$\text{NAD}^+ + \text{ethanol} \xrightarrow{\text{ADH}} \text{NADH} + \text{H}^+ + \text{acetaldehyde}.$$

Wavelength 339, Hg 334 or Hg 365 nm; light path 10 mm; incubation volume 1.00 ml; final volume 3.01 ml; 37°C.

Incubation mixture

0.20 ml	glycylglycine buffer (0.25 mol/l; pH 7.4)	50 mmol/l
0.10 ml	ATP, Na salt, (12 mg/ml; neutralized)	1.9 mmol/l
0.20 ml	NMN (20 mg/ml buffer)	10.8 mmol/l
0.10 ml	MgCl$_2$ (0.15 mol/l)	15 mmol/l
0.30 ml	water	
0.10 ml	enzyme in water	

incubate for 30 min at 37°C, place in a boiling water-bath for 1 min and centrifuge off precipitate.

Assay mixture

2.50 ml	pyrophosphate buffer (75 mmol/l; pH 8.7; 0.84 g semicarbazide-HCl, 0.17 g glycine, 1.00 ml 96% ethanol per 100 ml)	PP$_i$ semicarbazide glycine ethanol	62 mmol/l 62.5 mmol/l 19 mmol/l 165 mmol/l
0.50 ml	supernatant fluid		
0.01 ml	ADH yeast (30 mg/ml)		20 kU/l

Stability: the lyophilized enzyme stabilized with sucrose is stable for several months at 4°C; the enzyme solution in water is stable for a few days at 4°C.

Purity required: specific activity $\geq$ 0.15 U/mg protein (37°C). Contaminant (related to the specific activity of NAD-pyrophosphorylase) $\leq$ 0.5% "NADH oxidase".

Reference

[1] *A. Kornberg,* J. Biol. Chem. *182,* 779 (1950).

Neuraminidase

from *Clostridium perfringens*

Acylneuraminyl hydrolase, EC 3.2.1.18

N-Acetyl-neuraminyl glycoside $\xrightarrow{\text{neuramindase}}$ N-acetylneuraminic acid + aglycone

Michaelis constant (acetate buffer, pH 4.5; 37°C) [1]

N-acetylneuraminyl lactose 2.4×10^{-3} mol/l

Effectors: in contrast to neuraminidase from *V. cholerae,* the activity is not dependent on the presence of metal ions [1, 2]. The enzyme is inactivated by iodoacetate, arsenite, Fe^{3+}, Hg^{2+} and is competitively inhibited by N-acetylneuraminic acid. Sensitivity to atmospheric oxygen is observed in the presence of traces of heavy metals [3]. The enzyme is protected by chelating agents, e.g. serum albumin or other proteins, and KCN.

Molecular weight [4]: 56000 (by gel filtration).

Specificity: hydrolysis of 2,3- 2,6- and 2,8-glucosidic linkages joining terminal non-reducing N- or O-acylneuraminyl residues to galactose, N-acetylhexosamine, or N- or O-acylated neuraminyl residues in oligosaccharides, glycoproteins, glycolipids or colominic acid. According to sialic acid characterization, the enzymes from *C. perfringens* and the *V. cholerae* have wide specificity including both hydrophilic (e.g. glycoproteins) and amphiphilic (e.g. gangliosides) glycoconjugates [5,6]. The glycosidic linkage of N-acetyl-4-O-acetylneuraminic acid is resistant to their action [7].

pH Optimum [8]: pH 5.

Assay

$$NANA\text{-}lactose + H_2O \xrightarrow{\text{neuraminidase}} NANA + lactose$$

$$Lactose + H_2O \xrightarrow{\text{β-galactosidase}} glucose + galactose$$

$$Galactose + NAD^+ \xrightarrow{\text{Gal-DH}} galactono\text{-}lactone + NADH + H^+.$$

Wavelength 339, Hg 334 or Hg 365 nm; light path 10 mm; incubation volume 0.80 ml; incubation temperature 25°C; final volume 3.09 ml; temperature 20–25°C.

Incubation mixture

0.30 ml	acetate buffer (0.1 mol/l; pH 4.5)	39 mmol/l
0.40 ml	NANA-lactose, ammonium salt (10 mg/ml)	5.5 mmol/l
0.10 ml	enzyme in 0.3% bovine serum albumin	

incubate for exactly 10 min at 25°C; place in a boiling water-bath for 1 min and cool to room temperature.

Assay mixture

2.50 ml	phosphate buffer (0.1 mol/l; pH 7.6)	80 mmol/l
0.10 ml	NAD (10 mg/ml)	0.45 mmol/l
0.05 ml	MgSO$_4$ (0.05 mol/l)	0.8 mmol/l
0.40 ml	incubation mixture	
0.02 ml	Gal-DH (5 mg/ml)	0.16 U/l
0.02 ml	β-galactosidase (5 mg/ml)	0.97 kU/l

Stability: the powder is stable for many months at 4°C.

Purity required: activity $\geqslant$ 1 U/mg substance (25°C).

References

[1] *J. T. Cassidy, G. W. Jourdian, S. C. Roseman,* J. Biol. Chem. *240,* 3501 (1965).
[2] *E. A. Popenoe, R. M. Drew,* J. Biol. Chem. *228,* 673 (1957).
[3] *E. Mohr,* Z. Naturforsch. *15b,* 575 (1960).
[4] *E. Balke, R. Drzeniek,* Z. Naturforsch. *24b,* 599 (1969).
[5] *R. Drzeniek,* Histochem. J. *5,* 271 (1973).
[6] *A. Gottschalk, R. Drzeniek* in: *A. Gottschalk* (ed.), Glycoproteins – Their Composition, Structure and Function, Elsevier, Amsterdam 1972, p. 381.
[7] *R. Schauer, H. Faillard,* Hoppe Seyler's Z. Physiol. Chem. *349,* 961 (1968).
[8] *R. M. Burton,* J. Neurochem. *10,* 503 (1963).

Nucleosidediphosphate Kinase

from beef liver

ATP : nucleosidediphosphate phosphotransferase, EC 2.7.4.6

$$\text{ATP + UDP} \quad \underset{\xrightarrow{\hspace{1cm} \text{NDPK}}}{\rightleftharpoons} \quad \text{ADP + UTP}$$

Equilibrium constant

$$K = \frac{[\text{ADP}] \times [\text{UTP}]}{[\text{ATP}] \times [\text{UDP}]} = \text{ca. 1}$$

Molecular weight [1]*: 102000.

Specificity*: the enzyme is not specific. With ADP as phosphoryl acceptor UTP, ITP, GTP, ATP, dATP, dCTP and dGTP react as substrates; UDP, IDP, GDP, CDP and dTDP react with ATP or dATP as donor.

---

* For the enzyme from brewers' yeast.

Assay

$$dTDP + ATP \xrightarrow{\text{NDPK}} dTTP + ADP$$

$$ADP + PEP \xrightarrow{\text{PK}} ATP + pyruvate$$

$$Pyruvate + NADH + H^+ \xrightarrow{\text{LDH}} \text{L-lactate} + NAD^+.$$

Wavelength 339, Hg 334 or Hg 365 nm; light path 10 mm; final volume 3.00 ml; 25°C.

2.50 ml	triethanolamine buffer		
	(0.1 mol/l; pH 7.6)		83.3 mmol/l
0.20 ml	ATP, Na salt, (20 mg/ml)		2.2 mmol/l
0.10 ml	PEP, tricyclohexylammonium salt,	PEP	1.1 mmol/l
	(15 mg/ml; in solution of MgCl$_2$,	MgCl$_2$	16.7 mmol/l
	0.5 mol/l; KCl, 2 mol/l)	KCl	67 mmol/l
0.05 ml	NADH, Na salt, (10 mg/ml)		0.2 mmol/l
0.02 ml	PK (0.2 mg/ml)		266 U/l
0.02 ml	LDH (2 mg/ml)		7.3 kU/l
0.10 ml	dTDP, Na salt, (10 mg/ml)		0.7 mmol/l
0.01 ml	enzyme in buffer		

Prepare blank without sample.

Stability: solutions in 50% glycerol, pH ca. 6, are stable for several months at 4°C.

Purity required: specific activity $\geq$ 80 U/mg protein (25°C; with dTDP as substrate). Contaminant (related to the specific activity of NDPK) $\leq$ 0.1% LDH.

Reference

[1] *R. H. Yue, R. L. Ratliff, S. A. Kuby*, Biochemistry 6, 2923 (1967).

Nucleosidemonophosphate Kinase

from beef liver

ATP : nucleosidemonophosphate phosphotransferase, EC 2.7.4.4

$$ATP + \text{nucleoside monophosphate} \underset{}{\overset{\text{NMPK}}{\rightleftharpoons}} ADP + \text{nucleoside diphosphate}$$

Equilibrium constant [1]

$$K = \frac{[ADP] \times [CDP]}{[ATP] \times [CMP]} = 1.01$$

Specificity: the liver enzyme reacts with ATP and UMP, CMP, GMP. There is no activity with UTP, CTP, GTP, ITP. Mg^{2+} and Mn^{2+} are activators [1].

Assay

$$UMP + ATP \xrightarrow{\text{NMPK}} ADP + UDP$$

$$ADP + UDP + 2\,PEP \xrightarrow{\text{PK}} 2\,\text{pyruvate} + ATP + UTP$$

$$2\,\text{pyruvate} + 2\,NADH + 2\,H^+ \xrightarrow{\text{LDH}} 2\,\text{lactate} + 2\,NAD^+.$$

Wavelength 339, Hg 334 or Hg 365 nm; light path 10 mm; final volume 2.85 ml; 25°C.

2.00 ml	triethanolamine buffer (0.1 mol/l; pH 7.6)		70 mmol/l
0.10 ml	UMP, Na salt, (10 mg/ml)		0.7 mmol/l
0.30 ml	ATP, Na salt, (100 mg/ml)		17.4 mmol/l
0.20 ml	PEP, tricyclohexylammonium salt, (5 mg/ml)		0.7 mmol/l
0.10 ml	MgSO$_4$/KCl (0.5 mol/l; 2 mol/l; resp.)	MgSO$_4$	17.6 mmol/l
		KCl	70 mmol/l
0.08 ml	NADH, Na salt, (10 mg/ml)		0.42 mmol/l
0.01 ml	PK (10 mg/ml)		7 kU/l
0.02 ml	LDH (5 mg/ml)		19.3 kU/l
0.02 ml	enzyme in buffer		

prepare blank without UMP solution.

Stability: the lyophilized enzyme stabilized with sucrose is stable for several months at 4°C; the enzyme solution in water is stable for about 1 week at 4°C.

Purity required: specific activity $\geq$ 0.5 U/mg protein (25°C; measured with UMP and ATP as substrates). Contaminants (related to the specific activity of NMPK) $\leq$ 1% ATPase, $\leq$ 0.2% "NADH oxidase".

Reference

[1] *J. L. Strominger, L. A. Heppe, E. S. Maxwell,* Biochim. Biophys. Acta *32*, 412 (1959).

Nucleoside Phosphorylase

from calf spleen

Purine-nucleoside: orthophosphate ribosyltransferase, EC 2.4.2.1

$$\text{Purine nucleoside} + P_i \xrightarrow{\text{NP}} \text{D-ribose-1-P} + \text{purine}$$

Specificity: the enzyme from spleen catalyzes the phosphorolytic cleavage of adenosine, guanosine, inosine and their deoxy analogues [1].

Assay

$$\text{Inosine} + P_i \xrightarrow{\text{NP}} \text{hypoxanthine} + \text{ribose-1-P}$$

$$\text{Hypoxanthine} + 2\,H_2O + 2\,O_2 \xrightarrow{\text{XOD}} \text{urate} + 2\,H_2O_2 .$$

Wavelength 293 nm; light path 10 mm; final volume 3.13 ml; 25 °C;
$\varepsilon = 1.21 \times \text{mmol}^{-1} \times \text{mm}^{-1}$.

3.00 ml	phosphate buffer (0.1 mol/l; pH 7.4)	96 mmol/l
0.10 ml	inosine (2 mg/ml)	0.26 mmol/l
0.01 ml	XOD (10 mg/ml)	13 U/l
0.02 ml	enzyme in buffer	

Stability: crystalline suspensions in ammonium sulphate solution, 3.2 mol/l; pH ca. 6, are stable for several months at 4 °C.

Purity required: specific activity $\geq$ 20 U/mg protein (25 °C). Contaminants (related to the specific activity of NP) $\leq$ 0.01% AP and guanase, $\leq$ 0.5% ADA.

Reference

[1] *M. Friedkin, H. Kalckar,* in: *P. D. Boyer, H. Lardy, K. Myrbäck* (eds.), The Enzymes, Vol. *5,* Academic Press, New York 1961, p. 237.

OMP-Pyrophosphorylase

from brewer's yeast

Orotidine-5′-phosphate: pyrophosphate phosphoribosyl transferase, EC 2.4.2.10

$$\text{Orotate} + \text{PRPP} \xrightarrow{\text{OMP-pyrophosphorylase}} \text{OMP} + PP_i$$

Equilibrium constant [1]

$$K = \frac{[\text{orotate}] \times [\text{PRPP}]}{[\text{OMP}] \times [\text{PP}_i]} = 8.35$$

Michaelis constants [2]

Orotate	2×10^{-5} mol/l
PRPP	2×10^{-5} mol/l
5-Fluoro-orotate	2×10^{-5} mol/l
PP_i	1.2×10^{-4} mol/l

Inhibitor constants [2]

6-Uracil sulphonic acid	7×10^{-6} mol/l
6-Uracil sulphonamide	3.9×10^{-4} mol/l
6-Uracil methylsulphone	7.1×10^{-4} mol/l

Effectors [2]: Mg ions, 2×10^{-3} mol/l, give optimal activation.

pH Optimum: for the forward reaction (UMP formation) pH 7.5 – 8.5, for the reverse reaction pH 6.0 to 7.0.

Specificity: 5-fluoro-orotate reacts at approx. twice the rate obtained with orotate [4]. 5-Bromo, 5-chloro-, 5-amino-, 5-nitro- and 5-methyl derivates of orotate do not react. The enzyme is specific for PRPP and inorganic pyrophosphate. The enzyme preparation contains OMP-decarboxylase (Orotidine-5'-phosphate carboxy-lyase, EC 4.1.1.23).

Assay

$$\text{Orotate} + \text{PRPP} \xrightarrow{\text{OMP-pyrophosphorylase}} \text{OMP} + \text{PP}_i$$

$$\text{OMP} \xrightarrow{\text{OMP-decarboxylase}} \text{UMP} + \text{CO}_2.$$

Wavelength 295 nm; light path 10 mm; final volume 2.1 ml; 25°C; $\varepsilon = 0.395$ l $\times$ mmol^{-1} $\times$ mm^{-1}.

2.00 ml	glycylglycine buffer (50 mmol/l; pH 8.0; MgCl$_2$, 1 mmol/l)	glycylglycine MgCl$_2$	47.6 mmol/l 1 mmol/l
0.04 ml	orotate (10 mmol/l)		0.19 mmol/l
0.04 ml	PRPP (20 mmol/l)		0.38 mmol/l
0.02 ml	enzyme in buffer		

Stability: the lyophilized material is stable for several months at 4°C.

Purity required: specific activity $\geqq$ 7 U/g protein (25°C; orotate and PRPP as substrates).

References

[1] *H. Möllering,* in: *H. U. Bergmeyer,* Methods in Enzymatic Analysis, 2nd edn., Academic Press, New York 1974, p. 1959.
[2] *J. G. Flask,* in: *S. P. Colowick, N. O. Kaplan* (eds.), Methods in Enzymology, Vol. *VI,* Academic Press, New York 1963, p. 148.
[3] *W. L. Holmes,* J. Biol. Chem. *223,* 677 (1956).
[4] *J. L. Dahl, D. L. Way, R. E. Parks,* J. Biol. Chem. *234,* 2998 (1959).

Oxalate Decarboxylase

from *Aspergillus* species

Oxalate carboxy-lyase, EC 4.1.1.2

$$\text{Oxalate} \xrightarrow{\text{oxalate decarboxylase}} \text{formate} + CO_2$$

Michaelis constants (phosphate buffer, 0.1 mol/l, pH 6; 37°C) [1]

Oxalate 8×10^{-3} [mol/l]

Effectors [2]: sodium azide, thioglycolate, phenylhydrazine, hydroxylamine, HgCl, sulphite, dithiothreitol and mercaptoethanol inhibit.

Specificity [2]: the enzyme is highly specific for oxalate and does not react with formate, acetate, propionate, glyoxalate, glycollate, mesoxalate, pyruvate, oxalo-acetate, oxalosuccinate, succinate, fumarate, tartrate, itaconate, citrate, gluconate or any of the common amino acids.

pH Optimum [3]: pH 6 (phosphate buffer, 0.1 mol/l; 37°C).

Assay

$$\text{Oxalate} \xrightarrow{\text{oxalate-DC}} \text{formate} + CO_2$$

$$\text{Formate} + NAD^+ \xrightarrow{\text{formate-DH}} CO_2 + NADH + H^+.$$

Wavelength 339, Hg 334 or Hg 365 nm; light path 10 mm; incubation volume 0.6 ml; final volume 3.00 ml; 37°C.

0.30 ml	phosphate buffer (0.1 mol/l; pH 5.0)	50 mmol/l
0.10 ml	enzyme in phosphate buffer (pH 5.0)	
0.20 ml	oxalate, dipotassium salt (0.2 mol/l; pH 5.0)	67 mmol/l

incubate for exactly 2 min, add

2.00 ml	phosphate buffer (0.15 mol/l; pH 7.5)	110 mmol/l
0.20 ml	NAD, Li salt, (40 mg/ml)	3.8 mmol/l
0.20 ml	formate-DH (40 U/ml)	2.7 kU/l

Stability: suspensions in ammonium sulphate solutions, 3.2 mol/l; pH 6, are stable for several months at 4°C.

Purity required: specific activity $\geqslant 20$ U/mg protein (37°C).

References

[1] *H. E. Walter, Boehringer Mannheim*, unpublished.
[2] *E. Emiliani, P. Bekes,* Arch. Biochem. Biophys. *105*, 488 (1963).
[3] *H. O. Beutler et al.,* Fresenius Z. Anal. Chem. *301*, 186 (1980).

Oxalate Oxidase

from barley seedlings

Oxalate : oxygen oxidoreductase, EC 1.2.3.4

$$\text{Oxalate} + O_2 \xrightarrow{\text{oxalate oxidase}} 2\,CO_2 + H_2O_2$$

Michaelis constant [1]: oxalate 4×10^{-4} mol/l

Effectors: fluoride, iodoacetate inhibit at concentrations of 5×10^{-4} mol/l [1]. Mercaptoethanol and dithiothreitol interfere when assaying the enzyme with 3-methyl-2-benzothiazolinone hydrazone and N,N-dimethylaniline in the indicator reaction. Excess substrate concentrations (oxalate > 5 mmol/l) inhibit. The enzyme is rather insensitive to heat inactivation (e.g. 10 min/60°C).

Molecular weight [2]: 26000 (SDS-gel electrophoresis).

Specificity: the enzyme is specific for oxalate; reaction with other compounds has not so far been reported.

pH Optimum: pH 3.8 (succinate buffer, 0.05 mol/l).

Assay

$$\text{Oxalate} + O_2 \xrightarrow[\text{oxidase}]{\text{oxalate}} 2\,CO_2 + H_2O_2$$

$$H_2O_2 + \text{MBTH*} \xrightarrow{\text{POD}} \text{indamine dyestuff} + H_2O + OH^-.$$

Wavelength Hg 578 nm; light path 10 mm; final volume 3.02 ml; 37°C; $\varepsilon = 3.0\,l \times$ mmol^{-1} × mm^{-1}.

2.70 ml	succinate buffer/chromogen (50 mmol/l; pH 3.8 containing MBTH, 23 mg/ml;	succinate	45 mmol/l
	dimethylaniline 0.1 ml/l)	MBTH	20 mg/l
		dimethyl-	
		aniline	0.09 ml/l
0.01 ml	POD (1 mg/ml)		0.8 U/l
0.10 ml	EDTA-Na$_2$H$_2$ · 2 H$_2$O (0.1 mol/l)		3.3 mmol/l
0.01 ml	enzyme in water		
0.18 ml	water		
0.02 ml	oxalate, dipotassium salt, (0.2 mol/l)		1.3 mmol/l

Stability: lyophilized and stabilized enzymes are stable for several months at 4°C.

Purity required: activity $\geqslant$ 0.25 U/mg substance (37°C; ca. $\geqslant$ 5 U/mg enzyme protein).

References

[1] *J. Chiriboga,* Arch. Biochem. Biophys. *116*, 516 (1966).
[2] *H. E. Walter, Boehringer Mannheim*, unpublished.

Papain

from *Papaya* latex

EC 3.4.22.2

Papain hydrolyzes proteins, peptides, amides and esters, especially bonds with basic amino acids. Preferential cleavage: Arg-, Lys-, Phe-X- (the peptide bond next but one to the carboxyl group of phenylalanine); limited hydrolysis of native immunoglobulins.

---

* MBTH 3-Methyl-2-benzothiazolinone hydrazone hydrochloride.

Michaelis constants

		k_0 (s^{-1})	
Benzoyl-L-arginine ethyl ester	1.89×10^{-3} mol/l	12.0	pH 6; 38°C [1]
Benzoyl-L-argininamide	3.9×10^{-2} mol/l	11.0	

Hippuryl methyl ester	2.05×10^{-2} mol/l	2.7	pH 6; 35°C [2]
Thiohippuryl methyl ester	6.3×10^{-3} mol/l	0.14	

Hippurylamide	1.6×10^{-1} mol/l	0.6	
Hippuryl methyl ester	2.1×10^{-2} mol/l	3.1	pH 6; 37 – 40°C [1]
Carbobenzoxyhistidinamide	–	4.0	
Carbobenzoxyglycylglycine	2.7×10^{-1} mol/l	0.08	

Carbobenzoxyglycine-4-nitro-phenyl ester	9.3×10^{-6} mol/l	2.73	
Carbobenzoxyglycine-3-nitro-phenyl ester	1.89×10^{-5} mol/l	2.18	
Carbobenzoxyglycine-2-nitro-phenyl ester	1.52×10^{-4} mol/l	2.14	phosphate buffer, pH 6.8; 25°C [3]
Carbobenzoxyglycine phenyl ester	1.07×10^{-4} mol/l	2.45	
Carbobenzoxyglycine ethyl ester	5.14×10^{-3} mol/l	1.96	

Molecular weight [4, 5]: 21 000.

Specificity: amino acid amides, such as benzoyl-L-lysinamide, carbobenzoxy-L-glutamate diamide, carbobenzoxy-L-histidinamide, carbobenzoxy-L-leucinamide, L-leucinamide and hippurylamide are hydrolyzed at rates which decrease in the order given [5].

Assay

$$\text{Benzoyl-L-arginine ethyl ester} + H_2O \xrightarrow{\text{papain}} \text{benzoyl-L-arginine} + \text{ethanol} .$$

Potentiometric determination of the acid equivalents liberated. Final volume 7.52 ml; 25°C.

3.75 ml	substrate benzoyl-L-arginine ethyl ester, BAEE (80 mmol/l; EDTA, 0.52 mmol/l cysteine, 2.6 mmol/l; pH 6.2)	BAEE EDTA cysteine	40 mmol/l 0.26 mmol/l 1.3 mmol/l
2.50 ml	NaCl (3 mol/l)	NaCl	1.0 mmol/l
1.25 ml	water		
0.02 ml	enzyme in water		

Stability: suspensions in acetate buffer (50 mmol/l; pH 4.5, with NaCl, 0.2 mol/l) are stable for several months at 4°C.

Purity required: specific activity $\geqq$ 30 U/mg protein (25°C).

References

[1] *L. A. E. Sluyterman,* Biochim. Biophys. Acta *85,* 305, 316 (1964).
[2] *G. Lowe, A. Williams,* Biochem. J. *96,* 189 (1965).
[3] *J. F. Kirach, M. Igelstrom,* Biochemistry, *5,* 783 (1966).
[4] *L. Cunningham,* in: *M. Florkin, E. H. Stotz* (eds.), Comprehensive Biochemistry, Vol. *16,* Elsevier Publ. Comp., Amsterdam 1965, p. 85.
[5] *E. L. Smith, J. R. Kimmel,* in: *P. D. Boyer, H. Lardy, K. Myrbäck* (eds.), The Enzymes, Vol. *4,* New York 1960, p. 133.

Pepsin

from porcine gastric mucosa

EC 3.4.23.1

Pepsin hydrolyzes peptides including those linkages with adjacent aromatic or dicarboxylic L-amino acids; preferential cleavage: Phe-, Leu-.

Michaelis constants

Acetyl-L-phenylalanyl- diiodotyrosine	7.5×10^{-5} mol/l	glycine buffer, pH 2.0; 37°C [1]
Acetyl-L-phenylalanyl- L-phenylalanine	4.3×10^{-4} mol/l	citrate buffer, pH 1.85; 37°C [2]
Carbobenzoxy-L-glutamyl- L-tyrosine	1.89×10^{-3} mol/l	pH 4.0; 31.6°C [3]
Carbobenzoxy-L-glutamyl- L-tyrosine ethyl ester	1.78×10^{-3} mol/l	
Acetyl-L-phenylalanyl- L-tyrosine	2.4×10^{-3} mol/l	pH 2.0; 37°C [4]
Acetyl-L-tyrosyl-L-tyrosine	6.3×10^{-3} mol/l	

Inhibitor constants

Acetyl-D-phenylalanyl-L-diiodotyrosine	8×10^{-5} mol/l	} glycine buffer,
Acetyl-L-diiodotyrosine	8.8×10^{-4} mol/l	} pH 2.0; 37 °C [1]
Acetaminocinnamoyl-L-diiodotyrosine	9×10^{-5} mol/l	pH 2.0; 35 °C [5]
Acetyl-L-phenylalanyl-D-phenylalanine	1.4×10^{-3} mol/l	} citrate buffer,
Acetyl-D-phenylalanyl-L-phenylalanine	4.3×10^{-3} mol/l	} pH 1.85; 37 °C [2]
Methanol	6.08×10^{-4} mol/l [6]	
n-Amyl alcohol	1.4×10^{-2} mol/l [6]	

Molecular weight [6, 7]: 36 000.

Specificity: only peptide bonds are hydrolyzed. Esters and amides do not react. Both the amino acids must have the L-configuration. Acylation of the α-amino group increases the sensitivity of the dipeptide to pepsin. The rate of hydrolysis decreases if glutamate is replaced by glutamine [7, 8].

Assay: the rate of hydrolysis of denatured haemoglobin by pepsin is measured by the increase in absorption at 280 nm according to *Anson* [9]. A unit is the enzyme activity which causes an increase in absorption of 0.001 per min under the conditions of the assay.

Wavelength 280 nm; light path 10 mm; final volume 6.0 ml; 37 °C.

5.0 ml	haemoglobin solution	Hb	16.7 mg/ml
	(haemoglobin, 20 mg/ml; HCl, 60 mmol/l)	HCl	52 mmol/l
0.1 ml	enzyme in HCl (10 mmol/l)		

incubate for 10 min at 37 °C.

10.0 ml   trichloroacetic acid (5%)

filter, measure absorbance of filtrate.

Stability: lyophilized enzymes are stable for several months at $+4$ °C.

Purity required: activity ≥ 2500 units/mg substance according to *Anson* (37 °C).

References

[1] *W. T. Jackson, M. Schlamowitz, A. Shaw*, Biochemistry *4*, 1537 (1965).
[2] *A. J. Cornish-Bowden, J. R. Knowles*, Biochem. J. *96*, 71 P (1965).
[3] *E. J. Casey, K. J. Laidler*, J. Am. Chem. Soc. *72*, 2159 (1950).
[4] *L. E. Baker*, J. Biol. Chem. *211*, 701 (1954).

[5] *M. S. Silver,* J. Am. Chem. Soc. *87*, 1627 (1965).
[6] *J. Tang,* J. Biol. Chem. *240*, 3810 (1965).
[7] *F. A. Bovey, S. S. Yanari,* in: *P. D. Boyer, H. Lardy, K. Myrbäck* (eds.), The Enzymes, Vol. *4*, Academic Press, New York 1960, p. 63.
[8] *L. Cunningham,* in: *M. Florkin, E. H. Stotz* (eds.), Comprehensive Biochemistry, Vol. *16*, Elsevier Amsterdam 1965, p. 173.
[9] *M. L. Anson,* J. Gen. Physiol. *22*, 79 (1938).

Peroxidase

from horse radish

Donor:hydrogen-peroxide oxidoreductase, EC 1.11.1.7

$$\text{Donor} + H_2O_2 \xrightarrow{\text{POD}} \text{oxidized donor} + 2H_2O$$

Molecular weight [1, 2]: 40000.

Specificity and kinetic properties: POD is specific for the hydrogen acceptor; only H_2O_2, methyl- and ethyl peroxide are active. In contrast the enzyme is not specific for the hydrogen donor. A large number of phenols, aminophenols, diamines, indophenols, leuko-dyes, ascorbate and several amino acids react [3].

Rate constants [3, 4]

a) Hydrogen acceptors $1 \times \text{mol}^{-1} \times \text{s}^{-1}$

H_2O_2	9×10^8
Methyl peroxide	1.5×10^6
Ethyl peroxide	3.6×10^6

b) Hydrogen donors

4-Hydroxydiphenyl	8×10^7
Hydroquinone	3×10^6
Hydroquinone monomethyl ether	2×10^6
Catechol	2×10^6
Catechol monomethyl ether	2.5×10^5
Resorcinol	3×10^6
Pyrogallol	3×10^5
2-Phenylenediamine	5×10^7
3-Phenylenediamine	1×10^6

Aniline	7	$\times 10^4$
4-Aminobenzoate	1	$\times 10^3$
Ascorbate	2	$\times 10^4$
Dihydroxymaleate	2	$\times 10^4$
Urate	2	$\times 10^4$
NADH	3	$\times 10^3$
Nitrite	17	
Leukomalachite green	3	$\times 10^5$

Assay

$$H_2O_2 + DH_2 \xrightarrow{\text{POD}} 2 H_2O + D$$

(DH_2 = leuko-dye; D = dye).

Wavelength Hg 436 nm; light path 10 mm; final volume 3.11 ml; 25°C; ε = 2.55 $1 \times \text{mmol}^{-1} \times \text{mm}^{-1}$ (related to 1 mmol tetraguaiacol).

3.00 ml	phosphate buffer (0.1 mol/l; pH 7.0)	96.5 mmol/l
0.05 ml	guaiacol (2.2 mg/ml)	0.3 mmol/l
0.04 ml	H_2O_2 (1 µl 30% H_2O_2/ml;	
	A_{240nm} must be 0.400 ± 0.010)	0.12 mmol/l
0.02 ml	enzyme in buffer	

Stability: suspensions in ammonium sulphate solution, 3.2 mol/l; pH ca. 6, and lyophilized enzymes are stable for several months at 4°C.

Purity required: specific activity $\geq$ 180 U/mg protein (25°C, measured with guaiacol as substrate); purity number = 2.8 (definition according to *Theorell* [5]). For some purposes, e.g. determination of blood sugar (with glucose oxidase) a preparation with 50 U/mg lyophilized material is suitable; purity number = 0.6. "Amine oxidases" should be absent.

Purity required as enzyme label for enzyme-immunoassay: specific activity $\geq$ 250 U/mg protein (25°C; guaiacol as substrate).

References

[1] *R. Cecil, A. G. Ogston*, Biochem. J. *49*, 105 (1951).
[2] *L. M. Shannon, E. Kay, L. Y. Lew*, J. Biol. Chem. *241*, 2166 (1966).
[3] *K. G. Paul*, in: *P. D. Boyer, H. Lardy, K. Myrbäck* (eds.), The Enzymes, Vol. *8*, Academic Press, New York 1963, p. 227.
[4] *T. E. Barman*, Enzyme Handbook, Vol. *I*, Springer, Berlin 1969, p. 234.
[5] *H. Theorell*, Acta Chem. Scand. *4*, 22 (1950).

Phosphatase, acid

from potatoes

Orthophosphoric-monoester phosphohydrolase (acid optimum), EC 3.1.3.2

$$\text{Orthophosphate monoester} + H_2O \xrightarrow{\text{AcP}} \text{alcohol} + P_i$$

Specificity: acid phosphatase hydrolyzes a number of phosphomonoesters and phosphoproteins. Phosphodiesters are not hydrolyzed [1]. For kinetic properties of the potato enzyme, cf. [2].

Assay

$$\text{4-Nitrophenyl phosphate} + H_2O \xrightarrow{\text{AcP}} \text{4-nitrophenol} + P_i \,.$$

Wavelength Hg 405 nm; light path 10 mm; incubation volume 1.05 ml; final volume 3.05 ml; 25°C; $\varepsilon = 1.85 \; 1 \times \text{mmol}^{-1} \times \text{mm}^{-1}$.

1.00 ml	citrate buffer (0.1 mol/l; pH 5.6)	95 mmol/l
0.03 ml	4-nitrophenyl phosphate, Na salt, (0.6 mol/l)	18 mmol/l
0.02 ml	enzyme in 10% albumin solution	

incubate for 5 min at 25°C, add

2.00 ml   NaOH (0.5 mol/l) and read against blank.

Stability: lyophilized enzymes and suspensions in 1% serum albumin solution containing ammonium sulphate, 3.2 mol/l; pH ca. 6, are stable for several months at 4°C.

Purity required: specific activity $\geq$ 60 U/mg protein (25°C). For some purposes a preparation with a specific activity of 2 U/mg is sufficient.

References

[1] *G. Schmidt*, in: *P. D. Boyer, H. Lardy, K. Myrbäck* (eds.), The Enzymes, Vol. *5*, Academic Press, New York 1961, p. 37.
[2] *E. F. Alvarez*, Biochim. Biophys. Acta *59*, 663 (1962).

Phosphatase, alkaline

from calf intestine

Orthophosphoric-monoester phosphohydrolase (alkaline optimum), EC 3.1.3.1

$$\text{Orthophosphate monoester} + H_2O \xrightarrow{\text{AP}} \text{alcohol} + P_i$$

Effectors: alkaline phosphatase is inhibited by P_i, metal chelating agents and divalent heavy metal ions [1]. L-Phenylalanine, L-tryptophan and L-cysteine also inhibit [2].

Molecular weight (pH 8.7) [3]: 100 000.

Specificity: alkaline phosphatase catalyzes the hydrolysis of numerous phosphate esters, such as esters of primary and secondary alcohols, sugar alcohols, cyclic alcohols, phenols and amines. Phosphodiesters do not react. The enzyme hydrolyzes PP_i [1]. The kinetic properties of the enzyme depend on many factors, such as purity of enzyme, concentration of enzyme in the assay, buffer, pH, etc. For specificity and kinetic data of the enzyme from calf intestine, cf. [4, 5].

Assay

$$\text{4-Nitrophenyl phosphate} + H_2O \xrightarrow{\text{AP}} \text{4-nitrophenol} + P_i.$$

Wavelength Hg 405 nm; light path 10 mm; final volume 3.05 ml; 25°C; $\varepsilon = 1.85$ $1 \times \text{mmol}^{-1} \times \text{mm}^{-1}$.

3.00 ml	glycine buffer (0.1 mol/l; pH 10.5;	glycine	98.4 mmol/l
	MgCl$_2$, 1 mmol/l; ZnCl$_2$, 0.1 mmol/l)	MgCl$_2$	1 mmol/l
		ZnCl$_2$	0.1 mmol/l
0.03 ml	4-nitrophenyl phosphate, Na salt, (0.6 mol/l)		6 mmol/l
0.02 ml	enzyme in MgCl$_2$, 1 mmol/l		

Stability: dry powder and suspensions in ammonium sulphate solution, 3.2 mol/l; pH ca. 7, containing MgCl$_2$, 1 mmol/l, are stable for several months at 4°C. Solutions in NaCl (3 mol/l, MgCl$_2$; 1 mmol/l, ZnCl$_2$, 0.1 mmol/l, triethanolamine, 30 mmol/l; pH 7.6) with highly purified enzyme (for EIA) are stable for several months at 4°C.

Purity required: specific activity $\geq$ 300 U/mg protein (25°C) corresponds to approx. $\geq$ 1500 U/mg protein with diethanolamine as buffer system. Contaminants (related to the specific activity of AP) $\leq$ 0.001% PDE, $\leq$ 0.01% ADA and AMP deaminase. A preparation with a specific activity of 35 U/mg protein is sufficient for many purposes.

Purity required as enzyme label for enzyme-immunoassay: specific activity $\geqslant$ 2500 U/mg protein (37°C; 4-nitrophenyl phosphate as substrate, diethanolamine as buffer; pH 9.8).

References

[1] *T. C. Stadtman,* in: *P. D. Boyer, H. Lardy, M. Myrbäck* (eds.), The Enzymes, Vol. 5, Academic Press, New York 1961, p. 55.
[2] *R. P. Cox, P. Gilbert, M. J. Griffin,* Biochem. J. *105*, 155 (1967).
[3] *L. Engstrom,* Biochim. Biophys. Acta *52*, 36 (1961).
[4] *R. K. Morton,* Biochem. J. *61*, 232, 240 (1955).
[5] *H. N. Fernley, P. G. Walker,* Biochem. J. *104*, 1011 (1967).

Phosphodiesterase

from beef heart

3′:5′-Cyclic-AMP 5′-nucleotido-hydrolase, EC 3.1.4.17

$$\text{A-3:5-MP} + H_2O \xrightarrow{\text{PDE}} \text{AMP}$$

Michaelis constants*

		Relative rate [1]
A-3:5-MP	ca. 1×10^{-6} mol/l and 1×10^{-4} mol/l	100
dA-3:5-MP	−	59 − 63
I-3:5-MP	−	56 − 135
G-3:5-MP	−	54 − 136
U-3:5-MP	−	5 − 8
C-3:5-MP	−	0.6

Inhibitor constants: caffeine 5.0×10^{-2} mol/l [2]*.

Effectors: the enzyme is stimulated by imidazole and ammonium ions [3]. Ammonium ions inhibit at concentrations $\leqq 0.6$ mmol/l [1].

Specificity: PDE from beef heart specifically hydrolyzes the 3′:5′-diester bond in cyclic nucleotide phosphates. DNA, RNA, polyadenyl- and polyuridyl acids do not react. Nucleotide-2′:3′-phosphates are not hydrolyzed. Mg^{2+} can be replaced by Mn^{2+} or Co^{2+}.

Assay

$$\text{A-3:5-MP} + H_2O \xrightarrow{\text{PDE}} \text{AMP}$$

$$\text{AMP} + H_2O \xrightarrow{\text{AP}} \text{adenosine} + P_i$$

$$\text{Adenosine} + H_2O \xrightarrow{\text{ADA}} \text{inosine} + NH_3.$$

Wavelength 265 nm; light path 10 mm; final volume 3.28 ml; 25°C; $\varepsilon = 0.81$ $1 \times$ mmol$^{-1} \times$ mm^{-1}.

3.00 ml	glycylglycine buffer (0.1 mol; pH 7.5)	91.5 mmol/l
0.10 ml	MgSO$_4$ · 7 H$_2$O (10 mg/ml)	1.3 mmol/l
0.10 ml	A-3:5-MP (1 mg/ml; neutralized)	82 μmol/l
0.01 ml	ADA (2 mg/ml)	1.2 kU/l
0.04 ml	AP (1 mg/ml; $\geqq$ 350 U/mg)	4.5 kU/l
0.03 ml	enzyme in buffer	

---

* For the enzyme from dog heart.

Phosphodiesterase, free from calmodulin

Wavelength 265 nm; light path 10 mm; final volume 3.0 ml; 30°C; $\varepsilon = 0.81$ $1 \times \text{mmol}^{-1} \times \text{mm}^{-1}$.

2.68 ml	glycylglycine buffer (0.1 mol/l; pH 7.5; CaCl$_2$, 3 mmol/l)	glycylglycine CaCl$_2$	90 mmol/l 2.7 mmol/l
0.10 ml	MgSO$_4$ (41 mmol/l)		1.4 mmol/l
0.01 ml	ADA (2 mg/ml)		1.3 kU/l
0.04 ml	phosphatase, alkaline (1 mg/ml; $\geq$ 350 U/mg)		20 kU/l
0.02 ml	calmodulin (1 mg protein dissolved in 1 ml MES-buffer, 10 mmol/l; CaCl$_2$, 1 mmol/l; NaCl, 0.1 mol/l; pH 6.5; 1 part of this solution diluted with 99 parts of the MES-buffer before starting the assay)	calmodulin	67 mg/l
0.05 ml	enzyme in water		

allow to stand for 10 min at 30°C; add

0.10 ml	A-3:5-MP (1 mg/ml)	0.1 mmol/l

determine activity with and without calmodulin.

$$\text{Activation factor} = \frac{\text{activity with calmodulin}}{\text{activity without calmodulin}}$$

Stability: solutions in 50% glycerol containing Tris, 50 mmol/l and magnesium sulphate, 50 mmol/l; pH ca. 7, are stable for several months at 4°C.

Highly purified, calmodulin-free lyophilized enzymes stabilized with phosphate and other substances are stable for many months at -20°C.

Purity required: specific activity $\geq$ 0.25 U/mg protein (25°C). Contaminant (related to the activity of PDE) $\leq$ 3% phosphatases.

Purity requirements for the calmodulin-free enzyme: 10 fold activation after saturation with calmodulin at 30°C. One unit is the catalytic activity which will be achieved with A-3:5-MP as substrate and calmodulin saturation at 30°C.

References

[1] *G. Michal et al.,* Pharm. Res. Commun. *6,* 203 – 252 (1974).
[2] *K. G. Nair,* Biochemistry 5, 150 (1966).
[3] *R. W. Butcher, E. W. Sutherland,* J. Biol. Chem. *237,* 1244 (1962).

Phosphodiesterase

from calf spleen

Oligonucleate 3'-nucleotidohydrolase, EC 3.1.16.1

$$\text{Oligonucleotide} + (n-1)\, H_2O \xrightarrow{\ PDE\ } n\ 3'\text{-nucleotides}$$

Michaelis constant (ammonium acetate buffer, pH 5.7; 37°C) [1, 2]:

2'-deoxy-thymidine 3'-4-nitrophenyl phosphate 3×10^{-3} mol/l .

Effectors: EDTA and sulphydryl reagents are activators. Bivalent cations (Mg^{2+}, Mn^{2+} and, more effectively, Cu^{2+}, Hg^{2+} and Zn^{2+}) are inhibitors. Arsenite and fluoride are weak inhibitors [3].

Specificity: PDE from calf spleen attacks both polyribonucleotides and oligodeoxyribonucleotides, cleaving nucleoside 3'-phosphates from the 5'-OH end [2]. A 5'-phosphate terminus on the substrate inhibits hydrolysis. 4-Nitrophenyl esters of nucleoside 3'-phosphates and bis(4-nitrophenyl)phosphate are split only very slowly [3]. The enzyme is an exonuclease and non-specific with respect not only to the nature of the base but also to the sugar moiety.

pH Optimum [4]: 7.

Assay

$$\text{Thymidine-3'-4-NP} + H_2O \xrightarrow{\ PDE\ } \text{thymidine-3-MP} + 4\text{-nitrophenol} .$$

Wavelength Hg 405 nm; light path 10 mm; incubation volume 0.13 ml; final volume 3.03 ml; 25°C; $\varepsilon = 1.85\, l \times mmol^{-1} \times mm^{-1}$.

0.065 ml	citrate buffer (0.1 mol/l; pH 6.0)	50 mmol/l
0.015 ml	thymidine 3'-4-nitrophenyl phosphate (30 mg/ml)	7 mmol/l
0.05 ml	enzyme in citrate buffer	

incubate for exactly 5 min at 25°C, add

2.90 ml   NaOH (0.1 mol/l)

measure against a blank in which the sample has been added after the addition of NaOH.

Stability: in suspensions in ammonium sulphate solution, 3.2 mol/l; pH ca. 6, a decrease in activity of approx. 20% may occur within 6 months.

Purity required: specific activity $\geqslant$ 2 U/mg protein (25 °C). Contaminants related to the specific acitivity of PDE $\leqslant$ 0.2% phosphatases, $\leqslant$ 1.0% ADA.

References

[1] *W. E. Razell, H. G. Khorana,* J. Biol. Chem. *236*, 1144 (1961).
[2] *W. E. Razell* in: *S. P. Colowick, N. O. Kaplan* (eds.), Methods in Enzymology, Vol. *VI*, Academic Press, New York 1963, p. 245.
[3] *A. Bernardi, G. Bernardi,* in: *P. D. Boyer* (ed.), The Enzymes, Vol. *IV*, Academic Press, New York, 1971, p. 329.
[4] *H. C. Khorana* in: *P. D. Boyer, H. Lardy, K. Myrbäck* (eds.), The Enzymes, Vol. *5*, Academic Press, New York 1961, p. 91.

Phosphodiesterase

from snake venom

Oligonucleate 5′-nucleotido hydrolase, EC 3.1.4.1

$$\text{Oligonucleotide} + (n\text{-}1)\,H_2O \xrightarrow{\text{PDE}} n\ 5'\text{-nucleotides}$$

Michaelis constants [1, 2]*

4-Nitrophenyl-pT	5.0×10^{-4} mol/l	Tris buffer, pH 8.9; 37 °C
4-Nitrophenyl-pU	5.4×10^{-4} mol/l	
pTpT	2.1×10^{-4} mol/l	
TpT (3′,5′- or 5′,5′-diester)	5.3×10^{-4} mol/l	Tris buffer, pH 8.9; 37 °C
Methyl-4-nitrophenyl phosphate	1.2×10^{-2} mol/l	
Benzyl-4-nitrophenyl phosphate	6.8×10^{-3} mol/l	
di-4-Nitrophenyl-pT	7.7×10^{-4} mol/l	Tris buffer, pH 8.0; 37 °C
4-Nitrophenyl-pT	4.9×10^{-4} mol/l	
3′-O-Acetyl-4-nitrophenyl-pT	1.5×10^{-4} mol/l	

Molecular weight [3]: 115000.

Specificity: snake venom PDE attacks polyribonucleotides and oligodeoxyribonucleotides, cleaving nucleoside 5′-phosphates from the 3′-OH end.

---

* For the enzyme from *Crotalus adamanteus.*
  Abbreviations: esterified phosphate is designated by p; p is on the left of the base when it is bound to the 5′-OH group and right of the base, when it is bound to the 3′-OH group.

Assay

$$\text{bis-4-Nitrophenyl phosphate} + H_2O \xrightarrow{\text{PDE}}$$

$$\text{4-nitrophenol} + \text{4-nitrophenyl phosphate} \, .$$

Wavelength Hg 405 nm; light path 10 mm; final volume 3.02 ml; 25°C; $\varepsilon = 1.85$ $1 \times \text{mmol}^{-1} \times \text{mm}^{-1}$.

3.00 ml	Tris buffer (0.2 mol/l; pH 8.9; bis-4-nitrophenyl phosphate, Ca salt, 3.3 mg/ml)	Tris bis-4-nitrophenyl phosphate	0.2 mol/l 3.3 mg/ml
0.01 ml	enzyme in water		

Stability: solutions in 50% glycerol, pH ca. 6, are stable for several months at 4°C.

Purity required: specific activity $\geqq$ 1.5 U/mg protein (25°C). Contaminants (related to the specific activity of PDE) $\leqq$ 0.1% AP and 5'-nucleotidase, endonucleases are not detectable.

References

[1] *W. E. Razzell*, in: *S. P. Colowick, N. O. Kaplan* (eds.), Methods in Enzymology, Vol. *VI*, Academic Press, New York 1963, p. 236.
[2] *W. E. Razzell, H. G. Khorana*, J. Biol. Chem. *234*, 2105 (1959).
[3] *G. R. Philipps*, Hoppe-Seyler's Z. phys. Chem. *356*, 1085 (1975).

Phosphoenolpyruvate Carboxylase

from wheat

Orthophosphate: oxaloacetate carboxy-lyase (phosphorylating), EC 4.1.1.31

$$\text{Phosphoenolpyruvate} + CO_2 + H_2O \xrightarrow{\text{PEP carboxylase}} P_i + \text{oxaloacetate}$$

Equilibrium constant [1, 2]: the reaction is essentially irreversible.

Michaelis constants [2]

Phosphoenolpyruvate	$5 - 6 \times 10^{-4}$ mol/l
Hydrogen carbonate	3.1×10^{-4} mol/l

Effectors: 4-chloromercuribenzoate causes irreversible inhibition. Intermediates of the tricarboxylic acid cycle are inhibitory [3].

Molecular weight [2]: 350000.

Specificity: the enzyme catalyzes the fixation of carbon dioxide with phosphoenol-pyruvate to produce oxaloacetate and inorganic phosphate. The enzyme from *E. coli* is activated by several CoA-derivatives [4].

Assay

$$PEP + CO_2 + OH^- \xrightarrow{\text{PEP carboxylase}} \text{oxaloacetate} + P_i$$

$$\text{Oxaloacetate} + NADH + H^+ \xrightarrow{\text{MDH}} \text{L-malate} + NAD^+.$$

Wavelength 339, Hg 334 or Hg 365 nm; light path 10 mm; final volume 3.01 ml; 25°C.

2.50 ml	Tris buffer (0.1 mol/l; pH 8.0)	83 mmol/l
0.10 ml	KHCO$_3$ (0.3 mol/l)	10 mmol/l
0.10 ml	PEP, tricyclohexylammonium salt, (70 mg/ml)	5 mmol/l
0.10 ml	MgCl$_2$ (0.2 mol/l)	6.7 mmol/l
0.10 ml	GSH (45 mg/ml, neutralized)	4.9 mmol/l
0.05 ml	NADH, disodium salt, (10 mg/ml)	0.2 mmol/l
0.01 ml	MDH (5 mg/ml)	20 kU/l
0.05 ml	enzyme in water	

Stability: suspensions in ammonium sulphate solution, 3.2 mol/l; pH ca. 6, are stable for many months at 4°C.

Purity required: specific activity $\geqslant$ 5 U/mg protein (25°C). Contaminants (related to the specific activity of PEP carboxylase) $\leqslant$ 0.02% "NADH oxidase", $\leqslant$ 0.5% malic enzyme, $\leqslant$ 1% LDH.

References

[1] *M. F. Utter* in: *P. D. Boyer, H. Lardy, K. Myrbäck* (eds.), The Enzymes, Vol. *5*, Academic Press, New York 1961, p. 331.
[2] *H. Maruyama et al.,* J. Biochem. *241*, 2405 (1966).
[3] *M. Perl,* J. Biochem. *76*, 1095 (1975).
[4] *J. L. Canorvas, H. L. Kornberg,* Proc. R. Soc. *165* (B), 189 (1966).

Phosphoglucomutase

from rabbit muscle

α-D-Glucose 1,6-bisphosphate:α-D-glucose 1-phosphate phosphotransferase, EC 2.7.5.1

$$\text{D-Glucose-1-P} + \text{D-glucose-1,6-P}_2 \; \underset{}{\overset{PGluM}{\rightleftharpoons}}$$

$$\text{D-glucose-6-P} + \text{D-glucose-1,6-P}_2$$

Equilibrium constant (veronal buffer, pH 6.2 – 7.5; 30°C)

$$K = \frac{[\text{D-glucose-6-P}]}{[\text{D-glucose-1-P}]} = 17.2$$

Michaelis constants (imidazole buffer, pH 7) [1]

				Relative rate
Glucose-1-P	8	$\times 10^{-6}$ mol/l		1.00
Glucose-6-P	4.7	$\times 10^{-5}$ mol/l	25°C	0.35
Glucose-1,6-P$_2$	6	$\times 10^{-8}$ mol/l		–
Mannose-1-P	2.45	$\times 10^{-1}$ mol/l		0.06
Mannose-6-P	>5	$\times 10^{-4}$ mol/l		–
Ribose-1-P	9	$\times 10^{-4}$ mol/l	22°C	0.009
Ribose-5-P	4	$\times 10^{-4}$ mol/l		–
Galactose-1-P	>2	$\times 10^{-1}$ mol/l		0.0002
Galactose-6-P	>3	$\times 10^{-4}$ mol/l		–
Fructose-1-P	1	$\times 10^{-3}$ mol/l		–
Fructose-6-P	>2	$\times 10^{-3}$ mol/l	25°C	–
Fructose-1,6-P$_2$	2.7	$\times 10^{-6}$ mol/l		–

Molecular weight (pH 7.0): 62000.

Assay

$$\text{Glucose-1-P} \xrightarrow[\text{G-1,6-P}_2]{PGluM} \text{glucose-6-P}$$

$$\text{Glucose-6-P} + \text{NADP}^+ \xrightarrow{G6P\text{-}DH} \text{gluconate-6-P} + \text{NADPH} + \text{H}^+.$$

Wavelength 339, Hg 334 or Hg 365 nm; light path 10 mm; final volume 3.00 ml; 25°C.

2.50 ml	triethanolamine buffer (0.1 mol/l; pH 7.6)	84 mmol/l
0.20 ml	glucose-1-P, K salt, (20 mg/ml)	3.5 mmol/l
0.05 ml	glucose-1,6-P_2, tetracyclohexylammonium salt, (1 mg/ml)	0.02 mmol/l
0.10 ml	EDTA (10 mg/ml in buffer)	0.9 mmol/l
0.05 ml	$MgCl_2$ (0.1 mol/l)	1.7 mmol/l
0.05 ml	NADP, Na salt, (10 mg/ml)	0.19 mmol/l
0.01 ml	G6P-DH (1 mg/ml)	0.47 kU/l
0.02 ml	enzyme in EDTA solution (1 mmol/l)	

Stability: crystalline suspensions in ammonium sulphate solution, 3.2 mol/l; pH ca. 5, are stable for several months at 4°C.

Purity required: specific activity $\geq$ 200 U/mg protein (25°C). Contaminants (related to the specific activity of PGluM) $\leq$ 0.01% GR, HK and PGl.

Reference

[1] *O. H. Lowry, J. V. Passonneau,* J. Biol. Chem. *244*, 910 (1969).

6-Phosphogluconate Dehydrogenase

from yeast

6-Phospho D-gluconate: NADP$^+$ 2-oxidoreductase (decarboxylating), EC 1.1.1.44

$$Gluconate\text{-}6\text{-}P + NADP^+ \xrightarrow{\text{6-PGDH}} ribulose\text{-}5\text{-}P + NADPH + H^+ + CO_2$$

Equilibrium constant: the oxidation reaction is favoured.

Michaelis constants

	Form I	Form II
Gluconate-6-P	5.2×10^{-5} mol/l	3.1×10^{-5} mol/l [1]
NADP	2.0×10^{-5} mol/l	2.0×10^{-5} mol/l [1]
Gluconate-6-P	1.6×10^{-4} mol/l [2]	
NADP	2.6×10^{-5} mol/l [2]	

Inhibitor constant (phosphate buffer, pH 7.5) [3]

Pyridoxal-5-P	4.3×10^{-5} mol/l	competitive

Effectors: the inhibition by pyridoxal-5-P depends on the formation of a *Schiff's* base between the aldehyde group of the inhibitor and the ε-amino group of a lysine residue in the active site of the enzyme. The inhibition is reversed by the addition of gluconate-6-P.

Molecular weight (phosphate buffer, pH 6.2) [1]: Form I 101 000, Form II 111 000.

Specificity: the enzyme is specific for NADP; NAD does not react.

Assay

Wavelength 339, Hg 334 or Hg 365 nm; light path 10 mm; final volume 2.92 ml; 25°C.

2.50 ml	triethanolamine buffer (0.1 mol/l; pH 7.6)	85 mmol/l
0.20 ml	gluconate-6-P, Na salt, (15 mg/ml)	2.7 mmol/l
0.10 ml	$MgCl_2$ (0.1 mol/l)	3.4 mmol/l
0.10 ml	NADP, Na salt, (10 mg/ml)	0.4 mmol/l
0.02 ml	enzyme in buffer	

Stability: crystalline suspensions in ammonium sulphate solution, 3.2 mol/l; pH 7.5, are stable for several months at 4°C.

Purity required: specific activity $\geq$ 12 U/mg protein (25°C). Contaminants (related to the specific activity of 6-PGDH) $\leq$ 0.01% G6P-DH, GR and HK, $\leq$ 0.03% PGI.

References

[1] *M. Rippa, M. Signorini, S. Pontremoli*, Eur. J. Biochem. *1*, 170 (1967).
[2] *S. Pontremoli, E. Grazi*, in: *S. P. Colowick, N. O. Kaplan* (eds.), Methods in Enzymology, Vol. *IX*, Academic Press, New York 1966, p. 137.
[3] *M. Rippa, L. Spanio, S. Pontremoli*, Arch. Biochem. Biophys. *118*, 48 (1967).

Phosphoglucose Isomerase

from yeast

D-Glucose 6-phosphate ketol-isomerase, EC 5.3.1.9

$$\text{D-Glucose-6-P} \xrightleftharpoons{\text{PGI}} \text{D-fructose-6-P}$$

Equilibrium constant (Tris buffer, pH 8.0; 30°C)

$$K = \frac{[\text{D-fructose-6-P}]}{[\text{D-glucose-6-P}]} = 2.98 \times 10^{-1}$$

Michaelis constant [1]: Glucose-6-P 7×10^{-4} mol/l

Molecular weight (pH 7.0) [1]: 145 000.

Specificity: the enzyme only reacts with glucose-6-P and fructose-6-P.

Assay

$$\text{Fructose-6-P} \xrightarrow{\text{PGI}} \text{glucose-6-P}$$

$$\text{Glucose-6-P} + \text{NADP}^+ \xrightarrow{\text{G6P-DH}} \text{gluconate-6-P} + \text{NADPH} + \text{H}^+.$$

Wavelength 339, Hg 334 or Hg 365 nm; light path 10 mm; final volume 2.93 ml; 25°C.

2.50 ml	triethanolamine buffer (0.1 mol/l; pH 7.6)	85 mmol/l
0.10 ml	fructose-6-P, Na salt, (14 mg/ml)	1.4 mmol/l
0.20 ml	MgCl$_2$ (0.1 mol/l)	6.8 mmol/l
0.10 ml	NADP, Na salt, (10 mg/ml)	0.39 mmol/l
0.01 ml	G6P-DH (1 mg/ml)	460 U/l
0.02 ml	enzyme in buffer	

Stability: crystalline suspensions in ammonium sulphate solution, 3.2 mol/l; pH ca. 6, are stable for several months at 4°C.

Purity required: specific activity $\geq$ 350 U/mg protein (25°C). Contaminants (related to the specific activity of PGI) $\leq$ 0.01% F-6-PK, GR, 6-PGDH, PGluM, $\leq$ 0.2% β-fructosidase.

Reference

[1] *E. A. Noltmann, F. H. Bruns,* Biochem. Z. *331*, 436 (1959).

3-Phosphoglycerate Kinase

from yeast

ATP: 3-phospho-D-glycerate 1-phosphotransferase, EC 2.7.2.3

$$\text{Glycerate-3-P} + \text{ATP} \xrightleftharpoons{\text{PGK}} \text{glycerate-1,3-P}_2 + \text{ADP}$$

Equilibrium constant (phosphate buffer, pH 6.9; 25°C) [1]

$$K = \frac{[\text{glycerate-1,3-P}_2] \times [\text{ADP}]}{[\text{glycerate-3-P}] \times [\text{ATP}]} = 3.1 \times 10^{-4}$$

Michaelis constants

ATP	1.1×10^{-4} mol/l	
Glycerate-3-P	2.0×10^{-4} mol/l	(phosphate buffer,
ADP	2.0×10^{-4} mol/l	pH 6.9; 25°C) [2]
Glycerate-1,3-P_2	1.8×10^{-6} mol/l	
Mg^{2+}	2.5×10^{-4} mol/l	

Glycerate-3-P	6.2×10^{-4} mol/l	(Tris buffer, pH 7.8;
MgATP^{2-}	4.0×10^{-4} mol/l	25°C) [3]

Molecular weight [1]: 47 000.

Specificity: the enzyme does not react with AMP, PEP or glycerate-2,3-P_2.

Assay

$$\text{Glycerate-3-P} + \text{ATP} \xrightarrow{\text{PGK}} \text{glycerate-1,3-}P_2 + \text{ADP}$$

$$\text{Glycerate-1,3-}P_2 + \text{NADH} + H^+ \xrightarrow{\text{GAPDH}} \text{GAP} + \text{NAD}^+ + P_i.$$

Wavelength 339, Hg 334 or Hg 365 nm; light path 10 mm; final volume 3.14 ml; 25°C.

2.50 ml	triethanolamine buffer (0.1 mol/l; pH 7.6)	78.3 mmol/l
0.10 ml	EDTA (10 mg/ml buffer)	0.9 mmol/l
0.05 ml	NADH, Na salt, (10 mg/ml)	0.2 mmol/l
0.20 ml	ATP, Na salt, (10 mg/ml)	1.1 mmol/l
0.20 ml	glycerate-3-P, tricyclohexylammonium salt, (50 mg/ml)	6.5 mmol/l
0.05 ml	MgSO$_4$ (0.1 mol/l)	1.6 mmol/l
0.02 ml	GAPDH (5 mg/ml)	2.5 kU/l
0.02 ml	enzyme in buffer	

Stability: crystalline suspensions in ammonium sulphate solution, 3.2 mol/l; pH ca. 7, containing EDTA, 1 mmol/l, are stable for several months at 4°C.

Purity required: specific activity $\geq$ 400 U/mg protein (25°C). Contaminants (related to the specific activity of PGK) $\leq$ 0.001% "NADH-oxidase", $\leq$ 0.01% GAPDH, GDH and myokinase, $\leq$ 0.1% TIM.

References

[1] *W. K. G. Krietsch, Th. Bücher,* Eur. J. Biochem. *17,* 568 (1970).
[2] *Th. Bücher,* in: *S. P. Colowick, N. O. Kaplan* (eds.), Methods in Enzymology, Vol. *I,* Academic Press, New York 1955, p. 415.
[3] *M. Larsson-Raznikiewicz,* Biochim. Biophys. Acta *132,* 33 (1967).

Phosphoglycerate Mutase

from rabbit muscle

2,3-Bisphospho-D-glycerate : 2-phospho-D-glycerate phosphotransferase, EC 2.7.5.3

$$\text{D-Glycerate-2-P} \xrightleftharpoons[\text{D-glycerate-2,3-P}_2]{\text{PGM}} \text{D-glycerate-3-P}$$

Equilibrium constant (imidazole buffer, pH 6.8; 37 °C) [1]

$$K = \frac{[\text{D-glycerate-3-P}]}{[\text{D-glycerate-2-P}]} = 5.0$$

Michaelis constants (Tris buffer, pH 7; 30 °C) [1]:

Glycerate-2-P	$< 2 \quad \times 10^{-3}$ mol/l
Glycerate-3-P	$5 \quad \times 10^{-3}$ mol/l
Glycerate-2,3-P$_2$	1.25×10^{-4} mol/l

Effectors: the enzyme is inhibited by glycerate-2-P.

Molecular weight (pH 7.0): 57 000 [2], 64 000 [3].

Assay

$$\text{Glycerate-3-P} \xrightarrow{\text{PGM}} \text{glycerate-2-P}$$

$$\text{Glycerate-2-P} \xrightarrow{\text{enolase}} \text{PEP} + H_2O$$

$$\text{PEP} + \text{ADP} \xrightarrow{\text{PK}} \text{pyruvate} + \text{ATP}$$

$$\text{Pyruvate} + \text{NADH} + H^+ \xrightarrow{\text{LDH}} \text{L-lactate} + \text{NAD}^+.$$

Wavelength 339, Hg 334 or Hg 365 nm; light path 10 mm; final volume 2.93 ml; 25 °C.

2.50 ml	triethanolamine buffer (0.1 mol/l; pH 7.6)	85 mmol/l
0.03 ml	MgSO$_4$ (0.1 mol/l)	1 mmol/l
0.15 ml	glycerate-3-P, tricyclohexylammonium salt, (50 mg/ml)	4.7 mmol/l
0.05 ml	NADH, Na salt, (10 mg/ml)	0.2 mmol/l
0.10 ml	ADP, Na salt, (10 mg/ml)	0.57 mmol/l
0.05 ml	glycerate-2,3-P$_2$, tricyclohexylammonium salt, (6 mg/ml)	0.12 mmol/l
0.01 ml	LDH (5 mg/ml)	9.4 kU/l
0.01 ml	PK (2 mg/ml)	1.4 kU/l
0.01 ml	enolase (10 mg/ml)	1.4 kU/l
0.01 ml	enzyme in buffer	

Stability: crystalline suspensions in ammonium sulphate solution, 3.2 mol/l; pH ca. 6, are stable for several months at 4°C.

Purity required: specific activity $\geq$ 500 U/mg protein (25°C). Contaminants (related to the specific activity of PGM) $\leq$ 0.01% enolase and PGK.

References

[1] *S. Grisolia,* in: *S. P. Colowick, N. O. Kaplan* (eds.), Methods in Enzymology, Vol. *V*, Academic Press, New York 1962, p. 236.
[2] *L. I. Pizer,* J. Biol. Chem. *235*, 895 (1960).
[3] *H. Edelhoch, V. W. Rodwell, S. Grisolia,* J. Biol. Chem. *228*, 891 (1957).

Phospholipase A$_2$

from bee venom

Phosphatide 2-acylhydrolase, EC 3.1.1.4

$$\text{Glycerophospholipid} + H_2O \xrightarrow{\text{PL-A}_2} \text{lysoglycerophospholipid} + \text{fatty acid}$$

Effectors: the enzyme is activated by divalent metal cations, $Ca^{2+} > Mg^{2+} \gg Sr^{2+} \gg Ba^{2+}$. EDTA inhibits; however, the addition of both EDTA and Ca^{2+} in excess has an optimal activating effect. Organophosphorus compounds do not inhibit [1].

Molecular weight [1]: monomer 19000.

Specificity: phospholipase A$_2$ from bee venom hydrolyzes the acyl residues specifically at the 2-position of the glycerol skeleton. It is highly active towards the synthetic substrate 1.2-dihexanoyl-phosphatidylcholine [1]. In comparison with the other enzymes of the phospholipase A type activation by Ca^{2+} has been reported for the enzymes from *Crotalus adamanteus* [2] and from porcine pancreas [3]. The bee venom and porcine pancreatic enzymes show a striking degree of similarity: they both are highly basic, contain six disulphide bridges and are not inhibited by organophosphorus compounds. The enzyme from *Crotalus* venom differs from the other two in a number of respects, the most important of which is that the molecule is acidic rather than basic [1]. PL-A$_2$ from bee venom is much more agressive towards intact cells (e.g. erythrocytes) than the enzymes from pancreas and from *Crotalus*; in this respect the bee enzyme resembles the PL-A$_2$ from cobra (*Naja naja*) venom [4, 5].

pH Optimum [1]: 8.0.

Assay

Titrator with 1 ml burette; recorder; 50 ml reaction vessel; magnetic stirrer; microprocessor-controlled dispenser; final volume approx. 30.1 ml; 40°C.

30.00 ml   substrate emulsion (1 fresh egg yolk is homogenized with 100 ml H_2O; 125 ml $CaCl_2$, 0.01 mol/l, and 125 ml sodium deoxycholate, 4 mmol/l, are added and the pH adjusted with NaOH, 0.5 mol/l, to pH 8).

adjust mixture in reaction vessel to 40°C, add

0.05 ml   enzyme in water

titrate at constant pH. Run a blank without sample.

Stability: the lyophilized enzyme is stable for many months at 4°C.

Purity required: activity $\geqslant$ 2400 U/mg substance (40°C); ca. 3000 U/mg enzyme protein. Contaminant < 0.1 unit proteases/mg.

References

[1] *R. A. Shipolini et al.,* Eur. J. Biochem. *20,* 459 (1971).
[2] *M. A. Wells, D. I. Hanahan,* Biochemistry 8, 414 (1969).
[3] *G. H. de Haas et al.,* Biochim. Biophys. Acta *159,* 103 (1968).
[4] *R. F. A. Zwaal et al.,* Biochim. Biophys. Acta *233,* 474 (1971).
[5] *L. L. M. van Deenen et al., in: Y. Hatefi, L. Djavadi-Ohaniance* (eds.), The Structural Basis of Membrane Function, Academic Press, New York 1976, pp. 21 – 38.

Phospholipase A_2

from porcine pancreas

Phosphatide 2-acylhydrolase, EC 3.1.1.4

$$\text{Glycerophospholipid} + H_2O \xrightarrow{\text{PL-A}_2} \text{lysoglycerophospholipid} + \text{fatty acid}$$

Effectors [1, 2]: Ca ions are necessary for optimal activity. Zn^{2+}, Cd^{2+} and Pb^{2+} inhibit strongly at concentrations of 10^{-3} mol/l. Cu^{2+}, Ag^+, Ba^{2+}, F^- and CN^- give less than 20% inhibition at 10^{-3} mol/l. EDTA strongly inhibits by complexing with Ca^{2+}.

Specificity [1]: phospholipase A_2 from porcine pancreas acts stereospecifically on all common types of *sn*-3-phosphoglycerides, hydrolyzing exclusively fatty acid ester bonds at the glycerol C-2 position, regardless of chain length or degree of saturation. In this respect the porcine enzyme behaves very similarly to phospholipase A_2 from snake venom. Large differences between both enzymes are found by studying the velocity of the breakdown of various substrates. Whereas the snake venom enzyme slowly attacks negatively charged phospholipids and is more active on positively charged amphipathic molecules, the reverse situation is found with the pancreatic enzyme. The latter shows a marked preference for anionic phospholipids such as phosphatidic acid, cardiolipin and phosphatidyl glycerol.

pH Optimum [1]: $7.9 - 8.4$.

Assay

cf. phospholipase A_2 from bee venom.

Stability: suspensions in ammonium sulphate solution, 3.2 mol/l; pH ca. 5.5, are stable for many months at 4°C.

Purity required: specific activity $\geqslant$ 600 U/mg protein (40°C). Contaminants $\leqslant$ 0.001 unit protease/mg (haemoglobin as substrate), $\leqslant$ 1 U trypsin/mg (BAEE as substrate).

References

[1] *G. H. de Haas et al.,* Biochim. Biophys. Acta *159*, 103 (1968).
[2] *G. H. de Haas et al.,* Biochim. Biophys. Acta *239*, 252 (1971).

Phospholipase A_2

from *Crotalus durissus*

Phosphatide 2-acylhydrolase, EC 3.1.1.4

$$\text{Glycerophospholipid} + H_2O \xrightarrow{\text{PL-}A_2} \text{lysoglycerophospholipid} + \text{fatty acid}$$

Michaelis constant [1]: ovolecithin 3.9×10^{-2} mol/l

Effectors [2]: the enzyme is activated to varying degrees by Ca^{2+} and other divalent cations (Mg^{2+}, Mn^{2+}, Cd^{2+}), whereas certain divalent cations such as Zn^{2+} are potent inhibitors.

Molecular weight [3]: approx. 30000.

Specificity [4, 5]: the enzyme catalyzes the specific hydrolysis of the acyl group from the 2-position of glycerophospholipids such as phosphatidylcholine, phosphatidyl-ethanolamine and the vinyl ether analogues of phosphatidylcholine, phosphatidyl-ethanolamine, phosphatidylinositol, phosphatidylserine, phosphatidic acid and diphosphatidylglycerol.

pH Optimum: 7 – 8.

Assay

Wavelength 578 nm; light path 10 mm; final volume 3.05 ml; 25°C.

3.00 ml   buffer/substrate (mixture of 4 ml lecithin solution containing 50 mg lecithin; 0.5 ml methanol; 0.15 ml Tween 20 per ml; 7 ml cresol red (6 mg/ml methanol); 3 ml glycylglycine (13 mg/ml); 1 ml $CaCl_2$ ($CaCl_2 \cdot 2$ H_2O, 13.25 mg/ml); adjusted to pH 9.5, volume made up to 30 ml with water.)

0.050 ml   enzyme in water

the molar absorption coefficient has to be determined by addition of a HCl standard (0.030 ml HCl, 0.1 mol/l).

Stability: in solutions in glycerol, 50%; pH approx. 6, a decrease in activity of approx. 10% may occur within 6 months at 4°C.

Purity required: specific activity $\geqslant$ 200 U/mg protein (25°C).

References

[1] *K. Saito, D. I. Hanahan*, Biochemistry *1*, 521 (1962).
[2] *W. C. McMurray, W. L. Magee*, Annu. Rev. Biochem. *41*, 129 (1972).
[3] *M. A. Wells, D. I. Hanahan*, Biochemistry *8*, 414 (1969).
[4] *D. I. Hanahan*, The Enzymes *5*, 71 (1971).
[5] *M. A. Wells, D. I. Hanahan*, in: *S. P. Colowick, N. O. Kaplan* (eds.), Methods in Enzymology, Vol. *XIV*, Academic Press, New York 1978, p. 178.

Phospholipase C

from *Bacillus cereus*

Phosphatidylcholine cholinephosphohydrolase, EC 3.1.4.3

$$\text{Phosphatidylcholine} + H_2O \xrightarrow{\text{PL-C}} \text{1.2-diacylglycerol} + \text{phosphoylcholine}$$

Effectors: the enzyme is activated and stabilized by Zn^{2+} and Ca^{2+} [1]. 2-Phen-anthroline and EDTA inhibit [2].

Specificity: the enzyme hydrolyzes phosphatidylcholine, phosphatidylethanolamine and phosphatidylserine. Monoacylglycerylphosphoryl compounds such as lysolecithin are not hydrolyzed. Phosphatidylglycerol, phosphatidylinositol, diphosphatidyl-glycerol and phosphatidic acid do not react [3, 4].

The specificity of the enzyme can be altered by incubation with 2-phenanthroline and reactivation by addition of Co^{2+}. The enzyme pre-treated in such a way converts sphingomyelin at one third of its reaction rate with lecithin [5, 6].
Preparations containing sphingomyelinase activity are not suitable for the differentiation of choline-containing phospholipids.

pH Optimum [7]: 8.

Assay

$$Lecithin + H_2O \xrightarrow{PL-C} diglyceride + choline\ phosphate$$

$$Choline\ phosphate + H_2O \xrightarrow{AP} choline + P_i$$

$$Choline + O_2 \xrightarrow{choline\ oxidase} betainaldehyde + H_2O_2$$

$$Betainaldehyde + O_2 \xrightarrow{choline\ oxidase} betain + H_2O_2$$

$$2\,H_2O_2 + phenol + 4\text{-aminoantipyrine} \xrightarrow{POD}$$
$$4\,H_2O + 4\text{-(p-benzoquinoncmonoimino)-phenazone}\ .$$

Wavelength 500 nm; light path 10 mm; incubation volume for first reaction 0.95 ml; final volume 2.96 ml; incubation temperature 37°C; measuring temperature $20-25°C$; $\varepsilon = 0.61 \times mmol^{-1} \times mm^{-1}$.

0.40 ml	dimethylglutarate buffer (0.1 mol/l; pH 7.5)	42 mmol/l
0.30 ml	lecithin (20 mg/ml Triton X-100, 5%)	6.3 g/l
0.20 ml	H$_2$O	
0.05 ml	enzyme in buffer containing 1 mg bovine serum albumin per ml	

incubate 10 min at 37°C, stop reaction by adding

1.00 ml	SDS (2 mg/ml)	0.68 g/l

start colour reaction by adding

1.00 ml	chromogen (mixture of 4.5 ml Tris buffer, 50 mmol/l; pH 8.0; 0.05 ml 4-amino-antipyrine, 30 mg/ml; 0.05 ml phenol,	Tris 4-aminoantipyrine phenol	15 mmol/l 0.1 g/l 68 mg/l

| 20 mg/ml; 0.3 ml choline oxidase, 5 mg/ml; 0.02 ml POD, 10 mg/ml; 0.08 ml H_2O) | choline oxidase | 1000 U/l |
| | POD | 3400 U/l |

0.01 ml   alkaline phosphatase (10 mg/ml)   AP   4700 U/l

incubate for 20 min at 37°C; cool and read absorbance. Run a blank in which the sample is added after stopping the first reaction with SDS.

Stability: in ammonium sulphate solutions, 3.2 mol/l; pH approx. 6, a decrease of activity of approx. 10% may occur within 6 months at 4°C.

Purity required: specific activity $\geqslant$ 1600 U/mg protein (37°C).

References

[1] *C. Little, A. B. Otnaess,* Biochim. Biophys. Acta *391*, 326 (1975).
[2] *A. C. Ottolenghi,* in: *S. P. Colowick, N. O. Kaplan* (eds.), Methods in Enzymology, Vol. *XIV*, Academic Press, New York 1978, p. 188.
[3] *R. F. A. Zwaal et al.,* Biochim. Biophys. Acta *233*, 474 (1971).
[4] *A. B. Otnaess et al.,* Eur. J. Biochem. *79*, 459 (1977).
[5] *A. B. Otnaess,* FEBS Lett. *114*, 202 (1980).
[6] *H. Prydz et al.,* Scand. J. Clin. Lab. Invest. *42*, 57 (1982).
[7] *A. B. Otnaess et al.,* Eur. J. Biochem. *27*, 238 (1972).

Phospholipase D

from cabbage

Phosphatidylcholine phosphatidohydrolase, EC 3.1.4.4

$$\text{Phosphatidylcholine} + H_2O \xrightarrow{\text{PL-D}} \text{choline} + \text{phosphatidate}$$

Michaelis constant [1] (sodium acetate buffer, pH 5.8, 38°C)

lysophosphatidylcholine 1.5×10^{-4} mol/l.

Effectors: phospholipase D has an essential requirement for Ca^{2+}, 40 – 100 mmol/l. The enzyme is stimulated by ether and ether/chloroform mixtures, and inhibited by cationic detergents [1, 2].

Molecular weight [3]: 112500 ± 7500.

Specificity: phospholipase D shows a broad specificity towards glycerophospholipids. It hydrolyzes phosphatidylcholine, phosphatidylethanolamine, and phosphatidylserine with a decreasing order of reaction rate [1, 4]. It acts also on lysophosphatidylcholine [5], phosphatidylglycerol [6], and diphosphatidylglycerol [7].

Sphingomyelin is attacked only to a small extent [8]. Plasmalogens are not hydrolyzed by phospholipase D from cabbage [9, 10]. The substrate specificity depends strongly on the composition of the buffer used for incubation. For reviews see [11, 12].

pH Optimum: the pH optimum depends in a high degree on the reaction conditions. The addition of surfactants, such as SDS shifts the pH optimum from 4.9 to 6.6 [13]. In the same manner the concentration of Ca^{2+} will influence the pH optimum of the enzyme. The increase of the Ca^{2+} concentration from 0.5 mmol/l to 50 mmol/l results in obtaining two pH optima at 6.25 and 7.25 instead of one at 7.25 [3].

Assay

$$\text{Lecithin} + H_2O \xrightarrow{\text{PL-D}} \text{choline} + \text{phosphatidate}$$

$$\text{Choline} + 2\,O_2 + H_2O \xrightarrow[\text{oxidase}]{\text{choline}} \text{betain} + 2\,H_2O_2$$

$$2\,H_2O_2 + \text{4-aminoantipyrine} + \text{phenol} \xrightarrow{\text{POD}}$$
$$4\,H_2O + \text{4-(p-benzoquinonemonoimino)-phenazone}.$$

Wavelength 500 nm; light path 10 mm; incubation volume for first reaction 0.55 ml; final volume 3.05 ml; 37 °C; $\varepsilon = 1.21 \times mmol^{-1} \times mm^{-1}$.

0.40 ml	acetate buffer (0.05 mol/l; pH 5.6; CaCl$_2$, 10 mmol/l)	acetate	36 mmol/l
		CaCl$_2$	7.3 mmol/l
0.10 ml	lecithin (26 mmol/l, 0.5 g lecithin in 25 ml water ultrasonicated)	lecithin	3.6 g/l
0.05 ml	enzyme in bovine serum albumin, 100 mg/ml		

incubate for 10 min at 37 °C, add

2.50 ml	chromogen (mixture of 9 ml Tris buffer, 0.1 mol/l; pH 8.0; 0.8 ml 4-aminoanti-	Tris	37 mmol/l
		4-amino-	
	pyrine, 3 mg/ml; 0.8 ml phenol, 2 mg/ml;	antipyrine	89 mg/l
		phenol	66 mg/l
	0.8 ml EDTA, 0.06 mol/l; pH 8.0; 0.75 ml	EDTA	2 mmol/l
	POD, 100 U/ml; 0.5 ml choline oxidase,	POD	3750 U/l
	50 U/ml; made up to 20 ml with water)	choline-OD	1000 U/l

incubate for 20 min at 37 °C, then read absorbance; also run a blank in which the sample has been added after the first reaction has been stopped by the chromogen.

Stability: in the lyophilized state a decrease in activity of approx. 30% may occur within 6 months at 4 °C.

Purity required: activity $\geqslant$ 0.3 U/mg substance (37 °C). No contaminating enzymes which react with choline.

References

[1] *M. Kates, P. S. Sastry,* in: *S. P. Colowick, N. O. Kaplan* (eds.), Methods in Enzymology, Vol. *XIV,* Academic Press, New York 1978, p. 197 – 203.
[2] *S. J. Chen, P. G. Barton,* Can. J. Biochem. *40,* 1362 – 1375 (1971).
[3] *T. T. Allgyer, M. A. Wells,* Biochemistry *18,* 5348 – 5353 (1979).
[4] *F. M. Davidson, C. Long,* Biochem. J. *69,* 458 – 466 (1958).
[5] *C. Long, R. Odavic, E. J. Sargent,* Biochem. J. *85,* 33 (1962).
[6] *F. Haverkate, L. L. M. van Deenen,* Biochim. Biophys. Acta *84,* 106 – 108 (1964).
[7] *N. Z. Stanacer, L. Stuhne-Sekalec,* Biochim. Biophys. Acta *210,* 350 – 352 (1970).
[8] *R. Tzur, B. Shapiro,* Biochim. Biophys. Acta *280,* 290 – 296 (1972).
[9] *W. E. M. Lands, P. Hart,* Biochim. Biophys. Acta *98,* 532 – 538 (1965).
[10] *A. J. Slotboom, G. H. de Haas, L. L. M. van Deenen,* Chem. Phys. Lipids *1,* 192 – 208 (1967).
[11] *H. Brockerhoff, R. G. Jensen,* Lipolytic Enzymes, Academic Press, New York 1974.
[12] *M. Heller, R. Paoletti, D. Kritchevsky* (eds.), Advances in Lipid Research, Academic Press, New York 1978, pp. 292 – 293.
[13] *R. H. Quarles, R. M. C. Dawson,* Biochem. J. *112,* 795 – 799 (1969).

Phospholipase D

from *Streptomyces chromofuscus*

Phosphatidylcholine phosphatidohydrolase, EC 3.1.4.4

$$\text{Phosphatidylcholine} + H_2O \xrightarrow{\text{PL-D}} \text{choline} + \text{phosphatidate}$$

Michaelis constants [1]

Phosphatidylcholine	1.43×10^{-3} mol/l
Lysophosphatidylcholine	1.67×10^{-3} mol/l
Sphingomyeline	5.6×10^{-2} mol/l

Effectors [1]: Ca^{2+} stimulates the enzyme in the presence or absence of Triton X-100, Mg^{2+} stimulates in the absence of Triton X-100, Ba^{2+}, Mn^{2+}, Zn^{2+} and Co^{2+} inhibit. EDTA inhibits the enzyme activity completely. Activity is greatly affected by detergents such as Triton X-100 and deoxycholate. At maximal activity in the presence of 10 mmol Ca^{2+}/l both detergents are inhibitory at lower concentrations and stimulatory at higher concentrations. Ethyl ether activates in the presence of Triton X-100 and Ca^{2+}.

Molecular weight: 57000 [1], 50000 [2].

Specificity [1]: the enzyme hydrolyzes phosphatidylcholine, lysophosphatidylcholine, their corresponding ethanolamine compounds and sphingomyelin. The relative reaction rates are

Lysophosphatidylcholine	100
Phosphatidylcholine	87
Sphingosylphosphorylcholine	22

pH Optimum [1]: 8.

Assay

$$\text{Lecithin} + H_2O \xrightarrow{\text{PL-D}} \text{phosphatidate} + \text{choline}$$

$$\text{Choline} + 2\,O_2 + H_2O \xrightarrow{\text{choline oxidase}} \text{betain} + 2\,H_2O_2$$

$$2\,H_2O_2 + 4\text{-aminoantipyrine} + \text{phenol} \xrightarrow{\text{POD}}$$

$$4\,H_2O + 4\text{-(p-benzoquinonemonoimino)-phenazone}.$$

Wavelength 500 nm; light path 10 mm; incubation volume for first reaction 0.4 ml; final volume 2.90 ml; 37°C; $\varepsilon = 1.22\,1 \times mmol^{-1} \times mm^{-1}$.

0.10 ml	lecithin (13 mmol/l; mixture of 25 ml lecithin, 20 mg/ml, ultrasonicated, and 25 ml Triton X 100, 100 mg/ml)	lecithin Triton X-100	3.2 mmol/l 12 g/l
0.20 ml	Tris buffer (0.1 mol/l; pH 8.0)		50 mmol/l
0.05 ml	CaCl₂ (0.1 mol/l)		12 mmol/l
0.05 ml	enzyme in Tris buffer (0.05 mol/l; pH 8.0)		

incubate for 10 min at 37°C, add

2.50 ml	chromogen (mixture of 5 ml 4-aminoantipyrine, 3 mg/ml; 5 ml phenol, 2 mg/ml; 5 ml EDTA, 60 mmol/l, pH 8.0; 100 ml Tris buffer, 0.05 mol/l; pH 8.0; 5 ml POD, 90 kU/l; choline oxidase 150 U)	amino- antipyrin phenol EDTA Tris POD choline oxidase	108 mg/l 70 mg/l 2.1 mmol/l 36 mmol/l 3230 U/l 1080 U/l

incubate for 20 min at 37°C, then read absorbance; also run a blank with Tris buffer, 0.05 mol/l; pH 8.0, instead of sample.

Phosphomannose Isomerase

from yeast

D-**Mannose-6-phosphate ketol-isomerase, EC 5.3.1.8**

$$\text{D-Mannose-6-P} \xrightleftharpoons{\text{PMI}} \text{D-fructose-6-P}$$

Equilibrium constant (phosphate buffer, pH 6.8; 37 °C) [1]

$$K = \frac{[\text{D-fructose-6-P}]}{[\text{D-mannose-6-P}]} = 1.78$$

Michaelis constant (phosphate buffer, pH 6.8; 37 °C) [1]

Mannose-6-P 8×10^{-4} mol/l

Molecular weight [2]: 45000.

Specificity: the enzyme is specific for mannose-6-P.

Assay

$$\text{Mannose-6-P} \xrightarrow{\text{PMI}} \text{fructose-6-P}$$

$$\text{Fructose-6-P} \xrightarrow{\text{PGI}} \text{glucose-6-P}$$

$$\text{Glucose-6-P} + \text{NADP}^+ \xrightarrow{\text{G6P-DH}} \text{gluconate-6-P} + \text{NADPH} + \text{H}^+.$$

Wavelength 339, Hg 334 or Hg 365 nm; light path 10 mm; final volume 2.38 ml; 25 °C.

2.50 ml	triethanolamine buffer (0.1 mol/l; pH 7.6)	88.3 mmol/l
0.20 ml	mannose-6-P, K salt, (44.6 mmol/l)	3.2 mmol/l
0.10 ml	NADP, Na salt, (10 mg/ml)	0.4 mmol/l
0.01 ml	PGI (1 mg/ml)	1.2 kU/l
0.01 ml	G6P-DH (1 mg/ml)	0.5 kU/l
0.02 ml	enzyme in buffer	

Stability: suspensions in ammonium sulphate solution, 3.2 mol/l; pH ca. 6, containing EDTA, 1 mmol/l, are stable for several months at 4 °C.

Purity required: specific activity $\geq$ 150 U/mg protein (25 °C). Contaminants (related to the specific activity of PMI) $\leq$ 0.001% "NADPH oxidase", $\leq$ 0.01% GR, HK, 6-PGDH.

References

[1] *E. A. Noltmann, F. H. Bruns,* Biochem. Z. *330*, 514 (1958).
[2] *R. W. Gracy, E. A. Noltmann,* J. Biol. Chem. *243*, 3161 (1968).

Phosphorylase a

from rabbit muscle

1,4-α-D-Glucan: orthophosphate α-glucosyltransferase, EC 2.4.1.1

$$(\alpha\text{-}1,4\text{-Glucosyl})_n + P_i \xrightarrow{\text{phosphorylase a}} (\alpha\text{-}1,4\text{-glucosyl})_{n-1} + \alpha\text{-D-glucose-1-P}$$

Equilibrium constant (pH 7.3) [1]

$$K = \frac{[(\alpha\text{-}1,4\text{-glucosyl})_{n-1}] \times [\text{glucose-1-P}]}{[(\alpha\text{-}1,4\text{-glucosyl})_n] \times [P_i]} = 0.7$$

Michaelis constants [2]

	Constant ligand	Glucose (50 mmol/l)		V
Glucose-1-P	Glycogen	+	3.9×10^{-2} mol/l	60
Glucose-1-P	Glycogen + AMP	+	2.8×10^{-3} mol/l	60
Glucose-1-P	Glycogen	−	5.3×10^{-3} mol/l	57
Glucose-1-P	Glycogen + AMP	−	1.8×10^{-3} mol/l	65
P_i	Glycogen	+	5.4×10^{-2} mol/l	8.3
P_i	Glycogen + AMP	+	1.6×10^{-3} mol/l	8.3
P_i	Glycogen	−	1.4×10^{-3} mol/l	4.7
P_i	Glycogen + AMP	−	6.0×10^{-4} mol/l	6.4
Glycogen	P_i	+	3.6×10^{-4} mol/l	6.7
Glycogen	P_i + AMP	+	6.0×10^{-5} mol/l	6.7
Glycogen	P_i	−	4.0×10^{-5} mol/l	8.8
Glycogen	P_i + AMP	−	1.0×10^{-5} mol/l	9.1
AMP	P_i + Glycogen	+	2.0×10^{-6} mol/l	5.7
AMP	P_i + Glycogen	−	1.0×10^{-6} mol/l	6.8

Inhibitor constants [2]

Glucose (+ AMP)	9.0×10^{-2} mol/l
Glucose (− AMP)	7.9×10^{-3} mol/l

Molecular weights [3, 4]: phosphorylase a 370000, phosphorylase b 185000.

Effectors: heavy metals, SH-reagents and glucose inhibit whereas EDTA, mercapto-ethanol or dithiothreitol, AMP and glycogen activate or stabilize the enzyme.

Specificity: rabbit muscle phosphorylase exists in two forms, phosphorylase a and b. Phosphorylase b is converted by phosphorylase b kinase (EC 2.7.1.38), ATP and Mg^{2+} into phosphorylase a and the reverse reaction is carried out by phosphorylase phosphatase (EC 3.1.3.17). In resting muscle the phosphorylase is present in the b form [3, 5].

Phosphorylase is absolutely specific for α-D-glucose-1-P and α-1,4-glucosidic bonds in polysaccharides. AMP activates both forms of the enzyme. In the absence of AMP phosphorylase a has 60 – 70% of its maximum activity, whereas phosphorylase b is completely inactive.

Assay

$$P_i + (glucose)_n \xrightarrow{\text{phosphorylase a}} (glucose)_{n-1} + glucose\text{-}1\text{-}P$$

$$Glucose\text{-}1\text{-}P \xrightarrow{\text{PGluM}} glucose\text{-}6\text{-}P$$

$$Glucose\text{-}6\text{-}P + NADP^+ \xrightarrow{\text{G6P-DH}} gluconate\text{-}6\text{-}P + NADPH + H^+.$$

Wavelength 339, Hg 334 or Hg 365 nm; light path 10 mm; final volume 3.35 ml; 25 °C.

3.00 ml	potassium phosphate buffer (0.05 mol/l; pH 6.8)	45 mmol/l
0.03 ml	EDTA (3.7 mg/ml; pH 6.8)	0.1 mmol/l
0.10 ml	glycogen, AMP-free (6 mg/ml)	2 mg/ml
0.10 ml	NADP, Na salt, (10 mg/ml)	0.34 mmol/l
0.01 ml	glucose-1,6-P_2, tricyclohexylammonium salt, (1 mg/ml)	4 µmol/l
0.05 ml	$MgCl_2$ (1 mol/l)	15 mmol/l
0.02 ml	PGluM (5 mg/ml; NH_4^+-free dialyzed against acetate buffer, 50 mmol/l, pH 5.3)	0.8 kU/l
0.01 ml	G6P-DH (2 mg/ml; NH_4^+-free dialyzed against phosphate buffer, 50 mmol/l, pH 7.6)	6 kU/l
0.03 ml	enzyme in water (ca. 25 °C)	

Stability: the lyophilized enzyme stabilized with lactose, glycerol-2-P and EDTA is stable for several months at 4 °C.

Purity required: specific activity $\geq$ 20 U/mg protein (25 °C). Contaminants (related to the specific activity of phosphorylase a) $\leq$ 0.01% phosphatases.

References

[1] *S. Hestrin,* J. Biol. Chem. *179*, 943 (1949).
[2] *E. Helmreich, M. C. Michaelides, C. F. Cori,* Biochemistry *6*, 3695 (1967).
[3] *E. H. Fischer et al.,* in: *W. J. Whelan,* Control of Glycogen Metabolism, FEBS Meeting Oslo 1967, Academic Press, New York 1967, p. 19.
[4] *E. H. Fischer,* Biochemistry *6*, 3315 (1967).
[5] *N. B. Madsen, S. Shechosky,* J. Biol. Chem. *242*, 3301 (1967).

Phosphotransacetylase

from *Clostridium kluyveri*

Acetyl-CoA: orthophosphate acetyltransferase, EC 2.3.1.8

$$\text{CoA} + \text{acetyl-P} \quad \xrightleftharpoons{\text{PTA}} \quad \text{acetyl-CoA} + P_i$$

Equilibrium constant (Tris buffer, pH 7.4; 25 °C) [1]

$$K = \frac{[\text{acetyl-CoA}] \times [P_i]}{[\text{CoA}] \times [\text{acetyl-P}]} = 1.67 \times 10^2$$

Michaelis constants (25 °C) [1]

CoA	5.6×10^{-4} mol/l
Acetyl-P	6.6×10^{-4} mol/l

Effectors [1]: for full activity the enzyme requires NH_4^+, 7 mmol/l; K^+ and Mg^{2+} cannot replace NH_4^+ and even inhibit.

Molecular weight [1]: 38 000 – 41 000.

Specificity: the enzyme reacts with CoA (1.00), with dephospho-CoA (0.01), but not with oxidized CoA (in the absence of a reducing agent), nor with deamino-CoA nor CoA fragments (e.g. pantetheine-P).

Assay

Wavelength 233 nm; light path 10 mm; final volume 2.92 ml; 25 °C; $\varepsilon = 0.444$ $l \times mmol^{-1} \times mm^{-1}$.

2.50 ml	Tris buffer (0.1 mol/l; pH 7.4)	85.6 mmol/l
0.05 ml	GSH (30 mg/ml buffer)	1.7 mmol/l
0.20 ml	CoA (5 mg/ml)	0.4 mmol/l
0.10 ml	acetyl phosphate, K/Li salt, (40 mg/ml)	7.4 mmol/l
0.05 ml	ammonium sulphate (1 mol/l)	19 mmol/l
0.02 ml	enzyme in Tris buffer (25 mmol/l; pH 8.0; ammonium sulphate, 0.3 mol/l)	

Stability: crystalline suspensions in ammonium sulphate solution, 3.2 mol/l; pH ca. 8, are stable for several months at 4°C.

Purity required: specific activity $\geq$ 1000 U/mg protein (25°C). Contaminants (related to the specific activity of PTA) $\leq$ 0.01% acetyl-CoA deacylase, AK, ASAT, ALAT, LDH.

Reference

[1] *H. U. Bergmeyer et al.,* Biochem. Z. *338,* 114 (1963).

Plasmin (Fibrinolysin)

from human plasma

Plasmin, EC 3.4.21.7

Preferential cleavage: Lys > Arg-; higher selectivity than trypsin.

Michaelis constants: apparent *Michaelis* constants have been determined for the various types of plasmins on a variety of synthetic substrates at different pH and temperature. For further details see references cited in [1].

Enzyme preparation	Substrate	
	H-D-Val-Leu-Lys-4-NA	*Cbz-Lys-4-NPE
Glu^1-plasmin	–	13.5×10^{-6} mol/l
Lys^{77}-plasmin	210×10^{-6} mol/l	15 mol/l
Val^{442}-plasmin	280×10^{-6} mol/l	–
Val^{561}-plasmin	1260×10^{-6} mol/l	83 mol/l
	*TAME [2]	
Val^{442}-plasmin	4×10^{-3} mol/l	–

---

* Abbreviations:
Cbz-Lys-4-NPE N^α-carbobenzoxy-L-lysine-4-nitrophenyl ester
TAME   N-toluenesulphonyl-L-arginine methyl ester.

Effectors: α_1-globulin [3] and α_2-globulin [4] from human plasma inhibit. Furthermore there are inhibitors described from a number of mammalian tissues and a variety of plants [5]. Inhibiting are specific antibodies to plasmin [6] as well as amino-methylcyclohexane carboxylic acid [7], derivatives of L-lysine ε-aminocaproic acid ester and ω-guanidino acid ester [8]. ε-Aminocaproic acid inhibits non-competitively [9], the active site reagents DFP (diisopropyl fluorophosphate) and TLCK (L-1-chloro-3-tosylamido-7-amino-2-heptanone) inhibit irreversibly [10].

Molecular weight [11]: ca. 76500.

Specificity: plasmin is a trypsin-like serine protease formed from plasminogen by proteolysis which results in multiple forms of the active plasmin. It forms an activated complex with streptokinase and converts fibrin into soluble products. Preferential cleavage occurs at lysyl > arginyl peptide bonds of proteins and peptides. It hydrolyzes casein, denatured haemoglobin, basic amino acid esters and amides.

Assay

Wavelength 405 nm; light path 10 mm; final volume 2.40 ml; 25°C; $\varepsilon = 1.04\ l \times mmol^{-1} \times mm^{-1}$.

1.60 ml	Tris buffer (50 mmol/l; pH 8.2)	33 mmol/l
0.20 ml	NaCl (saline, 9 mg/ml)	13 mmol/l
0.40 ml	Chromozym® PL (1.9 mg Tos-Gly-Pro-Lys-4-NA-AcOH per ml glycine, 100 mmol/l)	17 mmol/l
0.20 ml	enzyme in glycine buffer, 50 mmol/l; (PEG 6000, 5 mg/ml) pH 2.5	

Stability: suspensions in ammonium sulphate solution, 3.2 mol/l; pH 8.6, are stable for many months at 4°C.

Purity required: specific activity $\geqslant$ 8 U/mg protein.

References

[1] *C. Kenneth et al.,* Human Plasmin, in: *S. P. Colowick, N. O. Kaplan* (eds.), Methods in Enzymology, Vol. *80,* Academic Press, New York 1981, pp. 379–386.

[2] *I. R. Rowell, F. J. Castellino,* J. Biol. Chem. *255,* 5329 (1980).

[3] *A. Rimon, Y. Shamash, B. Shapiro,* J. Biol. Chem. *241,* 5102 (1966).

[4] *H. G. Schwiek et al.,* Z. Gesamte Inn. Med. Grenzgeb. *21,* 193 (1966).

[5] *R. Vogel et al.,* Natural Proteinase Inhibitors, Academic Press, New York 1968.

[6] *K. C. Robins, L. Summaria,* Immunochemistry *3,* 29 (1966).

[7] *S. Okomoto et al.,* Ann. N.Y. Acad. Sci. *146,* 414 (1968).

[8] *M. Maramatu, S. Fuji,* J. Biochem. (Tokyo) *64,* 807 (1968).

[9] *M. Iwamoto et al.,* J. Biochem. (Tokyo) *64,* 759 (1968).

[10] *W. R. Groskopf et al.,* J. Biol. Chem. *242,* 5046 (1967) and *244,* 359 (1969).

[11] *K. C. Robbins et al.,* J. Biol. Chem. *250,* 4044 (1975).

Polyol Dehydrogenase

cf. Sorbitol Dehydrogenase, p. 309

Pyrophosphatase, inorganic

from yeast

Pyrophosphate phosphohydrolase, EC 3.6.1.1

$$\text{Pyrophosphate} + H_2O \xrightarrow{\text{inorg. pyrophosphatase}} 2\,P_i$$

Equilibrium constant: strongly in favour of hydrolysis

Molecular weight (pH 6.7): 63000.

Specificity: in the presence of Mg^{2+} the enzyme specifically hydrolyzes PP_i; organic phosphates such as ADP, ATP or TPP are not attacked. In the presence of Zn^{2+} the enzyme catalyzes the hydrolysis of some nucleoside di- and triphosphates [1].

Assay

Wavelength Hg 578 nm; light path 10 mm; final volume 10.00 ml; 25 °C.

3.00 ml	Tris buffer (50 mmol/l; pH 7.0)	48 mmol/l
0.05 ml	$MgCl_2$ (0.1 mol/l)	1.6 mmol/l
0.05 ml	$Na_4P_2O_7$ (0.1 mol/l)	1.6 mmol/l
0.02 ml	enzyme in buffer	

incubate for exactly 15 min at 25 °C and then pipette in 7.00 ml phosphate reagent (10 ml H_2SO_4, 2.5 mol/l; 10 ml ammonium molybdate solution, 2.5%; 10 ml sodium pyrosulphite solution, 2.7% and 4-methylaminophenol sulphate solution, 1%; 40 ml H_2O); mix, allow to stand for 10 min and measure against the blank. Calculate the activity of the pyrophosphatase by means of an orthophosphate standard.

Stability: crystalline suspensions in ammonium sulphate solution, 3.2 mol/l; pH ca. 6, are stable for several months at 4 °C.

Purity required: specific activity $\geq$ 200 U/mg protein (25 °C). Contaminants (related to the specific activity of pyrophosphatase) $\leq$ 0.01% phosphatase (pH 7), ATPase.

Reference

[1] *M. J. Schlesinger, M. J. Coon,* Biochim. Biophys. Acta *41*, 30 (1960).

Proteinase K

from *Tritirachium album*

EC 3.4.21.14

Hydrolysis of keratin and of native proteins

Effectors: metal-chelating agents such as EDTA and sulphydryl-reacting compounds such as mono-iodoacetic acid and 4-chloromercuribenzoate have no inhibitory effect. Phenylmethylsulphonyl fluoride and diisopropyl phosphofluoridate inhibit completely, suggesting the involvement of serine in the mechanism of action of proteinase K.

Specificity: native proteins and keratin are hydrolyzed at the carboxyl group of their aromatic or hydrophobic amino acid residues. The enzyme also shows high proteolytic activity in the presence of 0.5% (w/v) sodium dodecylsulphate and is suitable for the isolation of undegraded m-RNA and DNA.

Molecular weight [1]: 18500.

Assay

Wavelength Hg 578 nm; light path 10 mm; incubation volume 5.40 ml; incubation temperature 37°C; final volume 18.0 ml; measuring temperature 20–25°C.

5.00 ml	buffer/substrate solution (mixture of 2.0 g haemoglobin, 36 g urea; 8 ml NaOH, 1 mol/l; 10 ml boric acid, 1 mol/l; adjusted with HCl, 2 mol/l; to pH 7.5, made up to 100 ml with water)	borate haemoglobin urea	9.3 mmol/l 19 g/l 333 g/l
0.40 ml	enzyme in water		

incubate for exactly 10 min at 37°C; add

10.00 ml	trichloroacetic acid (0.3 mol/l)
0.40 ml	HCl (0.05 mol/l)

incubate for ca. 10 min at room temperature, filter, mix, add

5.00 ml	filtrate
10.00 ml	NaOH (0.5 mol/l)
3.00 ml	*Folin & Ciocalteu's* phenol reagent (commercial preparation diluted 1 + 2)

incubate for ca. 15 min at room temperature, read absorbance against blank, in which the sample has been added after the addition of TCA. Run a standard with 0.40 ml standard solution (tyrosine, 25 mmol/l) instead of sample. 0.40 ml sample is added to this assay mixture after addition of TCA.

Stability: the lyophilized enzyme is stable for many months at 4°C.

Purity required: activity $\geqslant$ 20 units/mg substance (37°C; haemoglobin as substrate).

Reference

[1] *W. Ebeling et al.,* Eur. J. Biochem. *47*, 91 (1974).

Pyroglutamate aminopeptidase

from calf liver

L-Pyroglutamyl-peptide hydrolase, EC 3.4.11.8

$$\text{Pyroglutamyl-peptide} + H_2O \xrightarrow{\text{pyroglutamate aminopeptidase}} \text{pyroglutamate} + \text{peptide}$$

Michaelis constants

Source	Substrate	K_m (mol/l)
Pseudomonas fluorescens [1 – 3]	pyr-Ala*	2×10^{-3} (pH 7.3; 30°C)
Bacillus subtilis [4]	pyr-β-naphthylamide	1.7×10^{-3} (pH 8; 37°C)
Bacillus amyloliquefaciens [5]	L-pyro-4-nitroanilide	0.69×10^{-3} (Tris-HCl buffer, 0.1 mol/l; pH 8.5; 30°C)
	L-pyro-4-methylcoumarinylamide	0.33×10^{-3} (Tris-HCl buffer, 0.1 mol/l; pH 8.5; 30°C)

Effectors: the enzyme is very sensitive to substances which interact with sulphydryl groups, such as 4-mercuriphenyl sulphonate and iodacetamide. EDTA and a sulphydryl compound are required for activity, but not divalent metal ions [6].

Specificity: the enzyme selectively removes amino-terminal pyroglutamate residues from polypeptide chains.

---

* Abbreviation: pyr = L-pyroglutamyl (L-pyrrolidinol) residue.

Substrate	Relative rate	
pyr-Ala	1.0	
pyr-Ile	0.5	
pyr-Val	0.22	for the enzyme from *P. fluorescens*
pyr-Leu	0.19	
pyr-Phe	0.14	
pyr-Tyr	0.09	
pyr-β-naphthylamide	1.0	
pyr-glutamic diethyl ester	3.0	for the enzyme from *B. subtilis*
pyr-anilide	0.1	

Assay

Wavelength Hg 578 nm; light path 10 mm; incubation volume 1.20 ml; incubation temperature 37°C; final volume 5.0 ml; temperature for colour reaction 37°C; $\varepsilon = 4.5\ l \times mmol^{-1} \times mm^{-1}$.

1.00 ml	phosphate buffer (0.1 mol/l; pH 8.0; EDTA, 0.01 mol/l; glycerol 5% (v/v); dithiothreitol, 5 mmol/l)	phosphate EDTA DTT	83 mmol/l 8.3 mmol/l 4.2 mmol/l 1.7 mmol/l
0.10 ml	L-pyroglutamic acid-β-naphthylamide (0.02 mol/l)		
0.10 ml	enzyme in buffer		

incubate for exactly 15 min at 37°C; add

1.00 ml   trichloroacetic acid (25% w/v)

mix, add

1.00 ml   incubation mixture
1.00 ml   nitrite solution (0.1%, w/v)

incubate for 3 min at 37°C, add

1.00 ml   ammonium sulphamate (0.5%, w/v)
2.00 ml   N-(naphthyl)-ethylene diamine (0.05%, w/v)

incubate for 20 min at 37°C, read absorbance against blank in which the sample has been added after the addition of TCA.

Stability: the lyophilized enzyme is stable for many months at 4°C.

Purity required: activity $\geq$ 4 mU/mg substance (37°C).

References

[1] *R. W. Armentrout, R. F. Doolittle,* Arch. Biochem. Biophys. *132,* 80 (1969).
[2] *R. W. Armentrout,* Biochim. Biophys. Acta *191,* 756 (1969).
[3] *J. A. Uliana, R. F. Doolittle,* Arch. Biochem. Biophys. *131,* 561 (1969).
[4] *A. Szewczuk, M. Mulczyk,* Eur. J. Biochem. *8,* 63 (1969).
[5] *K. Fujiwara, D. Tsuru,* J. Biochem. *83,* 1145 (1978).
[6] *R. F. Doolittle, R. W. Armentrout,* Biochemistry 7, 516 (1968).

Pyruvate Decarboxylase

from yeast

2-Oxo acid carboxy-lyase, EC 4.1.1.1

$$\text{Pyruvate} \xrightarrow{\text{PyDC}} \text{acetaldehyde} + CO_2$$

Michaelis constants (citrate buffer, pH 6; 30 °C) [1]

Pyruvate	3×10^{-2} mol/l
Mg^{2+}	2.5×10^{-5} mol/l
Thiamine pyrophosphate	2.5×10^{-6} mol/l

Molecular weight (pH 6.8) [2]: 175000.

Specificity: the yeast enzyme decarboxylates higher homologues up to 2-oxocaproate with decreasing rate; 2-oxoglutarate and acetoacetate do not react [3].

Assay

$$\text{Pyruvate} \xrightarrow{\text{PyDC}} \text{acetaldehyde} + CO_2$$

$$\text{Acetaldehyde} + NADH + H^+ \xrightarrow{\text{ADH}} \text{ethanol} + NAD^+.$$

Wavelength 339, Hg 334 or Hg 365 nm; light path 10 mm; final volume 3.00 ml; 25 °C.

2.79 ml	citrate buffer (0.2 mol/l; pH 6.0)	0.186 mol/l
0.10 ml	pyruvate, Na salt, (100 mg/ml)	30 mmol/l
0.08 ml	NADH, Na salt, (10 mg/ml)	0.32 mmol/l
0.01 ml	ADH, yeast, (10 mg/ml)	10 kU/l
0.02 ml	enzyme in buffer	

Stability: suspensions in ammonium sulphate solution, 2.5 mol/l; pH ca. 6.0, are stable up to one to two months at 4 °C.

Purity required: specific activity $\geqq$ 20 U/mg protein (25 °C).

References

[1] *M. F. Utter,* in: *P. D. Boyer, H. Lardy, K. Myrbäck* (eds.), The Enzymes, Vol. *5,* Academic Press, New York 1961, p. 320.
[2] *J. Ullrich, J. H. Wittorf, C. J. Gubler,* Biochim. Biophys. Acta *113,* 595 (1966).
[3] *T. P. Singer, J. Pensky,* J. Biol. Chem. *196,* 375 (1952).

Pyruvate Kinase

from rabbit muscle

ATP : pyruvate 2-*O*-phosphotransferase, EC 2.7.1.40

$$\text{Pyruvate} + \text{ATP} \xrightleftharpoons{\text{PK}} \text{PEP} + \text{ADP}$$

Equilibrium constant (Tris buffer, pH 7.4; 30 °C) [1]

$$K = \frac{[\text{PEP}] \times [\text{ADP}]}{[\text{pyruvate}] \times [\text{ATP}]} = 1.55 \times 10^{-4}$$

Michaelis constants

			Relative rate [1]
ATP	8.6×10^{-4} mol/l	Tris buffer, pH 7.4; 30 °C	1
Pyruvate	1.0×10^{-2} mol/l		1
ADP	3.0×10^{-4} mol/l		170
PEP	7.0×10^{-5} mol/l		170
ADP	2.1×10^{-4} mol/l	glycylglycine buffer, pH 8.5; 0 °C [2]	
PEP	3.2×10^{-5} mol/l		

Inhibitor constants [2]

ATP	1.2×10^{-4} mol/l	glycylglycine buffer, pH 8.5; 0 °C	(ADP, competitive)
	1.4×10^{-3} mol/l		(PEP, competitive)
Pyruvate	1.0×10^{-3} mol/l	ammonium buffer, pH 9.0	(PEP, competitive)

Molecular weight [3]: 237 000.

Specificity: ATP can be replaced by UTP, GTP, CTP, ITP or dATP. In the reverse reaction ADP (1.00), IDP (0.75), GDP (0.60) and the corresponding deoxy compounds can react (2'-dADP 0.11); AMP is inactive. The enzyme is specific for PEP and requires Mg^{2+} (Mn^{2+}, Co^{2+}) and K^+ (NH_4^+, Rb^+) for full activity.

Assay

$$PEP + ADP \xrightarrow{\text{PK}} pyruvate + ATP$$

$$Pyruvate + NADH + H^+ \xrightarrow{\text{LDH}} \text{L-lactate} + NAD^+.$$

Wavelength 339, Hg 334 or 365 nm; light path 10 mm; final volume 2.98 ml; 25 °C.

2.50 ml	triethanolamine buffer (0.1 mol/l; pH 7.6)	TEA	85.6 mmol/l
0.20 ml	PEP, tricyclohexylammonium salt, (3.75 mg/ml	PEP	0.54 mmol/l
	in MgSO$_4$, 0.05 mol/l; KCl, 0.2 mol/l)	MgSO$_4$	2.5 mmol/l
		KCl	10 mmol/l
0.20 ml	ADP, neutralized with KOH, (30 mg/ml)	ADP	4.7 mmol/l
0.05 ml	NADH, Na salt, (10 mg/ml)	NADH	0.2 mmol/l
0.01 ml	LDH (5 mg/ml)	LDH	9.2 kU/l
0.02 ml	enzyme in buffer		

Stability: crystalline suspensions in ammonium sulphate solution, 3.2 mol/l; pH ca. 6, and solutions in 50% glycerol, pH ca. 6, are stable for several months at 4 °C.

Purity required: specific activity $\geq$ 200 U/mg protein (25 °C). Contaminants (related to the specific activity of PK) $\leq$ 0.01% enolase, LDH, myokinase.

References

[1] *J. T. McQuate, M. F. Utter*, J. Biol. Chem. *234*, 2151 (1959).
[2] *A. M. Raynard, L. F. Hass, D. D. Jacobsen, P. D. Boyer*, J. Biol. Chem. *236*, 2277 (1961).
[3] *G. L. Cottam, P. F. Hollenberg, M. Coon*, J. Biol. Chem. *244*, 1481 (1969).

Pyruvate Oxidase

of bacterial origin

Pyruvate: oxygen oxidoreductase (phosphorylating), EC 1.2.3.3

$$Pyruvate + P_i + O_2 + H_2O \xrightarrow{\text{pyruvate oxidase}} \text{acetyl-P} + CO_2 + H_2O_2$$

Equilibrium constant: the reaction is irreversible.

Michaelis constants [1]

Pyruvate	1.7×10^{-3} mol/l
Phosphate	4.0×10^{-3} mol/l

Effectors [1]: divalent cations such as Ca^{2+}, Co^{2+}, Mn^{2+} and Mg^{2+} are necessary for catalytic activity. EDTA, $ZnCl_2$ and $BaCl_2$ inhibit at concentrations of 5 mmol/l, also P_i at high concentrations.

Molecular weight [1]: approx. 190000.

Specificity: the enzyme reacts with pyruvate; 2-oxobutyrate is oxidized at 3% of the rate with pyruvate. There is no reaction with oxaloacetate and 2-oxoglutarate [1]. Methylene blue, ferricyanide and tetrazolium salts can substitute for oxygen as electron acceptors [2, 3]. Arsenate can partially substitute for P_i [2].

pH Optimum [1]: $6.5 - 8$.

Assay

Wavelength 546 nm; light path 10 mm; final volume 2.49 ml; 25°C; $\varepsilon = 0.457\,1 \times$ mmol$^{-1} \times$ mm^{-1}.

1.80 ml	potassium phosphate buffer (0.1 mol/l; pH 7.0)	72 mmol/l
0.20 ml	pyruvate, Na salt, (0.5 mol/l)	40 mmol/l
0.10 ml	$MgCl_2$ (0.2 mol/l)	8 mmol/l
0.05 ml	thiamine pyrophosphate (10 mmol/l)	0.2 mmol/l
0.20 ml	4-aminoantipyrine (20 mg/ml)	1.6 g/l
0.05 ml	dichlorophenol sulphonic acid (0.34 mol/l)	6.8 mmol/l
0.02 ml	FAD, Na salt, (0.83 mg/ml)	8 µmol/l
0.02 ml	POD (2 mg/ml phosphate buffer)	4 kU/l
0.05 ml	enzyme in phosphate buffer (0.1 mol/l; pH 7.0; FAD 10 µmol/l)	

Stability: the lyophilized enzyme is stable for several months at -20°C.

Purity required: activity $\geqslant 10$ U/mg substance (25°C); $\triangleq$ ca. 30 U/mg protein. Contaminants (related to the specific activity of pyruvate oxidase) $\leqslant 0.05\%$ ALAT and ASAT.

References

[1] *H. Möllering, Boehringer Mannheim,* unpublished.
[2] *L. P. Hager, D. M. Geller, F. Lipmann,* Fed. Proc. *13,* 734 (1954).
[3] *A. P. Nygaard* in: *Hoppe-Seyler, Thierfelder,* Handbuch der Phys. und Path.-chemischer Analyse, 10th ed. Vol. V A, Enzyme, Springer, Berlin 1964, p. 887.

Ribonuclease

from bovine pancreas

Ribonucleate 3′-pyrimidino-oligo-nucleotidohydrolase, EC 3.1.4.22

Reaction, see under "Specificity".

Michaelis constants

(3′-5′)Cytidylyl-adenosine	1.4×10^{-3} mol/l	
(3′-5′)Cytidylyl-guanosine	1.4×10^{-3} mol/l	
(3′-5′)Cytidylyl-purine-9-riboside	5.0×10^{-3} mol/l	
(3′-5′)Cytidylyl-N^6-methyl-adenosine	1.5×10^{-3} mol/l	
(3′-5′)Cytidylyl-N^6-dimethyl-adenosine	5.0×10^{-3} mol/l	
(3′-5′)Cytidylyl-3-isoadenosine	5.0×10^{-3} mol/l	
(3′-5′)Cytidylyl-cytidine	3.3×10^{-3} mol/l	dimethylglutarate buffer, pH 7.0; 25 °C [1]
(3′-5′)Cytidylyl-uridine	3.7×10^{-3} mol/l	
(3′-5′)Cytidylyl-thymidine	1.1×10^{-3} mol/l	
(3′-5′)Uridylyl-adenosine	1.3×10^{-3} mol/l	
(3′-5′)Uridylyl-adenosine-N^1-oxide	1.4×10^{-3} mol/l	
(3′-5′)Uridylyl-guanosine	2.0×10^{-3} mol/l	
(3′-5′)Uridylyl-cytidine	1.7×10^{-3} mol/l	
(3′-5′)Uridylyl-uridine	3.7×10^{-3} mol/l	
(3′-5′)Adenylyl-3′-cytidylate	9×10^{-4} mol/l	
U-2:3-MP, cycl.*	6.9×10^{-3} mol/l	pH 7.0; 25 °C [2]
C-2:3-MP, cycl.*	3.5×10^{-3} mol/l	

Inhibitor constants (Tris buffer, pH 7.5; 25 °C) [3]: Uridine-3′-phosphate (U-2:3-MP, cycl. as substrate) 2.16×10^{-3} mol/l.

Molecular weight [4]: 13 680.

Specificity: RNase catalyzes the hydrolysis of 3′- and 5′-phosphate diester linkages of ribose in RNA with the formation of oligonucleotides with end-terminal 2′:3′-cyclic phosphate diester groups, which are then hydrolyzed to the corresponding 3′-mono-ester. The enzyme only hydrolyzes phosphodiester linkages which originate from pyri-

---

* The hydrolysis is activated by adenine, adenosine, cyclic A-2:3-MP, or adenylyl-adenosine.

midine 3'-phosphates. RNase hydrolyzes a large number of 2': 3'-cyclo compounds of cytidine and uridine derivates [5 – 7].

Assay

$$\text{Ribonucleate} \xrightarrow[\text{H}_2\text{O}]{\text{RNase}} \text{oligonucleotides}$$

A unit is the enzyme activity which under the test conditions causes an absorbance change from A_0 to A_1 (= final absorbance), indicating complete reaction, in one minute.

Wavelength 300 nm; light path 10 mm; final volume 3.00 ml; 25°C.

1.5 ml	substrate/acetate buffer (0.1 mol/l; pH 5.0;	50 mmol/l
	RNA, 1 mg/ml)	RNA 0.5 g/l
1.4 ml	water	
0.1 ml	enzyme in water	

Stability: lyophilized enzymes are stable for several months at 4°C.

Purity required: activity $\geq$ 40 units/mg substance (25°C; according to *Kunitz* [8]). The preparation should be completely free from DNase.

References

[1] *H. Follmann, H. J. Wieker, H. Witzel,* Eur. J. Biochem. *1,* 243 (1967).
[2] *H. J. Wieker, H. Witzel,* Eur. J. Biochem. *1,* 251 (1967).
[3] *C. S. Cheung, H. I. Abrash,* Biochemistry *3,* 1883 (1964).
[4] *D. G. Smith, W. H. Stein, S. Moore,* J. Biol. Chem. *238,* 227 (1963).
[5] *C. B. Anfinsen, F. H. White,* in: *P. D. Boyer, H. Lardy, K. Myrbäck* (eds.), The Enzymes, Vol. 5, Academic Press, New York 1961, p. 95.
[6] *J. P. Hummel, G. Kalnitzky,* Annu. Rev. Biochem. *33,* 25 (1964).
[7] *H. G. Gassen, H. Witzel,* Eur. J. Biochem. *1,* 36 (1967).
[8] *M. Kunitz,* J. Gen. Physiol. *24,* 15 (1940).

Sphingomyelinase

from *Bacillus cereus*

Sphingomyelin cholinephosphohydrolase, EC 3.1.4.12

$$\text{Sphingomyelin} + \text{H}_2\text{O} \xrightarrow{\text{sphingomyelinase}} \text{N-acylsphingosine} + \text{choline phosphate}$$

Effectors [1]: Mg^{2+} activates the enzyme, Ca^{2+} inhibits completely at 5 mmol/l. Deoxycholate stimulates at concentrations $> 0.1\%$. EDTA inhibits completely at 0.25

mmol/l, 2-phenanthroline is without effect. Zn^{2+} and Ca^{2+} partly restore the EDTA-inhibited enzyme activity, SH-reagents do not affect the enzyme.

Molecular weight [2]: 30000.

Specificity [1]: the enzyme exhibits a high specificity for sphingomyelin; phosphatidylcholine, phosphatidylethanolamine and phosphatidylinositol are not hydrolyzed. Lysophosphatidylcholine is hydrolyzed to a small extent by a side activity of the enzyme.

pH Optimum: 6.0 – 7.0 in borate buffer [1], 9.0 in diethylbarbiturate buffer [2].

Assay

$$\text{Sphingomyelin} + H_2O \xrightarrow{\text{sphingomyelinase}} \text{acylsphingosine} + \text{choline phosphate}$$

$$\text{choline phosphate} + H_2O \xrightarrow{\text{AP}} \text{choline} + P_i$$

$$P_i + NH_4VO_3 + (NH_4)_6MO_7O_{24} \longrightarrow \text{coloured complex}.$$

Wavelength Hg 365 nm; light path 10 mm; incubation volume for enzyme reaction 0.57 ml; final volume 2.20 ml; incubation temperature 37°C; measuring temperature 20 – 25°C.

0.50 ml	sphingomyelin (3 mg/ml barbiturate buffer, 50 mmol/l; pH 9.0; ultrasonicated)	sphingomyelin barbiturate	2.6 g/l 44 mmol/l
0.01 ml	$MgCl_2$ (0.1 mol/l)		1.7 mmol/l
0.01 ml	alkaline phosphatase (10 mg/ml)		6100 U/l
0.025 ml	enzyme in barbiturate buffer		

incubate for exactly 15 min at 37°C and then add 0.025 ml trichloroacetic acid, 3 mol/l, and heat for 5 min in a boiling water-bath, cool and centrifuge for ca. 9 min. Use 0.20 ml of the supernatant (also run a blank in which the sample was added after the precipitation with TCA) and mix with 2.0 ml phosphate reagent (10 ml ammonium molybdate solution, 10%; 10 ml ammonium vanadate solution, 0.235%; 5.7 ml HNO_3, conc.; make up to 100 ml with water), mix, allow to stand for 10 min and measure against a blank.
Calculate the activity of the sphingomyelinase by means of an orthophosphate standard curve.

Stability: in suspensions in ammonium sulphate solutions, 3.4 mol/l; pH ca. 7.3; $MgCl_2$, 2 mmol/l, a decrease in activity of approx. 10% may occur within 6 months at 4°C.

Purity required: specific activity $\geqslant 800$ U/mg protein (37 °C). Contaminant $\leqslant 0.1\%$ PL-C.

References

[1] *H. Ikezawa et al.,* Biochim. Biophys. Acta *528*, 247 (1978).
[2] *S. Schiefer, Boehringer Mannheim,* unpublished.

Sorbitol Dehydrogenase (Polyol Dehydrogenase)

from sheep liver

L-Iditol: NAD 5-oxidoreductase, EC 1.1.1.14

$$\text{D-Sorbitol} + \text{NAD}^+ \xrightleftharpoons{\text{SDH}} \text{D-fructose} + \text{NADH} + \text{H}^+$$

Equilibrium constant (25 °C) [1]

$$K = \frac{[\text{D-fructose}] \times [\text{NADH}] \times [\text{H}^+]}{[\text{D-sorbitol}] \times [\text{NAD}^+]} = 2.05 \times 10^{-9} \text{ mol/l}$$

Michaelis constants

Sorbitol	7×10^{-4} mol/l [1]
Fructose	$0.25 - 0.30$ mol/l
L-Erythrulose	2.5×10^{-2} mol/l

Inhibitor constants [2]:

	concentration (mmol/l)	inhibition
4-Chloromercuribenzoate	0.1	100%
Cysteine	2	50%
Glutathione	2	28%
KCN	1.0	5%
	1.0	50%
2-Phenanthroline	0.83	50%
	0.083	13%
EDTA	1	12%

Molecular weight [2]: 115000.

Specificity: the enzyme catalyzes the following reactions in the presence of NAD or NADH:

Ribitol	⇌	D-ribulose
Xylitol	⇌	D-xylulose
D-Sorbitol	⇌	D-fructose
L-Iditol	⇌	L-sorbose
Allitol	⇌	allulose
L-Gala-D-glucoheptide	⇌	perseulose (L-galaheptulose)
D-Altro-D-glucoheptide	⇌	sedoheptulose (D-altroheptulose)

In addition, L-erythrulose is reduced and, at ca. 3% of this rate, a mixture of glyceraldehyde phosphate and dihydroxyacetone phosphate. The relative rates of oxidation are as follows: D-sorbitol (1.00), L-iditol (0.96), altro-D-glucoheptide (0.87), xylitol (0.85), L-gala-D-glucoheptide (0.52), ribitol (0.49) and allitol (0.45). The following are not oxidized: erythritol, D- and L-arabitol, dulcitol, D-gulitol, D-iditol, D-mannitol, D-talitol, perseitol, volemitol and cyclitol (myo-inositol, D- and L-inositol, etc.).

The following are not reduced: D-tagatose, D-mannoheptulose, D-glucoheptulose, D-glucose, fructose-6-P, DL-glyceraldehyde, dihydroxyacetone, pyruvate, hydroxypyruvate, 2-oxoglutarate, acetaldehyde and glycolaldehyde. NADP(H) cannot replace NAD(H).

Assay

Wavelength 339, Hg 334 or Hg 365 nm; light path 10 mm; final volume 2.97 ml; 25°C.

2.50 ml	triethanolamine buffer (0.1 mol/l; pH 7.6)	83.3 mmol/l
0.40 ml	D-fructose (200 mg/ml)	150 mmol/l
0.05 ml	NADH, Na salt, (10 mg/ml)	0.2 mmol/l
0.02 ml	enzyme in water	

(at this fructose concentration SDH is not saturated with substrate).

Stability: lyophilized enzymes containing 83% maltose are stable for several months at 4°C.

Purity required: specific activity ≥ 25 U/mg protein (25°C). Contaminants (related to the specific activity of SDH) ≤ 0.01% ADH, ≤ 0.02% GlDH, glucose dehydrogenase, ≤ 0.05% LDH, MDH.

References

[1] *S. Hollmann,* in: *Hoppe-Seyler/Thierfelder,* Handbuch der physiol. und path.-chem. Analyse, Vol. *VI* A, Springer, Berlin 1964, p. 704.
[2] *G. Smith,* Biochem. J. *83*, 135 (1962).

Subtilisin

from *Bacillus subtilis*

EC 3.4.21.14

Hydrolysis of proteins and peptide amides

Michaelis constants [1]

Substrate, methyl ester of	K_m (mol/l) for subtilisin from various strains of *B. subtilis*			
	Amylo-saccha-riticus	Novo	Carlsberg	BPN
N-Acetyl-L-tyrosine*	146×10^{-3}	70×10^{-3}	90×10^{-3}	
N-Acetyl-L-tyrosine	46×10^{-3}	90×10^{-3}	70×10^{-3}	
N-Acetyl-L-tryptophan	31×10^{-3}	90×10^{-3}	50×10^{-3}	
N-Acetyl-L-phenylalanine	44×10^{-3}	60×10^{-3}	30×10^{-3}	
N-Benzoyl-L-arginine*	12×10^{-3}	7×10^{-3}	7×10^{-3}	
N-Tosyl-L-arginine	42×10^{-3}	70×10^{-3}	44×10^{-3}	
N-Acetyl-valine		280×10^{-3}	190×10^{-3}	
N-Acetyl-L-alanine				123×10^{-3}
N-Acetyl-L-leucine				66×10^{-3}
N-Acetyl-L-phenyl-alanine*				16.6×10^{-3}
N-Acetyl-L-tryptophan*				23.8×10^{-3}
N-Acetyl-L-lysine				91×10^{-3}

* ethyl ester

Inhibitor constants [1]

Inhibitor	K_i (mol/l) for subtilisin from various strains of *B. subtilis*			
	Carlsberg	Novo	BPN	conditions*:
Indole	50×10^{-3}	50×10^{-3}	30×10^{-3}	pH 7.5; 37°C
Phenol	100×10^{-3}	110×10^{-3}	100×10^{-3}	pH 7.4; 37°C
Hydrocinnamate	140×10^{-3}	300×10^{-3}	340×10^{-3}	pH 8.0; 37°C
Methyl butyrate	170×10^{-3}	170×10^{-3}	210×10^{-3}	pH 7.0; 37°C

* Hydrolysis of ATEE in 0.1 mol/l KCl, containing 8% dioxane (v/v).

Effectors [1]: various phenyl arsonates inhibit; benzyloxycarbonyl-L-phenylalanine bromoethyl ketone causes loss of hydrolytic activity. Certain aromatic compounds (see e.g. inhibitor constants) are competitive inhibitors. In contrast to their action on chymotrypsin and trypsin, chloromethylketones, 4-nitrobenzene sulphonate and proflavin do not affect the proteolytic activity of subtilisin.

Molecular weight [1]: determination by sequence studies

Strain of *B. subtilis*

BPN	27 537
Carlsberg	27 277
Novo	27 537
Amylosacchariticus	27 671

Specificity [1]: the subtilisins manifest a wide specificity with both esterase and proteinase activity and are suitable for hydrolyzing a variety of proteins, including casein, haemoglobin, gelatine and oxidized lysozyme. The amide bond is more resistant to the enzyme's attack than the ester bond. Cleavage appears to occur mainly at the NH_2- and COOH-terminal sides of neutral and acidic amino acid residues.

pH Optimum: subtilisins act optimally at alkaline pH values.

Assay

Wavelength Hg 578 nm; light path 10 mm; incubation volume 2.40 ml; incubation temperature 37°C; final volume 7.50 ml; measuring temperature 37°C.

1.00 ml	borate buffer (0.1 mol/l; pH 8.0)	42 mmol/l
1.00 ml	substrate (20 mg casein/ml, adjusted to pH 8.0)	8.3 g/l
0.20 ml	HCl (0.05 mol/l)	4.2 mmol/l
0.20 ml	enzyme in $CaCl_2$ (2 mmol/l)	1.7 mmol/l

incubate for exactly 10 min at 37°C, add

2.00 ml	trichloroacetic acid (0.1 mol/l)
0.20 ml	$CaCl_2$ (2 mmol/l)

incubate for 10 min at 37°C, filter, mix

1.50 ml	filtrate
5.00 ml	Na_2CO_3 (0.4 mol/l)
1.00 ml	*Folin & Ciocalteu's* phenol reagent (commercial preparation diluted 1 + 2)

incubate for ca. 20 min at 37°C, read absorbance against blank, in which the sample has been added after the addition of TCA. Run a standard with 0.20 ml standard solution (tyrosine, 5.0 mmol/l) instead of sample. 0.20 ml sample is added to this assay mixture after the addition of TCA.

Stability: the lyophilized enzyme is stable for many months at 4°C.

Purity required: activity $\geq$ 5 units/mg substance (37°C, casein as substrate).

Reference

[1] *F. S. Markland, E. C. Smith,* in: *P. D. Boyer* (ed.), The Enzymes, Vol. *III*, Academic Press, New York 1971, p. 561.

Thermolysin

from *Bacillus thermoproteolyticus*

EC 3.4.24.4

Preferential cleavage of -Val, -Leu, -Ile, -Phe.

Effectors: thermolysin is a zinc-containing metallo-endoproteinase and is inhibited by metal chelating agents, but is insensitive to inhibitors of thiol- and serine proteases [1, 2]. The enzyme is reversibly inactivated by diethyl pyrocarbonate [3]. The activity of the enzyme may be enhanced by its covalent modification with N-hydroxysuccin-imide esters of aliphatic and aromatic amino acids [3].

Specificity [4 – 8]: the enzyme hydrolyzes (especially at high temperatures from 70°C – 80°C) proteins and peptides, preferably at bonds involving the NH_2-groups of leucine, isoleucine, valine and phenylalanine. Thermolysin has elastase activity which can be supressed in the presence of NaCl, 0.1 mol/l, without loss of proteolytic activity.

Assay

Wavelength Hg 578 nm; light path 10 mm; incubation volume 2.20 ml; incubation temperature 37°C; final volume 7.50 ml; measuring temperature 37°C.

1.00 ml	borate buffer (0.1 mol/l; pH 7.2)	4 mmol/l
1.00 ml	substrate solution (2 g casein dissolved in some water and 10 ml NaOH, 1 mol/l, diluted and pH adjusted to 7.2, made up with water to 100 ml)	4.5 mg/ml
0.20 ml	enzyme in water	

incubate for exactly 10 min at 37°C; add

2.00 ml   trichloroacetic acid (0.1 mol/l)
0.20 ml   CaCl$_2$ (2 mmol/l)

incubate for ca. 10 min at 37°C, filter, mix, use

1.50 ml   filtrate
5.00 ml   Na$_2$CO$_3$ (0.4 mol/l)
1.00 ml   *Folin & Ciocalteu's* phenol reagent (commercial preparation diluted 1 + 2)

incubate for ca. 20 min at 37°C, read absorbance against blank in which the sample has been added after the addition of TCA. Run a standard with 0.20 ml standard solution (tyrosine, 50 mmol/l) instead of sample. 0.2 ml sample is added to this assay mixture after addition of TCA.

Stability: the lyophilized enzyme is stable for many months at 4°C.

Purity required: activity $\geq$ 40 units/mg substance (37°C; casein as substrate).

References

[1] *S. A. Latt et al.,* Biochem. Biophys. Res. Commun. *37,* 333 (1969).
[2] *H. Matsabura, J. Feder, in: P. D. Boyer* (ed.), The Enzymes, Vol. *III,* Academic Press, New York 1971, p. 721.
[3] *S. Blumberg, B. L. Vallee,* Biochemistry *14,* 2410 (1975).
[4] *H. Matsabura et al.,* Biochem. Biophys. Res. Commun. *21,* 242 (1965).
[5] *H. Matsabura et al.,* Arch. Biochem. Biophys. *115,* 324 (1966).
[6] *K. Morihara, H. Tsuzuki,* Biochim. Biophys. Acta *118,* 215 (1966).
[7] *H. Matsabura,* Biochem. Biophys. Res. Commun. *24,* 427 (1966).
[8] *K. Morihara, M. Ebata,* J. Biochem. *61,* 149 (1967).

Thrombin (Fibrinogenase)

from human plasma

Thrombin, EC 3.4.21.5

Preferential cleavage: Arg-; activates fibrinogen to fibrin.

Michaelis constants [1]

TAME (4-toluenesulphonyl-L-arginine methyl ester)	$4-6 \times 10^{-3}$ mol/l
BAME (benzoyl-L-arginine methyl ester)	$1.2-1.3 \times 10^{-3}$ mol/l

TLME (4-toluenesulphonyl-L-lysine methyl ester)	$5.3 - 25 \times 10^{-3}$ mol/l
AlME (acetyl-L-lysine methyl ester)	$17 - 40 \times 10^{-3}$ mol/l
AAME (acetyl-L-arginine methyl ester)	$10 - 21 \times 10^{-3}$ mol/l
Carbobenzoxy-L-arginine methyl ester	$5.2 - 13 \times 10^{-3}$ mol/l
Carbobenzoxy-L-lysine ester	$13 - 14 \times 10^{-3}$ mol/l
Fibrin [2]	4.3×10^{-7} mol/l

Effectors: thrombin is produced from its blood plasma precursor prothrombin by a variety of agents [3]. The enzyme is inactivated by several naturally occurring specific thrombin inactivators [4]. The most important is antithrombin of blood plasma. DFP (diisopropyl fluorophosphate) and PMSF (phenylmethylsulphonyl fluoride) [5], 4,3-phenoxyethoxy benzamidine and toluenesulphonyl-lysyl-chloromethyl ketone inactivate. Heparin and the combination of antithrombin and heparin inhibit. Hirudin [6] was shown to be one of the most selective natural inhibitors of thrombin.

Molecular weight [7]: multiple forms of human, bovine and porcine thrombins exist whose molecular weights are generally agreed to be ca. 34000.

Specificity: thrombin converts fibrinogen to fibrin and hydrolyzes peptides, amides and esters of L-arginine. With regard to polypeptide substrates, thrombin is far more specific than trypsin even though both enzymes catalyze the hydrolysis of similar low molecular-weight ester and amide substrates. Factor V and VIII of blood coagulation are modified and result in initially enhanced, and then diminished biological activity.

Assay

Wavelength 405 nm; light path 10 mm; final volume 1.10 ml; 25 °C; $\varepsilon = 1.04 \, l \times mmol^{-1} \times mm^{-1}$.

0.90 ml	Tris buffer (50 mmol/l; pH 8.3; NaCl, 0.227 mol/l)	Tris	41 mmol/l
		NaCl	0.2 mol/l
0.10 ml	Chromozym® TH (1.9 mmol/l)		0.17 mmol/l
0.10 ml	enzyme in buffer		

Stability: the lyophilized enzyme stabilized with buffer substances (pH 6.9) is stable for several months at 4 °C.

Purity required: specific activity $\geqslant$ 120 U/mg protein (25 °C; $\triangleq$ approx. 755 NIH units). Contaminant (related to the specific activity of thrombin) $\leqslant$ 3% factor Xa.

References

[1] *S. Magnusson,* Thrombin and Prothrombin, in: *P. D. Boyer* (ed.), The Enzymes, Vol. *III*, Academic Press, New York 1971, pp. 278 – 320.
[2] *G. Y. Shinowara,* Biochim. Biophys. Acta *113*, 359 (1966).

[3] *B. J. Davis,* Ann. N. Y. Acad. Sci. *121*, 404 (1964).

[4] *A. C. Allison, J. Humphrey,* Nature *183*, 1590 (1959).

[5] *R. L. Lundblad,* Biochemistry *10*, 2501 (1971) and Biophys. Res. Commun. *66*, 482 (1975).

[6] *F. Markwardt,* Hirudin as an Inhibitor of Thrombin, in: *S. P. Colowick, N. O. Kaplan* (eds.), Methods in Enzymology, Vol. *XIX*, Academic Press, New York 1970, pp. 924–932.

[7] *W. H. Seegers,* Blood Clotting Enzymology, Academic Press, New York 1967.

Transaldolase

from yeast

Sedoheptulose-7-phosphate: D-glyceraldehyde-3-phosphate dihydroxyacetonetransferase, EC 2.2.1.2

$$\text{Sedoheptulose-7-P} + \text{D-glyceraldehyde-3-P} \xrightleftharpoons{\text{TA}}$$

$$\text{D-fructose-6-P} + \text{D-erythrose-4-P}$$

Equilibrium constants: the reaction is reversible. $K = 0.82$ (25 °C) [1]; $K = 1.05$ (37 °C) [2].

Michaelis constants (of substrates of highest affinity)

Sedoheptulose-7-P	1.7×10^{-4} mol/l
Fructose-6-P	4.3×10^{-4} mol/l
Erythrose-4-P	6.17×10^{-5} mol/l
Glyceraldehyde-3-P	2.2×10^{-4} mol/l
Fructose-6-P	$8 \quad \times 10^{-4}$ mol/l ⎫ for the isolated isoenzymes [3]
Erythrose-4-P	$2 \quad \times 10^{-5}$ mol/l ⎭

Inhibitor constants [4]: orthophosphate 5×10^{-2} mol/l. The inhibition is competitive with fructose-6-P, non-competitive to sedoheptulose-7-P, glyceraldehyde or glyceraldehyde-3-P.

Effectors: pyrophosphate or sulphate inhibit the enzyme in the same manner as orthophosphate; chloride and carbonate show no inhibitory effect.

Molecular weights [4]

αα	+	ββ	⇌	2αβ
isoenzyme I		isoenzyme II		isoenzyme II
76 100		65 200		68 100

Specificity

Donor substrates   *Acceptor substrates*

D-Fructose-6-P   D-Erythrose-4-P
D-Sedoheptulose-7-P   D-Glyceraldehyde-3-P
D-Fructose   Formaldehyde
L-Sorbose-6-P   D-Glyceraldehyde-3-P
2,5-D-*threo*-Diketohexose   L-Glyceraldehyde-3-P
Sedoheptulose   Hydroxypyruvate aldehyde
Octulose-8-P   D-Erythrose
                                        Ribose-5-P

No reaction with: Glycolaldehyde
                                        Glycolaldehyde phosphate

Assay

$$\text{Fructose-6-P + erythrose-4-P} \xrightarrow{\text{TA}} \text{sedoheptulose-7-P + GAP}$$

$$\text{GAP} \xrightarrow{\text{TIM}} \text{DAP}$$

$$\text{DAP + NADH + H}^+ \xrightarrow{\text{GDH}} \text{glycerol-3-P + NAD}^+.$$

Wavelength 339, Hg 334 or Hg 365 nm; light path 10 mm; assay volume 2.98 ml; 25°C.

2.50 ml	triethanolamine buffer (0.1 mol/l; pH 7.6;	triethanolamine	84 mmol/l
	EDTA, 10 mmol/l)	EDTA	8.4 mmol/l
0.20 ml	erythrose-4-P (10 mmol/l)		0.7 mmol/l

mix thoroughly

0.20 ml	fructose-6-P, Na-salt, (30 mg/ml)		5.4 mmol/l
0.05 ml	NADH, Na salt, (10 mg/ml)		0.2 mmol/l
0.01 ml	GDH/TIM (1 mg and 0.1 mg/ml, resp.)	GDH	134 U/l
		TIM	16.7 kU/l
0.02 ml	enzyme in buffer		

Stability: suspensions in ammonium sulphate solution, 3.2 mol/l; pH ca. 6, are stable for several months at +4°C.

Purity required: specific activity $\geq$ 15 U/mg protein (25°C). Contaminant (related to the specific activity of transaldolase) $\leq$ 0.01% "NADH oxidase".

References

[1] *B. L. Horecker, P. Z. Smyrniotis,* J. Biol. Chem. *212*, 811 (1955).
[2] *R. Venkataraman, E. Racker,* J. Biol. Chem. *236*, 1876 (1961).
[3] *O. Tsolas, B. L. Horecker,* Arch. Biochem. Biophys. *136*, 303 (1970).
[4] *O. Tsolas, B. L. Horecker,* in: *P. D. Boyer* (ed.), The Enzymes, Vol. *VII*, Academic Press, New York 1972, p. 259.

Triosephosphate Isomerase

from rabbit muscle

D-Glyceraldehyde-3-phosphate ketol-isomerase, EC 5.3.1.1

$$GAP \overset{TIM}{\rightleftharpoons} DAP$$

Equilibrium constant (veronal/acetate buffer, pH 8.0; 25 °C) [1]

$$K = \frac{[DAP]}{[GAP]} = 22.$$

Molecular weight [2]: 53 000.

Assay

$$GAP \overset{TIM}{\longrightarrow} DAP$$

$$DAP + NADH + H^+ \overset{GDH}{\longrightarrow} glycerol\text{-}3\text{-}P + NAD^+.$$

Wavelength 339, Hg 334 or Hg 365 nm; light path 10 mm; final volume 3.07 ml; 25 °C.

2.50 ml	triethanolamine buffer (0.3 mol/l; pH 7.6)	0.243 mol/l
0.50 ml	GAP (4 mg/ml)	3.8 mmol/l
0.05 ml	NADH, Na salt, (10 mg/ml)	0.2 mmol/l
0.01 ml	GDH (10 mg/ml)	1.3 kU/l
0.01 ml	enzyme in buffer	

Stability: crystalline suspensions in ammonium sulphate, 3.2 mol/l; pH ca. 6, are stable for several months at 4 °C.

Purity required: specific activity $\geqq$ 5000 U/mg protein (25 °C). Contaminants (related to the specific activity of TIM) $\leqq$ 0.001% GAPDH, $\leqq$ 0.01% aldolase, GDH.

References

[1] *P. Oesper, O. Meyerhof,* Arch. Biochem. Biophys. *27*, 223 (1950).
[2] *F. C. Hartmann,* Biochemistry *10*, 146 (1971).

Trypsin

from bovine pancreas

EC 3.4.21.4

Preferential cleavage: Arg-, Lys-

Michaelis constants

Benzoyl-L-arginine ethyl ester	5 $\times 10^{-5}$ mol/l	
Benzoyl-L-argininamide	2.1 $\times 10^{-3}$ mol/l	pH 7.8 – 8; 25 °C [1]
4-Toluenesulphonyl-L-arginine methyl ester	5 $\times 10^{-5}$ mol/l	

L-Lysine-4-nitroanilide	3.64 $\times 10^{-4}$ mol/l	
Benzoyl-L-argininc-4-nitro-anilide	9.39 $\times 10^{-4}$ mol/l	pH 8 – 8.5; 15 °C [2]

Inhibitor constants

4-Aminobenzamidine	8.25 $\times 10^{-6}$ mol/l	
Benzamidine	1.84 $\times 10^{-5}$ mol/l	pH 8.15; 15 °C [3]
Acetamidine	3.65 $\times 10^{-2}$ mol/l	
Phenylacetamidine	1.51 $\times 10^{-2}$ mol/l	

Ethylamine	6.2 $\times 10^{-2}$ mol/l	pH 6.6; 25 °C [4]
Butylamine	1.7 $\times 10^{-3}$ mol/l	

Effectors: trypsin is inhibited by organophosphorus compounds such as diisopropyl fluorophosphate and natural trypsin inhibitors from pancreas. Soybean, lima bean and egg white are further sources of inhibitors. Inhibition by benzamidine is reversible [3]. Calcium ions retard trypsin hydrolysis.

Molecular weight [5]: ca. 23 300.

Specificity: the protease activity of trypsin is specific towards positively charged side chains with lysine and arginine [7].

Assay [7]

$$\text{N-Benzoyl-L-arginine ethyl ester} + H_2O \xrightarrow{\text{trypsin}}$$

$$\text{N-benzoyl-L-arginine} + \text{ethanol}.$$

Wavelength 255 nm; light path 10 mm; final volume 3.00 ml; 25°C; $\varepsilon = 0.081$ $1 \times \text{mmol}^{-1} \times \text{mm}^{-1}$.

2.80 ml	Tris buffer/substrate (Tris, 50 mmol/l; pH 8.0;	Tris	46.7 mmol/l
	0.34 mg BAEE/ml; CaCl$_2$, 2.22 mg/ml)	BAEE	0.9 mmol/l
		CaCl$_2$	19 mmol/l
0.20 ml	enzyme in HCl, 1 mmol/l		

Stability: the lyophilized enzyme is stable for several months at 4°C.

Purity required: activity $\geqslant$ 40 U/mg substance (25°C).

References

[1] *L. Cunningham,* in: *M. Florkin, E. H. Stotz* (eds.), Comprehensive Biochemistry, Vol. *16*, Elsevier, Amsterdam 1965, p. 85.
[2] *B. F. Erlanger et al.,* Arch. Biochem. Biophys. *95*, 271 (1961).
[3] *M. Mares-Guia, E. Shaw,* J. Biol. Chem. *240*, 1579 (1965).
[4] *T. Inagami, T. Murachi,* J. Biol. Chem. *238*, 1905 (1963).
[5] *T. Hofmann,* Biochemistry *3*, 356 (1964).
[6] *V. Keil-Dlouha et al.,* FEBS Lett. *16*, 287 (1971).
[7] *G. W. Schwert, Y. Takenaka,* Biochim. Biophys. Acta *16*, 570 (1955).

Urease

from jackbeans, *Canavalia ensiformis*

Urea amidohydrolase, EC 3.5.1.5

$$\text{Urea} + H_2O \xrightarrow{\text{urease}} CO_2 + 2\,NH_3$$

Michaelis constant (phosphate buffer, pH 7; 25°C) [1]: urea 1.05×10^{-2} mol/l.

Effectors: Na^+, K^+ and NH_4^+ inhibit, while P_i activates; Suramin and thiourea are competitive inhibitors [2].

Molecular weight [3]: 480000.

Specificity: the enzyme is specific for urea.

Assay

$$\text{Urea} + H_2O \xrightarrow{\text{urease}} CO_3^{--} + 2\,NH_4^+$$

$$2\,NH_4^+ + 2\text{-oxoglutarate} + 2\,NADH \xrightarrow{\text{GlDH}}$$

$$2\,\text{L-glutamate} + 2\,NAD^+ + 2\,H_2O\,.$$

Wavelength 339, Hg 334 or Hg 365 nm; light path 10 mm; final volume 3.02 ml; 25 °C.

2.50 ml	triethanolamine buffer (0.1 mol/l; pH 8.0)	83 mmol/l
0.10 ml	2-oxoglutarate, Na salt, (76 mg/ml)	13 mmol/l
0.10 ml	ADP, Na salt, (20 mg/ml)	1.4 mmol/l
0.05 ml	NADH, Na salt, (10 mg/ml)	0.2 mmol/l
0.05 ml	GlDH (10 mg/ml)	20 kU/l
0.20 ml	urea (450 mg/ml)	0.5 mol/l
0.02 ml	enzyme in water	

Stability: lyophilized enzymes are stable for several months at 4 °C.

Purity required: activity $\geq 5-100$ U/mg substance (25 °C).

References

[1] *J. Peterson, K. Harmon, C. Niemann*, J. Biol. Chem. *176*, 1 (1948).
[2] *T. E. Barman*, Enzyme Handbook, Vol. *2*, Springer, Berlin 1969, p. 648.
[3] *W. N. Fishbein, K. Nagarajan*, Arch. Biochem. Biophys. *144*, 700 (1971).

Uricase

from *Aspergillus flavus*

Urate : oxygen oxidoreductase, EC 1.7.3.3

$$\text{Urate} + 2\,H_2O + O_2 \xrightarrow{\text{uricase}} \text{allantoin} + CO_2 + H_2O_2$$

Equilibrium constant [1, 2]: the reaction is irreversible.

Michaelis constant (TEA buffer, 0.05 mol/l, pH 8.5; 30°C) [3]

Ureate 6×10^{-5} mol/l

Effectors: Hg^{2+}, Cu^{2+} and CN^- are strong inhibitors [3]. Excess concentrations of the substrate, urate, inhibit. Phosphate and acetate activate at higher concentrations.

Molecular weight [3]: 93000.

Specificity: the enzyme is specific for urate. Caffeine, theobromine, theophylline, xanthine, 8-chloroxanthine and 2,8-dithio-6-hydroxypurine do not react.

pH Optimum [3]: pH 8.5 (TEA buffer, 0.05 mol/l; 30°C).

Assay

Wavelength 293 nm; light path 10 mm; final volume 3.15 ml; 25°C;
$\varepsilon = 1.26\,l \times mmol^{-1} \times mm^{-1}$.

3.00 ml	borate buffer (20 mmol/l; pH 8.5; saturated with oxygen)	19 mmol/l
0.10 ml	uric acid (1.2 mmol/l borate buffer)	0.04 mmol/l
0.05 ml	enzyme in 0.1% bovine serum albumin	

Stability: a decrease in the activity of lyophilized enzymes of approx. 10% may occur within 6 months.

Purity required: specific activity $\geqslant$ 15 U/mg protein (25°C).

References

[1] *P. Laboureur, C. Langlois,* Proc. Natl. Acad. Sci. *264*, 224 (1967).
[2] *P. Laboureur et al.,* Soc. d'Etudes Appl. Biochim., France, Dos 1642656 (1971).
[3] *P. Laboureur, C. Langlois,* Bull. Soc. Chim. Biol. *50* (4), 811 and 827 (1968).

Uricase

from hog liver

Urate: oxygen oxidoreductase, EC 1.7.3.3

$$Urate + 2\,H_2O + O_2 \xrightarrow{\text{uricase}} allantoin + H_2O_2 + CO_2$$

Michaelis constant (Tris buffer, pH 8; 20°C) [1]: urate 1.7×10^{-5} mol/l

Inhibitor constants (Tris borate buffer, pH 8.5; 20°C) [1, 2]

2,6,8-Trichloropurine	8.0×10^{-7} mol/l
2,6-Dichloro-8-hydroxypurine	1.3×10^{-6} mol/l
2-Hydroxy-6,8-diaminopurine	1.8×10^{-6} mol/l
2,6-Dihydroxypurine	1.2×10^{-5} mol/l
2,6-Dihydroxy-8-amino-purine (xanthine)	2.3×10^{-5} mol/l
2-Chloro-6-amino-8-hydroxypurine	4.0×10^{-5} mol/l
2,8-Dihydroxy-6-amino-purine	1.5×10^{-4} mol/l
2,6-Diamino-8-hydroxypurine	5.0×10^{-4} mol/l

Molecular weight (pH 10.5) [3]: 125000.

Specificity: the enzyme is specific for urate.

Assay

Wavelength 293 nm; light path 10 mm; final volume 3.12 ml; 25°C; $\varepsilon = 1.26$ $l \times mmol^{-1} \times mm^{-1}$.

3.00 ml	borate buffer (20 mmol/l; pH 9.5)	19.2 mmol/l
0.10 ml	uric acid (2 mg/ml buffer)	0.4 mmol/l
0.02 ml	enzyme in buffer	

Stability: solutions in 50% glycerol containing glycine, 50 mmol/l, and Na_2CO_3, 0.13 mol/l; pH 10.2, are stable for several months at 4°C.

Purity required: specific activity $\geqq$ 8 U/mg protein (25°C).

References

[1] *H. R. Mahler,* in: *C. A. Lamb, O. G. Bentley, J. M. Beattie* (eds.), Trace Elements, Academic Press, New York 1958, p. 311.
[2] *H. R. Mahler,* in: *P. D. Boyer, H. Lardy, K. Myrbäck* (eds.), The Enzymes, Vol. *VIII*, Academic Press, New York 1963, p. 285.
[3] *O. M. Pitts, W. W. Fish,* Biochemistry *13*, 888 (1974).

Uridinediphosphoglucose Dehydrogenase

from beef liver

UPDglucose: NAD 6-oxidoreductase, EC 1.1.1.22

$$UDPglucose + 2\,NAD^+ + H_2O \xrightarrow{\text{UDPG-DH}}$$
$$UDPglucuronate + 2\,NADH + 2\,H^+$$

Equilibrium constant: the reaction is irreversible.

Michaelis constants (Tris buffer, pH 7.5) [1]

UDPglucose	8.8×10^{-6} mol/l
NAD	1.57×10^{-4} mol/l
5-Fluoro-UDPglucose	2.69×10^{-5} mol/l
6-Aza-UDPglucose	7.32×10^{-5} mol/l

Specificity: the enzyme is specific for UDPglucose.

Assay

Wavelength 339, Hg 334 or Hg 365 nm; light path 10 mm; final volume 2.82 ml; 25°C.

2.50 ml	Tris buffer (50 mmol/l; pH 8.7)	44.3 mmol/l
0.10 ml	UDPglucose, Na salt, (10 mg/ml)	0.5 mmol/l
0.20 ml	NAD (20 mg/ml)	2 mmol/l
0.02 ml	enzyme in buffer	

Stability: suspensions in ammonium sulphate solution, 3.2 mol/l; pH ca. 6, are stable for several months at 4°C.

Purity required: specific activity $\geq$ 0.6 U/mg protein (25°C). Contaminant (related to the specific activity of UDPG-DH) $\leq$ 1% LDH.

Reference

[1] *N. D. Goldberg, J. L. Dahl, R. E. Parks,* J. Biol. Chem. *238*, 3109 (1963).

Uridinediphosphoglucose Pyrophosphorylase

from beef liver

UTP : α-D-glucose-1-phosphate uridylyltransferase, EC 2.7.7.9

$$\text{UTP + glucose-1-P} \xrightleftharpoons{\text{UDPG-PP}} \text{UDPglucose + PP}_i$$

Equilibrium constant (TEA buffer, pH 7.8; 25°C) [1]

$$K = \frac{[\text{PP}_i] \times [\text{UDPglucose}]}{[\text{UTP}] \times [\text{glucose-1-P}]} = 0.28 \text{ to } 0.34$$

Michaelis constants (TEA buffer, pH 7.8; 25 °C) [1]*

		Relative rate
UDPglucose	6.0×10^{-5} mol/l	1.000
dTDPglucose	3.5×10^{-4} mol/l	0.027
CDPglucose	–	0.008
GDPglucose	–	0.001
UDPgalactose	–	0.035
UDPxylose	–	0.039
UDPmannose	–	0.003
PP_i	8.4×10^{-5} mol/l	–
UTP	2.0×10^{-4} mol/l	–
Glucose-1-P	5.0×10^{-5} mol/l	–

Molecular weight [1]: 480000.

Specificity: the enzyme does not react with IDPglucose, ADPglucose, UDP-N-acetylglucosamine or UDPglucuronate [1]*.

Assay

$$UDPglucose + PP_i \xrightarrow{\text{UDPG-PP}} UTP + glucose\text{-}1\text{-}P$$

$$Glucose\text{-}1\text{-}P \xrightarrow{\text{PGluM}} glucose\text{-}6\text{-}P$$

$$Glucose\text{-}6\text{-}P + NADP^+ \xrightarrow{\text{G6P-DH}} gluconate\text{-}6\text{-}P + NADPH + H^+.$$

Wavelength 339, Hg 334 or Hg 365 nm; light path 10 mm; final volume 3.00 ml; 25 °C.

2.50 ml	Tris-HCl (0.05 mol/l; pH 7.8)	42 mmol/l
0.10 ml	$Na_4P_2O_7$ (45 mmol/l)	1.5 mmol/l
0.10 ml	UDPglucose, Na salt, (20 mg/ml)	1.0 mmol/l
0.10 ml	$MgCl_2$ (0.2 mol/l)	6.6 mmol/l
0.10 ml	NADP, Na salt, (10 mg/ml)	0.38 mmol/l
0.05 ml	glucose-1,6-P_2, tricyclohexylammonium salt, (1 mg/ml)	0.02 mmol/l
0.02 ml	PGluM (5 mg/ml; NH_4^+-free dialyzed against acetate buffer, 50 mmol/l; pH 5.3)	6.6 kU/l
0.01 ml	G6P-DH (2 mg/ml; NH_4^+-free dialyzed against phosphate buffer, 50 mmol/l, pH 7.6)	930 U/l
0.02 ml	enzyme in water	

Stability: the lyophilized enzyme is fairly stable for several months at 4 °C.

---

* For enzyme from calf liver.

Purity required: activity $\geqq$ 4 U/mg substance (approx. 100 U/mg protein) (25 °C).

References

[1] *R. G. Hansen et al.,* in: *S. P. Colowick, N. O. Kaplan* (eds.), Methods in Enzymology, Vol. *VIII*, Academic Press, New York 1966, p. 248.
[2] *R. L. Turnquist, R. Gaurth Hansen,* in: *P. D. Boyer* (ed.), The Enzymes, Vol. *VIII*, Academic Press, New York 1973, p. 51.

Uridylyltransferase

from calf liver

UDPglucose : α-D-galactose-1-phosphate uridylyltransferase, EC 2.7.7.12

$$\text{UDPglucose} + \text{galactose-1-P} \xrightarrow[\hspace{1cm}]{\text{UT}} \text{glucose-1-P} + \text{UDPgalactose}$$

Specificity: the rate of reaction with TDPglucose is 8.7% and with CDPglucose 0.3% of the rate with UDP-glucose. There is no reaction with ADPglucose, GDPglucose or IDPglucose. Mannose-1-P, xylose-1-P, galactose-6-P and fructose-1-P do not react [1]. For the properties of the enzyme from *E. coli*, cf. [2].

Assay

$$\text{Galactose-1-P} + \text{UDPglucose} \xrightarrow{\text{UT}} \text{UDPgalactose} + \text{glucose-1-P}$$

$$\text{Glucose-1-P} \xrightarrow{\text{PGluM}} \text{glucose-6-P}$$

$$\text{Glucose-6-P} + \text{NADP}^+ \xrightarrow{\text{G6P-DH}} \text{gluconate-6-P} + \text{NADPH} + \text{H}^+.$$

Wavelength 339, Hg 334 or Hg 365 nm; light path 10 mm; final volume 2.94 ml; 25 °C.

2.50 ml	glycine buffer (0.1 mol/l; pH 8.0)	85 mmol/l
0.10 ml	NADP, Na salt, (10 mg/ml)	0.4 mmol/l
0.10 ml	UDPglucose, Na salt, (10 mg/ml)	0.5 mmol/l
0.15 ml	galactose-1-P, K salt, (40 mg/ml)	4.6 mmol/l
0.03 ml	glucose-1,6-P_2, tricyclohexylammonium salt, (2 mg/ml)	25 µmol/l
0.03 ml	PGluM (10 mg/ml)	20 kU/l
0.01 ml	G6P-DH (5 mg/ml)	2.4 kU/l
0.02 ml	enzyme in buffer	

Stability: lyophilized enzymes stabilized with sucrose are stable for several months at 4°C. The enzyme solution in water is stable for a few days at 4°C.

Purity required: specific activity $\geq$ 1 U/mg protein (25°C). Contaminants (related to the specific activitiy of UT) $\leq$ 0.01% 6-PGDH, $\leq$ 0.1% "NADH oxidase", $\leq$ 1% glucose dehydrogenase, $\leq$ 5% UDPG-pyrophosphorylase.

References

[1] *J. S. Mayes, R. G. Hansen,* in: *S. P. Colowick, N. O. Kaplan* (eds.), Methods in Enzymology, Vol. *IX*, Academic Press, New York 1966, p. 708.
[2] *T. E. Barman,* Enzyme Handbook, Vol. *I*, Springer, Berlin 1969, p. 473.

Xanthine Oxidase

from milk

Xanthine: oxygen oxidoreductase, EC 1.2.3.2

$$\text{Xanthine} + H_2O + O_2 \xrightarrow{\text{XOD}} \text{urate} + H_2O_2$$

Molecular weight [4]: 275000.

Michaelis constants (phosphate buffer, 25°C)

Purine	3.0×10^{-6} mol/l	
Hypoxanthine	1.3×10^{-6} mol/l	
Xanthine	1.7×10^{-6} mol/l	
Acetaldehyde	1.6×10^{-2} mol/l	pH 7.8
Salicylaldehyde	1.1×10^{-3} mol/l	
Benzaldehyde	8.3×10^{-4} mol/l	
Xanthine	3.6×10^{-6} mol/l	pH 8.3 [2]
O_2	2.4×10^{-4} mol/l	
O_2	2.4×10^{-5} mol/l	pH 7.8 [3]

Specificity [5]: the enzyme is not specific, as it can react as an electron donor with aldehydes, purines, pteridines, pyrimidines, azapurines and other heterocyclic compounds.

Ferricyanide, cytochrome c and various dyes can replace O_2 as electron acceptor.

Assay

$$\text{Hypoxanthine} + 2\,H_2O + 2\,O_2 \xrightarrow{\ \text{XOD}\ } \text{urate} + 2\,H_2O_2 \ .$$

Wavelength 293 nm; light path 10 mm; final volume 3.05 ml; 25 °C; $\varepsilon = 1.20$ $l \times mmol^{-1} \times mm^{-1}$.

3.00 ml   substrate/phosphate buffer   50 mmol/l
           (50 mmol/l; pH 7.4; EDTA, 0.1 mmol/l;   EDTA   0.1 mmol/l
           hypoxanthine, 0.1 mmol/l;   hypoxanthine 0.1 mmol/l
           O$_2$ saturated)
0.05 ml   enzyme suspension

Stability: suspensions in ammonium sulphate solution, 3.2 mol/l; pH ca. 8, EDTA 10 mmol/l, are stable for few months at 4 °C.

Purity required: specific activity ≥ 0.4 U/mg protein (25 °C). Contaminants (related to the specific activity of XOD) $\leq 0.01\%$ guanase, NP, uricase, $\leq 0.1\%$ ADA, AP.

References

[1] L. Greenlee, P. Handler, J. Biol. Chem. 239, 1090 (1964).
[2] E. Ackermann, A. S. Brill, Biochim. Biophys. Acta 56, 397 (1962).
[3] I. Friedovich, P. Handler, J. Biol. Chem. 237, 916 (1962).
[4] P. Andrews et al., Biochem. J. 93, 627 (1964).
[5] R. C. Bray, in: P. D. Boyer, H. Lardy, K. Myrbäck (eds.), The Enzymes, Vol. 7, Academic Press, New York 1963, p. 533.

2.2.2   Coenzymes, Metabolites, and Other Biochemical Reagents

Hans-Otto Beutler and Martin Supp

Of the many substances required for enzymatic analysis, we present only a selection here with the most important data for each substance.

Most substances are included which are mentioned in the assay procedures of the volumes III – VII and which are essential for the determinations according to the equations of the reactions underlying them. Only physical constants necessary for the analysis are given; other constants (e.g. rotation) are added if they are of practical importance. Contaminants are mentioned under "Purity required" if they may interfere in the use of the preparation in enzymatic analysis.

Common reagents which are available in sufficient quality as recommended by the National Bureau of Standards (NBS) have been omitted, as have amino acids and simple sugars (a wide range of excellent quality is available commercially).

Nomenclature, notation, and abbreviations largely follow the recommendations of the International Union of Biochemistry (IUB) (cf. p. 102). Where appropriate, the synonyms and the usual abbreviations have also been given in the heading of each section. In many cases the application of the substances in analysis is mentioned.

References

Apart from a few original works, we do not give a detailed bibliography, but refer the reader to some basic surveys.

E. Eberius: Wasserbestimmung mit *Karl-Fischer*-Lösung, Monographie, Angewandte Chemie *66*, 121 (1954).

S. P. Colowick, N. O. Kaplan: Methods in Enzymology, Academic Press, New York 1957, Vol. *III*; 1963, Vol. *VI*; 1975, Vol. *XLI*.

H. M. Rauen, Biochemisches Taschenbuch, Springer-Verlag, Berlin-Göttingen-Heidelberg 1964; 2nd edit.

Hoppe-Seyler/Thierfelder: Handbuch der physiologisch- und pathologisch-chemischen Analyse. 10th ed., Vol. *II*, Springer-Verlag, Berlin-Göttingen-Heidelberg 1955.

R. M. C. Dawson, D. C. Elliott, W. H. Elliott, K. M. Jones, Data for Biochemical Research, Oxford University Press, Oxford 1959.

Publication No. 719, National Academy of Sciences – National Research Council. Washington D.C. 1960.

Specifications and Criteria for Biochemical Compounds, Publication No. 1344, National Academy of Sciences, National Research Council, Washington D.C., 1967.

Acetyl-coenzyme A (Acetyl-CoA)

S-Acetyl-coenzyme A

$C_{23}H_{38}N_7O_{17}P_3S$

$CoA - S - CO - CH_3$
(structural formula of CoA, see p. 341)

Molecular weights:

acetyl-CoA 809.6
acetyl-CoA-Li$_3$ 827.4

$\lambda_{max} = 260\ nm\ (pH\ 7)$
$\varepsilon_{233} = 0.44\ l \times mmol^{-1} \times mm^{-1}$
$\varepsilon_{260} = 1.6\ \ l \times mmol^{-1} \times mm^{-1}$

Stability: acetyl-CoA should be stored dry at $-20\,°C$. No detectable decomposition of the lithium salt occurs within 6 months. Slightly acid solutions are recommended because of the sensitivity of the thioester bond to alkali. The pyrophosphate bond is hydrolyzed in strong acid.

Analysis: enzymatic determination with CS; Li by flame photometry; water according to *K. Fischer.*

Purity required: 83% acetyl-CoA.

N-Acetylneuraminosyl-D-lactose

NANA-lactose

$C_{23}H_{39}NO_{19}$ (free acid)

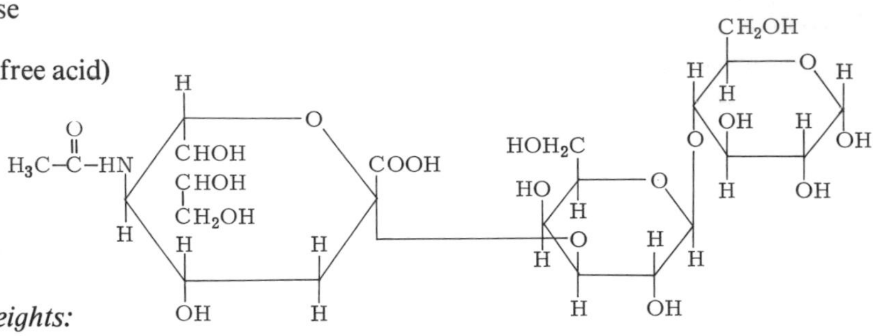

Molecular weights:

NANA-lactose   633.6
NANA-lactose-NH$_4$ 650.6

Melting point: 158 °C

$[\alpha]_D^{25}$: $+28°$

Stability: the substance is stable at $+4°C$. In acid solution it is hydrolyzed.

Analysis: enzymatic determination of lactose after hydrolysis with neuraminidase; water according to *K. Fischer.*

Purity required: 80% NANA-lactose; water < 10%.

Application: for the determination of neuraminidase activity.

Acetyl-L-phenylalanyl-L-arginine-O-ethyl ester, acetate

Ac-Phe-ArgOEt · HAc

$C_{21}H_{33}N_5O_6$ $CH_3CO\text{-}Phe\text{-}Arg\text{-}O\text{-}C_2H_5, CH_3COOH$

Molecular weight: 451.5

Melting point: 81 − 85 °C

$[\alpha]_D^{27}$: $−9.8°$ (water)

Stability: aqueous solutions of the substance are stable in water for 1 week, stored at +4°C.

Analysis: enzymatic determination of ethanol using alcohol dehydrogenase after alkaline hydrolysis with human kallikrein [1].

Purity required: 95% Ac-Phe-ArgOEt · HAc; <1% Ac-Phe-Arg; chromatographically pure (TLC).

Application: for the determination of the kallikrein (kininogenin) and urokinase activities.

Reference

[1]   *F. Fiedler, R. Geiger, Ch. Hirschauer, G. Leysath,* Peptide Esters and Nitroanilides as Substrates for the Assay of Human Urinary Kallikrein, Hoppe-Seyler's Z. Physiol. Chem. *259,* 1667 – 1673 (1978).

Acetylphosphate

$C_2H_5O_5P$ (free acid)

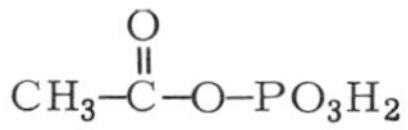

$$CH_3-\overset{\overset{\displaystyle O}{\|}}{C}-O-PO_3H_2$$

Molecular weights:

acetyl-P 140.0
acetyl-P-K,Li 184.1

Stability: acetyl-P as the crystalline potassium-lithium salt is stable at 4°C. It is rapidly hydrolyzed under acid or alkaline conditions. It is recommended to prepare only the daily requirement as a neutral solution. Because of the ease of hydrolysis it is not possible to avoid contamination with acetate and phosphate.

Analysis: enzymatic determination with PTA; K,Li by flame photometry.

Purity required: 90% acetyl-P-K,Li (68% acetyl-P).

Acetylpyridine-adenine dinucleotide (APAD)

$C_{22}H_{28}N_6O_{14}P_2$

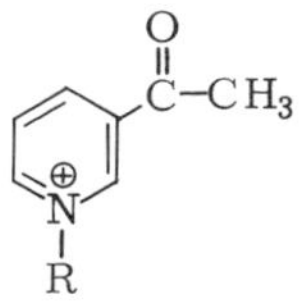

Molecular weight:

APAD 662.5

$\varepsilon_{260} = 1.65 \ 1 \times mmol^{-1} \times mm^{-1}$

R = Adenosinediphosphoribose

reduced form:

$$\varepsilon_{365} = 0.91 \; 1 \times mmol^{-1} \times mm^{-1}$$

Stability: APAD is stable for several days at 4°C in weakly acid solution (at pH of free acid); it is very alkali-labile. The solid can be stored in the cold and dry state for 6 months without an apparent loss of activity.

Analysis: enzymatic determination with ADH; water according to *K. Fischer*.

Purity required: 88% APAD.

Acetylthiocholine iodide

S-Acetylthiocholine iodide

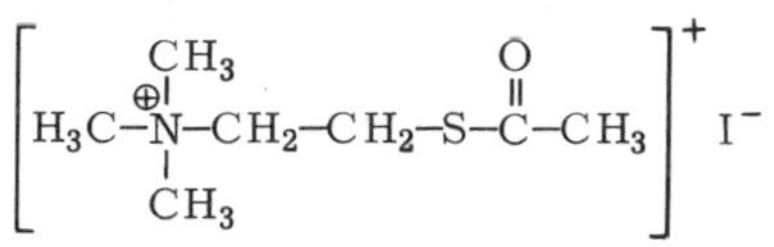

$C_7H_{16}INOS$

Molecular weight: 289.2

Melting point: 203 – 208°C

$\lambda_{max} = 226$ nm

Stability: in the solid state the substance is stable for 24 months at +4°C (light protected).

Analysis: titration of I or colorimetric determination with hydroxylamine.

Purity required: 98%; <0.1% thiocholine iodide (free).

Application: for the determination of acetylcholine esterase activity.

Adenosine

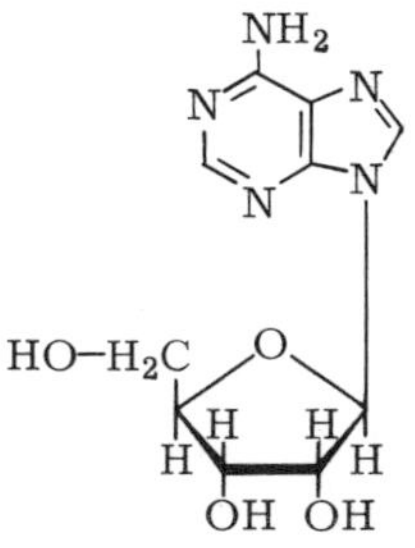

$C_{10}H_{13}N_5O_4$

Molecular weight: 267.2

$\varepsilon_{260} = 1.5 \; 1 \times mmol^{-1} \times mm^{-1}$

Stability: in the solid state the substance is stable at room temperature (at least 24 months).

Analysis: enzymatic determination with ADA; spectrophotometric determination.

Purity required: 98%; chromatographically pure.

Application: for the enzymatic determination of ADA activity.

Adenosine 5'-diphosphate (ADP)

$C_{10}H_{15}N_5O_{10}P_2$ (free acid)

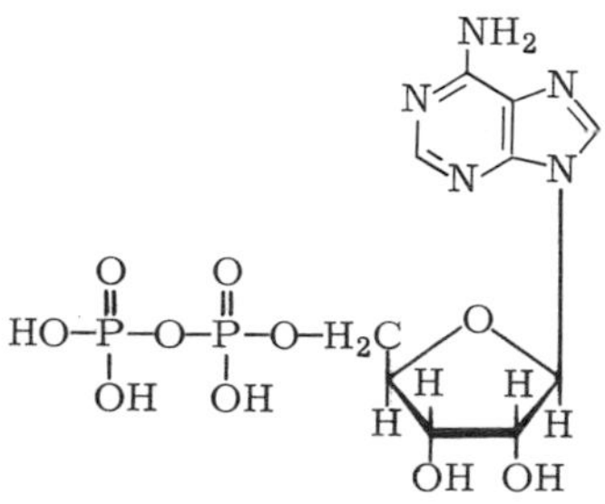

Molecular weights:

ADP 427.2
ADP-Na$_2$ 471.2

$\lambda_{max} = 259$ nm;
$\varepsilon_{259} = 1.54 \ l \times mmol^{-1} \times mm^{-1}$
$\varepsilon_{260} = 1.50 \ l \times mmol^{-1} \times mm^{-1}$

Stability: during storage the solid ADP slowly disproportionates to AMP and ATP. This reaction is faster in the case of the Na salt (up to 2% ATP in ca. 3 – 6 months). Neutral solutions are stable for months at $-15°C$, and several days at $4°C$. The optimum stability in solution is at pH ca. 9.

Analysis: enzymatic determination with PK/LDH; Na by flame photometry; water according to *K. Fischer.*

Purity required: disodium salt 82% ADP; 9% sodium; 6% water; 3% AMP; <1% ATP;
free acid 97% ADP; <3% AMP; <0.1% ATP.

Adenosine 3':5'-monophosphoric acid, cyclic (A-3:5-MP)

$C_{10}H_{12}N_5O_6P$ (free acid)

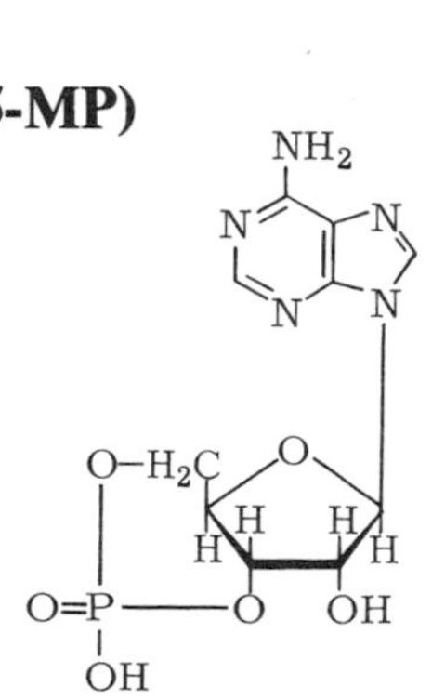

Molecular weights:

A-3:5-MP 329.2
A-3:5-MP · H$_2$O 347.2

$\varepsilon_{260} = 1.50 \ l \times mmol^{-1} \times mm^{-1}$
(phosphate buffer, 0.1 mol/l; pH 7.0)

Stability: the free acid shows no detectable decomposition when stored dry and at room temperature. Neutral aqueous (sterile) solutions are stable at $0\,°C$ for at least 1 month.

Analysis: enzymatic determination with PDE/AP/ADA; determination of organically bound phosphate according to *Fiske & Subbarow*; water according to *K. Fischer*.

Purity required: crystalline free acid 99% A-3:5-MP $\cdot$ H_2O (94% A-3:5-MP).

Adenosine 5′-monophosphate (AMP)

$C_{10}H_{14}N_5O_7P$ (free acid)

Molecular weights:

AMP 347.2;
AMP $\cdot$ H_2O 365.2;
AMP-Na_2 $\cdot$ 6 H_2O 499.2

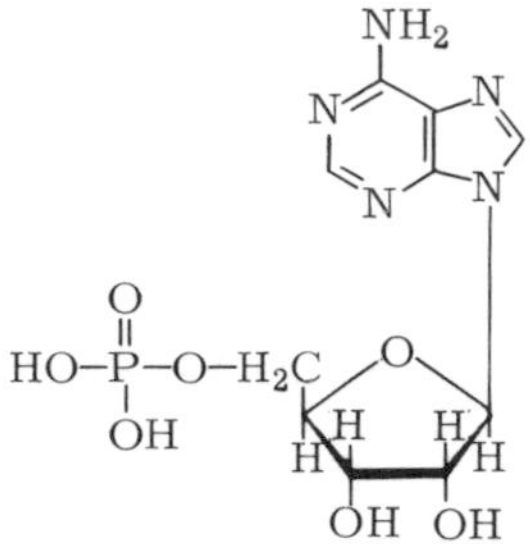

λ_{max} = 259 nm
ε_{259} = 1.54 $l \times mmol^{-1} \times mm^{-1}$
ε_{260} = 1.50 $l \times mmol^{-1} \times mm^{-1}$
(phosphate buffer, 0.1 mol/l; pH 7.0)

Stability: the free acid and its salts are stable for long periods in the dry state; neutral solutions are also stable.

Analysis: enzymatic determination with MK/PK/LDH; Na by flame photometry; water according to *K. Fischer*.

Purity required: crystalline free acid 98% AMP $\cdot$ H_2O (93% AMP); disodium salt 98% AMP-Na_2 $\cdot$ 6 H_2O (69% AMP).

Adenosine 5′-triphosphate (ATP)

$C_{10}H_{16}N_5O_{13}P_3$ (free acid)

Molecular weights:

ATP 507.2
ATP-Na_2H_2 $\cdot$ 3 H_2O 605.2

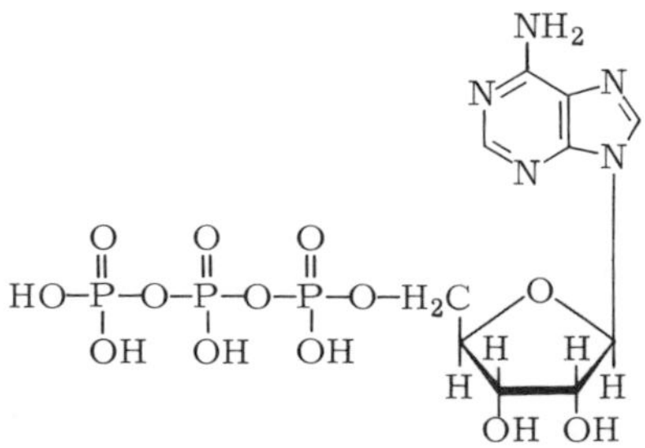

λ_{max} = 259 nm
ε_{259} = 1.54 $l \times mmol^{-1} \times mm^{-1}$
ε_{260} = 1.50 $l \times mmol^{-1} \times mm^{-1}$

Stability: the alkali salts slowly decompose on storage in the cold and dry state. Neutral solutions of the alkali salts are stable for months at $-15°C$, but only for ca. 1 week at $4°C$. The optimum stability in solution is observed at ca. pH 9. ATP-Tris (even in solution) is more stable. Magnesium ions and other divalent cations accelerate the hydrolysis of ATP.

Analysis: by enzymatic determination with PGK/GAP-DH; Na by flame photometry; water according to *K. Fischer*.

Purity required: 98% ATP-$Na_2H_2 \cdot 3 H_2O$ (82% ATP), by HPLC: sum of ADP and AMP <0.5 area %.

δ-Aminolaevulinic acid

5-Amino-4-oxo-pentanoic acid

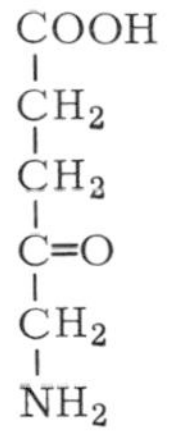

$C_5H_9NO_3 \cdot HCl$

Molecular weights:

aminolaevulinic acid 131.1
aminolaevulinic acid · HCl 167.57

```
COOH
|
CH2
|
CH2
|
C=O
|
CH2
|
NH2
```

Melting point: $147-148°C$ (HCl-compound; decomposition)

Stability: in the solid state the substance is stable at $+4°C$.

Analysis: determination of nitrogen.

Purity required: 98%.

Application: for the determination of porphobilinogen synthase.

Aminoantipyrine

4-Amino-1-phenyl-2,3-dimethyl-pyrazolone-(5)

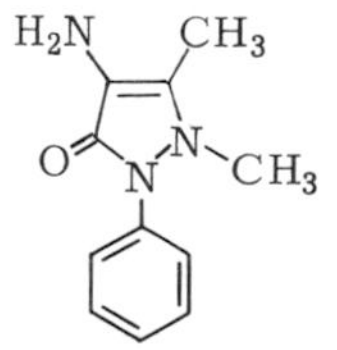

$C_{10}H_{13}N_3O$

Molecular weight: 203.2

Melting point: 106 – 109 °C

$\lambda_{max} = 241$ nm

Stability: in the solid state the substance is stable at room temperature.

Analysis: titration in a water-free medium with perchloric acid.

Purity required: 98%; chromatographically pure.

Application: for the enzymatic determination (with POD) of H_2O_2 formed by oxidase reactions e.g. ChOD, GOD, urate oxidase.

Aminophthalic acid hydrazide (Luminol)

5-Amino-2,3-dihydro-1,4-phthalazinedione

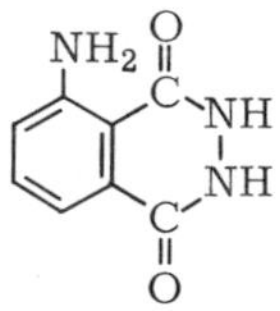

$C_8H_7N_3O_2$

Molecular weight: 177.16

$\lambda_{max} = 424$ nm (water)
$= 485$ nm (dimethylsulphoxide, DMSO)

Stability: luminol is stable at room temperature, if stored dry. Solutions in DMSO are stable for 1 week (room temperature).

Analysis: determination by HPLC.

Purity required: 80 area %.

Application: luminol is used as activator for the measurement of the phagocytose-associated chemiluminescence [1, 2]; it is a substrate of chemiluminescence for reactions with peroxidases [3].

References

[1] *R. C. Allen, L. B. Loose,* Phagocytic Activation of a Luminoldependent Chemiluminescence in Rabbit Alveolar and Peritoneal Macrophages, Biochem. Biophys. Res. Commun. *69,* 245 – 252 (1976).
[2] *T. Kato* et al., Measurement of Chemiluminescence in Freshly Drawn Human Blood, Klin. Wochenschr. *59,* 203 – 211 (1981).
[3] *K. Puget, A. M. Michelson,* Microestimation of Glucose and Glucose Oxidase, Biochimie *58,* 757 – 758 (1976).

Azino-benzthiazoline sulphonate (ABTS)

2,2'-Azino-bis-[3-ethylbenzthiazoline-sulphonate-(6)], diammonium salt

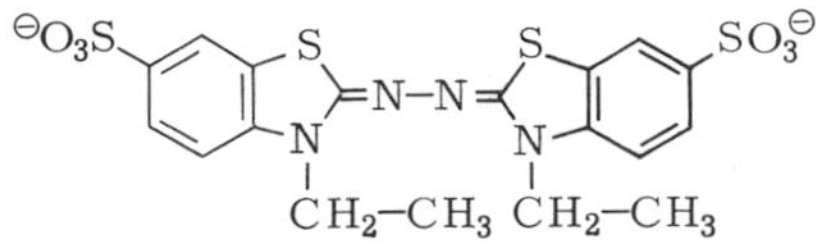

$C_{18}H_{16}N_4O_6S_4$ (anion)

Molecular weights:

ABTS 514.6
ABTS-$(NH_4)_2$ 548.7

λ_{max} = 341 nm (pH 7)

Stability: in the solid state the substance is stable at room temperature for at least 24 months.

Analysis: spectrophotometric determination.

Purity required: 99%; chromatographically pure.

Application: for the enzymatic determination (with POD) of H_2O_2 formed by oxidase reactions e.g. ChOD, GOD, urate oxidase.

Benzoyl-prolyl-phenylalanyl-arginine-4-nitroanilide acetate

Chromozym® PK (cf. p. 340)

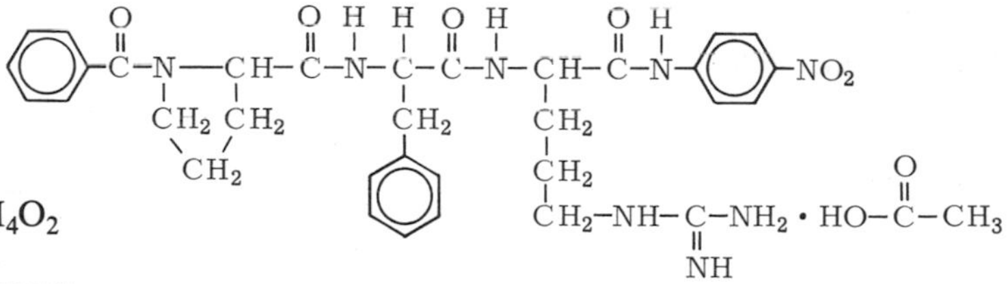

$C_{33}H_{38}N_8O_6 - C_2H_4O_2$

Molecular weight: 702.8

Melting point: 145°C

$[\alpha]_D^{20}$: $-65°$

λ_{max} = 315 nm (pH 7)

Stability: Chromozym® PK is stable at room temperature.

Analysis: spectrophotometric determination of 4-nitroaniline after acid hydrolysis, and after enzymatic hydrolysis using kallikrein; determination of Chromozym® PK from nitrogen.

Purity required: 90% Chromozym® PK; < 0.5% 4-nitroaniline.

Application: for the determination of serine protease activity, especially of prekallikrein and kallikrein.

S-Benzyl-L-cysteine-4-nitroanilide

$C_{16}H_{17}N_3O_3S$

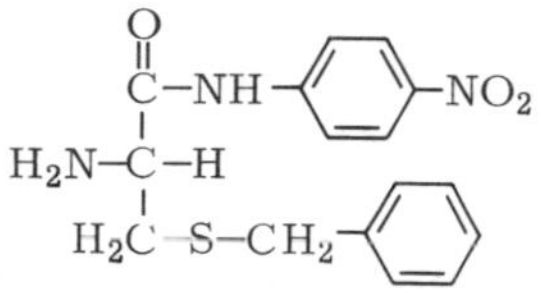

Molecular weight: 331.4

Melting point: 96 – 99 °C

Stability: the substance should be protected from light. It is stable for many months, stored at room temperature.

Analysis: spectrophotometric determination of 4-nitroaniline after acid hydrolysis; determination of nitrogen.

Purity required: 98%; 4-nitroaniline < 0.2%.

Application: for the determination of oxytocinase activity.

Butyrylthiocholine iodide

S-Butyrylthiocholine iodide

$C_9H_{20}INOS$

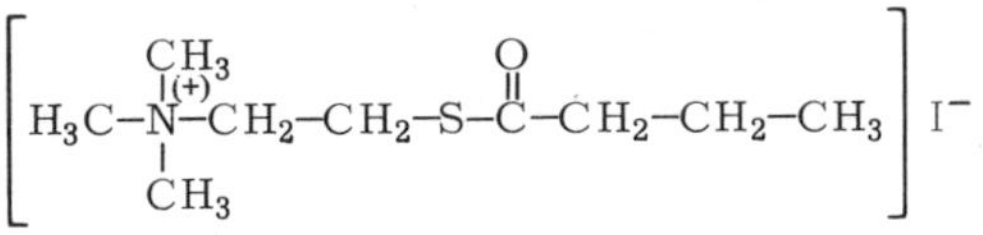

Molecular weight: 317.2

Melting point: 172 – 174 °C

λ_{max} = 226 nm

Stability: the substance should be stored dry and protected from light. It is stable for many months, if stored at +4 °C.

Analysis: titration of I or colorimetric measurement with hydroxylamine.

Purity required: 99%; < 0.15% thiocholine iodide (free).

Application: for the determination of the catalytic activities of acetylcholine esterase and butyrylcholine esterase.

Carbamoylphosphate

CH_4NO_5P (free acid)

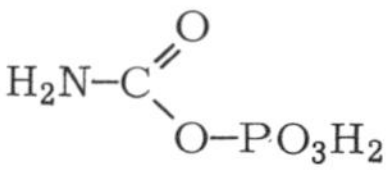

$$H_2N-C \overset{\displaystyle O}{\underset{\displaystyle O-PO_3H_2}{\big\langle}}$$

Molecular weights:

carbamoyl-P 141.0
carbamoyl-P-Li$_2$ 152.9
carbamoyl-P-Na$_2$ · 3 H$_2$O 239.0

Stability: the alkali salts should be stored in the dry state at 4°C. In alkaline solution inorganic phosphate and cyanate are formed, while in acid solution hydrolysis occurs [1]. As there is a rapid hydrolysis even in neutral solution (at 37°C the half-life is only 50 min) it is recommended to keep the solution cold or to store frozen.

Analysis: enzymatic determination of acid-labile phosphate according to [1]; Li, Na by flame photometry; water according to *K. Fischer.*

Purity required: crystalline disodium salt 95% carbamoyl-P-Na$_2$ · 2 H$_2$O (60% carbamoylphosphate);
dilithium salt 80% carbamoyl-P; < 3% inorganic phosphate.

References

[1] *R. L. Metzenberg, M. Marshall, P. P. Cohen,* in: *H. A. Lardy,* Biochemical Preparations, Vol. 7, John Wiley and Sons, New York-London 1959, p. 23.

Carbobenzoxy-valyl-glycyl-arginine-4-nitroanilide acetate

Chromozym® TRY (cf. p. 340)

$C_{27}H_{36}N_8O_7 - C_2H_4O_2$

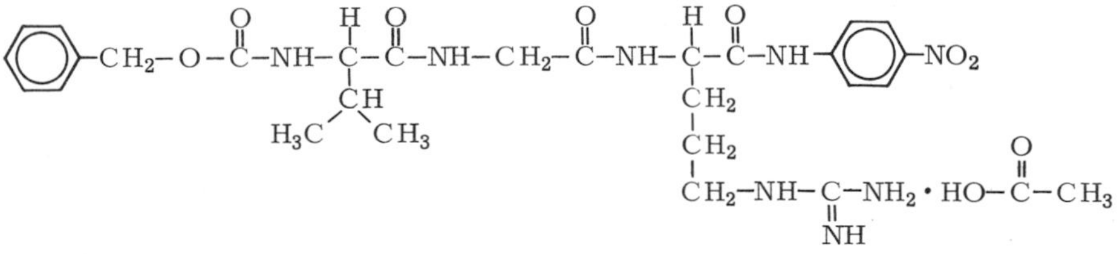

Molecular weight: 644.7

Melting point: 130°C

$[\alpha]_D^{20}$: $-27°$

λ_{max} = 315 nm (pH 7)

Stability: Chromozym® TRY is stable at room temperature.

Analysis: spectrophotometric determination of 4-nitroaniline after acid hydrolysis, and after enzymatic hydrolysis using trypsin; determination of Chromozym® TRY from nitrogen.

Purity required: 90% Chromozym® TRY; < 0.5% 4-nitroaniline.

Application: for the determination of serine proteases activity, especially of trypsin.

Chromozym®*

Chromozym® GK

cf. D-Valyl-cyclohexylalanyl-arginine-4-nitroanilide, diacetate p. 392.

Chromozym® PK

cf. Benzoyl-prolyl-phenylalanyl-arginine-4-nitroanilide acetate p. 337.

Chromozym® PL

cf. Tosyl-glycyl-prolyl-lysine-4-nitroanilide acetate p. 388.

Chromozym® TH

cf. Tosyl-glycyl-prolyl-arginine-4-nitroanilide acetate p. 388.

Chromozym® TRY

cf. Carbobenzoxy-valyl-glycyl-arginine-4-nitroanilide acetate p. 339.

---

* Chromozym® = registered trade mark of *Pentapharm AG,* Bâle, Switzerland, supplied by *Boehringer Mannheim.*

Coenzyme A (CoA)

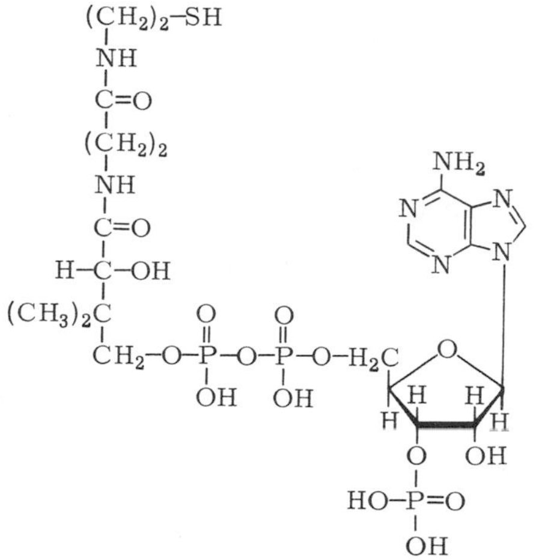

$C_{21}H_{36}N_7O_{16}P_3S$ (free acid)

Molecular weights:

CoA 767.6
CoA-Li$_3$ 785.4

$\lambda_{max} = 260$ nm (pH 7);
$\varepsilon_{260} = 1.60$ l $\times$ mmol$^{-1} \times$ mm^{-1}

$(CH_2)_2$–SH
|
NH
|
C=O
|
$(CH_2)_2$
|
NH
|
C=O
|
H–C–OH
|
$(CH_3)_2$C
|
CH$_2$–O–P–O–P–O–H$_2$C
  OH   OH

NH$_2$

HO–P=O
|
OH

Stability: in the solid state CoA is very labile and should be stored at $-20\,°$C. A weakly acid CoA solution as well as neutralized solutions should be used up within a day. At strongly acid pH the pyrophosphate bond is cleaved. The SH group is very sensitive to oxidizing agents, even to oxygen at alkaline and neutral pH.

Analysis: enzymatic determination with PTA; water according to *K. Fischer*.

Purity required: CoA free acid 85% CoA; 6% water; <5% other UV-absorbing compounds; <1% glutathione;
trilithium salt 82% CoA; 2.5% Li; 6% water; <2% glutathione.

Creatine phosphate

Phosphocreatine

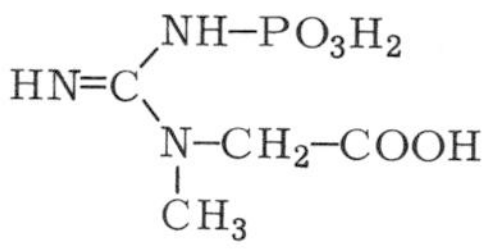

$C_4H_{10}N_3O_5P$ (free acid)

Molecular weights:

creatine-P 211.1
creatine-P-Na$_2 \cdot$ 4 H$_2$O 372.0

HN=C
NH–PO$_3$H$_2$
N–CH$_2$–COOH
|
CH$_3$

Stability: creatine-P is stable in the solid, crystalline state. It is very labile in acid solution, and somewhat more stable in almost neutral solution (pH 7.8); however, even at this pH a slow hydrolysis occurs.

Analysis: enzymatic determination with CK/HK/G6P-DH; Na by flame photometry; water according to *K. Fischer*.

Purity required: 98% creatine-P-Na$_2$ · 4 H$_2$O (63% creatine-P).

Creatinine

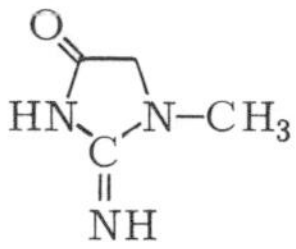

C$_4$H$_7$N$_3$O

Molecular weight: 113.2

Melting point: 250 – 280°C

λ_{max} = 234 nm (pH 7)

Stability: stable in the solid state at +4°C.

Analysis: determination of nitrogen.

Purity required: 98%.

Cytidine 2′ : 3′-monophosphoric acid, cyclic (C-2 : 3-MP)

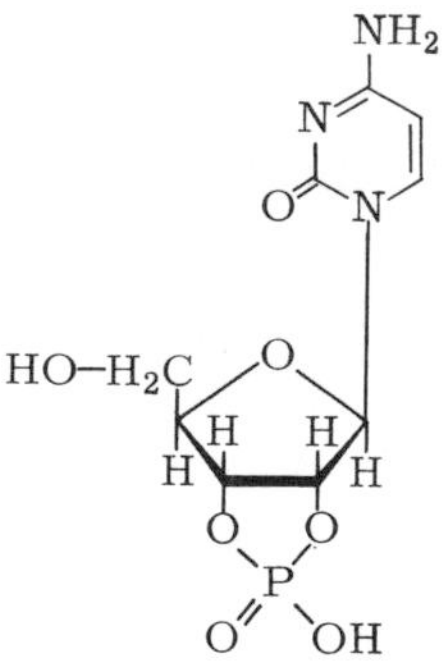

C$_9$H$_{12}$N$_3$O$_7$P (free acid)

Molecular weight: 305.2

λ_{max} = 260 nm
ε_{260} = 0.80 l × mmol^{-1} × mm^{-1}

Stability: the substance shows no decomposition, if stored dry and at +4°C.

Analysis: spectrophotometric determination and determination of nitrogen.

Purity required: crystalline free acid 96%; P$_i$ < 0.1%; chromatographically pure.

Deoxycholate

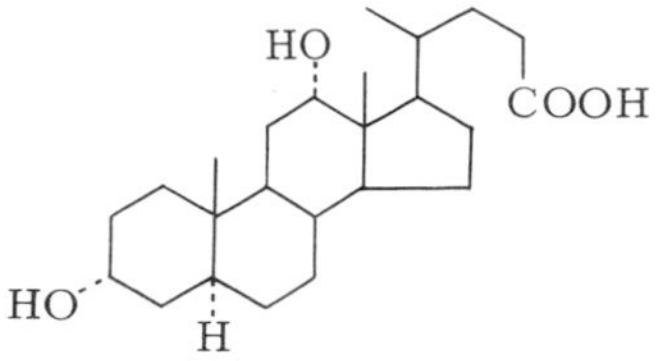

$C_{24}H_{40}O_4$ (free acid)

Molecular weights:

deoxycholic acid 392.57
deoxycholate-Na 414.57

Stability: deoxycholate, Na salt, is stable at room temperature.

Analysis: determination by HPLC.

Purity required: 98% deoxycholate, Na salt; < 1.5% cholate, Na salt; chromatographically (TLC) pure.

Application: for the determination of lipase activity.

Dihydroxyacetone phosphate (DAP)

$C_3H_7O_6P$ (free acid)

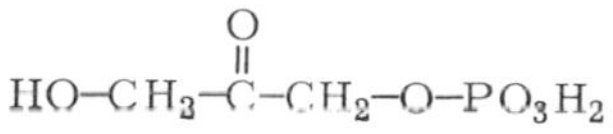

Molecular weights:

DAP 170.1
DAP-dimethylketal-$(CHA)_2 \cdot H_2O$ 432.5

Stability: the dicyclohexylammonium salt of DAP-dimethylketal is stable indefinitely at $+4\,^{\circ}C$, while in weak acid it is stable for several days in the frozen state. The phosphate group is split off in a few minutes by alkali, 1 mol/l, at room temperature.

Analysis: enzymatic determination with GDH; CHA by titration in anhydrous medium; water according to *K. Fischer*.

Purity required: 99% cyclohexylammonium salt of DAP-dimethylketal $\cdot H_2O$ (39% DAP).

Application: for the determination of triosephosphate isomerase activity.

Dimethylamino-naphthalene-dicarbonic acid hydrazide

7-Dimethylamino-naphthalene-1,2-dicarbonic acid hydrazide

$C_{14}H_{13}N_3O_2$

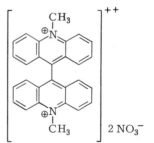

Molecular weight: 255.3

λ_{max} = 514 nm (chemiluminescence with H_2O_2)

Stability: the substance is stable at room temperature, if stored dry. Solutions in dimethylsulphoxide are stable for 1 week (room temperature).

Analysis: determination by HPLC.

Purity required: 90 area %.

Application: the compound is used as substrate for chemiluminescence reactions in presence of H_2O_2. The turnover of light quanta is approx. 2.3% [1].

Reference

[1] *K. D. Gundermann, W. Horstmann, G. Bergmann,* Synthese und Chemilumineszenz-Verhalten von 7-Dialkylamino-naphthalin-dicarbonsäure-(1.2)-hydraziden, Liebigs Ann. Chem. *682*, 127 – 141 (1965).

Dimethyl-bisacridinium-dinitrate (Lucigenine)

[9,9′-Bis-(N-methylacridinium)] dinitrate

$C_{28}H_{22}N_4O_6$

Molecular weight: 510.51

Melting point: 290 – 300°C (decomposition)

λ_{max} = 261 nm; 369 nm; 430 nm (NH$_4$OH, 0.1 mol/l)

Stability: lucigenine is stable at room temperature, if protected from light. Aqueous solutions of lucigenine are stable for 1 week (+4°C, pH 7). It decomposes in alkaline solutions.

Analysis: determination of nitrogen.

Purity required: 90%.

Application: lucigenine is used for chemiluminescence studies of redox activities of leukocytes.

References

[1] *R. C. Allen,* in: *M. De Luca, W. D. McElroy* (eds.), Bioluminescence and Chemiluminescence, Academic Press, New York 1981, pp. 63 – 73.

[2] *R. C. Allen,* in: *W. Adam, G. Cilento* (eds.), Chemical and Biological Generation of Excited States, Academic Press, New York (1982).

Dimethylthiazolyl-tetrazolium bromide (MTT)

$C_{18}H_{16}N_5SBr$

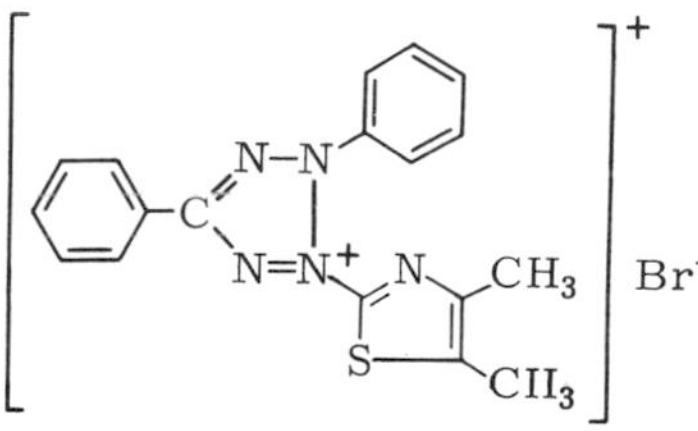

Molecular weight: 414.33

$\lambda_{max} = 242$ nm

Stability: MTT is stable at $+4°C$, protected from light.

Analysis: determination of nitrogen.

Purity required: 98%.

Application: for transferring hydrogen to tetrazolium salts to form coloured formazans, especially for the determination of ascorbic acid (ε_{578nm}, MTT-formazan $= 1.69 \pm 0.02$ l $\times$ mmol^{-1} $\times$ mm^{-1}; pH 3.5).

Dithio-bis-(2-nitrobenzoic acid) (Ellman's reagent, DTNB)

5,5'-Dithio-bis-(2-nitrobenzoic acid)

$C_{14}H_8N_2O_8S_2$

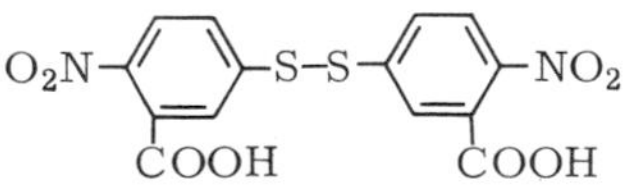

Molecular weight: 396.4

Melting point: approx. 240°C

$\lambda_{max} = 323$ nm

Stability: Ellman's reagent is stable at room temperature.

Analysis: colorimetric determination with cysteine.

Purity required: 93%.

Application: for the determination of compounds with sulphydryl groups.

Dithiothreitol (DTT)

threo-1,4-Dimercapto-2,3-butandiol

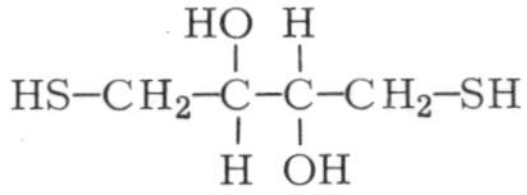

$$HS-CH_2-\overset{\overset{\displaystyle HO}{|}}{\underset{\underset{\displaystyle H}{|}}{C}}-\overset{\overset{\displaystyle H}{|}}{\underset{\underset{\displaystyle OH}{|}}{C}}-CH_2-SH$$

$C_4H_{10}O_2S_2$

Molecular weight: 154.3

Melting point: 40 – 44 °C

Stability: DTT is stable at +4 °C, protected from light.

Analysis: determination by the reaction with Ellman's reagent (dithio-bis-nitrobenzoic acid); DTT by iodometric titration.

Purity required: 95%; chromatographically pure.

Application: for reducing oxidized compounds (pH 7.5), especially for the determination of dehydroascorbic acid.

D-Erythrose 4-phosphate

$C_4H_9O_7P$ (free acid) $H_2O_3P-O-CH_2-CH(OH)-CH(OH)-CHO$

Molecular weights:

erythrose-4-P 198.1
erythrose-4-P-diethylacetal-Ba 409.5

Stability: the barium salt of the diethylacetal is stable at 4 °C. It is soluble in water. After removal of the acetal group the solution must be kept acid. It can be stored frozen for several weeks. Decomposition occurs in neutral, or even more rapidly in alkaline solution.

Analysis: enzymatic determination of erythrose-4-P with transaldolase after hydrolysis; Ba by complexometry; water according to *K. Fischer*.

Purity required: 38% D-erythrose-4-P; 33% Ba; 13% ethanol bound as acetal; <2% GAP; <3.5% glucose-6-P.

Application: for the enzymatic determination of transaldolase activity.

Ethylsulphonyl-toluidene (EST)

N-Ethyl-3-methyl-N(2-sulphoethyl) aniline, sodium salt

$C_{11}H_{16}NO_3SNa$

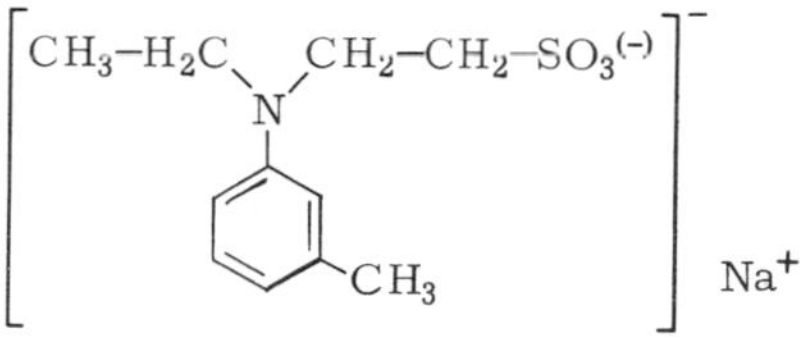

Molecular weights:

EST 242.3
EST-Na · H_2O 283.3

Melting point: 187 °C

Stability: EST is stable at room temperature.

Analysis: determination by HPLC.

Purity required: 99 area %; chromatographically pure.

Application: for the determination of uric acid.

Flavin-adenine dinucleotide (FAD)

$C_{27}H_{33}N_9O_{15}P_2$ (free acid)

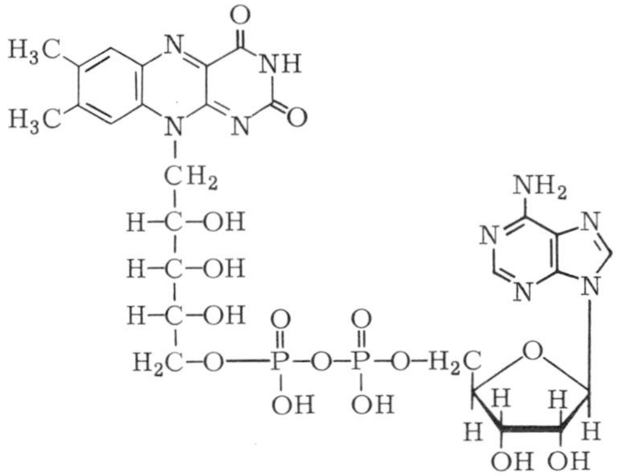

Molecular weights:

FAD 785.6
FAD-Na_2 829.6

λ_{max} = 264, 448 and 373 nm (pH 7.0)
ε_{260} = 4.62 l × mmol^{-1} × mm^{-1}
ε_{450} = 1.13 l × mmol^{-1} × mm^{-1}

Stability: the neutral aqueous solution of the Na salt is stable at $4\,°C$. The solutions are unstable below pH 3 and above pH 10, at high temperatures and on exposure to light.

Analysis: spectrophotometric determination; Na by flame photometry; water according to *K. Fischer.*

Purity required: 87% FAD; 4% sodium; 6% water; $P_i < 0.5\%$.

Flavin mononucleotide (FMN)

Lactoflavin phosphate, riboflavin 5'-monophosphate

$C_{17}H_{21}N_4O_9P$ (free acid)

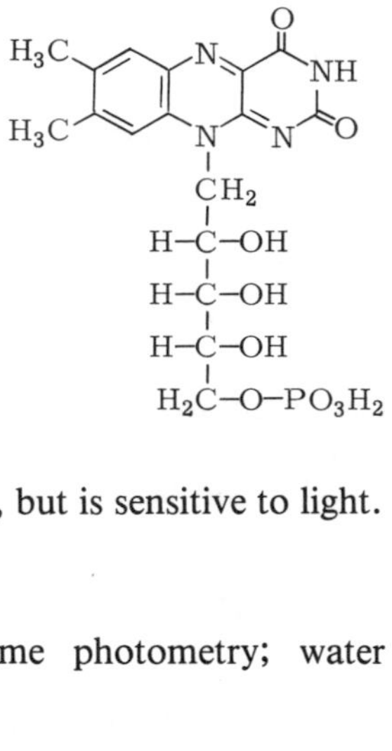

Molecular weights:

FMN 456.4
FMN-Na 478.4

$E_0 = -0.19\,V$ (pH 7)
E_0 (with enzyme protein) $= -0.06\,V$
$\lambda_{max} = 450\,nm;$
$\varepsilon_{450} = 1.22\ l \times mmol^{-1} \times mm^{-1}$

Stability: in the solid state FMN is stable for months at $4\,°C$, but is sensitive to light. Solutions of FMN are also unstable when exposed to light.

Analysis: spectrophotometric determination; Na by flame photometry; water according to *K. Fischer.*

Purity required: 87% FMN; 6% sodium; 6% water; $P_i < 0.5\%$.

Application: for the determination of NAD (P) H : FMN-oxidoreductase activity and for the determination of NAD (P) H by luciferase from *Photobacterium fischeri* [1].

Reference

[1] *W. Duane, J. W. Hastings,* Flavin Mononucleotide Reductase of luminous Bacteria, Mol. Cell. Biochem. *6,* 53 (1975).

D-Fructose 1,6-bisphosphate

α-D-(−)-Fructofuranose 1,6-bisphosphate

$C_6H_{14}O_{12}P_2$ (free acid)

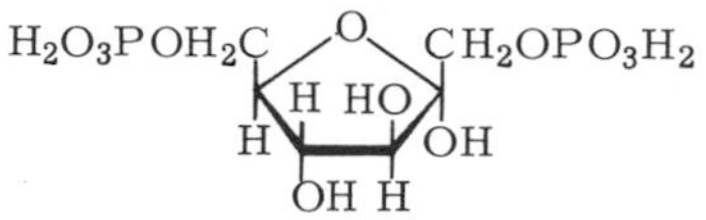

Molecular weights:

fructose-1,6-P_2 340.1
fructose-1,6-P_2 · $(CHA)_4$ · 10 H_2O 917.0
fructose-1,6-P_2-Na_3 · 8 H_2O 550.2

Stability: salts are stable at room temperature in the dry state. After long storage (6 − 12 months) a slight yellow colour can occur with the neutral salts. The sodium salt is extremely hygroscopic and therefore should be protected from moisture. Under sterile conditions neutral solutions of alkali salts are stable for several weeks. Fructose-1,6-P_2 is sensitive to acid. The C-1-phosphate bond is hydrolyzed about 10 times faster than the C-6-phosphate; in weakly alkaline solution there is an increase in colour.

Analysis: enzymatic determination with aldolase; Na by flame photometry; CHA titrimetrically in anhydrous medium; water according to *K. Fischer.*

Purity required: cyclohexylammonium salt 98% fructose-1,6-P_2-$(CHA)_4$ · 10 H_2O (36% fructose-1,6-P_2); free from other ester phosphates; trisodium salt 98% fructose-1,6-P_2-Na_3 · 8 H_2O (66% fructose-1,6-P_2); 12% Na; 26% water; <2% inorganic phosphate (as PO_4^{3-}). Other ester phosphates by enzymatic determination <2%.

Application: for the determination of fructose-1,6-bisphosphate aldolase activity.

D-Fructose 1-phosphate

α-D-(−)-Fructofuranose 1-phosphate

$C_6H_{13}O_9P$ (free acid)

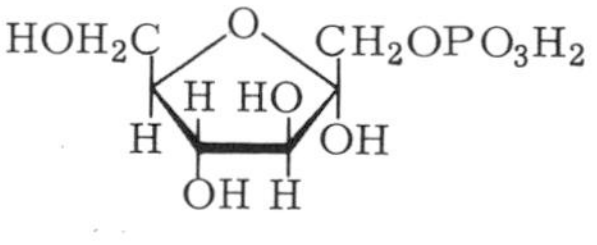

Molecular weights:

fructose-1-P 260.1
fructose-1-P-$(CHA)_2$ 458.5
fructose-1-P-Na_2 304.1

Stability: salts are stable at room temperature in the dry state. After storage for 6 – 12 months a slight yellow colour is often seen. Solutions should be prepared freshly. Fructose-1-P is sensitive to acid.

Analysis: enzymatic determination with F-6-PK; Na by flame photometry; CHA titrimetrically in anhydrous medium; water according to *K. Fischer*.

Purity required: crystalline dicyclohexylammonium salt 98% fructose-1-P-(CHA)$_2$ (56% fructose-1-P);
sodium salt 78% fructose-1-P; 13% Na; 5% water.

Application: for the determination of fructose-1,6-bisphosphate aldolase activity.

D-Fructose 6-phosphate

α-D-(−)-Fructofuranose 6-phosphate

C$_6$H$_{13}$O$_9$P (free acid)

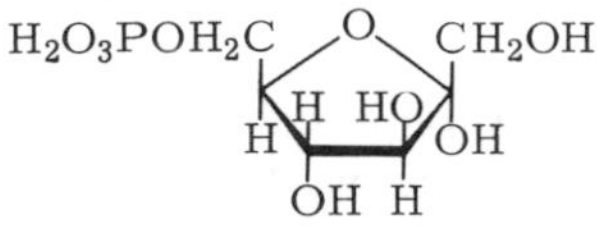

Molecular weights:

fructose-6-P 260.1
fructose-6-P-Na$_2$ 304.1
fructose-6-P-Ba 395.4

Stability: salts are stable at room temperature indefinitely. After prolonged storage (6 – 12 months) a slight yellow colour can occur with neutral salts. Neutral (sterile) solutions are stable for several weeks. Fructose-6-P is fairly stable to acid hydrolysis, but less stable to alkali.

Analysis: enzymatic determination with F-6-PK; Na by flame photometry; Ba by complexometry; water according to *K. Fischer*.

Purity required: barium salt 52% fructose-6-P; 30% Ba; 7% H$_2$O; < 3% of other ester phosphates;
sodium salt 75% fructose-6-P; 12% Na; 7% H$_2$O; < 3% of other ester phosphates.

Application: for the determination of phosphoglucose isomerase activity.

D-Galactose 1-phosphate

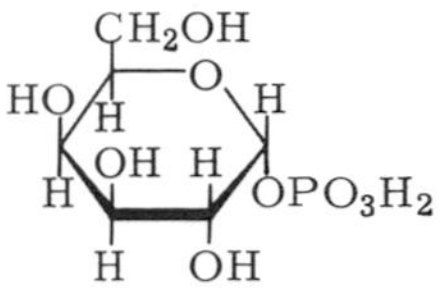

$C_6H_{13}O_9P$ (free acid)

Molecular weights:

galactose-1-P 260.2
galactose-1-P-K_2 · 5 H_2O 426.4

Stability: at room temperature the usual salts are very stable in the dry state. Neutral solutions of the salts are stable for some days. The phosphate group is cleaved by the action of acid.

Analysis: enzymatic determination with AP/Gal-DH; K by flame photometry; water according to *K. Fischer*.

Purity required: 95% galactose-1-P-K_2 · 5 H_2O (85% galactose-1-P).

Application: for the determination of uridyltransferase activity.

D-Gluconate 6-phosphate

6-Phosphogluconate

$C_6H_{13}O_{10}P$ (free acid)

Molecular weights:

gluconate-6-P 276.1
gluconate-6-P-Na_3 · 2 H_2O 378.1

Stability: the substance is stable at room temperature.

Analysis: enzymatic determination with 6-PGDH; Na by flame photometry; determination of free and organically bound phosphate according to *Fiske & Subbarow*; water according to *K. Fischer*.

Purity required: 98% gluconate-6-P-Na_3 · 2 H_2O; $< 0.05\%$ P_i; $< 0.1\%$ glucose-6-P.

Application: for the determination of 6-PGDH activity.

D-Glucose 1,6-bisphosphate

α-D-(+)-Glucopyranose 1,6-bisphosphate

$C_6H_{14}O_{12}P_2$ (free acid)

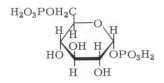

Molecular weights:

glucose-1,6-P_2 340.1
glucose-1,6-P_2-(CHA)$_4$ · 4 H_2O 808.9

Stability: the crystalline cyclohexylammonium salt shows no loss of content even after a year at room temperature. It is easily soluble in water, giving a solution of neutral reaction. The phosphate on the C_1 position, in contrast to that on the C_6, is easily split off by acid.

Analysis: enzymatic determination with G6P-DH after hydrolysis with HCl, 1 mol/l, (7 min; 100°C); water according to *K. Fischer.*

Purity required: 95% glucose-1,6-P_2-(CHA)$_4$ · 4 H_2O (40% glucose-1,6-P_2).

Application: for the determination of phosphoglucomutase activity.

D-Glucose 1-phosphate

α-D-(+)-Glucopyranose 1-phosphate

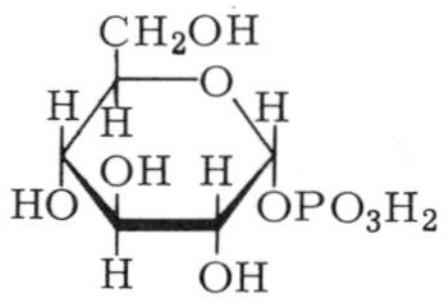

$C_6H_{13}O_9P$ (free acid)

Molecular weights:

glucose-1-P 260.2
glucose-1-P-K_2 · 2 H_2O 372.4
glucose-1-P-Na_2 · 4 H_2O 376.2

Stability: the crystalline potassium and sodium salts are stable. They dissolve to give a solution of neutral reaction. The phosphate group is easily hydrolyzed by acid.

Analysis: enzymatic determination with PGluM and G6P-DH; Na, K by flame photometry; water according to *K. Fischer.*

Purity required: dipotassium and disodium salt 98% (68% glucose-1-P). Preparations chemically prepared contain no oligosaccharides and only traces of glucose-1,6-P_2.

For this reason these preparations are used in preference to enzymatically prepared material in the assay of phosphoglucomutase or phosphorylase.

Application: for the determination of phosphoglucomutase activity.

D-Glucose 6-phosphate

α-D-(+)-Glucopyranose 6-phosphate

$C_6H_{13}O_9P$ (free acid)

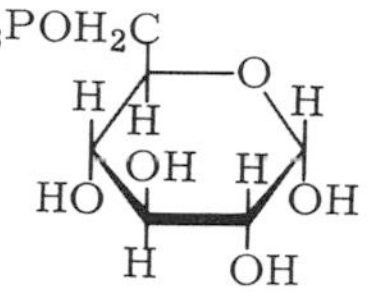

Molecular weights:

glucose-6-P 260.2
glucose-6-P-Ba · 7 H_2O 521.6
glucose-6-P-Na$_2$ 304.2

Stability: the salts are stable in the dry state and at room temperature. The sodium salt is very hygroscopic and therefore should be protected from moisture. Neutral solutions of the alkali salts are stable under sterile conditions. Glucose-6-P is very stable to dilute acid, but is less resistant to alkali.

Analysis: enzymatic determination with G6P-DH; Na by flame photometry; Ba by complexometry; water according to *K. Fischer*.

Purity required: barium salt 98% glucose-6-P-Ba · 7 H_2O (49% glucose-6-P), free from other ester phosphates;
sodium salt 76% glucose-6-P; 13% Na; 8% water; free from other ester phosphates.

L-γ-Glutamyl-3-carboxy-4-nitroanilide (Glupa-carboxylate)

$C_{12}H_{13}N_3O_7$

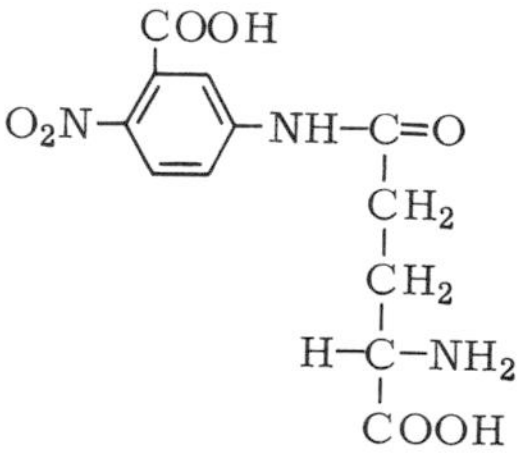

Molecular weights:

glupa-carboxylate 311.3
glupa-carboxylate · 2 H_2O 347.3

Melting point: 174 – 178 °C
$[\alpha]_D^{20}$: + 20°

Stability: stable at $+4°C$, if protected from light.

Analysis: photometric determination of 5-amino-2-nitrobenzoic acid after enzymatic hydrolysis with γ-glutamyl transferase, or after alkaline hydrolysis. Determination by HPLC.

Purity required: 99% glupa-carboxylate $\cdot$ 2 H_2O, 99.5% by HPLC.

Application: for the determination of γ-glutamyl transferase activity.

Glutathione, reduced (GSH)

γ-L-Glutamyl-L-cysteinylglycine

$C_{10}H_{17}N_3O_6S$

Molecular weight: 307.3

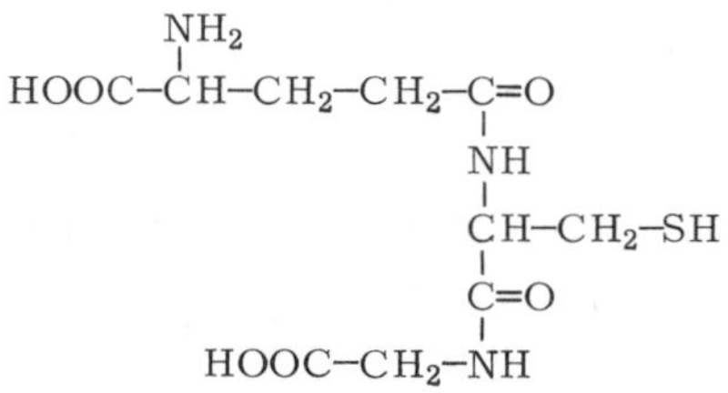

$$
\begin{array}{l}
\overset{\displaystyle NH_2}{|} \\
HOOC-CH-CH_2-CH_2-C{=}O \\
\qquad\qquad\qquad\quad | \\
\qquad\qquad\qquad\ \ NH \\
\qquad\qquad\qquad\quad | \\
\qquad\qquad\qquad\ \ CH-CH_2-SH \\
\qquad\qquad\qquad\quad | \\
\qquad\qquad\qquad\ \ C{=}O \\
\qquad\qquad\qquad\quad | \\
\qquad\qquad HOOC-CH_2-NH
\end{array}
$$

Stability: GSH is stable indefinitely in the crystalline and dry state. In solution GSH is easily oxidized to the disulphide, especially at alkaline pH and in the presence of atmospheric oxygen.

Analysis: enzymatic determination with glyoxylase; iodometric determination.

Purity required: 98% GSH; < 2% GSSG.

Glutathione, oxidized (GSSG)

$C_{20}H_{32}N_6O_{12}S_2$

Molecular weight: 612.6

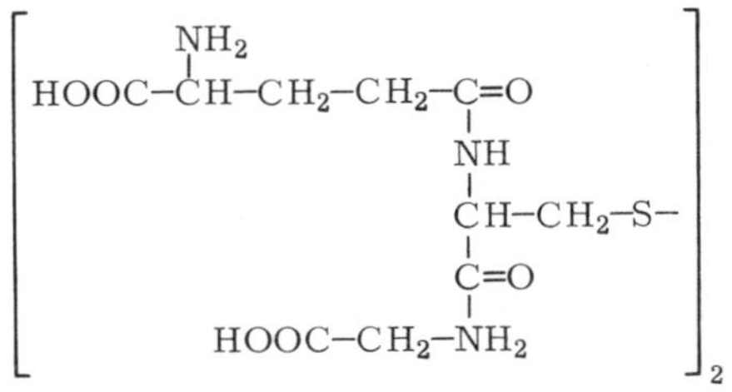

$$
\left[
\begin{array}{l}
\overset{\displaystyle NH_2}{|} \\
HOOC-CH-CH_2-CH_2-C{=}O \\
\qquad\qquad\qquad\quad | \\
\qquad\qquad\qquad\ \ NH \\
\qquad\qquad\qquad\quad | \\
\qquad\qquad\qquad\ \ CH-CH_2-S- \\
\qquad\qquad\qquad\quad | \\
\qquad\qquad\qquad\ \ C{=}O \\
\qquad\qquad\qquad\quad | \\
\qquad\qquad HOOC-CH_2-NH_2
\end{array}
\right]_2
$$

Stability: GSSG is stable at room temperature.

Analysis: enzymatic determination with glutathione reductase; water according to *Karl Fischer.*

Purity required: GSSG $> 90\%$; GSH $< 0.5\%$; $H_2O < 5\%$; organic solvents $< 5\%$.

DL-Glyceraldehyde 3-phosphate (GAP)

$C_3H_7O_6P$ (free acid)

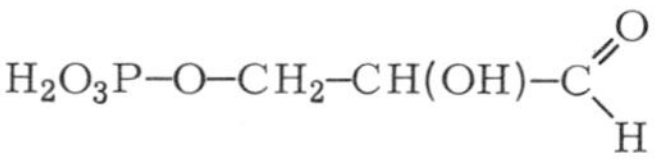

$$H_2O_3P{-}O{-}CH_2{-}CH(OH){-}C\overset{O}{\underset{H}{\big\backslash}}$$

Molecular weights:

GAP 170.1
GAP-diethylacetal-Ba 379.5
GAP-diethylacetal-$(CHA)_2 \cdot H_2O$ 460.6

Stability: the crystalline barium salt of the diethylacetal is stable for months at $4\,°C$. The free acid, which is obtained by treatment with cation exchange resin followed by hydrolysis of the acetal group [1], is easily soluble in water, but rapidly decomposes in alkali. The acid reacting solution is stable for some weeks in the frozen state.

Analysis: enzymatic determination with TIM/GDH.

Purity required: barium salt of GAP-diethylacetal 95% (42% DL-GAP or 21% D-GAP);
dicyclohexylammonium salt 98% DL-GAP-diethylacetal-$(CHA)_2 \cdot H_2O$ (18% D-GAP);
$< 0.1\%$ DAP.

Reference

[1] *C. E. Ballou, H. O. L. Fischer,* The Synthesis of D-Glyceraldehyde-3-phosphate, J. Amer. Chem. Soc. *77*, 3329 (1955).

D-Glycerate 2,3-bisphosphate

2,3-Diphosphoglycerate

$C_3H_8O_{10}P_2$ (free acid)

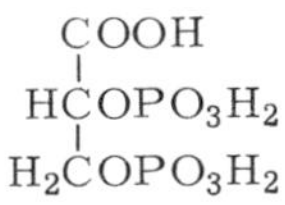

$$\begin{array}{l} COOH \\ | \\ HCOPO_3H_2 \\ | \\ H_2COPO_3H_2 \end{array}$$

Molecular weights:

glycerate-2,3-P_2 266.0
glycerate-2,3-P_2-$(CHA)_5 \cdot 4\ H_2O$ 834.0

Stability: the crystalline penta-cyclohexylammonium salt is stable at room temperature. It dissolves in water to give a solution of neutral reaction. The phosphate groups are only hydrolyzed with difficulty by acid.

Analysis: enzymatic determination with PGM; water according to *K. Fischer.*

Purity required: 98% glycerate-2,3-P_2-(CHA)$_5$ · 4 H_2O (31% glycerate-2,3-P_2).

D-Glycerate 3-phosphate

D-(−)-3-Phosphoglycerate (3-PG)

$C_3H_7O_7P$ (free acid) $H_2O_3P – O – CH_2 – CH(OH) – COOH$

Molecular weights:

glycerate-3-P 186.1
glycerate-3-P-BA · 2 H_2O 357.4
glycerate-3-P-(CHA)$_3$ · 3 H_2O 537.6
glycerate-3-P-Na$_3$ 252.0

Stability: salts of glycerate-3-P are stable at room temperature. Sodium salts are very hygroscopic. Neutral aqueous solutions are good media for micro-organisms; their stability is therefore mainly a question of sterility. Glycerate-3-P is very stable to acid hydrolysis.

Analysis: enzymatic determination with PGK; Na by flame photometry; CHA by titration in anhydrous medium; Ba by complexometry; water according to *K. Fischer.*

Purity required: barium salt 98% glycerate-3-P-Ba · 2 H_2O (51% glycerate-3-P); cyclohexylammonium salt 98% glycerate-3-P-(CHA)$_3$ · 3 H_2O (34% glycerate-3-P); trisodium salt 63% glycerate-3-P; 21% Na; 13% water.

Glycogen

α-1,4-α-1,6-Glucan

$(C_6H_{10}O_5)_n$

Molecular weight: $(162.14)_n$; 20×10^6 to 25×10^6

Stability: glycogen is stable in the dry state at 4°C. It is soluble in water. Strong acid causes hydrolysis of 1,4- and 1,6-glycosidic linkages with the formation of glucose.

Analysis: enzymatic determination with amyloglucosidase; water according to *K. Fischer.*

Purity required: AMP-free preparations 90%; other preparations 80% enzymatically pure glycogen.

Glycollate 2-phosphate

2-Phosphoglycollate

$C_2H_5O_6P$ (free acid) $H_2O_3P - O - CH_2 - COOH$

Molecular weights:

glycollate-2-P 156.0
glycollate-2-P-(CHA)$_3$ · 2 H$_2$O 489.6

Stability: the cyclohexylammonium salt is very stable at room temperature. Solutions in weak acetic acid or sulphuric acid (pH 4 6) undergo only about 1% hydrolysis even at 100°C (7 min).

Analysis: glycollate-2-P from phosphorus; CHA by titration in anhydrous medium; water according to *K. Fischer;* determination of free and organically bound phosphate according to *Fiske & Subbarow.*

Purity required: tricyclohexylammonium salt 98% glycollate-2-P-(CHA)$_3$ · 2 H$_2$O (31% glycollate-2-P); 7% water; < 2% inorganic phosphate.

Guanine

$C_5H_5N_5O$

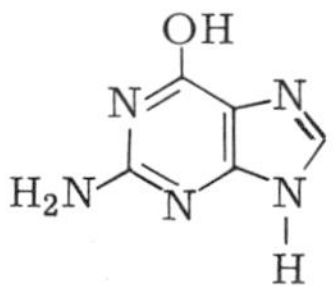

Molecular weight: 151.1

λ_{max} = 246 and 275 nm (pH 7.0)
ε_{246} or ε_{275} = 1.04 or 0.815 $1 \times mmol^{-1} \times mm^{-1}$ (pH 7.0)
ε_{260} = 0.73 $1 \times mmol^{-1} \times mm^{-1}$ (pH 7.0)

Stability: guanine is stable at room temperature.

Analysis: enzymatic determination with guanase.

Purity required: 98% guanine.

Guanosine

Guanine-9-D-ribofuranoside

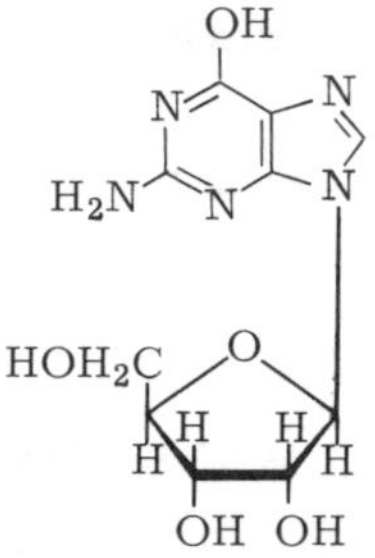

$C_{10}H_{13}N_5O_5$

Molecular weight: 283.2

$\lambda_{max} = 252$ nm (pH 7.0)
$\varepsilon_{252} = 1.36$ l $\times$ mmol^{-1} $\times$ mm^{-1}
$\varepsilon_{260} = 1.18$ l $\times$ mmol^{-1} $\times$ mm^{-1}

Stability: guanosine is stable in the crystalline state at room temperature. In acid solution guanosine is slowly hydrolyzed in the cold to guanine and D-ribose, while on heating the rate of hydrolysis is more rapid.

Analysis: guanosine by enzymatic and spectrophotometric determination; water according to *K. Fischer.*

Purity required: 98% guanosine; chromatographically pure.

Guanosine 5′-triphosphate (GTP)

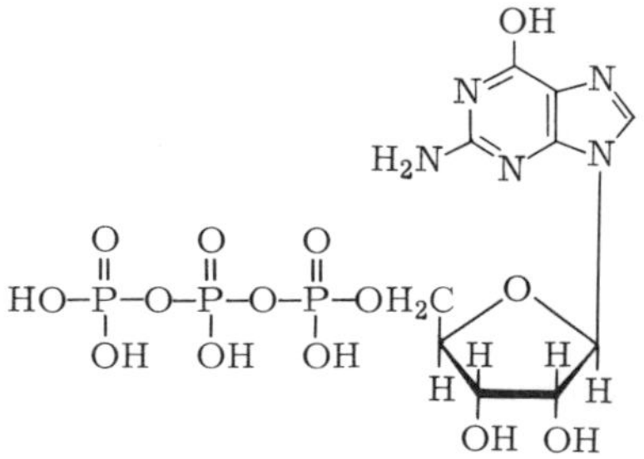

$C_{10}H_{16}N_5O_{14}P_3$ (free acid)

Molecular weights:

GTP 523.2
GTP-Li$_2$ 535.1
GTP-Na$_2$ 567.1

$\lambda_{max} = 252$ nm (pH 7.0)
$\varepsilon_{252} = 1.37$ l $\times$ mmol^{-1} $\times$ mm^{-1} (pH 7.0)
$\varepsilon_{260} = 1.18$ l $\times$ mmol^{-1} $\times$ mm^{-1} (pH 7.0)

Stability: alkali salts stored at 4°C in the dry state can lose 10% of their content in 6 months. Neutral solutions, stored frozen or in the cold, can be kept for several days without significant loss. At pH 3 and below there is a rapid hydrolysis of the pyrophosphate group. Lithium salts are more stable than sodium salts, but cannot be used in many of the reactions involved in protein synthesis.

Analysis: enzymatic determination with PGK; Li, Na by flame photometry; water according to *K. Fischer.*

Purity required: disodium salt > 83% GTP-Na$_2$ (75% GTP); < 5% GDP and GMP; dilithium salt 83% GTP-Li$_2$ (80% GTP).

Hyaluronic acid

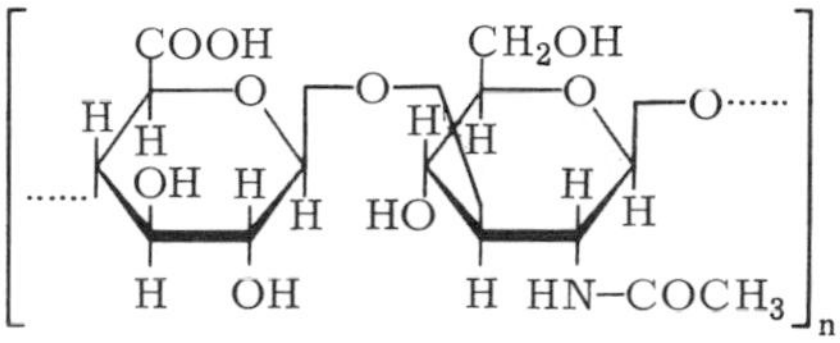

Molecular weight: 2×10^5 to 8×10^6
(according to method of preparation)

Stability: the free acid and the potassium salt, as well as their solutions, are stable at 4°C.

Analysis: enzymatic determination with hyaluronidase.

Purity required: hyaluronic acid or its salts should not contain more than 5% protein.

Application: for the determination of hyaluronidase activity.

DL-3-Hydroxybutyrate

$C_4H_8O_3$ (free acid)

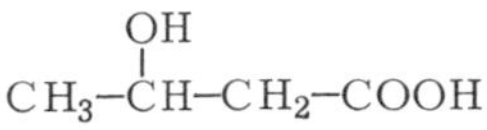

Molecular weights:

3-hydroxybutyric acid 104.1
3-hydroxybutyrate-Na 126.1

Stability: stable in the solid state and in solution.

Analysis: enzymatic determination with β-hydroxybutyrate dehydrogenase.

Purity required: free acid 95%;
sodium salt 99% (41% D-3-hydroxybutyrate).

Hydroxymethyl-glutaryl-coenzyme A

Hydroxymethyl-glutaryl-CoA

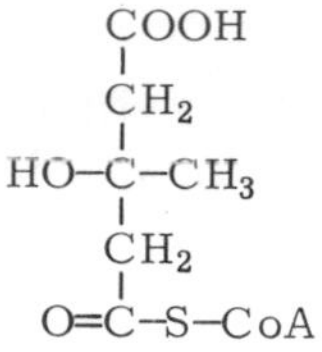

$C_{27}H_{44}N_7O_{20}P_3S$

Molecular weight: 911.7

$\lambda_{max} = 259$ nm
$\varepsilon_{260} = 1.46$ l $\times$ mmol^{-1} $\times$ mm^{-1}

```
        COOH
         |
         CH2
         |
    HO-C-CH3
         |
         CH2
         |
      O=C-S-CoA
```

Stability: hydroxymethyl-glutaryl-CoA should be stored at $-20\,°C$. At this temperature it is stable for some months.

Analysis: spectrophotometric determination; water according to *K. Fischer.*

Purity required: $>95\%$; $<5\%$ water.

Application: for the determination of hydroxymethyl-glutaryl-CoA reductase activity.

Hypoxanthine

6-Hydroxypurine

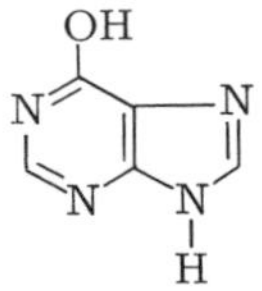

$C_5H_4N_4O$

Molecular weight: 136.1

$\lambda_{max} = 250$ nm (pH 7.0)
$\varepsilon_{250} = 1.06$ l $\times$ mmol^{-1} $\times$ mm^{-1}

Stability: crystalline hypoxanthine is stable in the dark at room temperature.

Analysis: enzymatic and spectrophotometric determination.

Purity required: 99% hypoxanthine; chromatographically pure.

Application: for the determination of xanthine oxidase activity.

Inosine

Hypoxanthine-9-D-ribofuranoside

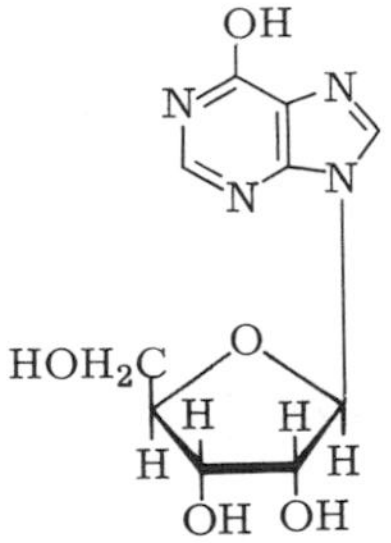

$C_{10}H_{12}N_4O_5$

Molecular weight: 268.2

λ_{max} = 248.5 nm
$\varepsilon_{248.5}$ = 1.225 l $\times$ mmol^{-1} $\times$ mm^{-1} (pH 6.0)

Stability: stable in the solid state at 4°C in the dark. In acid solution inosine is slowly hydrolyzed in the cold to hypoxanthine and D-ribose, while on heating the rate of hydrolysis is more rapid. Irradiation with UV-light causes decomposition of inosine.

Analysis: enzymatic and spectrophotometric determination.

Purity required: 98% inosine.

myo-Inositol

Cyclohexanhexol, *meso*-inositol

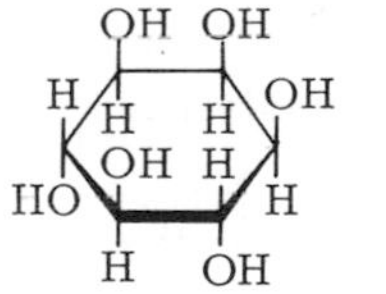

$C_6H_{12}O_6$

Molecular weights:

myo-inositol 180.2
myo-inositol · 2 H_2O 216.2

Stability: *myo*-inositol is stable in crystalline state and in solution at room temperature.

Analysis: enzymatic determination with inositol-dehydrogenase.

Purity required: 95% *myo*-inositol;
myo-inositol · 2 H_2O (75% *myo*-inositol).

Iodonitrotetrazolium chloride (INT)

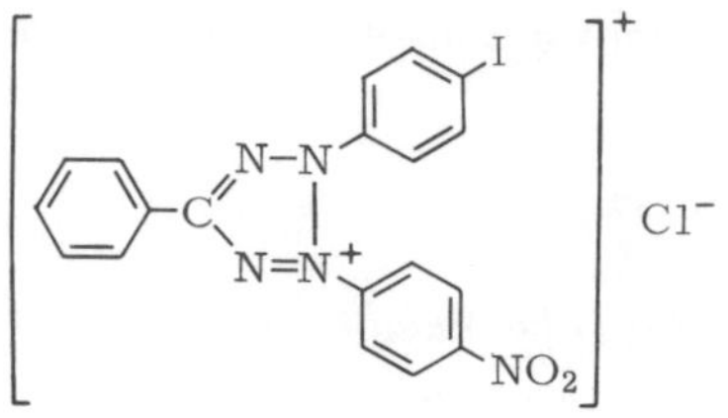

$C_{18}H_{13}N_5O_2ICl$

Molecular weight: 505.7

$\lambda_{max} = 248$ nm

Stability: INT is stable at $+4°C$, protected from light.

Analysis: colorimetric determination of the reduced INT (INT-formazan).

Purity required: 98% INT.

Application: for transferring hydrogen to tetrazolium salts to form coloured formazans, INT-formazan ($\varepsilon_{492} = 1.99 \pm 0.02$ l $\times$ mmol^{-1} $\times$ mm^{-1}, pH 8.6).

DL-Isocitrate

1-Hydroxy-1,2,3-propanetricarboxylate

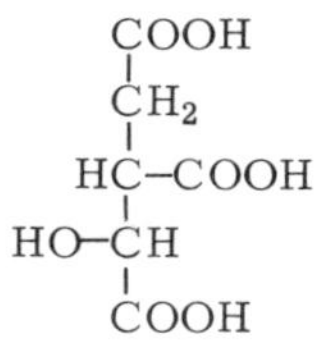

$C_6H_8O_7$ (free acid)

Molecular weights:

isocitric acid 192.12
isocitric acid-Na$_3$ · 2 H$_2$O 294.13

Analysis: enzymatic determination with isocitrate dehydrogenase.

Purity required: 99% DL-isocitric acid-Na$_3$ · 2 H$_2$O (32.8% D-isocitric acid).

Application: for the determination of isocitric dehydrogenase activity.

Lecithin (Phosphatidyl cholines)

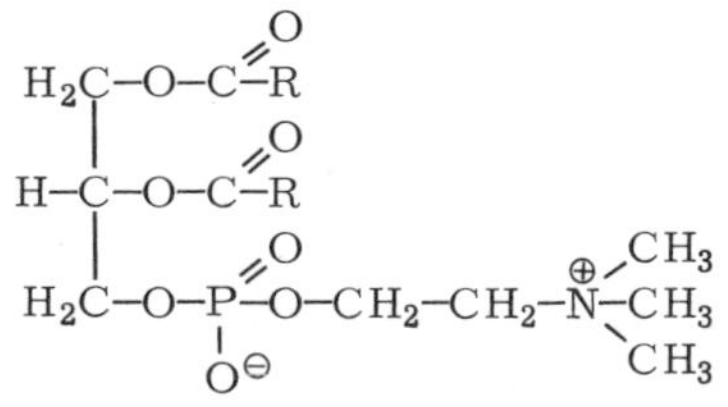

[(R-1,2-O-Diacylglycerol)phosphoryl]-choline

Molecular weight: 650 – 850
(according to the kind of fatty acids)

dipalmitoyl-R-1,2-lecithin: 733
distearoyl-R-1,2-lecithin: 789

Stability: the most synthetic compounds are stable at $+4°C$ and in the dark. After dissolving in organic solvents (methanol, acetone, chloroform) they are stable for 4 weeks (at $+4°C$). They can be stabilized by antioxidants such as butylyzed hydroxytoluene.

Properties: all lecithins can be used for preparing emulsions. By using water, it is possible to produce homogeneous suspensions (ultrasonic). For determination by enzymatic reaction lecithins can most favourably be dissolved in mixtures of t-butanol and water.

Analysis: enzymatic determination with PL-C/AP/choline kinase/PK/LDH; content from organically bound phosphorus according to *Fiske & Subbarow* and from nitrogen.

Purity required: lecithin from egg 70%; synthetic lecithin (dipalmitoyl-lecithin) 98%.

Application: for the determination of the activities of phospholipases C and D.

Leucine-4-nitroanilide (Leupa)

$C_{12}H_{17}N_3O_3$

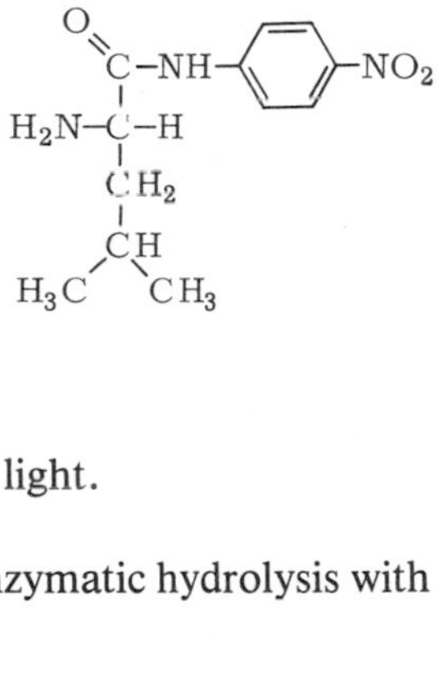

Molecular weight: 251.3

Melting point: $87-89°C$

$[\alpha]_D^{20} = +129°$

$\lambda_{max} = 312$ nm

Stability: very stable at room temperature, if protected from light.

Analysis: photometric determination of 4-nitroaniline after enzymatic hydrolysis with aminopeptidase M or after hydrolysis with HCl (1 mol/l).

Purity required: 95% Leupa; $<0.5\%$ 4-nitroaniline.

D-(−)-Luciferin (Photinus pyralis-luciferin)

D-(−)-2-(6'-Hydroxy-2'-benzothiazolyl-)Δ^2-thiazoline-4-carboxylic-acid

$C_{11}H_8N_2O_3S_2$

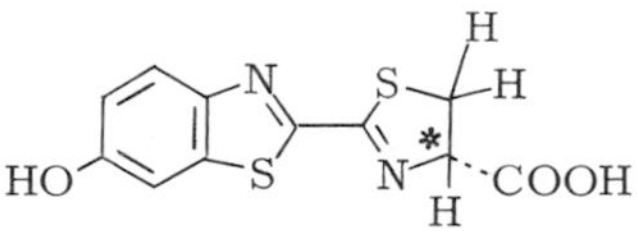

Molecular weight: 280.31

Melting point: 200 – 204 °C

$[\alpha]_D^{20}$: $-29°$

$\lambda_{max} = 265$ nm; 327 nm (pH 7)

Stability: luciferin is stable at $-20°C$, if protected from light. In aqueous solutions it is oxidized in presence of light. In buffer solutions luciferin is stable for 8 hours (0°C).

Analysis: determination of nitrogen.

Purity required: 95%; chromatographically pure.

Application: for the determination of luciferase activity (*Photinus pyralis*).

D-Maltoheptaose

$C_{42}H_{72}O_{36}$

Molecular weight: 1152

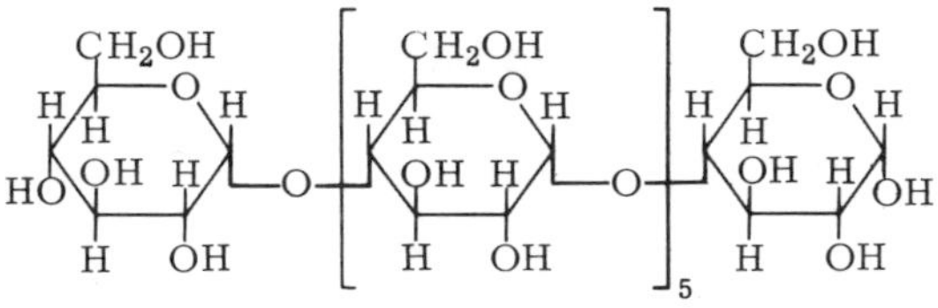

Stability: the substance is stable at room temperature. The aqueous solution is stable for 3 hours at room temperature (pH 7), and for 8 hours at $+4°C$.
The substance is hydrolyzed in acid solutions.

Analysis: enzymatic determination with amylase, α-glucosidase, hexokinase and G6P-DH; determination by HPLC; water according to *K. Fischer.*

Purity required: 92% (by enzymatic determination); 90% by HPLC (area %); < 5% hexaose; < 1% pentaose; < 0.2% other oligomers; < 3% water.

Application: for the determination of α-amylase activity.

D-Maltohexaose

$C_{36}H_{62}O_{31}$

Molecular weight: 990.86

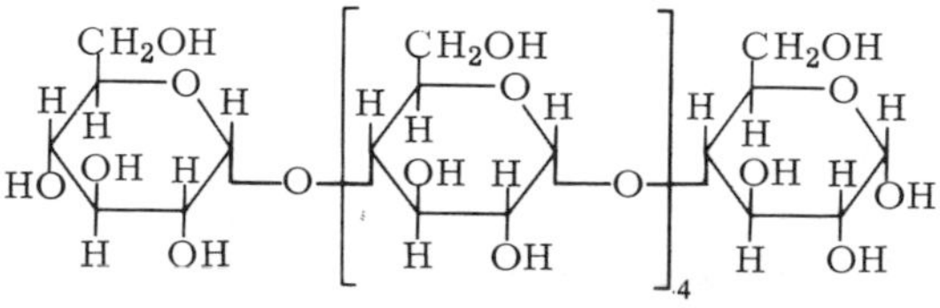

Stability: the substance is stable at room temperature. The aqueous solution is stable for 3 hours at room temperature (pH 7), and for 8 hours at $+4\,^\circ$C. The substance is hydrolyzed in acid solutions.

Analysis: enzymatic determination with amylase, α-glucosidase, hexokinase and G6P-DH; determination by HPLC; water according to *K. Fischer*.

Purity required: 92% (by enzymatic determination); 95% by HPLC (area %); $< 5\%$ other oligomers; $< 3\%$ water.

Application: for the determination of α-amylase activity.

D-Maltopentaose

$C_{30}H_{52}O_{26}$

Molecular weight: 828.72

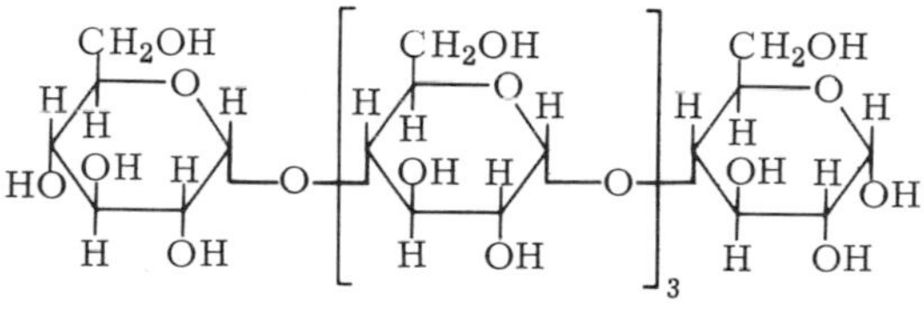

Stability: the substance is stable at room temperature. The aqueous solution is stable for 3 hours at room temperature (pH 7), and for 8 hours at $+4\,^\circ$C. The substance is hydrolyzed in acid solutions.

Analysis: enzymatic determination with amylase, α-glucosidase, hexokinase and G6P-DH; determination by HPLC; water according to *K. Fischer*.

Purity required: 92% (by enzymatic determination); 95% by HPLC (area %); $< 5\%$ other oligomers; $< 3\%$ water.

Application: for the determination of α-amylase activity.

D-Maltotetraose

$C_{24}H_{42}O_{21}$

Molecular weight: 666.58

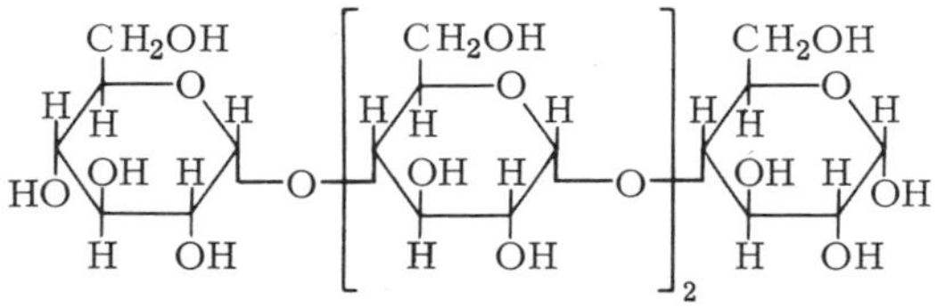

Stability: the substance is stable at room temperature. The aqueous solution is stable for 3 hours at room temperature (pH 7), and for 8 hours at $+4\,^\circ$C. The substance is hydrolyzed in acid solutions.

Analysis: enzymatic determination with amylase, α-glucosidase, hexokinase and G6P-DH; determination by HPCL; water according to *K. Fischer*.

Purity required: 92% (by enzymatic determination); 95% by HPLC (area %); < 5% other oligomers; < 3% water.

Application: for the determination of α-amylase activity.

Methylbenzothiazolone-hydrazonesulphonate (MBTHS)

$C_8H_9N_3O_3S_2$ (free acid)

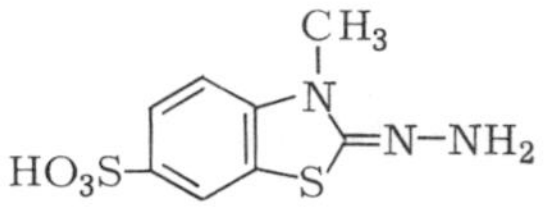

Molecular weights:

MBTHS 259.0
MBTHS-K 297.1

Stability: MBTHS-K is stable at room temperature. In aqueous solutions the substance is stable at room temperature for 2 days, at +4°C for 10 days.

Analysis: determination by HPLC.

Purity required: 90%.

Application: for the determination of H_2O_2 formed from oxidizable substances by oxygen-dependent oxidases.

4-Methylumbelliferyl-acetamido-deoxy-β-galactoside

4-Met-um-β-galactosamide

$C_{18}H_{21}NO_8$

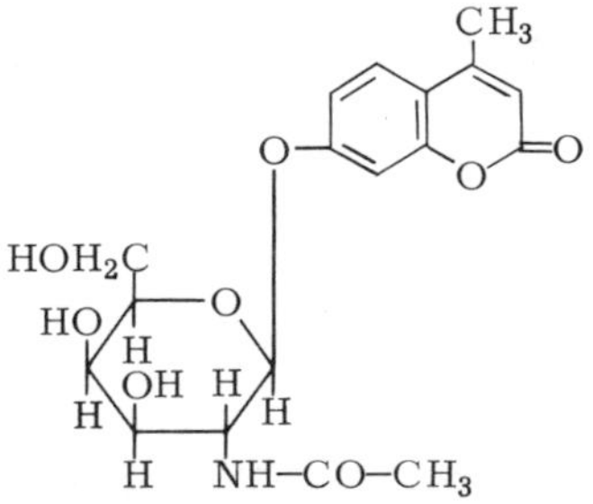

Molecular weight: 379.4

Melting point: 195°C

Stability: the substance is stable at −20°C. In acid solution it is hydrolyzed.

Analysis: fluorimetric determination of 4-methylumbelliferone after enzymatic hydrolysis with β-glucosaminidase. Content of 4-met-um-β-galactosaminide from carbon.

Purity required: 98%; < 2% 4-methylumbelliferone.

4-Methylumbelliferyl-β-D-galactopyranoside

4-Met-um-β-galactoside

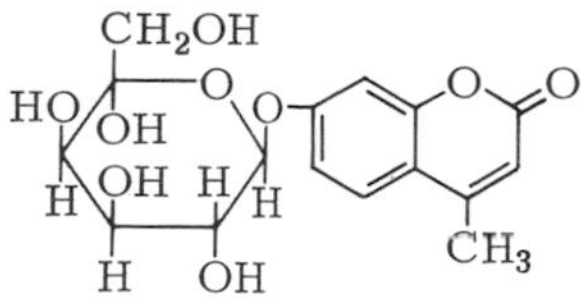

$C_{16}H_{18}O_8$

Molecular weight: 338.3

Melting point: 260°C (decomp.)

$[\alpha]_D^{20}$: − 64°

Stability: the substance is stable at − 20°C. In acid solution it is hydrolyzed.

Analysis: fluorimetric determination of 4-methylumbelliferone after enzymatic hydrolysis with β-galactosidase.

Purity required: 95% 4-met-um-β-galactoside; < 0.5% 4-methylumbelliferone; chromatographically pure (TLC).

Application: for the fluorimetric determination of β-galactosidase activity.

4-Methylumbelliferyl-α-D-glucopyranoside

4-Met-um-α-glucoside

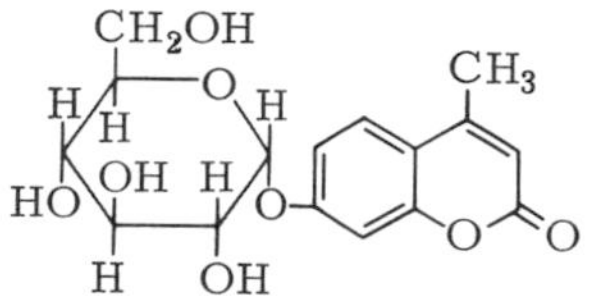

$C_{16}H_{18}O_8$

Molecular weight: 338.3

Melting point: 214°C

$[\alpha]_D^{20}$: + 157°

Stability: the substance is stable at − 20°C. In acid solution it is hydrolyzed.

Analysis: fluorimetric determination of 4-methylumbelliferone after enzymatic hydrolysis with α-glucosidase.

Purity required: 95% 4-met-um-α-glucoside; < 0.5% 4-methylumbelliferone; chromatographically pure (TLC).

Application: for the fluorimetric determination of α-glucosidase activity.

4-Methylumbelliferyl-β-D-glucopyranoside

4-Met-um-β-glucoside

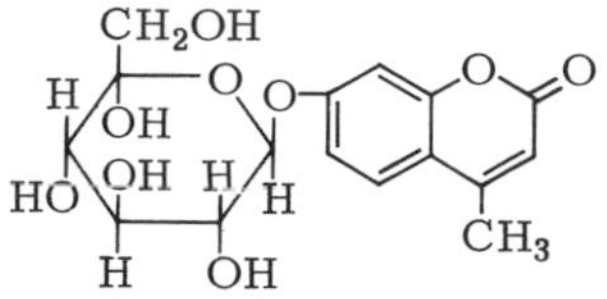

$C_{16}H_{18}O_8$

Molecular weight: 338.3

$[\alpha]_D^{20}$: −94.5°

Stability: the substance is stable at −20°C. In acid solution it is hydrolyzed.

Analysis: fluorimetric determination of 4-methylumbelliferone after enzymatic hydrolysis with β-glucosidase.

Purity required: 95% 4-met-um-β-glucoside; < 0.5% 4-methylumbelliferone; chromatographically pure (TLC).

Application: for the fluorimetric determination of β-glucosidase activity.

4-Methylumbelliferyl-α-D-mannopyranoside

4-Met-um-α-mannoside

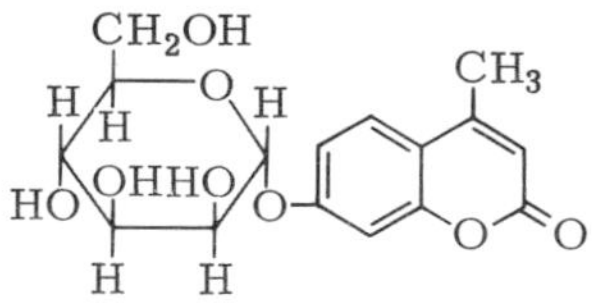

$C_{16}H_{18}O_8$

Molecular weight: 338.3

Melting point: 223°C

$[\alpha]_D^{20}$: +168°

Stability: the substance is stable at −20°C. In acid solution it is hydrolyzed.

Analysis: fluorimetric determination of 4-methylumbelliferone after enzymatic hydrolysis with α-mannosidase.

Purity required: 75% 4-met-um-α-mannoside; < 0.5% 4-methylumbelliferone; chromatographically pure (TLC).

Application: for the fluorimetric determination of α-mannosidase activity.

Myristic aldehyde

n-Tetradecanal

$C_{14}H_{28}O$

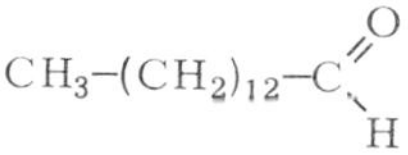

$CH_3-(CH_2)_{12}-C\overset{O}{\underset{H}{\diagup}}$

Molecular weight: 212.38

Melting point: 32 – 42 °C

Stability: myristic aldehyde is stable at room temperature. In methanol and other alcohols myristic aldehyde is stable for 1 day at +4 °C.

Analysis: titration of myristic aldehyde by hydroxylamine (as oxime).

Purity required: 95%.

Application: as substrate for bioluminescence measurements, especially for the determination of luciferase activity (*Photobacterium fischeri*) and for the determination of NAD(P)H.

β-Nicotinamide-adenine dinucleotide (NAD)

$C_{21}H_{27}N_7O_{14}P_2$ (free acid)

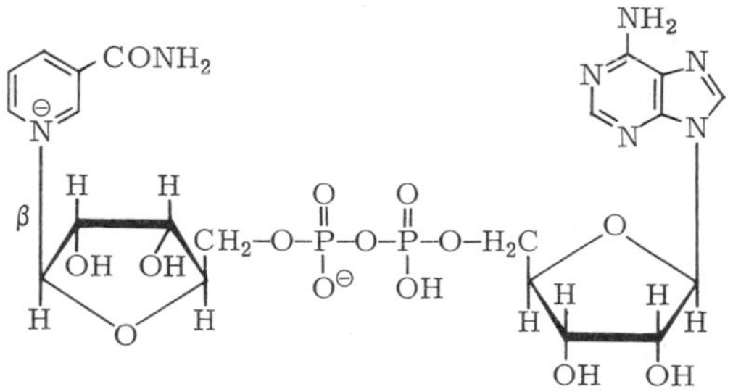

Molecular weights:

NAD 663.4
NAD-Li · 2 H_2O 705.4

$E_0 = -0.316$ V (pH 7.0)

α- and β-NAD differ sterically in the glycosidic linkage between the nicotinamide and ribose. Only β-NAD is active as coenzyme.

$\lambda_{max} = 260$ nm
$\varepsilon_{260} = 1.76$ $l \times mmol^{-1} \times mm^{-1}$

Stability: NAD decomposes slowly even if stored dry and in the cold (2 to 4% per year). NAD-Li salts are more stable than the free acid. Aqueous solutions at slightly acid pH are moderately stable, e.g. 1% solution keeps for about one week at 4°C. Above pH 7 the rate of decomposition increases with pH. For enzyme assays the formation of enzyme inhibitors is very critical.

Analysis: enzymatic determination. Determination of impurities: the difference between the content enzymatically determined and that calculated from the absorption at 260 nm is a measure of the contaminating nucleotides.

The sum of the pyridine coenzymes can be obtained by measurements of the cyanide complex at 327 nm; water according to *K. Fischer;* water is not quantitatively measured by either the *K. Fischer* method or dry weight determination. Accompanying nucleotides are best determined (also α-NAD) by HPLC.

Purity required: 96% β-NAD (enzymatically active. calculated as free acid); 2% other UV-absorbing compounds (calculated as NAD).
NAD-Li · 2 H$_2$O 99%; 1% UV-absorbing contaminants. The presence of organic solvents, even in traces (especially ethanol), is undesirable.

β-Nicotinamide-adenine dinucleotide, reduced (NADH)

C$_{21}$H$_{29}$N$_7$O$_{14}$P$_2$ (free acid)

Molecular weights:

β-NADH 665.4
β-NADH-Na$_2$ 709.4

λ_{max} = 259 nm and 339 nm
ε see Table 1

Table 1. Absorption coefficients (ε) of nicotinamide-adenine dinucleotide (NAD), of its cyanide complex (NAD-CN) and of reduced nicotinamide-adenine dinucleotide (NADH).

	ε l × mmol^{-1} × mm^{-1}				
	260 nm	327 nm	334 nm	339 nm	365 nm
NAD (pH 7)	1.76 ± 0.01 [1]				
NAD-CN (pH 10)	1.41	0.59		0.493	
NADH (pH 8.7)	1.43 ± 0.01 [1]		0.618	0.63	0.34

Stability: stored at 4°C, dry under nitrogen and protected from light, the alkali salts of NADH are stable for several months. NADH preparations, especially in solution, are partially oxidized to NAD. Light, and in particular heavy metals, can considerably accelerate this process.

With aged preparations the amount of NADH corresponding to the absorption at 339 nm can only be partly oxidized. Such preparations should not be used for measurements of enzyme activity, but are still suitable for substrate analysis (end-point determinations). NADH is strongly hygroscopic. In moist preparations inhibitors for dehydrogenases are formed and at the same time the preparations become coloured (yellow to brown) [1].

Analysis: enzymatic determination. Determination of impurities: the difference between the enzymatically determined content and that calculated from the absorption at 260 nm is a measure of the contaminating nucleotides. Contamination of NAD is measured by absorbance ratio A_{260}/A_{339} and by HPLC.

Purity required: disodium salt 80% β-NADH (enzymatically active, calculated as free acid); 6% Na; <1% α-NAD; <3% other UV-absorbing compounds (calculated as NADH), 5% water; traces of organic solvents generally do not interfere. Ratio of absorbances $A_{260}/A_{339} < 2.30$.

Reference

[1] *E. Haid, P. Lehmann, J. Ziegenhorn,* Clin. Chem. *21,* 884 – 887 (1975).

β-Nicotinamide-adenine dinucleotide phosphate (NADP)

$C_{21}H_{28}N_7O_{17}P_3$ (free acid)

Molecular weights:

β-NADP 743.4
β-NADP-Na$_2$ 787.4

$E_0 = -0.317$ V (pH 7.0; 30 °C)

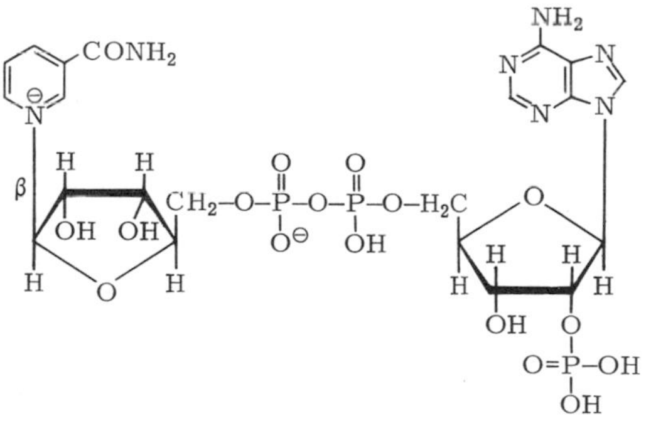

As for NAD the α- and β-forms of NADP differ sterically in the glycosidic linkage between nicotinamide and ribose. The β-form only is enzymatically active.

$\lambda_{max} = 260$ nm
$\varepsilon_{260} = 1.80$ $l \times mmol^{-1} \times mm^{-1}$

Stability: NADP is stable if stored at $+4°C$ in the dry state. In aqueous solution at weakly acid pH it is stable for about 1 week at $+4°C$. Above pH 7 the rate of decomposition increases with increasing pH.

Analysis: enzymatic determination. Determination of impurities: the difference between the amount determined enzymatically and that found from the absorption at 260 nm is a measure of the contaminating nucleotides.
The sum of the pyridine coenzymes can be determined by measurement of the cyanide complex at 327 nm.
The best way to determine contaminating nucleotides is by HPLC. Water according to *K. Fischer*.

Purity required: 85% NADP (enzymatically active, calculated as the free acid); 2% UV-absorbing compounds (calculated as NADP); 5% Na; 5% water; absence of organic solvents is desirable.

β-Nicotinamide-adenine dinucleotide phosphate, reduced (NADPH)

$C_{21}H_{30}N_7O_{17}P_4$ (free acid)

Molecular weights:

β-NADPH 745.4
β-NADPH-Na$_4$ 833.4

$\lambda_{max} = 259$ and 339 nm
ε as for NAD and derivatives, see Table 1, instead of $\varepsilon_{365} = 0.350$ l $\times$ mmol^{-1} $\times$ mm^{-1}

Stability: the alkali salts of NADPH are stable for several months if stored at $+4°C$, dry and protected from light. Aqueous solutions at pH 8 – 9 and stored cool are stable for several days. Acid conditions result in a rapid loss of activity. Aged preparations can contain inhibitors for dehydrogenases (e.g. GlDH), but these do not interfere in substrate analysis (end-point measurements).

Analysis: enzymatic determination. Determination of impurities: the difference between the amount determined enzymatically and that calculated on the basis of the absorption at 260 nm is a measure of the contaminating nucleotides; also, measurements of absorbance ratio A_{260}/A_{339} and HPLC. Na by flame photometry; water according to *K. Fischer*.

Purity required: tetrasodium salt 80% β-NADPH; < 1% NADP; Na 11%; water 6%; other nucleotides < 2%.

Nitro-blue-tetrazolium salt (NBT)

$C_{40}H_{30}N_{10}O_6Cl_2$

Molecular weights:

NBT 817.66
NBT · 3 H$_2$O 871.66

λ_{max} = 257 nm (water)

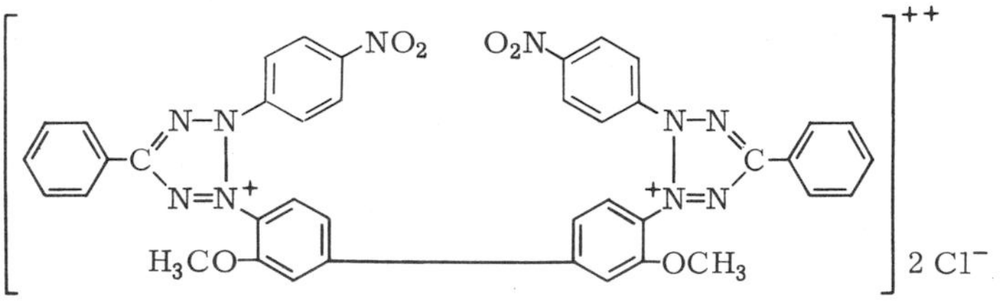

Stability: NBT is stable at $+4°C$, protected from light.

Analysis: Content of NBT from nitrogen.

Purity required: 97% NBT, at most some impurities chromatographically detected (TLC). Water according to *K. Fischer.*

Application: for transferring hydrogen to tetrazolium salts to form coloured formazans. (ε_{540}, NBT-formazan = 0.72 l $\times$ mmol^{-1} $\times$ mm^{-1}).

4-Nitrophenyl-α-fucoside (4-NP-α-fucoside)

4-Nitrophenyl-α-L-fucopyranoside

$C_{12}H_{15}NO_7$

Molecular weight: 285.25

Melting point: 195 – 197 °C

$[\alpha]_D^{20}$: $-262°$ (ethanol)

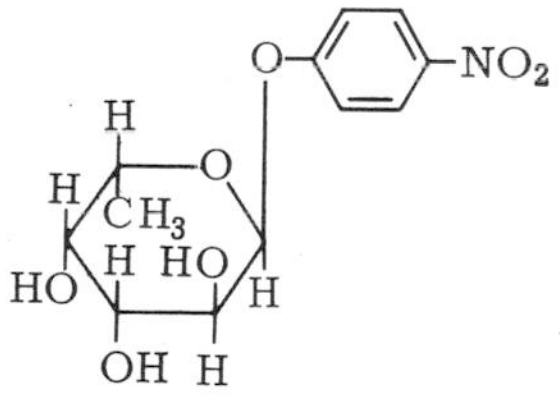

Stability: in the solid state 4-NP-α-fucoside is stable for more than two years ($+4°C$; dry). In aqueous solutions (acid) it is unstable.

Analysis: spectrophotometric determination of 4-nitrophenol after enzymatic hydrolysis with α-fucosidase.

Purity required: 95% 4-NP-α-fucoside; chromatographically pure (TLC); $<0.5\%$ 4-nitrophenol.

4-Nitrophenyl-β-galactoside (4-NP-β-gal)

4-Nitrophenyl-β-D-galactopyranoside

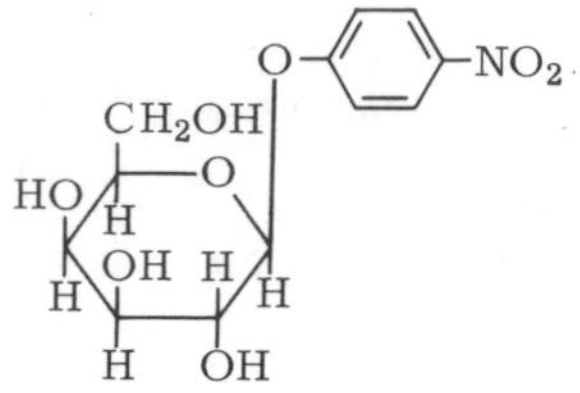

$C_{12}H_{15}NO_8$

Molecular weight: 301.26

Melting point: 173 – 174 °C

$[\alpha]_D^{20}$: – 84°

Stability: in the solid state 4-NP-β-gal is stable for more than 2 years (+ 4 °C, dry, protected from light). In aqueous and acid solutions it is unstable.

Analysis: spectrophotometric determination of 4-nitrophenol after enzymatic hydrolysis with β-galactosidase, or after hydrolysis with HCl (1 mol/l).

Purity required: 99% 4-NP-β-gal; chromatographically pure (TLC); < 0.2% 4-nitrophenol.

4-Nitrophenyl-α-glucoside (4-NP-α-gluc)

4-Nitrophenyl-α-D-glucopyranoside

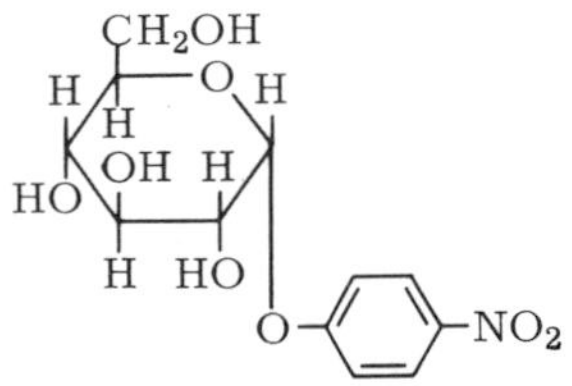

$C_{12}H_{15}NO_8$

Molecular weight: 301.26

Melting point: 214 °C

$[\alpha]_D^{20}$: + 259°

Stability: in the solid state 4-NP-α-gluc is stable for more than 2 years (+ 4 °C, dry). In aqueous and acid solutions it is unstable.

Analysis: spectrophotometric determination of 4-nitrophenol after enzymatic hydrolysis with α-glucosidase or after acid hydrolysis.

Purity required: 99% 4-NP-α-gluc; chromatographically pure (TLC); < 0.2% 4-nitrophenol.

4-Nitrophenyl-β-glucoside (4-NP-β-gluc)

4-Nitrophenyl-β-D-glucopyranoside

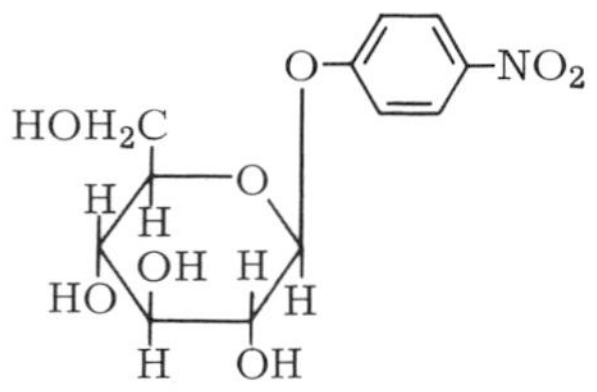

$C_{12}H_{15}NO_8$

Molecular weight: 301.26

Melting point: 164 – 166°C

$[\alpha]_D^{20}$: – 104°

Stability: in the solid state 4-NP-β-gluc is stable for more than 2 years (+ 4°C; dry). In aqueous and acid solutions it is unstable.

Analysis: spectrophotometric determination of 4-nitrophenol after enzymatic hydrolysis with β-glucosidase or after acid hydrolysis.

Purity required: 99% 4-NP-β-gluc; chromatographically pure (TLC); < 0.2% 4-nitrophenol.

4-Nitrophenyl-β-glucuronide (4-NP-β-glucuronide)

4-Nitrophenyl-β-D-glucopyranosido uronic acid

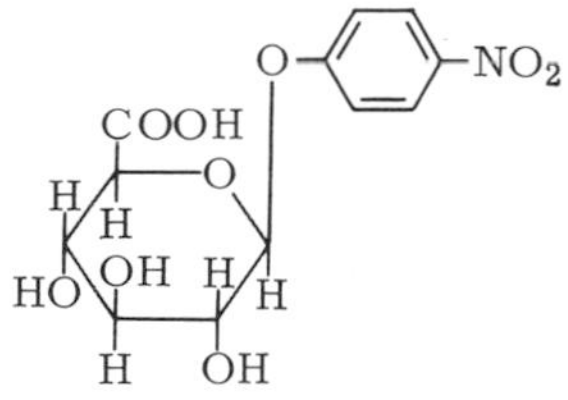

$C_{12}H_{13}NO_9$ (free acid)

Molecular weight: 315.2

Melting point: 138 – 140°C

$[\alpha]_D^{20}$: – 98°

Stability: in the solid state 4-NP-β-glucuronide is stable at + 4°C (dry). In acid solutions it is unstable.

Analysis: spectrophotometric determination of 4-nitrophenol after enzymatic hydrolysis with glucuronidase.

Purity required: 83% 4-NP-β-glucuronide; < 0.5% 4-nitrophenol.

4-Nitrophenyl-α-D-maltoheptaoside

$C_{48}H_{75}O_{38}N$

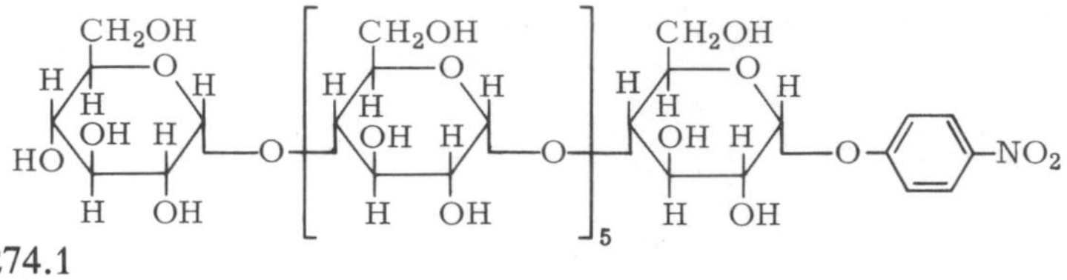

Molecular weight: 1274.1

$[\alpha]_D^{20}$: 194°

λ_{max} = 303 nm

Stability: the substance is stable at $+4°C$ (dry). The solutions are stable for 5 days at room temperature (pH 7), and for two weeks at $+4°C$. The substance is hydrolyzed in acid solution.

Analysis: spectrophotometric determination of 4-nitrophenol after enzymatic hydrolysis with α-amylase and α-glucosidase, or after acid hydrolysis with HCl (1 mol/l). Purity by HPLC.

Purity required: 90% 4-NP-maltoheptaoside; < 5% 4-NP-maltohexaoside.

Application: for the determination of α-amylase activity.

4-Nitrophenyl-α-D-maltohexaoside

$C_{42}H_{65}O_{33}N$

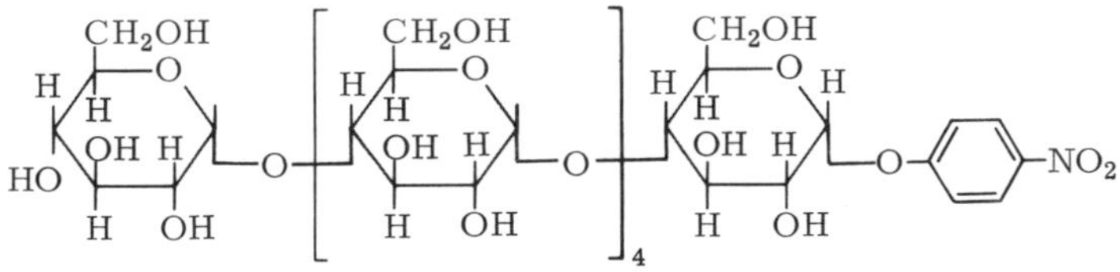

Molecular weight: 1112.1

λ_{max} = 303 nm

Stability: the substance is stable at $+4°C$ (dry). The solutions are stable for 5 days at room temperature (pH 7), and for two weeks at $+4°C$. The substance is hydrolyzed in acid solution.

Analysis: spectrophotometric determination of 4-nitrophenol after enzymatic hydrolysis with α-amylase and α-glucosidase, or after acid hydrolysis with HCl (1 mol/l). Purity by HPLC.

Purity required: 90% 4-NP-maltohexaoside; < 6% other oligomers.

Application: for the determination of α-amylase activity.

4-Nitrophenyl-α-D-maltopentaoside

$C_{36}H_{55}O_{28}N$

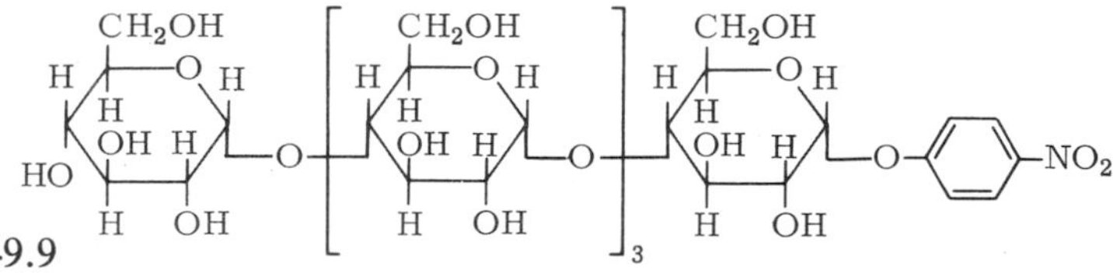

Molecular weight: 949.9

$\lambda_{max} = 303$ nm

Stability: the substance is stable at $+4\,^{\circ}C$ (dry). The solutions are stable for 5 days at room temperature (pH 7), and for two weeks at $+4\,^{\circ}C$. The substance is hydrolyzed in acid solution.

Analysis: spectrophotometric determination of 4-nitrophenol after enzymatic hydrolysis with α-amylase and α-glucosidase, or after acid hydrolysis with HCl (1 mol/l). Purity by HPLC.

Purity required: 90% 4-NP-maltopentaoside; < 6% other oligomers.

Application: for the determination of α-amylase activity.

4-Nitrophenyl-α-D-maltotetraoside

$C_{30}H_{45}O_{23}N$

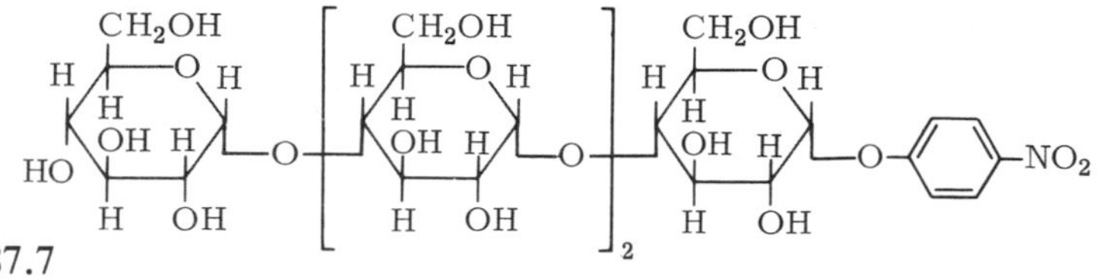

Molecular weight: 787.7

$\lambda_{max} = 303$ nm

Stability: the substance is stable at $+4\,^{\circ}C$ (dry). The solutions are stable for 5 days at room temperature (pH 7), and for two weeks at $+4\,^{\circ}C$. The substance is hydrolyzed in acid solution.

Analysis: spectrophotometric determination of 4-nitrophenol after enzymatic hydrolysis with α-amylase and α-glucosidase, or after acid hydrolysis with HCl (1 mol/l). Purity by HPLC.

Purity required: 90% 4-NP-maltotetraoside; < 6% other oligomers.

Application: for the determination of α-amylase activity.

4-Nitrophenyl-α-mannoside (4-NP-α-man)

4-Nitrophenyl-α-D-mannopyranoside

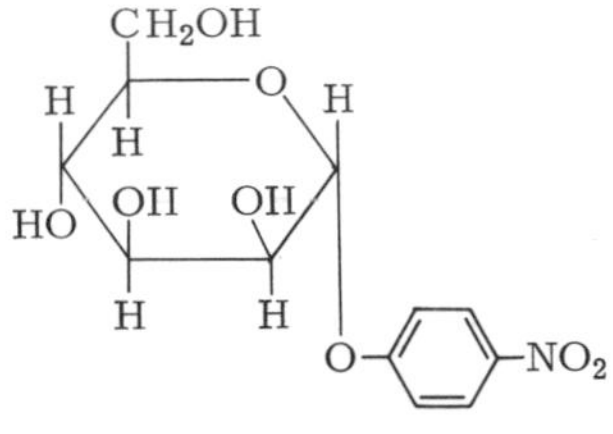

$C_{12}H_{15}NO_8$

Molecular weight: 301.26

Melting point: 174 – 176 °C

$[\alpha]_D^{20}$: + 165°

Stability: in the solid state 4-NP-α-man is stable at room temperature (dry; protected from light). In acid solution it is unstable.

Analysis: spectrophotometric determination of 4-nitrophenol after enzymatic hydrolysis with α-mannosidase or after hydrolysis with HCl (1 mol/l).

Purity required: 99% 4-NP-α-man; chromatographically pure (TLC); < 0.2% 4-nitrophenol.

4-Nitrophenyl-β-mannoside (4-NP-β-man)

4-Nitrophenyl-β-D-mannopyranoside

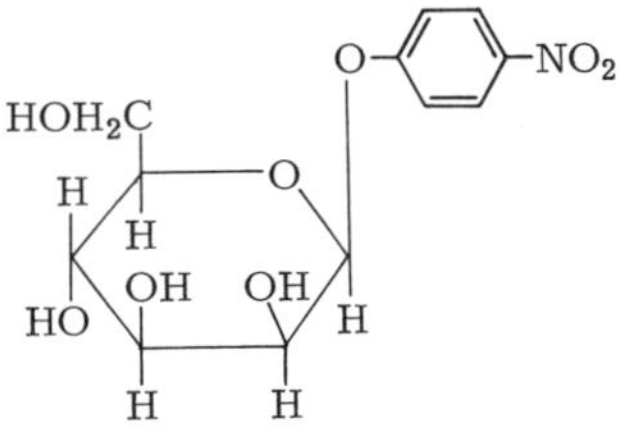

$C_{12}H_{15}NO_8$

Molecular weight: 301.26

Melting point: 202 °C

$[\alpha]_D^{20}$: 105°

Stability: in the solid state 4-NP-β-man is stable for more than 2 years (+ 4 °C; dry; protected from light). In acid solutions it is unstable.

Analysis: spectrophotometric determination of 4-nitrophenol after hydrolysis with HCl (1 mol/l).

Purity required: 96% 4-NP-β-man; chromatographically pure (TLC); < 0.2% 4-nitrophenol.

4-Nitrophenyl phosphate (4-NPP)

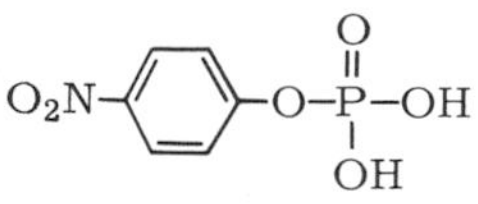

$C_6H_6NO_6P$ (free acid)

Molecular weights:

4-NPP 219.1
4-NPP-Na$_2$ · 6 H$_2$O 371.1

λ_{max} = 308 nm; 220 nm

Stability: 4-NPP is stable at +4°C, if stored dry and protected from light. In aqueous solutions 4-NPP is stable at room temperature for some days (at pH 9). The 4-NPP content decreases at higher or lower pH.

Analysis: spectrophotometric determination of 4-nitrophenol after enzymatic hydrolysis with alkaline phosphatase or after acid hydrolysis. Content from organically bound phosphorus according to *Fiske & Subbarow*; water according to *K. Fischer*.

Purity required: 98% 4-NPP-Na$_2$ · 6 H$_2$O (56% 4-NPP); < 0.07% 4-nitrophenol.

Application: for the determination of alkaline phosphatase activity.

4-Nitrophenyl sulphate (4-NP-sulphate)

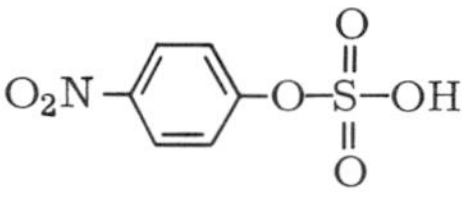

$C_6H_5NO_6S$ (free acid)

Molecular weights:

4-NP-sulphate 219.3
4-NP-sulphate-K 257.3

Stability: the compound is stable, if stored at +4°C.

Analysis: spectrophotometric determination of 4-nitrophenol after enzymatic cleavage with arylsulphatase. Determination of 4-NP-sulphate from nitrogen.

Purity required: 98%; < 0.5% 4-nitrophenol.

Application: for the determination of arylsulphatase activity.

Orotic acid

Uracil-4-carboxylic acid

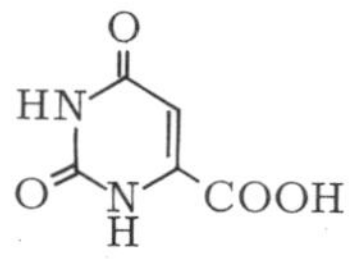

$C_5H_4N_2O_4$

Molecular weights:

Orotic acid 156.1
Orotic acid $\cdot$ H_2O 174.1

λ_{max} = 205 and 282 nm (pH 6)
ε_{205} = 1.09 l $\times$ mmol^{-1} $\times$ mm^{-1}
ε_{282} = 0.752 l $\times$ mmol^{-1} $\times$ mm^{-1}

Stability: orotic acid is stable indefinitely in the crystalline state.

Analysis: spectrophotometrie determination.

Purity required: 98% orotic acid or orotic acid $\cdot$ H_2O (88% orotic acid).

Application: for the determination of the activities of orotate-P-ribosyl transferase and orotidine-5'-P decarboxylase.

Orotidine 5'-monophosphate (OMP)

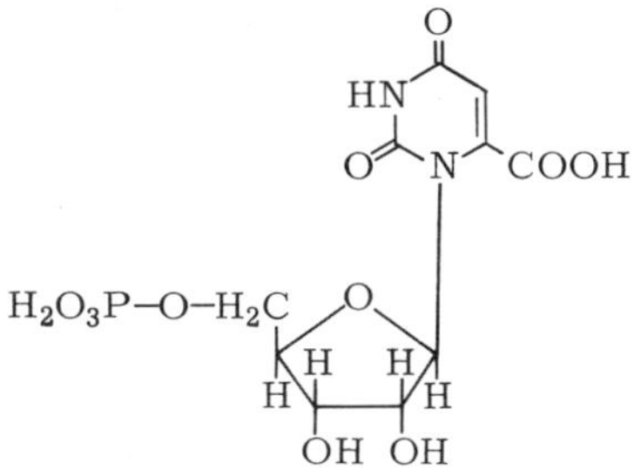

$C_{10}H_{13}N_2O_{11}P$ (free acid)

Molecular weights:

OMP 368.2
OMP-Na$_3$ 434.2

λ_{max} = 205 nm (pH 6.0)
ε_{205} = 1.09 l $\times$ mmol^{-1} $\times$ mm^{-1}

Stability: in the solid state the substance is stable at +4°C.

Analysis: enzymatic determination of OMP with orotidine-5'-P decarboxylase; spectrophotometric determination; Na by flame photometry; water according to *K. Fischer.*

Purity required: 98% OMP-Na$_3$ (83% OMP); < 2% other nucleoside phosphates.

Application: for the determination of OMP decarboxylase activity.

2-Oxobutyric acid

α-Ketobutyric acid

C$_4$H$_6$O$_3$

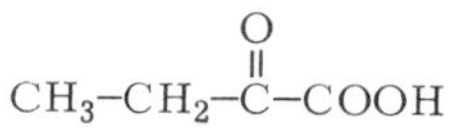

$$CH_3-CH_2-\overset{\displaystyle O}{\overset{\displaystyle \|}{C}}-COOH$$

Molecular weight: 102.1

Melting point: 31 – 33 °C

Stability: very stable, if stored dry at +4 °C.

Analysis: enzymatic determination with LDH.

Purity required: 96% 2-oxobutyric acid.

2-Oxoglutarate

α-Ketoglutarate

C$_5$H$_6$O$_5$

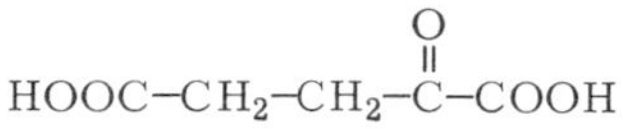

$$HOOC-CH_2-CH_2-\overset{\displaystyle O}{\overset{\displaystyle \|}{C}}-COOH$$

Molecular weights:

2-oxoglutaric acid 146.1
2-oxoglutarate-Na$_2$ · 2 H$_2$O 226.1

Stability: the free acid and the disodium salt are stable at room temperature in the solid state. Neutral aqueous solutions (stored sterile) have a limited stability at +4 °C.

Analysis: enzymatic determination with GlDH.

Purity required: 98% 2-oxoglutaric acid;
95% 2-oxoglutarate-Na$_2$ · 2 H$_2$O.

Phenazine methosulphate (PMS)

5-Methylphenazinium methylsulphate

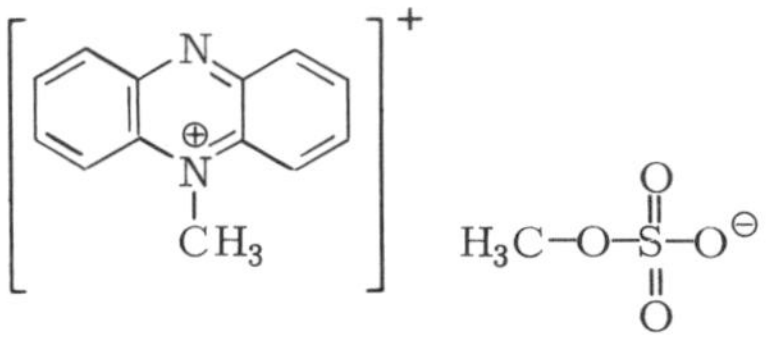

C$_{14}$H$_{14}$N$_2$O$_4$S

Molecular weight: 306.34

λ_{max} = 258 nm

Stability: PMS should be protected from light. It is stable at +4°C.

Analysis: determination of PMS from nitrogen.

Purity required: 95% PMS.

Application: for transferring hydrogen of reducing substances to tetrazolium salts forming formazans, especially for determining ascorbic acid.

PZ-Pro-Leu-Gly-Pro-D-Arg; ("Wünsch substrate")

Phenylazobenzyl-oxycarbonyl-L-prolyl-L-leucyl-glycyl-L-prolyl-D-arginine

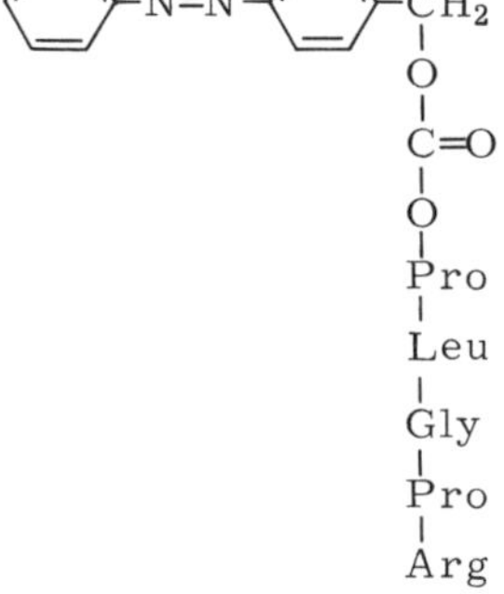

C$_{38}$H$_{52}$N$_{10}$O$_8$

Molecular weight: 776.9

Melting point: 157 – 159°C

$[\alpha]_D^{20}$: −89°

Stability: the "Wünsch substrate" is stable at +4°C.

Analysis: spectrophotometric determination;
determination of the "Wünsch substrate" from nitrogen.

Purity required: 90%; < 2% impurities (TLC).

Application: for the determination of collagenase activity.

Reference

E. Wünsch, H. G. Heidrich, Z. Physiol. Chem. *333*, 149–151 (1963).

Phosphoenolpyruvate (PEP)

$C_3H_5O_6P$ (free acid)

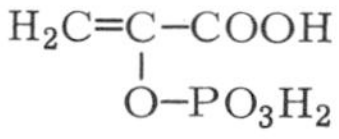

$$H_2C=\underset{|}{C}-COOH$$
$$O-PO_3H_2$$

Molecular weights:

PEP 168.0
PEP-$(CHA)_3$ 465.3
PEP-K 206.1
PEP-Na · H_2O 208.0

Stability: the tricyclohexylammonium, sodium, and potassium salts are stable at +4°C, and are not hygroscopic. Neutral aqueous solutions should be prepared freshly each week. PEP is sensitive to acid hydrolysis and heavy metal ions. The acid potassium and sodium salts should be neutralized immediately after dissolving.

Analysis: enzymatic determination with PK; K, Na by flame photometry; CHA by titration in anhydrous medium; water according to *K. Fischer*.

Purity required: cyclohexylammonium salt 98% PEP-$(CHA)_3$ (35.5% PEP), free from pyruvate;
potassium salt 95% PEP-K (77% PEP);
sodium salt 95% PEP-Na · H_2O (77% PEP).

5-Phospho-α-D-ribose 1-diphosphate (PRPP)

$C_5H_{13}O_{14}P_3$ (free acid)

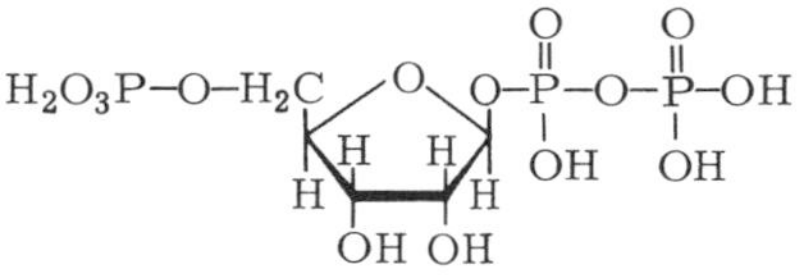

Molecular weights:

PRPP 390.1
PRPP-Na_4 478.1

Stability: if kept dry and protected from light at +4°C, the salts of PRPP are stable for several months. Aqueous solutions in glycine buffer are stable for a few days if stored in a cool place at pH 8.

Analysis: enzymatic determination with OMP pyrophosphorylase and OMP decarboxylase; determination of free and organically bound phosphorus according to *Fiske & Subbarow*; Na by flame photometry; water according to *K. Fischer*.

Purity required: sodium salt 64% PRPP-Na$_4$ (45% PRPP), 15% organically bound phosphorus; 9% water; < 1% inorganic phosphate.

Application: for the determination of OMP pyrophosphorylase and OMP decarboxylase activities.

Pyridoxal phosphoric acid (PALP)

C$_8$H$_{10}$NO$_6$P

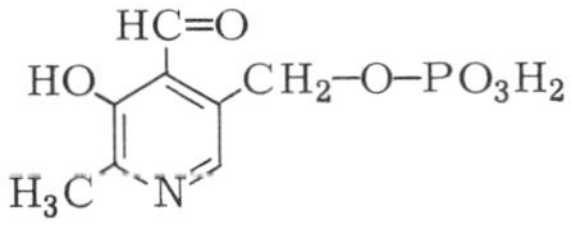

Molecular weights:

PALP 247.1
PALP · H$_2$O 265.2

$\lambda_{max} = 388$ nm
$\varepsilon_{388} = 0.65$ l $\times$ mmol^{-1} $\times$ mm^{-1} } NaOH, 1 mol/l
$\lambda_{max} = 295$ nm
$\varepsilon_{295} = 0.67$ l $\times$ mmol^{-1} $\times$ mm^{-1} } HCl, 1 mol/l

Stability: in the solid state pyridoxal phosphate is stable, while in solution there is 2 – 4% decomposition at + 4°C and 5 – 7% at room temperature per month. In acid solution at 100°C pyridoxal phosphate is hydrolyzed within a few hours.

Analysis: enzymatic determination with aminotransferase; water according to *K. Fischer*.

Purity required: 98% PALP or PALP · H$_2$O.

Application: for the determination of apo-aminotransferases.

Pyruvate

C$_3$H$_4$O$_3$ (free acid)

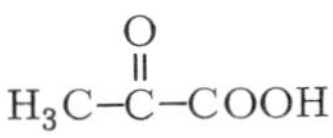

Molecular weights:

Pyruvic acid 88.1
Pyruvate-Na 110.0

Stability: the sodium salt in the solid state and in (sterile) solution is stable even at room temperature. Freezing of solutions often causes a slight decomposition.

Analysis: enzymatic determination with LDH. Check for the presence of the dimeric form (γ-hydroxy-γ-methyl-α-oxoglutaric acid) by chromatography.

Purity required: The crystalline sodium salt (98%) must be free from the dimer.

D-Ribose 5-phosphate

D-Ribofuranose 5-phosphate

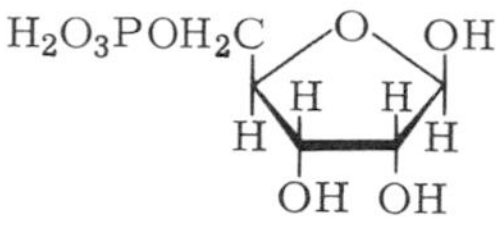

$C_5H_{11}O_8P$ (free acid)

Molecular weights:

ribose-5-P 230.0
ribose-5-P-Ba · 6 H₂O 473.5
ribose-5-P-Na₂ · 2 H₂O 312.1

Stability: the salts are stable in the solid state at room temperature. The sodium salt is hygroscopic. Neutral aqueous solutions are stable if stored sterile. Ribose-5-P is rather stable to weak acid.

Analysis: enzymatic determination with ribose-5-P pyrophosphorylase; Na by flame photometry; Ba complexometrically; determination of free and organically bound phosphate according to *Fiske & Subbarow*; water according to *K. Fischer*.

Purity required: barium salt 98% ribose-5-P-Ba · 6 H₂O (48% ribose-5-P); 28% Ba; 23% water; < 1% inorganic phosphate;
disodium salt 75% ribose-5-P; 15% Na; 10% water; traces of inorganic phosphate.

Application: for the determination of transketolase activity.

Sphingomyelins

Molecular weights: 720 – 860
(according to the kind of fatty acids)
$[\alpha]_D^{20}$: +5.3° (chloroform/methanol)

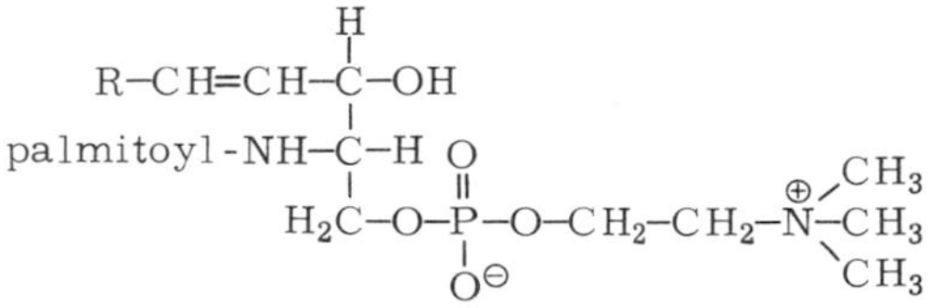

Stability: sphingomyelins are stable at $+4\,^{\circ}$C, if kept dry and protected from light. In solutions of chloroform they are stable at $+4\,^{\circ}$C for approx. 6 months.

Analysis: enzymatic determination with sphingomyelinase/AP; colorimetric measurement of the inorganic phosphorus formed according to *Fiske & Subbarow*.

Purity required: synthetic sphingomyelins (e.g. palmitoylsphingomyelin) 96%.

Application: for the determination of sphingomyelinase activity.

Succinyl-L-alanyl-L-alanyl-L-alanine-4-nitroanilide

Suc-(Ala)$_3$-4-NA

$C_{19}H_{25}N_5O_8$

$$COOH-CH_2-CH_2-CO-Ala$$
$$|$$
$$Ala$$
$$|$$
$$Ala-NH-\langle\rangle-NO_2$$

Molecular weight: 451.4

Melting point: 246 $^{\circ}$C

$[\alpha]_D^{19}$: -24.8° (dimethylformamide)

$\lambda_{max} = 315$ nm (pH 8.0)
$\varepsilon_{315} = 1.46 \; l \times mmol^{-1} \times mm^{-1}$

Stability: the substance is instable in aqueous solutions. The solution of Suc-(Ala)$_3$-4-NA in N-methylpyrrolidone is stable for months if stored at $+4\,^{\circ}$C in the dark. For use it is recommended to dilute this solution with Tris buffer (0.2 mol/l, pH 8.0) [1].

Analysis: spectrophotometric determination of 4-nitroaniline after acid hydrolysis, and/or after enzymatic hydrolysis using elastase; determination of Suc-(Ala)$_3$-4-NA from nitrogen.

Purity required: 90% Suc-(Ala)$_3$-4-NA; $<0.5\%$ 4-nitroaniline; chromatographically pure (TLC).

Application: for the determination of elastase activity in biological materials.

Reference

[1] *J. Bieth, B. Spiess, C. G. Wermuth,* The Synthesis and Analytical Use of a Highly Sensitive and Convenient Substrate of Elastase, Biochemical Medicine *11*, 350–357 (1974).

Tetrahydrofolic acid (THF)

Tetrahydro-pteroylglutamic acid

$C_{19}H_{23}N_7O_6$

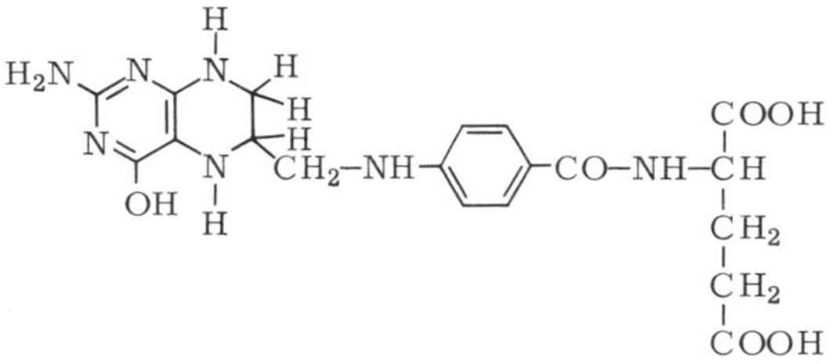

Molecular weight: 445.4

λ_{max} = 286 nm (in NaOH, 1 mol/l)
ε_{286} = 1.85 l $\times$ mmol^{-1} $\times$ mm^{-1} (NaOH, 1 mol/l)
λ_{max} = 297 nm (pH 7.0)
ε_{297} = 2.2 l $\times$ mmol^{-1} $\times$ mm^{-1} (pH 7.0)

Stability: stored in sealed ampoules under oxygen-free nitrogen or argon at +4°C THF is stable for months. In the presence of atmospheric oxygen there is a rapid oxidation to dihydrofolic acid, which has an absorption maximum at 282 nm.

Analysis: enzymatic determination with excess FH$_4$-formylase.

Purity required: the product is very difficult to obtain in a pure state due to the ease with which it is oxidized. Details on purity and content are therefore less reliable.

Application: for the determination of tetrahydrofolate formylase activity.

Thiamine-pyrophosphoric acid (TPP)
(Cocarboxylase)

$C_{12}H_{19}N_4O_7P_2S$

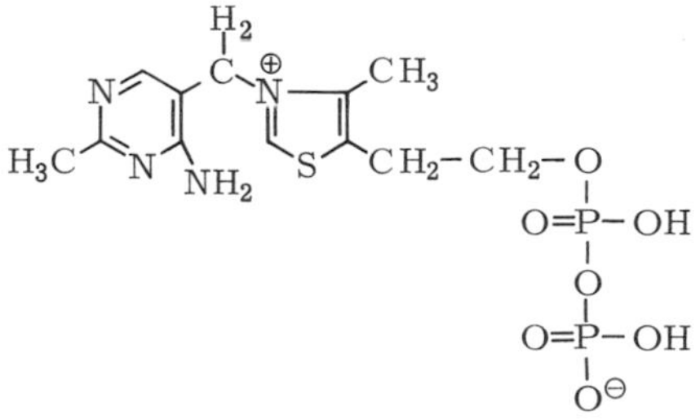

Molecular weights:

TPP 425.3

TPP · HCl 461.9

λ_{max} = 235 and 267 nm (pH 5)
ε_{235} = 1.01 l $\times$ mmol^{-1} $\times$ mm^{-1}
ε_{267} = 0.92 l $\times$ mmol^{-1} $\times$ mm^{-1}

Stability: in the solid and crystalline state the substance is stable indefinitely, in neutral aqueous solution at +4°C for 4 weeks. When heating the acid solution the pyrophosphate bond is slowly cleaved. In alkaline solution decomposition occurs with increasing yellow colour.

Analysis: spectrophotometric determination, determination of nitrogen.

Purity required: 98% TPP · HCl (91% TPP).

Tosyl-glycyl-prolyl-arginine-4-nitroanilide acetate

Chromozym®* TH (cf. p. 340)

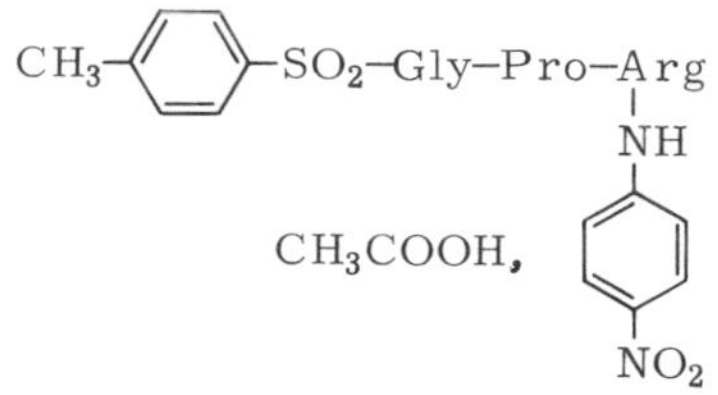

$C_{26}H_{34}NO_7S - C_2H_4O_2$

Molecular weight: 662.6

Melting point: 133 – 136 °C

$[\alpha]_D^{20}$: $-40°$
$\lambda_{max} = 315$ nm (pH 7)

CH_3—⟨benzene⟩—SO_2–Gly–Pro–Arg
|
NH

CH_3COOH, ⟨benzene⟩
|
NO_2

Stability: Chromozym® TH is stable at room temperature.

Analysis: spectrophotometric determination of 4-nitroaniline after acid hydrolysis, and after enzymatic hydrolysis using thrombin; determination of Chromozym® TH from nitrogen.

Purity required: 90% Chromozym® TH; <0.5% 4-nitroaniline.

Application: for the determination of serine proteases activity, especially of thrombine.

Tosyl-glycyl-prolyl-lysine-4-nitroanilide acetate

Chromozym®* PL (cf. p. 340)

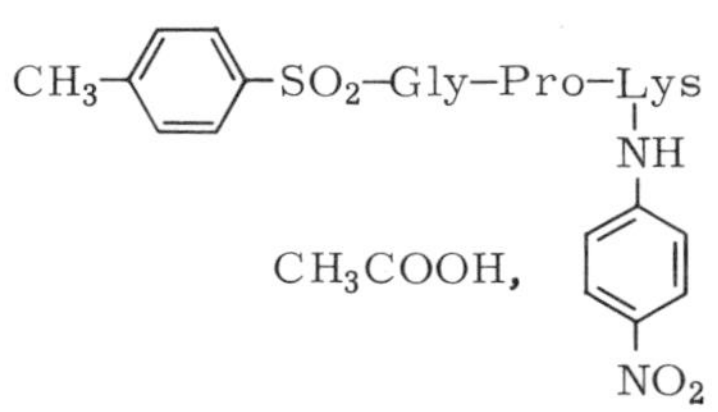

$C_{28}H_{38}O_9N_6S - C_2H_4O_2$

Molecular weight: 634.7

Melting point: 125 °C (decomposition)

$[\alpha]_D^{20}$: $-45°$
$\lambda_{max} = 313$ nm (pH 7)

CH_3—⟨benzene⟩—SO_2–Gly–Pro–Lys
|
NH

CH_3COOH, ⟨benzene⟩
|
NO_2

---

* Chromozym® = registered trade-mark of Pentapharm AG, Bâle, Switzerland.

Stability: Chromozym® PL is stable at room temperature.

Analysis: photometric determination of 4-nitroaniline after acid hydrolysis, and after enzymatic hydrolysis using plasmin; content of Chromozym® PL from nitrogen.

Purity required: 90% Chromozym® PL; <0.5% 4-nitroaniline; chromatographically pure (TLC).

Application: for the determination of the activity of serine proteases, especially of plasmin.

Uric acid

2,6,8-Trihydroxypurine

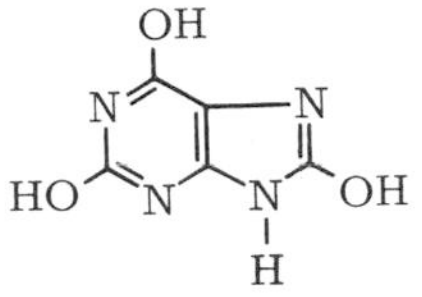

$C_5H_4N_4O_3$

Molecular weight: 168.1

λ_{max} = 231 and 283 nm (pH 1), 293 nm (pH 9)
ε_{231} = 0.85 $1 \times mmol^{-1} \times mm^{-1}$
ε_{283} = 1.15 $1 \times mmol^{-1} \times mm^{-1}$
ε_{293} = 1.26 $1 \times mmol^{-1} \times mm^{-1}$

Stability: uric acid is stable in the solid state and in solution.

Analysis: enzymatic determination with uricase; spectrophotometric determination.

Purity required: 98% uric acid.

Application: for the determination of uricase activity.

Uridine 5'-diphosphate (UDP)

$C_9H_{14}N_2O_{12}P_2$ (free acid)

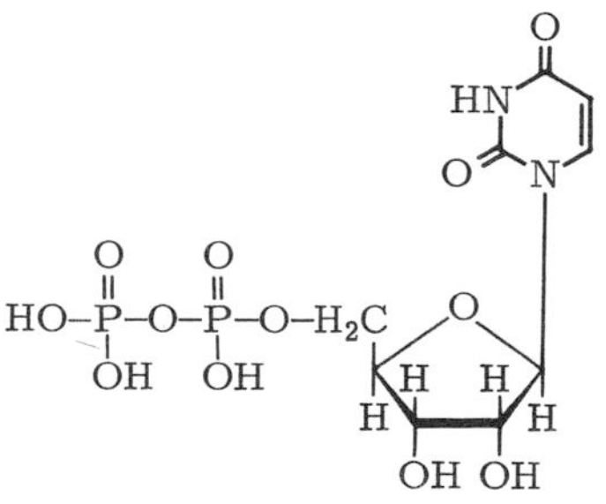

Molecular weights:

UDP 404.2
UDP-K_2 · 3 H_2O 534.4

λ_{max} = 260 nm
ε_{260} = 0.99 $1 \times mmol^{-1} \times mm^{-1}$
(phosphate buffer, 0.1 mol/l; pH 7.0)

Stability: in the solid state the substance is stable at +4°C for 24 months.

Analysis: enzymatic determination with PK/LDH; K by flame photometry; water according to *K. Fischer*.

Purity required: 97% UDP-K_2 · 3 H_2O (72% UDP); <1% UMP; <3% UTP.

Uridine 5′-diphosphoglucose (UDPglucose)

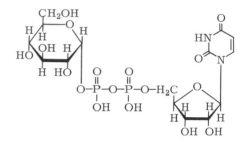

$C_{15}H_{24}N_2O_{17}P_2$ (free acid)

Molecular weights:

UDPglucose 556.1
UDPglucose-Na_2 610.3

λ_{max} = 262 nm (pH 2)
ε_{262} = 1.00 l × mmol^{-1} × mm^{-1}
ε_{260} = 0.99 l × mmol^{-1} × mm^{-1}

Stability: the salts are stable; after 5 min in HCl, 1 mol/l, at 100°C glucose is split off and after 15 min 1 phosphate group.

Analysis: enzymatic determination with UDPglucose-DH; spectrophotometric determination; Na by flame photometry; water according to *K. Fischer*.

Purity required: sodium salt 83% UDPglucose; 2% other uridine phosphates; 7% sodium; 5% water.

Application: for the determination of the activities of UDPG-DH, UDPG-pyrophosphorylase and uridyltransferase.

Uridine 5′-diphosphoglucuronate (UDPglucuronate)

$C_{15}H_{22}N_2O_{18}P_2$ (free acid)

Molecular weights:

UDPglucuronic acid 580.3
UDPglucuronate-Na_2 624.2

λ_{max} = 258 nm
ε_{258} = 1.00 l × mmol^{-1} × mm^{-1}
ε_{260} = 0.99 l × mmol^{-1} × mm^{-1}

Stability: in the solid state the neutral salts are stable. In mineral acids (below pH 3 – 4) UDPglucuronate is hydrolyzed to UMP, glucuronic acid and inorganic phosphate.

Analysis: enzymatic determination with excess UDPglucuronyl-transferase; Na by flame photometry; water according to *K. Fischer*.

Purity required: sodium salt 78% UDPglucuronate; 8% sodium; 10% water.

Application: for the determination of UDPglucuronate pyrophosphorylase activity.

Uridine 5'-monophosphate (UMP)

$C_9H_{13}N_2O_9P$ (free acid)

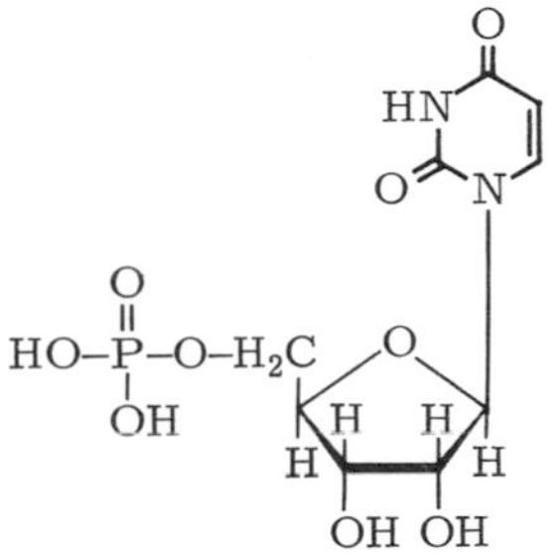

Molecular weights:

UMP 324.2
UMP-Na$_2$ · 3 H$_2$O 422.2

λ_{max} = 260 nm
ε_{260} = 0.99 l × mmol^{-1} × mm^{-1}
(phosphate buffer, 0.1 mol/l; pH 7.0)

Stability: in the solid state the substance is stable at room temperature.

Analysis: enzymatic determination with NMP-kinase/PK/LDH;
Na by flame photometry; water according to *K. Fischer*.

Purity required: 99% UMP-Na$_2$ · 3 H$_2$O (76% UMP).

Uridine 5'-triphosphate (UTP)

$C_9H_{15}N_2O_{15}P_3$ (free acid)

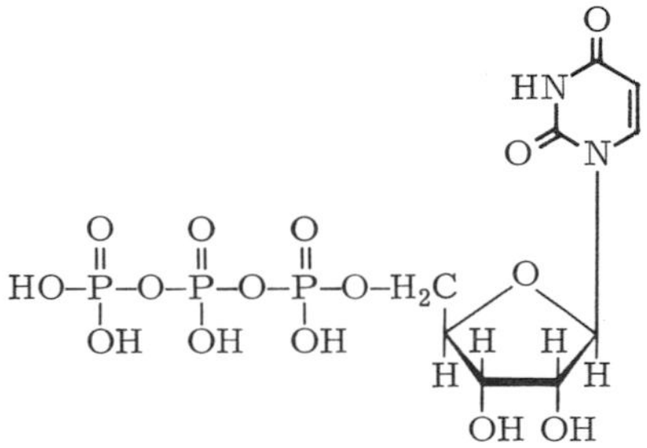

Molecular weights:

UTP 484.2
UTP-Na$_2$ 550.1

λ_{max} = 260 nm
ε_{260} = 0.99 l × mmol^{-1} × mm^{-1}
(phosphate buffer, 0.1 mol/l; pH 7.0)

Stability: in the solid state the substance is stable for 12 months at −20°C.

Analysis: enzymatic determination with PGK/GAPDH; Na by flame photometry; water according to *K. Fischer.*

Purity required: trisodium salt 75% UTP; 11% sodium; 8% water; <4% UDP; <1% UMP.

D-Valyl-cyclohexylalanyl-arginine-4-nitroanilide, diacetate

Chromozym® GK (cf. p. 340)

$C_{30}H_{50}N_8O_9 \cdot 2\ CH_3COOH$

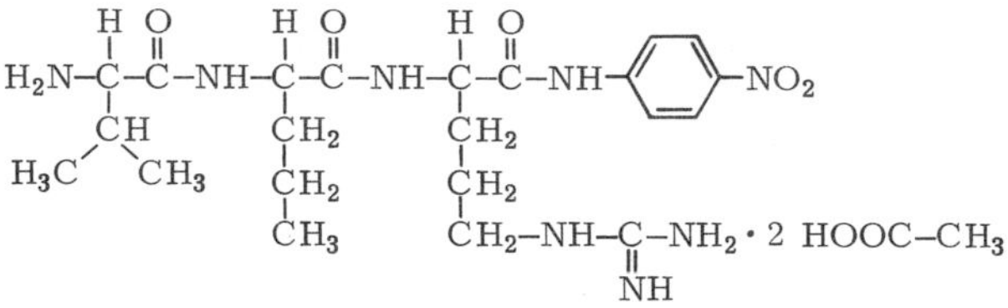

Molecular weight: 786.9

Melting point: 110°C

$[\alpha]_D^{20}: -59°$

$\lambda_{max} = 315$ nm (pH 7)

Stability: Chromozym® GK is stable at room temperature.

Analysis: spectrophotometric determination of 4-nitroaniline after acid hydrolysis, and/or after enzymatic hydrolysis using kallikrein; determination of Chromozym® GK from nitrogen.

Purity required: 90% Chromozym® GK; <0.5% 4-nitroaniline.

Application: for the determination of glandular kallikrein activity.

H-D-Valyl-L-leucyl-L-arginine-4-nitroanilide, dihydrochloride

Val-Leu-Arg-4-NA · 2 HCl

$C_{23}H_{38}N_8O_5 \cdot 2\ HCl$

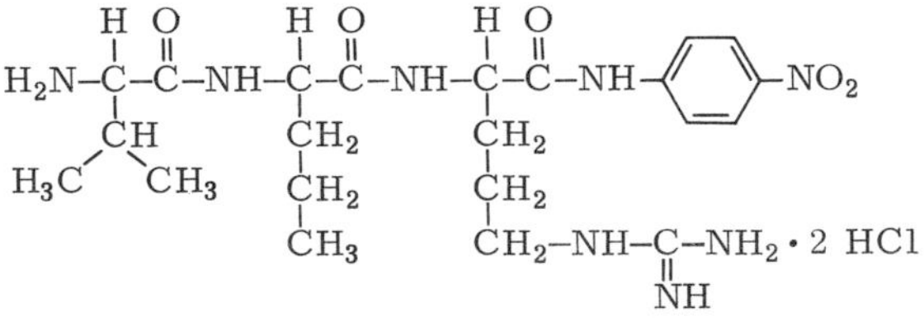

Molecular weight: 579.5

$\lambda_{max} = 316$ nm (water)
$\varepsilon_{316} = 1.3\ l \times mmol^{-1} \times mm^{-1}$

Stability: the substance is stable at room temperature for more than one year if it is stored dry. In aqueous solution it is stable for at least 6 months if kept at $+4\,^{\circ}C$ [1].

Analysis: spectrophotometric determination of 4-nitroaniline after acid hydrolysis, and/or after enzymatic hydrolysis using pancreas kallikrein; determination of Val-Leu-Arg-4-NA from nitrogen.

Purity required: 90% Val-Leu-Arg-4-NA; < 0.5% 4-nitroaniline; chromatographically pure (TLC).

Application: for the determination of glandular kallikrein activity [2].

References

[1] *P. Friberger, L. Aurell, G. Claeson,* Chromogenic Substrates for Kallikreins and Related Enzymes, Agents Actions Suppl. *9*, 83 – 90 (1982).
[2] *E. Amundsen, J. Putter, P. Friberger, M. Knos, M. Larsbraten, M. Claeson,* Methods for the Determination of Glandular Kallikrein by means of a Chromogenic Tripeptide Substrate, Adv. Exp. Med. Biol. *120*, 83 – 95 (1979).

2.3   Standard and Reference Materials

Rudolf Portenhauser

Highly sensitive and precise methods have been developed for analytical measurements in many areas of biological science, e.g. in the fields of environmental analysis, medical diagnosis, and physiological research. The application of these methods requires analytical practice of high quality, particularly with respect to the accuracy of individual analyses, as well as the proficiency of inter-laboratory networks.

Besides the need to develop reliable analytical techniques, the need to create reliable reference materials was also early recognized. The use of reference and/or standard materials is essential for effective communication among scientists, from laboratory to laboratory and from country to country across language barriers, and, last but not least, is a prerequisite for the comparability of analytical data resulting from related measurement problems.

This chapter gives a brief review of the role of standard and reference materials and the availability of standard substances in biological chemical analysis.

394 2 Reagents for Enzymatic Analysis

2.3.1 General

2.3.1.1 Fields of Application

The field of application is large and only a few examples of the consistency and composition of standard materials can be given.

Many substances important in clinical chemistry are stable, simple in chemical structure and easily purified, and are therefore readily obtainable as reference material; in contrast, other clinically relevant compounds are very complex and variable in composition in the native state, and their measurement may be influenced by the tissue in which they occur. In such cases the reference materials consist of mixtures of these compounds in lyophilisates, or solutions in their native matrices. For measurements of blood coagulation, international reference preparations of heparins, thromboplastin, plasmin and other clotting factors are available. In the area of determination of enzyme activity a number of purified enzyme preparations are held and distributed by international institutions, national societies, or industrial companies.

The application of standard and reference materials is very important in the verification of the wavelength and/or absorbance scales of spectrophotometers, and in the calibration of thermometers and thermostatted vessels and cuvette holders in manual or automated instruments. Reference materials for these applications are available from national authorities as listed in the following table. In the field of clinical methodology, for example the collection of blood specimens for diagnosis, quantitative measurement of clinically relevant metabolites, or the development and use of chromatographic methods for drug monitoring, including sample preparation, are also objects of standardization procedures and recommendations of professional organizations, such as the National Committee for Clinical Laboratory Standards, NCCLS, or European Committee for Clinical Laboratory Standards, ECCLS, and others (see Table 1).

2.3.1.2 Definitions

The following terms are defined within the framework and according to current practices of the National Bureau of Standards (NBS) and NCCLS and are commonly accepted in the fields of chemical analyses (cf. chapter 3.1.1.4 "Standards").

Primary reference materials

Generic class of well characterized, stable, homogeneous materials, produced in quantity and having one or more physical or chemical properties experimentally determined within stated analytical uncertainties. Primary reference materials are those having properties certified by a recognized national standard institution (NBS in USA, National Physical Laboratory in the United Kingdom, etc.). These substances are certified with the most accurate and reliable analytical techniques available consistent with end-use requirements.

Secondary reference materials

Primary and secondary reference materials are mainly distinguished in terms of their end-use. In general, secondary reference materials are prepared by commercial organizations or by industrial laboratories for immediate use as working standards. In many cases the secondary reference materials are directly related to primary reference materials. Many suppliers of medical materials use primary reference materials as calibration standards to control the production of secondary reference materials. Normally, a primary reference material is produced by a national standards laboratory having legal authority to issue such material, using definitive analytical methods whenever possible, and is supplied with an appropriate certificate of authenticity.

Definitive methods

These are defined as the most accurate methods available to measure a given chemical property and have a valid and well described theoretical foundation. The results obtained have negligible systematic errors and high levels of precision.

2.3.1.3   Sources of Materials

High-quality reference materials usually carry the authority of an international professional society, a national administration, a national standard laboratory or similar institution. They are entitled, for example, "International Reference Preparations" by WHO, and "Standard Reference Materials" by NBS. A few thousand reference materials are now available, not only from these institutions, but also from commercial industrial sources.

   The following table shows a list of the most important national and international organizations from which information is available, and/or from which standard reference materials, reference preparations or standard reagents can be purchased. The list contains the central address of the organization, the national laboratory responsible for the respective standard material, and the commonly used abbreviations.

Table 1. National and international organisations representing initiators and distributors of standard and reference materials.

Institution	Address; availability of information/substances	Abbreviations for materials	Nature or application of materials	Materials relevant in enzymatic analysis*
Community Bureau of References, Brussels (CBR or BCR)	Community Bureau of References, 200, Rue de la Loi, B-1049 Brussels, Belgium	BCR- or CBR-Standards	Enzymes	γ-GT; (in preparation: CK, ALT, AP)
European Committee for Clinical Laboratory Standards (ECCLS)			(Guidelines and Recommendations for working procedures in clinical chemistry)	(several in preparation)
European Pharmacopoeia (Ph. Eur.)	Technical Secretariat European Pharmacopoeia Commission, Council of Europe, F-67006 Strasbourg-Cedex, France	CRS, Chemical Reference Substances	Pure chemicals for identification in chromatographic procedures (TLC, GL, HPLC), infrared spectrophotometry Drugs Enzymes Hormones	Chymotrypsin Insulin Thyroxin
Fédération Internationale de Pharmaceutique (FIP)[a]	Centre des Standards de la Commission Internationale des Enzymes Pharmaceutiques Dept. of General Biochemistry and Physical Pharmacy Faculty of Pharmaceutical Sciences, State University Gent Wolterslaan 16, B-9000 Gent, Belgium	FIP Standards and Substrates	Pancreatic enzymes Proteins Bile salts	Bromelain Cellulase Chymotrypsin Enterokinase Lysozyme Pancreas Lipase Pancreas Protease Papain Pepsin Plasmin Trypsin

Table 1 (continued)

Institution	Address; availability of information/substances	Abbreviations for materials	Nature or application of materials	Materials relevant in enzymatic analysis*
International Federation of Clinical Chemistry (IFCC)	Council of the International Federation of Clinical Chemistry	IFCC-Standards	Multicomponent biological (human) materials	Human serum albumin Human serum 74/1 (containing: IgA, G, M α_1-Antitrypsin, Orosomucoid, Transferrin)
National Bureau of Standards (NBS)[b]	Office of Standard Reference Materials National Bureau of Standards, Washington D.C. 20234, USA	Standard or Certified Reference Materials (SRM or CRM resp.)	Pure chemicals; Complex biological materials; Glass filters and transmittance standards for spectrophotometry; Quartz cuvettes; Thermometers; Melting point standard	Bilirubin Buffer substances Cholesterol Cortisol Creatinine Electrolytes Glucose Metals Protein standards Urea Uric acid Freeze-dried human serum, containing calcium, chloride, enzymes (acid phosphatase, alkaline phosphatase, ALAT, ASAT, CK, LDH, γ-GT), Lithium, Potassium
National Committee for Clinical Laboratory Standards (NCCLS)	National Committee for Clinical Laboratory Standards, National Office 771, East Lancaster Avenue, Villanove, PA. 19085 USA	AS ~ (Approved Standards) ~C ~H ~I ~L ~M NRSCC (Nat. Ref. Syst. in Clin. Chem.)	Clinical chemistry Haematology Instrumentation Labelling of clin. laboratory material Microbiology Test procedures Reagents	Standardized Protein Solution (bovine albumin)

Table 1 (continued)

Institution	Address; availability of information/substances	Abbreviations for materials	Nature or application of materials	Materials relevant in enzymatic analysis* (cf. Tab. 2)
United States Pharmacopeial Convention, Inc. (USP)[c]	United States Pharmacopeial Convention, Inc. 12061 Twinbrook Parkway Rockville, MD. 20852, USA available in the FRG: Promochem GmbH, PO Box 1246, 4230 Wesel; Zentrallaboratorium Deutscher Apotheker e.V. PO Box 5360 6236 Eschborn	USP-Reference Standards USP-AS (Authentic Substances NF – (National Formulary) Ref. Stds. FCC – (Food Chemicals Codex) Ref. Stds.	Biological substances, Metabolites, Drugs, Alkaloids, Enzymes, Antibiotics, Hormones, Vitamins, etc. Melting point standards	Amino acids (cf. Tab. 2) Ascorbic acid Cholesteryl caprylate Dextrose Fructose Fumaric acid Urokinase
World Health Organization (WHO)	World Health Organization Distribution and Sales Service CH-1211 Geneva 27 Switzerland	IRP (International Reference Preparations) IRR (International Reference Reagents) IS (International Standards)	Antibiotics[2,4] Antibodies[2,4,6] Antigens[2,3,4,6] Blood products and related substances[1,4,5,6,8,9] Endocrinological and related substances[4] Miscellaneous[4,7] Reference reagents[1,5,6,7,10]	AFP CEA Human chorionic gonadotrophin (HCG) Salmonella-Antibodies Insulin Interferons IgA, D, E, G, M Lecithins Heparin Thrombin Thromboplastin Antithrombin III Anti-*Trichinella*-Serum (in preparation: Apolipo-proteins AI, B[1])

Table 1 (continued)

Institution	Address; availability of information/substances	Abbreviations for materials	Nature or application of materials	Materials relevant in enzymatic analysis*
WHO Collaborating Centre for Chemical Reference Substances	WHO collaborating Centre for Chemical Reference Substances, Apoteksbolaget AB, Centrallaboratorium, Box 3045, S-17103 Solna 3, Sweden	International Chemical Reference Substances for Pharmaceuticals	Pharmaceutical products Antibiotics Drugs Hormones Vitamins	Cortisone Digitoxin Digoxin Folic acid Riboflavin Vitamin A

* only selected examples of a variety of reference materials available from the respective organization.
a The described specifications of FIP Standard Reference Substances, respectively assay principles and FIP unit definitions in this chapter are quoted from "Pharmaceutical Enzymes, Properties and Assay Methods", *R. Ruyssen, A. Lauwers* (eds.), E. Story-Scientia P.V.B.A. Scientific Publishing Company, Gent/Belgium, 1978.
b *R. W. Seward & R. Mavrodineanu* (1981) "Standard Reference Materials: Summary of the Clinical Laboratory Standards Issued by the National Bureau of Standards"
  NBS Special Publication 260-71.
  National Bureau of Standards, Dept. of Commerce, Washington D.C. 20234, USA.
c The United States Pharmacopeia, 20th Revision, USP XX; The National Formulary, 15th Edition, NF XV; (1980), USP Convention, Inc. 12601 Twin-brook Parkway, Rockville, MD. 20852, USA.

WHO-standards and reference preparations are held and distributed by:

[1] Centre for Disease Control, Atlanta, GA. 30333, USA

[2] FAO/WHO International Laboratory for Biological Standards, Central Veterinary Laboratory, Weybridge, Surrey, England

[3] International Agency for Research on Cancer, 150 Cours Albert-Thomas, F-69008 Lyon, France

[4] International Laboratory for Biological Standards, National Institute for Biological Standards and Control, Hampstead, London NW3 6RB, England

[5] International Laboratory for Biological Standards, Central Laboratory, Netherlands
Red Cross Blood Transfusion Service, Plesmanlaan 125, NL-1000 Amsterdam, Netherlands

[6] International Laboratory for Biological Standards, Statens Seruminstitut, 80 Amager Boulevard, DK-2000 Copenhagen, Denmark

[7] Research Resources Branch, National Institutes of Allergy and Infectuous Diseases, National Institute of Health, Bethesda, MD., USA

[8] Rijksinstituut voor de Volksgezondheid, Postbus 1, NL-3720 BA, Bilthoven, Netherlands

[9] WHO, CH-1211 Geneva 27, Switzerland

[10] WHO Collaborating Centre for Reference and Research on Viral Hepatitis, Centres for Disease Control, Phoenix, Arizona, USA

2.3.2   Standard and Reference Materials for Enzymatic Analysis

In clinical laboratories in particular the accuracy of analyses has to be a paramount objective, because of the consequences in diagnosis and treatment of disease which depend upon such data.

Therefore, a summary of primary reference materials, for clinically relevant parameters as well as for the calibration of instruments, is given. The list refers to the certified primary reference materials currently available. It contains information about stability, proper storage, impurities, and instructions for the preparation of stock solutions and working standards of the respective materials.

2.3.2.1   Materials for Method Calibration

Albumin

cf. Bovine Serum Albumin (Total Protein Standard), p. 403, Human Serum Albumin Standard, p. 417, and Standardized Protein Solution (Bovine Serum Albumin), p. 427.

Amino Acids

A variety of L-amino acids listed in Table 2 is available as USP-Reference Standards from the United States Pharmacopeial Convention, Inc.

Table 2. Amino acids available as USP-Reference Standards

Amino acid	USP-Catalogue-No.	Amino acid	USP-Catalogue-No.
L-Alanine (200 mg)	125	L-Methionine (200 mg)	4115
L-Arganine (200 mg)	425	L-Phenylalanine (200 mg)	5305
Arginine · HCl (200 mg)	426	L-Proline (200 mg)	5685
L-Cysteine · HCl (200 mg)	1615	L-Serine (200 mg)	6125
L-Histidine (200 mg)	3085	L-Threonine (200 mg)	6672
L-Isoleucine (200 mg)	3495	L-Tryptophan (200 mg)	7005
L-Leucine (200 mg)	3570	L-Tyrosine (500 mg)	7050
L-Lysine Acetate (200 mg)	3715	L-Valine (200 mg)	7085
L-Lysine · HCl (200 mg)	3720		

Anticonvulsant Drug Level Assay Standard for Valproic Acid and Carbamazepin

NBS-No SRM 1599

This SRM consists of four vials of freeze dried human serum, three of which contain the two drugs of different concentration levels: near, above, and lower than the concentrations usually used to control convulsions caused by epilepsy. The fourth vial is a serum blank. Drugs of known purity were added by weight to the human serum base.

Certified values: they were determined from analyses of the reconstituted SRM 1599 by gas chromatography and liquid chromatography.

Antiepilepsy Drug Level Assay Standard

NBS-No SRM 900

This SRM is certified for concentrations of four drugs – phenytoin, ethosuximide, phenobarbital, and primidone – in a processed human serum base. It is supplied as a set of four different freeze-dried preparations with four different concentrations of the respective drug: toxic, therapeutic, sub-therapeutic, and a serum blank.

Storage and stability: the SRM 900 should be stored at $2-8\,^{\circ}$C, and should not be exposed to sunlight or ultraviolet radiation. Under such storage, the material is expected to be stable for at least 2 years.

After reconstitution the contents should be used within one day.

Certified values: they are determined from analyses of the reconstituted material by gas chromatography and liquid chromatography.

Antithrombin III, Human

WHO, Code 72/1

Storage: the unopened ampoules containing the freeze-dried residues derived from the original 1 ml aliquots of normal human plasma should be stored at $-20\,^{\circ}$C. The predicted loss of activity when stored for 10 years at $-20\,^{\circ}$C is 1 %; on storage at $+20\,^{\circ}$C the loss is 0.1% in 1 week.

Standard solution: the total contents of one ampoule (0.9 IU) are reconstituted with 1.0 ml distilled water at room temperature, dissolved by gentle swirling, and transferred immediately to a plastic tube. Do not weigh out any portion of the freeze-dried material. The reconstituted plasma should be kept on ice during the assay and used as soon as possible. Unused ampoules must be discarded, not frozen for later use.

This Antithrombin III Reference Preparation was calibrated by collaborative study involving 12 laboratories.

Assay: immunological methods (*Mancini, Laurell*), clotting tests, and amidolytic procedures.

Ascorbic Acid

USP-RS; Cat. No 430

Storage: the material is preserved in air-tight, light-resistant containers.

Assay: 400 mg of ascorbic acid are dissolved in a mixture of 100 ml carbon dioxide-free water and 25 ml of H_2SO_4, 1 mol/l. The solution is titrated at once with iodine solution, 0.1 mol/l, adding a few ml of starch solution as the end-point is approached. Each ml of iodine, 0.1 mol/l, is equivalent to 8.806 mg ascorbic acid.

Purity: generally ascorbic acid contains not less than 99% and not more than 100.5% of $C_6H_8O_6$. Heavy metals $\leqslant 0.002\%$; ash $< 0.1\%$.

Specific rotation determined in a solution in carbon dioxide-free water containing 1 g/10 ml is between $+20.5°$ and $+21.5°$.

Bilirubin

NBS-No SRM 916

Stability: stored under conditions that totally exclude light, kept in tightly closed vials in a desiccator at $+4°C$, the SRM 916 material is stable for at least 3 years. Standard solutions may be preserved for 1 week at $-20°C$.

Standard solutions: transfer 20.2 mg SRM 916 to a 100 ml flask and dissolve in 2 ml sodium carbonate, 0.1 mol/l, and 1.5 ml sodium hydroxide, 0.1 mol/l. This solution is diluted to 100 ml with pooled serum [1] and well mixed.

Purity: 99.0 ± 2.0% (thin-layer chromatography [2]); chloroform 0.8%; insoluble (in chloroform) < 0.01%; ash 0.01%.

References

[1] Recommendations on a Uniform Bilirubin Standard. Clin. Chem. 8, 405 – 407 (1962).
[2] Z. J. Petrika, C. J. Watson, J. Chromatogr. 37, 76 (1964).

Bovine Serum Albumin (Total Protein Standard)

NBS-No SRM 926

Stability: the material should be stored in a well closed container below $+4°C$. Storage in a desiccator protected from sunlight and UV radiation is recommended. Under these conditions SRM 926 is expected to be stable for at least 3 years.
The lyophilized SRM is extremely hygroscopic. For application, the container should be opened and samples weighed under controlled atmospheric conditions, e.g. in dry nitrogen atmosphere in a glove box etc.

Standard solutions: albumin solutions may be prepared from the lyophilized SRM by adding the powder to the surface of the liquid and allowing the powder to dissolve by diffusion into the liquid. Avoid disturbing the solution during this process.

Purity: see NBS-No SRM 927.

Bovine Serum Albumin (Solution, 7%) (Total Protein Standard)

NBS-No SRM 927

Stability: material is supplied in sealed ampoules and must be stored so that the temperature does not exceed $+4\,^{\circ}$C, nor may the solution be allowed to freeze. Opened ampoules should be used promptly. Under proper storage in sealed ampoules this SRM is expected to be stable for 3 years.

Standard solutions: solutions of lower protein concentration may be prepared by transfer of the appropriate aliquot to a 25 ml volumetric flask and dilution with an aqueous NaCl solution, 0.15 mol/l.

Purity: peptide mass 70.45 ± 0.10 g/l, biuret method [1] 70.48 ± 0.50 g/l, method using optical density at 278 nm 70.77 ± 0.23 g/l; pH 6.66 ± 0.01; Na^+ 0.0291 ± 0.0004 mol/l; Cl^- 0.0210 ± 0.0002 mol/l; volume $2.06 - 2.28$ ml.

Reference

[1] *B. T. Doumas,* Standards for Total Serum Protein Assays − A Collaborative Study, Clin. Chem. *21* (8), 1159−1166 (1957).

Bromelain

FIP-Standard Reference Substance

Working standard solution, about 20 µg/ml, corresponding to $5 - 6$ FIP units/mg: The necessary amount of bromelain is dissolved in a solution containing cysteine, 5 mmol/l, EDTA-Na_2H_2 · 2 H_2O, 1.0 mmol/l, KCl, 10 mmol/l, and Tris buffer, 50 mmol/l, adjusted to pH 7.15 at $25\,^{\circ}$C. The solution is stable for 1 day.
   Immediately before use this solution is equilibrated to $35\,^{\circ}$C in a water-bath.

Assay principle: a casein solution is incubated with bromelain for 10 min at $35\,^{\circ}$C and pH 7.0. The reaction is stopped by adding a protein precipitating reagent (containing trichloroacetic acid, 0.11 mol/l, sodium acetate 0.22 mol/l, and acetic acid, 0.33 mol/l) and the undigested casein is removed by filtration. The amount of peptides remaining in solution is determined photometrically at 275 nm.
   Casein, *Merck* No. 2244 is used; supplier *E. Merck,* Darmstadt (FIP controlled material is available from the Fed. Internat. Pharmaceutique, Gent).

Unit definition: 1 FIP unit of bromelain activity is contained in that amount of a standard preparation, which hydrolyzes a suitable preparation of casein (FIP controlled) under the standard conditions at an initial rate such that an amount of peptides is liberated which is not precipitated by the protein precipitation reagent given above and which gives the same absorbance as 1 µmol tyrosine per ml at 275 nm.

Calcium Carbonate

NBS-No SRM 915

Stability: the material is stable for at least 10 years when stored at room temperature in the tightly closed original bottle. Solutions prepared from SRM 915 are stable indefinitely when stored in glass stoppered bottles.

Standard solutions: for atomic absorption prepared according to ref. [1]. For titrimetric procedures, 100 mg/l (5 milliequivalents): add to 0.25 g dried SRM 915 9 ml de-ionized water and 1 ml concentrated HCl, dissolve and fill up to the 1000 ml-mark with de-ionized water. Store in a pyrex bottle.

Purity: 99.99 ± 0.03% (emission and atomic absorption spectrometry neutron activation analysis); water 0.01 + 0.005%.

Reference

[1] *J. P. Cali, G. N. Bowes Jr., D. S. Young,* Clin. Chem. *19,* 1208–1213 (1973).

Calcium Pantothenate

USP-RS; Cat. No. 870.

Storage: the material must be stored in air-tight containers and dried at 105 °C for 3 h before use.

Standard stock solution, 50 µg/ml: 50 mg calcium pantothenate are dissolved in about 500 ml water in a 1000 ml volumetric flask. 10 ml acetic acid, 0.2 mol/l, and 100 ml sodium acetate solution (1 in 60) are added and diluted with water to volume. Store in a refrigerator.

Standard solutions containing 0.01 µg/ml to 0.04 µg/ml are prepared on the day of assay by diluting an aliquot of the stock solution with water.

Purity (calculated on the dried basis): loss on drying less than 5% of its weight. The material contains not less than 90% and not more than 110% of dextrorotatory calcium pantothenate.

Specific rotation of a solution of 500 mg in 10 ml: between $+25°$ and $+27.5°$; heavy metals $< 0.002\%$; nitrogen content (*Kjeldahl*) $5.7\% - 6.0\%$; calcium content (titration) $8.2\% - 8.6\%$.

Cellulase

FIP-Standard Reference Substance

Working standard solution: the amount of cellulase corresponding to about 0.03 FIP units/ml is dissolved in ice-cold water.

Substrate: anhydrous hydroxyethylcellulose (*Natrosol Hercules* Type 250 H), FIP controlled.

Assay principle: the activity of cellulase preparations is preferably determined by following the changes in the degree of polymerization of linear polymers during the degradation of the substrate.

Unit definition: one FIP unit of endocellulase activity is contained in that amount of a standard preparation that produces under the given experimental conditions a hydrolysis of hydroxyethylcellulose (FIP controlled) with an initial rate such that an apparent quantity of one micromole glucosidic linkage is hydrolyzed per minute.

Chloride

cf. Sodium Chloride, p. 426.

Cholesterol (5-Cholesten-3β-ol)

NBS-No SRM 911a

Stability: stored under inert gas, as delivered, and at $-15°C$ the material is stable for 10 years. Cholesterol should be stored in a refrigerator or freezer, protected from sunlight or UV radiation.

If stored in the dark at room temperature the material should not be used after 6 months from the date of purchase. The ethanol solution, containing 5 mmol/l, is stable for 4 months at $0°C$.

Standard solutions: stock standard solutions are prepared by dissolving 194 mg cholesterol in 100 ml absolute ethanol (5 mmol/l).

Purity: 99.8% (liquid chromatography, gravimetric recovery, thin-layer chromatography, gas chromatography, neutron activation, IR and UV spectroscopy).

Chymotrypsin

CRS

Storage: store in sealed containers at 2 °C to 8 °C, protected from light and moisture.

Standard solution: 25 mg chymotrypsin (CRS) is dissolved in 250 ml HCl, 1 mmol/l, and the solution is stored at 0 °C to 5 °C.

For each titration 1 ml is warmed to about 25 °C over 15 min and 50 μl, corresponding to about 27 nkat, is used. Hydrolyzed substrate: acetyltyrosine ethyl ester. The assay procedure is described in [1].

Enzyme activity: chymotrypsin CRS is crystallized from an extract of the pancreas of beef. It contains not less than 4.0 μkat/mg.

Reference

[1] Eur. Pharm. 1977, Suppl. to Vol. *III*, p. 88 – 91. Malmaisonneuve S.A., 57 Sainte-Ruffine, France.

Chymotrypsin

FIP-Standard Reference Substance

The standard material chymotrypsin is a proteolytic enzyme from an extract of the pancreas of oxen prepared by activation of chymotrypsinogen and purified by crystallization.

Storage: store in a refrigerator in well-closed containers, protected from light and moisture.

Standard solution, 0.1 mg/ml, corresponding to 24 units/ml: 25 mg chymotrypsin are accurately weighed and dissolved in 250 ml HCl, 1 mmol/l. The solution is stored at 0° – 5 °C.

Working solution: a portion of 1 ml of the standard solution is brought to room temperature over 15 min and for each titration an aliquot of 50 μl is used.

Assay principle: the activity of chymotrypsin is measured by the hydrolysis of N-acetyl-L-tyrosine ethyl ester at pH 8.0 and 25 °C. The consumption of base necessary to neutralize the acid liberated during hydrolysis is recorded as a function of time.

Unit definition: one FIP unit of chymotrypsin activity is contained in that amount of the standard preparation which under the specified conditions hydrolyzes 1 micromole of the substrate per minute (Int. FIP unit) or per second (μkat).

Purity: specific activity not less than 240 FIP units/mg. Loss on drying (*in vacuo*, 2 h at 60 °C) <5%; histamine not more than 1 μg calculated as histamine chloride per 240 FIP units (= 4 μkat) of chymotrypsin activity.

Cortisol (Hydrocortisone)

NBS-No SRM 921

Stability: material should be stored in a well closed container at or below 30 °C and should be protected from heat and direct sunlight. When stored refrigerated the cortisol is stable for at least 5 years.

Standard solutions: stock solution, 1 mg/ml: 50 mg SRM 921 are transferred to a 50 ml volumetric flask and 35 ml absolute ethanol are added. When the material is dissolved, dilute to the mark with absolute ethanol.

Standard solution, 5 μg/ml: dilute 1 ml ethanol solution to 200 ml with water.

Both solutions are stable for 6 months when stored in a well stoppered glass container at 4 °C.

Purity: cortisol 98.9% ± 0.2% (liquid and thin-layer chromatography, proton magnetic resonance spectroscopy); 21-dehydrocortisol 0.6%; 21-acetylcortisol 0.2%; 21-dehydrocortisone 0.1%; cortisone 0.2%; ash 0.002%; insoluble matter 0.001%; loss on drying 0.08%.

Creatinine

NBS-No SRM 914

Stability: protected from sunlight and stored at or below 30 °C, preferably refrigerated, the material is stable for 5 years. The stock standard solution, 1 mg/ml, was found to be stable indefinitely when stored at +4 °C in an all glass container.

Standard solution: a stock standard solution is prepared by dissolving 0.1 g creatinine in 100 ml HCl, 0.1 mol/l.

Purity: 99.8 ± 0.1% (paper, thin-layer, and gas-liquid chromatography, phase-solubility analysis [1]); volatile matter 0.03%; chloride 0.07%; ash 0.003%; insoluble matter 0.001%.

Reference

[1] *W. J. Madors,* Phase Solubility Analysis, in Organic Analysis, Vol. *II*, p. 253, Interscience Publishers, Inc., New York 1954.

Cyanocobalamin (Vitamin B$_{12}$)

USP-RS; Cat. No 1520

Storage: USP-cyanocobalamin-RS consists of 1.5 g of mixture with mannitol 8.66 µg/mg. The material must be preserved in air-tight, light-resistant containers. Before using it is dried over silica gel for 4 h.

Standard stock solution, 1 µg/ml: to an accurately weighed, suitable quantity of USP-cyanocobalamin-RS, 25% ethanol is added to make a solution with a concentration of 1 µg/ml. The solution is stored in the refrigerator.

Standard solution: dilutions of the standard stock solution can be made with water to give final vitamin B$_{12}$ concentrations between 0.01 ng/ml and 0.04 ng/ml. These standard solutions should be freshly prepared for each assay.

Purity: 96.0 − 100.5% cyanocobalamin, $C_{63}H_{88}CoN_{14}O_{14}P$, calculated on the dried basis. Loss on drying < 12%; absorbance ratio A_{361}/A_{550} between 3.15 and 3.40.

Digitoxin

USP-RS; Cat. No 1990

Storage: the material is preserved in well-closed containers. Before use it is dried *in vacuo* for 1 h at 105 °C.

Standard solution: an accurately weighed quantity of USP-digitoxin RS is dissolved in ethanol and quantitatively diluted stepwise with ethanol to obtain a solution with a known concentration of about 40 µg/ml.

Purity (dry material): 93.0 − 103.0% digitoxin, $C_{41}H_{64}O_{13}$ (chromatographic procedure); ash negligible from 100 mg.

Typical analysis [1]: 99.7%. Loss on drying 0.6%, total amount of impurities (5 spots on thin-layer chromatography) < 0.2%; gitoxin not detected <0.05%, (chromatographic procedure).

Reference

[1] WHO-Collaborating Center, Solna, Sweden. WHO/Pharm/78.494, App. *7*, 14 – 15.

Digoxin

USP-RS; Cat. No 2000

Storage: the material is preserved in tight containers. Dry *in vacuo* at 105 °C for 1 h before using.

Standard solution: an accurately weighed quantity of USP-digoxin-RS is dissolved in hot ethanol, cooled, and diluted quantitatively and stepwise with ethanol to obtain a solution with a known concentration of about 25 μg/ml.

Purity: loss on drying $\leqslant$ 1.0% of its weight; specific rotation, determined at 546.1 nm in a solution containing 1 g per 10 ml in pyridine is between +13.6° and +14.3°; ash negligible from 100 mg.

Typical analysis [1]: 100.0%; loss on drying 0.1%; total solid impurities 0.6%; digoxigenin-bis-digitoxoside 0.1% (HPLC); gitoxin 0.1% (HPLC); digoxigenin 0.05% (HPLC); digitoxin 0.05% (HPLC).

Reference

[1] WHO-Collaborating Center, Solna, Sweden. WHO/Pharm/78.494, App. *9*, 18 – 19.

Enterokinase

FIP-Standard Reference Substance

Assay principle: pure trypsinogen is incubated with enterokinase under standard conditions at pH 6.0 – 6.2 and 35 °C. The trypsin activity generated is measured by using the accepted method for trypsin.

Unit definition: one FIP unit of enterokinase activity is contained in that amount of the standard preparation which under the specified standard conditions forms one FIP unit of trypsin per minute.

Folic Acid

USP-RS, Cat. No 2860

Storage: the material is preserved in well-closed light-resistant containers and should not be dried. The water content should be determined at the time of use.

Standard solution: about 30 mg of the reference standard material is weighed accurately, corrected for water content, and dissolved in an aqueous solvent containing 2 ml NH_4OH and 1 g $NaClO_3$ per 100 ml.

Purity: 97 – 102% folic acid, $C_{19}H_{19}N_7O_6$, calculated as the anhydrous base; loss on drying (3 h at 105 °C) 5.0 – 8.5%.

Typical analysis [1]: A_{256}/A_{365} 2.80 – 3.00; total impurities (sum of peak areas) 0.3% of the area of the main peak (HPLC); free amines 0.1%; water 8%; ash < 0.3%.

Reference

[1] WHO/Pharm/78.494, Appendix *6*, 12 – 13.

Fructose

USP-RS; Cat. No 2865

Storage: the material is preserved in well-closed containers. Before using it is dried *in vacuo* at 70 °C for 4 h.

Standard solution: 25 mg of USP-fructose-RS is transferred to a 10 ml volumetric flask, water is added to volume and mixed.

Purity: 98.0 – 102.0% fructose, $C_6H_{12}O_6$; loss on drying < 0.5%; ash < 0.5%; chloride < 0.018%; sulphate < 0.025%; arsenic ⩽ 1 ppm; Ca^{2+}, Mg^{2+}, sum < 0.005%; heavy metals < 5 ppm.

D-Glucose (Dextrose)

NBS-No SRM 917

Stability: material should be stored in a well-closed container protected from heat or sunlight at or below 30 °C. When stored refrigerated, the material is stable for at least 5 years.

Standard solution: 1.000 g SRM 917 is weighed into a 100 ml volumetric flask, solubilized with 0.2% benzoic acid solution and filled to the mark with 0.2% benzoic acid solution. 1 ml of this solution contains 0.01 g glucose. When stored in a refrigerator, this solution is stable indefinitely.

Purity: 99.9% (gas-liquid chromatography, differential scanning calorimetry, proton magnetic resonance spectrometry); α-D-glucose $> 99.0\%$; β-D-glucose $< 1.0\%$; moisture 0.06%; ash 0.002%; insoluble matter 0.001 to 0.006%; nitrogen $< 0.001\%$.

γ-Glutamyltransferase (γ-GT)

CBR-Enzyme Reference

γ-GT isolated from pig kidney, dissolved in a 6% solution of bovine serum albumin (without addition of sodium azide), and lyophilized in amounts of 1.00148 g in glass ampoules [1, 2].

Storage and stability: the freeze-dried material can be stored at $+2$ to $8\,^{\circ}$C, but preferably at $-20\,^{\circ}$C. Long-term stability experiments are in progress.

In the reconstituted state the γ-GT reference material is stable for at least $10-15$ days, even when stored at $+37\,^{\circ}$C.

Standard solution: the contents of one ampoule are reconstituted with 1.0 ml distilled water, giving an enzymic γ-GT activity of about 217 ± 1 U/l (proposed IFCC-method at $30\,^{\circ}$C, pH 8.0, γ-L-glutamyl-3-carboxy-4-nitroanilide, 6 mmol/l, glycylglycine, 150 mmol/l).

Purity: electrophoresis of the enzyme concentrate in 7% acrylamide gels shows one protein band corresponding to the γ-GT activity. ALT, AST, LDH not detectable; AP and LAP (as percentage of γ-GT activity) 0.47% and 0.37%, respectively. LDH in the used bovine serum albumin 0.054 U/g BSA.

References

[1] "Purified γ-Glutamyltransferase as an Enzyme Reference Material" (1982), CEIMED, Centre d'Etude des Interferences des Medicaments et autres Xenobiotiques sur les Membranes et les Systemes Biologiques. 7, Rue Albert Lebrun, 54002 Nancy, France.
[2] "Stability Study of Lyophilized γ-Glutamyltransferase" National Institute for Biological Standards and Control, (1982), Holly Hill, Hampstead, London, NW3 6RB.

Heparin (Sodium Salt)

WHO, 3rd International Standard for Heparin, Code No 65/69

Lyophilized sodium salt of heparin, extracted from porcine intestinal mucosa.

Storage: the sealed ampoules containing the freeze-dried heparin standard should be stored at $-20\,^{\circ}$C in the dark.

Standard solution: after opening the ampoule with a new glass-file, a measured quantity of water is added and, after a few minutes, the solid is dissolved by gentle mixing. One ampoule contains 7.9 mg solid material, corresponding to 1370 IU with confidence limits (P = 0.95) of 1342 – 1402 IU [1].

Do not attempt to weigh out any portion of the freeze-dried material.

Analytical data on the ampouled material: wet weight 1.010 g ± 1%; dry weight of the freeze-dried plug 7.899 mg; oxygen content 0.13%; moisture content 1.41%.

Reference

[1] Bull. WHO 1970, *42*, p. 129 – 149.

Human Serum

NBS-No SRM 909

Storage and stability: the freeze-dried serum should be stored between 2 and 8 °C. It should not be frozen nor exposed to sunlight or ultraviolet radiation. Under the recommended storage conditions, SRM 909 is expected to be stable for at least one year.

Reconstitution (diluent water included with SRM 909):

Procedure A, reconstitution with weighing of the freeze-dried serum: the mass of dry serum is determined by the difference in weight between the vial with lyophilisate and the vial after removal of the reconstituted lyophilisate. The reconstituted serum is stored at 2 – 8 °C and is preferably used within 8 h. The concentration of an analyte, after the contents of a vial is weighed and reconstituted with 10.00 ± 0.02 ml diluent water, is calculated by multiplying the mass of freeze-dried serum, in grams, by the certified concentration of the analyte per gram of freeze-dried serum. For example, if the mass of freeze-dried serum in a vial is 0.8703 g, the concentration of uric acid in this vial would be:

$$0.5681 \text{ mmol} \times l^{-1} \times g^{-1} \times 0.8703 \text{ g} = 0.4944 \text{ mmol/l}.$$

The uncertainty is also calculated similarly and for this example would be:

$$0.0050 \text{ mmol} \times l^{-1} \times g^{-1} \times 0.8703 \text{ g} = 0.0044 \text{ mmol/l}.$$

Certified values: the analyte concentrations and uncertainties per gram of freeze-dried serum after reconstitution of SRM 909 according to procedure A (analytical method: isotope dilution mass spectrometry) are:

calcium 3.560 ± 0.013 mmol × l^{-1} × g^{-1}; chloride 128.0 ± 1.5 mmol × l^{-1} × g^{-1}; glucose 7.811 ± 0.095 mmol × l^{-1} × g^{-1}; lithium 1.945 ± 0.033 mmol × l^{-1} × g^{-1}; potassium 4.1546 ± 0.0098 mmol × l^{-1} × g^{-1}; uric acid 0.5681 ± 0.0050 mmol × l^{-1} × g^{-1}; cholesterol 4.346 ± 0.030 mmol × l^{-1} × g^{-1}; magnesium 1.425 ± 0.072 mmol × l^{-1} × g^{-1}.

Procedure B, reconstitution without weighing the freeze-dried serum: reconstitution is performed with 10.00 ± 0.02 ml of the diluent water; the reconstituted serum is stored at 2 – 8 °C and is recommended to be used within 8 h.

Certified values: the certified concentrations of the constituents and the tolerance limits for the use with procedure B (analytical method: isotope dilution mass spectrometry) are:

calcium 3.02 (+0.17, −0.06) mmol/l; chloride 108 (+7, −3) mmol/l; glucose 6.62 (+0.44, −0.20) mmol/l; lithium 1.65 (+0.12, −0.06) mmol/l; potassium 3.52 (+0.19, −0.06) mmol/l; uric acid 0.481 (+0.031, −0.012) mmol/l; cholesterol 3.68 (+0.22, −0.08) mmol/l; magnesium 1.21 (+0.14, −0.10) mmol/l.

Information on enzymes determined in SRM 909: the catalytic concentrations (U/l) of seven enzymes were measured in an interlaboratory study. The values given in the table are consensus values, not certified by the National Bureau of Standards. The enzyme values listed below are dependent on materials and methods. A detailed description of the preparation of test solutes, methods employed, and participants of the study is given in the NBS-Special publication 260-83 [1].

Reconstitution procedure for enzyme determinations: the procedure is different from those given for the determinations of parameters with certified values. Measurements were performed after the lyophilized serum had been reconstituted with 10 ml ice-cold diluent water (0 – 1 °C) by gently swirling the vial at 10 min intervals for one hour, always returning the vial to an ice-water-bath.

Table: Catalytic concentrations of enzymes in reconstituted SRM 909

Enzyme	Overall mean	Within-laboratory standard deviation	Combined within- and between-laboratory standard deviation
	U/l	U/l	U/l
Acid phosphatase	0.23	0.02	0.04
Alkaline phosphatase	75.4	1.9	2.2
Alanine aminotransferase	24.2	0.8	1.8
Aspartate aminotransferase	30.7	0.4	0.9
Creatine kinase	123.0	9.7	10.2
Lactate dehydrogenase	229.2	5.0	9.6
γ-Glutamyltransferase	16.4	0.3	0.4

Brief description of enzyme methods:

All methods use a reaction temperature of 29.77 °C which can be verified by using a gallium melting point cell, NBS-No SRM 1968.

1. Acid phosphatase, EC 3.1.3.2:

Modified method of *Ewen & Spitzer* [2].

Final reaction conditions: pH 5.4; acetate buffer, 0.15 mol/l; thymolphthalein monophosphate, 1.0 mmol/l; Brij-35, 1.5 g/l; volume fraction (sample/total) 0.083 (1 : 12).

2. Alkaline phosphatase, EC 3.1.3.1:

Method of *Bowers & McComb* [3] with modified reaction temperature.

Final reaction conditions: pH 10.5; 4-nitrophenylphosphate, 16.0 mmol/l; 2-amino-2-methyl-1-propanol, 1.0 mol/l; magnesium acetate, 1.0 mmol/l; volume fraction (sample/total) 0.0164 (1 : 61).

3. Alanine aminotransferase, EC 2.6.1.2:

Modified method according to [4, 5].

Final reaction conditions: pH 7.3; L-alanine, 0.5 mol/l; 2-oxoglutarate, 15 mmol/l; pyridoxal 5'-phosphate, 0.11 mmol/l; Tris buffer, 89 mmol/l; NADH (as NADH-Na_2, $2H_2O$), 0.16 mmol/l; LDH (EC 1.1.1.27), 2.2 kU/l (25 °C); volume fraction (sample/total) 0.083 (1 : 12).

4. Aspartate aminotransferase, EC 2.6.1.1:

Modified method according to [6, 7].

Final reaction conditions: pH 7.8; L-aspartate, 175 mmol/l; 2-oxoglutarate, 15 mmol/l; pyridoxal 5'-phosphate, 0.11 mmol/l; Tris buffer, 89 mmol/l; NADH (as NADH-Na_2, $2H_2O$), 0.16 mmol/l; MDH (EC 1.1.1.37), 950 U/l (25 °C); volume fraction (sample/total) 0.083 (1 : 12).

5. Creatine kinase, EC 2.7.3.2:

Method according to [8, 9] with modifications: adenylate kinase inhibitors are removed; a blank reaction is run; EDTA-Na_2H_2 is added; reaction temperature is 29.77 °C.

Final reaction conditions: pH 6.6; imidazole acetate, 100 mmol/l; creatine phosphate, 30 mmol/l; ADP, 2 mmol/l; D-glucose, 20 mmol/l; NAD, 2 mmol/l; HK (EC 2.7.1.1), 2500 U/l, G-6-PDH (EC 1.1.1.49) (from *Leuconostoc mesenteroides*), 1500 U/l, magnesium acetate, 10 mmol/l; N-acetyl cysteine, 20 mmol/l; EDTA-Na_2H_2, 2 mmol/l; volume fraction (sample/total) 0.043 (1 : 23).

6. Lactate dehydrogenase, EC 1.1.1.27:

Modified method according to *Bowers* [10].

Final reaction conditions: pH 7.35; sodium pyruvate, 1.2 mmol/l; NADH, 0.15 mmol/l; Tris buffer, 96.8 mmol/l; volume fraction (sample/total) 0.016 (1 : 61).

7. γ-Glutamyltransferase, EC 2.3.2.2:

Modified method according to [11].

Final reaction conditions: pH 7.90; L-γ-glutamyl-3-carboxy-4-nitroanilide, 6 mmol/l; glycylglycine, 150 mmol/l; volume fraction (sample/total) 0.091 (1 : 11).

References

[1] G. N. Bowers, Jr., R. Alvarez, J. P. Cali, K. R. Eberhardt, D. J. Reeder, R. Schaffer, G. A. Uriano (1983), Standard Reference Materials: The Measurement of the Catalytic (Activity) Concentration of Seven Enzymes in NBS Human Serum SRM 909. NBS Special Publication 260-83. National Bureau of Standards, Dept. of Commerce, Washington D.C. 20234, USA.
[2] L. M. Ewen, R. W. Spitzer, Improved Determination of Prostatic Acid Phosphatase (Sodium Thymolphthalein Monophosphate Substrate), Clin. Chem. 22, 627 – 632 (1976).
[3] G. N. Bowers, Jr., R. B. McComb, Clin. Chem. 21, 1988 – 1995 (1975).
[4] IFCC Methods for the Measurement of Catalytic Concentrations of Enzymes, Part 3. IFCC Method for Alanine Aminotransferase (EC 2.6.1.2) (Stage 2, Draft 1, 1979-11), Clin. Chim. Acta 105, 147F – 172F (1980).
[5] Recommendations of the Alanine Aminotransferase Study Group, Subcommittee on Enzymes, Committee on Standards, American Association for Clinical Chemistry – R. Rej, Clin. Chem. 26, 1023 (1980).
[6] IFCC Methods for the Measurement of Catalytic Concentrations of Enzymes. Part 2. IFCC Method for Aspartate Aminotransferase (EC 2.6.1.1) (Stage 2, Draft 1977), Clin. Chim. Acta 70, F19 – F42, following page 470 (1976).
[7] Recommendations of the Aspartate Aminotransferase Study Group, Subcommittee on Enzymes, Committee on Standards, American Association for Clinical Chemistry – Clin. Chem. 26, 1023 – 1024 (1980).
[8] IFCC Methods for the Measurement of Catalytic Concentrations of Enzymes, Part 7. Method for Creatine Kinase (EC 2.7.3.2) (Stage 1, Draft 1981), obtained from IFCC Expert Panel of Enzymes, M. Hørder (Chairman), Department of Clinical Chemistry, University of Odense, Odense, Denmark.
[9] Recommendation of the Study Group on Creatine Kinase, Subcommittee on Enzymes, Committee on Standards, American Association for Clinical Chemistry, 1979 – personal communication with R. Elser.
[10] G. N. Bowers, Jr., Lactic Dehydrogenase, in: D. Seligson (ed.), Standard Methods of Clinical Chemistry, Vol. 4, Academic Press, New York 1963, p. 163.
[11] IFCC Methods for the Measurement of Catalytic Concentrations of Enzymes, Part 4. IFCC Method for γ-Glutamyltransferase (EC 2.3.2.2) (Stage 1, Draft 1981-3) – personal communication from L. M. Shaw.

Human Serum Standard

IFCC 74/1 (Provisional recommendation, [1])

Storage and stability: 1 ml aliquots of human serum containing 0.1% sodium azide are sealed under nitrogen in ampoules. It is recommended that the ampoules are stored at $-20°C$, or better, at $-70°C$. The components in sealed ampoules remain stable for 4 years with the exception of lipoproteins and complement.

Specification: pooled sera collected from healthy male adults aged $20-30$ years. To avoid excess lipoprotein, all subjects had fasted for 18 h prior to the collection of the blood.

Human albumin content 39.8 to 40.7 g/l (method: *Laurell* rocket). No polymer within the limit of detection (0.4%).

IgG   108 ± 2.6 IU/ml (radial immunodiffusion),
IgA   102 ± 1.6 IU/ml (radial immunodiffusion),
IgM   146 ± 3.0 IU/ml (radial immunodiffusion).

From these, the arbitrary weight units can be assigned as IgG 8.68; IgA 1.45; IgM 1.24 g/l.

The contents of α_1-antitrypsin, transferrin, orosomucoid and complement have not yet been determined.

Reference

[1] *J. R. Hobbs, N. Harboe, C. Alper, B. G. Johansson,* International Fed. Clin. Chem. Committee on Standards; Expert Panel on Proteins, J. Clin. Chem. Clin. Biochem. *18*, 99–103 (1980).

Human Serum Albumin Standard

IFCC, Provisional Recommendation [1]

Purpose: international liquid standard for the comparison of albumin assays.

Proposed storage: the material should be filled in 1.1 ml aliquots in ampoules, and should be kept at $+2°C - +6°C$ in the dark.

Proposed specification: (source: plasmapheresis of normal persons); no detectable quantities of other proteins should be found by immunochemical methods; dimer or higher polymers should not exceed 1.0% (molecular sieving with polyacrylamide gel); nitrogen content *(Kjeldahl)* should be at least 15.7%; ash ≤ 0.1% w/w; hexose ≤ 0.1%; lipid (as fatty acid) $1-2$ mol/mol, i.e. $0.4-0.9\%$. Haematin absorbance at 405 nm in the final product diluted to 10 g/l ≤ 0.025; non-protein amino compounds (as norleucine) ≤ 0.05%; Ca^{2+} ≤ 8.75 μmol/g protein.

Reference

[1] *J. R. Hobbs, N. Harboe, C. Alper, B. G. Johansson, Th. Peters,* IFCC Committee on Standards; Expert Panel on Proteins, Provisional Recommendation (1978) on Specification for Human Serum Albumin Standard, J. Clin. Chem. Clin. Biochem. *18*, 96–98 (1980).

4-Hydroxy-3-methoxy-D,L-mandelic Acid (VMA)

NBS-No SRM 925

Stability: the material should be kept in the well-closed original bottle and stored in a desiccator. Refrigeration at 4°C, or, preferably −20°C is recommended. The SRM should be protected from sunlight and UV radiation. The material should not be used after 5 years from the date of purchase.

Standard stock solution, 1 mg/ml: 100 mg SRM 925 are transferred to a 100 ml volumetric flask, dissolved in HCl, 0.01 mol/l, and diluted to the mark. The contents are transferred to a glass-stoppered brown bottle. When stored in a refrigerator at 4°C the solution is stable for about 3 months.

Working standard solution, 50 µg/ml, (diluted with HCl, 0.01 mol/l) should be discarded after 1 week [1, 2].

Purity: 99.4 ± 0.4% (gas-liquid chromatography; thin-layer chromatography); uncharacterized compound max. 0.5%; keto-VMA max. 0.1%; volatile matter 0.02%; ash 0.004%.

References

[1] *F. W. Sunderman Jr.,* Colorimetric Determination of VMA in Urine, in Standard Methods of Clinical Chemistry Vol. *6, R. P. McDonald* (ed.), Academic Press, New York 1970, pp. 99−106.
[2] *N. W. Tietz,* Fundamentals of Clinical Chemistry, W. B. Saunders, Co., Philadelphia, PA. 1970, pp. 577−580.

Iron Metal (Clinical Standard)

NBS-No SRM 937

Storage and stability: the material should be stored in the tightly-closed original bottle under normal laboratory conditions.

Standard stock solution, 20 mmol/l: approximately 1 g SRM 937 is weighed to the nearest 0.1 mg and transferred to a 1 l volumetric flask. The metal is dissolved in 100 ml HCl, 6 mol/l, and diluted to 1 l with water. The exact concentration of this stock solution is expressed by

$$c = \frac{m_{SRM\,937} \times 0.999}{55.847 \times 1000} \quad \text{mol/l} .$$

Working stock standard solution, 0.2 mmol/l: transfer 10 ml stock standard solution to a 1 l volumetric flask and add HCl, 0.2 mol/l, to the graduation line.

Working standard solutions: more diluted solutions may be prepared by pipetting known volumes of the working stock standard solution into volumetric flasks and diluting to the graduation line with HCl, 0.2 mol/l. These solutions should be prepared daily.

Purity: assay 99.90 ± 0.02% by weight; Ni 0.001%; Si 0.008%; C, Cr, Co, each 0.007%; Cu, Mn, O_2, S, each 0.006%; Mo 0.005%; P 0.003%; Ge, N_2 0.001%; other elements < 0.003%.

Lead Nitrate

NBS-No SRM 928

Storage: the dry material should be stored in the tightly-closed original bottle under normal laboratory conditions.

Stability of solutions: because of the instability of non-acidified aqueous lead solutions it is recommended that three levels of concentration be used.

a) a stock standard solution (50 mmol/l) is prepared by dissolving 1.6561 g SRM 928 in ion-free water. If the solution is cloudy add a few drops of NH_4OH. Mix, dilute to 100 ml in a volumetric flask and transfer immediately to a dry polyethylene bottle previously washed with acid and rinsed with water. This solution is stable for 6 months.

b) an intermediate solution (500 µmol/l) is prepared by a 1 : 100 dilution of the above stock solution. This solution may be stored in a polyethylene bottle at room temperature for one month.

c) working standard solutions of 0.5, 1.0, 2.5, and 5.0 µmol/l should be prepared each time an analysis is performed.

Purity: 100.00 ± 0.03% (determination of lead as lead chromate using a slight excess of $K_2Cr_2O_7$; excess chromate is determined spectrophotometrically [1]).

Reference

[1] *E. J. Catanzaro, T. J. Murphy, W. R. Shields, E. L. Garner,* J. Res. NBS *72A*, 26 – 267 (1968).

Lipase

cf. Pancreas Lipase, p. 422.

Lithium Carbonate

NBS-No SRM 924

Stability: the material should be stored in the well-closed original bottle under normal laboratory conditions. Stable indefinitely.

Standard solution 1.0 mmol/l: SRM 924 is dried for 4 h at 200 °C then cooled to room temperature in a dessicator. Dissolve 73.91 mg SRM 924 in 50 ml of de-ionized water and 20 ml HCl, 0.1 mol/l. Dilute to the mark with de-ionized water and mix well in a 2 l volumetric flask. The solution is stable indefinitely when stored in a well stoppered all-glass container.

Purity: 100.05 ± 0.02% (CO_3^{2-}-determination by coulometric acidimetry).

Lysozyme

FIP-Standard Reference Substance

White crystalline powder, water soluble, available in sealed ampoules.

Working standard solution: 25 mg lysozyme hydrochloride is dissolved in 50 ml bi-distilled and gas-free water. The solution is kept in a thermostat at 25 °C.

Substrate: killed and dried germs (non-pathogenic) of *Micrococcus luteus* ATCC 4698 (FIP controlled). The substrate must be stored at −20 °C and used within one year. It is recommended to keep small quantities in sealed ampoules *in vacuo* or under nitrogen.

Working substrate suspension: 50 mg *Micrococcus luteus* ATCC 4698 is suspended in 100 ml phosphate buffer, 0.15 mol/l, pH 7.0, containing NaCl, 0.017 mol/l. The suspension is filtered over glass wool, the resulting solution is maintained at 25 °C and the absorbance is adjusted to 0.66 at 450 nm.

Assay principle: the decrease of absorbance at 450 nm, 10 mm path length, of a mixture of 3 ml working substrate suspension and 10 μl of the working standard lysozyme solution is followed at 25 °C for 2 minutes.

Unit definition: one conventional unit of enzyme represents a decrease of absorbance of 0.001 per minute.

Specification of FIP-lysozyme:
Electrophoretically pure.
Nitrogen content: *Kjeldahl* method 18%; formaldehyde titration of amino acids 0.4%. Water content < 1%; ash negligible.

Magnesium Gluconate Dihydrate

NBS-No SRM 929

Stability: the material should be stored in the tightly-closed original bottle under normal laboratory conditions. The material is hygroscopic and must be dried before use; such drying will not remove water of hydration. Stored under these conditions, the material will show no significant change in properties. If not stated otherwise by NBS, the user should not use the material beyond 5 years after the purchase date.

Standard solution: a standard solution of magnesium, 5 mmol/l, is prepared by placing 1.125 g dried SRM 929 in a 500 ml volumetric flask and dissolving the material with reagent-grade water [1]. Lower concentrations required for analysis may be prepared by dilution. Solutions prepared as above are stable for at least 60 days under normal laboratory conditions.

Purity: the material is highly purified magnesium gluconate dihydrate, $Mg(C_6H_{11}O_7)_2$ $\cdot$ 2 H_2O. Magnesium, weight percent 5.403 $\pm$ 0.022 (thermal ionization isotope-dilution mass spectrometry [2]).

References

[1] National Committee for Clinical Laboratory Standards (NCCLS), Type I.
[2] *E. J. Catanzaro, T. J. Murphy, E. L. Garner, W. R. Shields*, J. Res. NBS *70A*, No 6, 553 – 558 (1966).

D-Mannitol

NBS-No SRM 920

Stability: when stored in a well-closed container at 30 °C or less, protected from direct sunlight, the material is stable for at least 5 years.

Standard solution, 1 μmol/ml: 182.2 mg SRM 920 is transferred to a 1 l volumetric flask, dissolved in sulphuric acid, 0.2 mol/l, and diluted to the mark with sulphuric acid, 0.2 mol/l. This standard solution and the working solutions (dilutions with sulphuric acid, 0.2 mol/l, [1]) should be stored in glass-stoppered brown bottles at 4 °C. Under such conditions these standards should be stable for 6 months.

Purity: D-mannitol 99.8 $\pm$ 0.1% (differential scanning calorimetry; polarimetry, gas-liquid chromatography, thin-layer chromatography); D-glucitol 0.1%; total alditol 99.9%; loss on drying < 0.02%; ash < 0.001%; insoluble matter < 0.001%.

Reference

[1] *S. C. Kanter,* Mannitol as a Primary Standard in the Determination of Triglycerides, Clin. Chim. Acta *16*, 177 – 178 (1967).

Pancreatic Lipase

FIP-Standard Reference Substance

Unit definition: one unit of lipase activity corresponds to that amount of the standard reference substance which under the conditions of the assay liberates one micro-equivalent of fatty acid per minute.

Pancreatic Protease

FIP-Standard Reference Substance

Unit definition: one unit of protease activity corresponds to that amount of the standard reference substance which under the conditions of the assay hydrolyzes the substrate at an initial rate such that an amount of peptides is liberated per minute which is not precipitated by trichloroacetic acid and which gives the same absorbance at 275 nm as one micromole of tyrosine per ml.

Pantothenic Acid

cf. Calcium Pantothenate, p. 405.

Papain

FIP-Standard Reference Substance

Working standard solution: the amount of papain containing about 3 FIP units/ml is dissolved in water and the pH is adjusted to 7.0. The solution is kept in ice-water during the assay.

Assay principle: the specific activity of papain is determined following the hydrolysis of the synthetic substrate N-benzoyl-L-arginine ethyl ester, at pH 7.0 and 25 °C. The consumption of base necessary to neutralize the liberated acid is recorded as a function of time.

Unit definition: one FIP unit of papain is defined as the enzyme activity which, under the specified conditions, hydrolyzes 1 micromole of N-benzoyl-L-arginine ethyl ester per minute.

Pepsin

FIP-Standard Reference Substance

White or slightly yellowish, crystalline, hygroscopic powder with a faint odour, soluble in water, insoluble in alcohol or ether.

Storage: the material should be stored in a desiccator.

Standard reference working solution: a solution of 0.5 units/ml in HCl, 60 mmol/l, is prepared and should be used within 15 min.

Assay principle: the most widely used assay method for pepsin is described by *Anson* [1] using haemoglobin (FIP controlled) in a 2% solution (w/v) as substrate.

Unit definition: one FIP unit of pepsin activity is contained in that amount of the standard preparation which upon incubation at 25 °C for 1 min with a suitable preparation of pure haemoglobin causes the decomposition of the haemoglobin to such an extent that the amount of hydroxyaryl substances liberated reacts with *Folin-Ciocalteu* reagent to form a coloured solution of intensity equal to that resulting from the reaction of 1 μmol of tyrosine with the reagent.

Standard tyrosine solution: weigh 60.4 mg pure tyrosine, dissolve in HCl, 0.1 mol/l, and make up to 1000 ml with HCl. This solution contains 1 μmol/3 ml. Pipette 3 ml aliquots into 20 ml distilled water and mix. Add 1 ml NaOH, 3.85 mol/l, and 1 ml *Folin-Ciocalteu* reagent. After 15 min read the absorbance at 540 nm against a blank of 23 ml distilled water, 1 ml NaOH, 3.85 mol/l, and 1 ml *Folin-Ciocalteu* reagent.

Purity: characterized by determination of specific activity, between 0.5 and 0.7 units/mg. Water content < 5%. A sample of 10 g pepsin may not contain any germ of *Salmonellae*, nor a sample of 1 g pepsin any germ of *Escherichia coli*.

Reference

[1] *M. L. Anson,* J. Gen. Physiol. *22*, 79, 1938.

Plasmin

FIP-Standard Reference Substance

Stability of plasmin solution: stable for several days at pH 1 to pH 4. After addition of lysine to plasmin a product is obtained which is soluble at pH 4 to pH 9. This solution is stable for several hours at room temperature.

Working standard solution, 0.04 – 0.09 FIP units/ml: dissolve the appropriate amount of plasmin in distilled water, add 10% phosphate/lysine buffer (66.6 mmol/l, 100 mmol/l, pH 7.5) and dilute with distilled water to volume. The enzyme solution loses approx. 10% of its activity per hour at 25°C.

Assay principle: casein (FIP controlled) is hydrolyzed by plasmin at pH 7.5 and 35.5°C for 2 min (blank) and 22 min (sample). The reaction is stopped by addition of perchloric acid and the undigested precipitated casein is removed by filtration. The amount of peptides remaining in solution is determined spectrophotometrically at 275 nm.

Unit definition: one FIP unit of plasmin is defined as the enzyme activity which, under the specified conditions, gives rise in 1 min to the formation of perchloric acid soluble peptides with an absorbance at 275 nm equal to that of a tyrosine solution, 1 mmol/l.

Plasmin, Human

WHO, 1st International Reference Preparation (IRP) of Plasmin Human, Code No 72/379.

The standard consists of a partially purified, glycerol-activated preparation of human plasma plasminogen [1].

Storage: the material should be stored at 0°C or below.

Standard solution: each ampoule of the batch of standard contains approximately 1 ml of a solution of the plasmin preparation in 50% glycerol solution. One ampoule contains 8 IU [2].

References

[1] *J. T. Sgouris, J. K. Inman, K. B. McCall, L. A. Heyndman, H. D. Anderson* (1960), Vox Sang., *5*, 357.
[2] *T. B. L. Kirkwood, P. I. Campbell, P. I. Guffney*, Thrombos., Diathes. haemorrh. (Stuttgart), *34*, 20 (1975).

Potassium Chloride

NBS-No SRM 918

Stability: the material should be stored in the well-closed original container under normal laboratory conditions. It is stable indefinitely. It is recommended that weighing and other manipulations of the solid SRM 918 not be made when the relative humidity exceeds 75%.

Standard solution, 100 mmol/l: 7.46 g SRM 918 are transferred to a 1 l volumetric flask, 3 ml concentrated nitric acid and 100 ml de-ionized water are added. After all the salt is dissolved, the flask is filled up to the mark with de-ionized water. This solution is stable indefinitely when stored in a well-stoppered all-glass container.

Purity: 99.9 ± 0.0% (potassium, gravimetry and isotope dilution analysis, chloride by a coulometric argentimetric procedure).

Protease

cf. Pancreatic Protease, p. 422.

Pyruvate

cf. Sodium Pyruvate, p. 426.

Retinol

cf. Vitamin A, p. 431.

Riboflavin

WHO-Collaborating Center; Contr. No 382035

Proposed ICRS (International Chem. Ref. Substance).

Storage: the substance can be stored at ambient temperature, protected from light. It is not hygroscopic.

Purity: 99.4 ± 0.3% (HPLC); total impurities 0.9%; lumiflavin not detected; loss on drying (105 °C) 0.3%; ash 0.0 mg/g; ratio of absorbance $A_{374}/A_{267} = 0.320$; ratio of absorbance $A_{444}/A_{267} = 0.376$; specific rotation (dried substance) $-126.7°$ (at 20 °C in a 5 mg/ml solution of NaOH, 0.05 mol/l), $+57.9°$ (at 25 °C in a 5 mg/ml solution of HCl, 420 g/l).

Serum

cf. Human Serum, p. 413, and Human Serum Standard, p. 416.

Sodium Chloride

NBS-No SRM 919

Stability: the material should be stored in the well closed original container under normal laboratory conditions. It is stable indefinitely. It is recommended that weighing and other manipulations should not be made when the relative humidity exceeds 60%.

Standard solution, 100 mmol/l: (suitable for both, sodium and chloride determinations): 5.85 g SRM 919 (dried at 110°C) are placed in a 1 l volumetric flask, 3 ml concentrated nitric acid and 100 ml de-ionized water are added. After NaCl is dissolved, dilute to the mark with de-ionized water. This solution is stable indefinitely when stored in a well-stoppered all-glass container.

Purity: 99.9 ± 0.0% (sodium assayed by gravimetric procedure via conversion to sodium sulphate).

Sodium Pyruvate

NBS-No SRM 910

Storage and stability: the material should be stored in the tightly capped bottle at 2–6°C. Before opening, it should be allowed to warm to room temperature. Under proper storage this material should be stable for at least 5 years.

Standard solutions: stock solutions (250 mg/ml) and dilutions thereof should be prepared daily with distilled water. The pH of a solution of 250 mg/ml in distilled water is 5.90 ± 0.06. Such solutions are stable for several weeks at −20°C. However, at +20°C and at an initial pH of 5.90, the concentration of para-pyruvate doubles in 5 days, and triples within 24 h if the pyruvate is dissolved in potassium phosphate, 0.25 mol/l, at pH 8.0 and +20°C.

Purity: sodium pyruvate 98.7 ± 0.2 weight % (HPLC); sodium 2-hydroxy-4-keto-2-methylpentanedioate (para-pyruvate) 0.9 ± 0.2% (HPLC, high field proton NMR spectroscopy); moisture 0.28%; methanol 0.21%; water, insoluble matter 0.004%.

Sorbitol

USP-RS; Cat. No 6170

Storage: the material is preserved in air-tight containers. It is to be used without drying.

Standard solution: 75 mg USP-sorbitol-RS is dissolved in 5 drops of water, and diluted with methanol to 50.0 ml to obtain a solution with a concentration of about 1.5 mg/ml.

Purity: sorbitol $91.0\% - 100.5\%$, $C_6H_{14}O_6$, calculated on the anhydrous basis. It may contain small amounts of other polyhydric alcohols. Water $\leqslant 1.0\%$; ash $\leqslant 0.1\%$; chloride $\leqslant 0.005\%$; sulphate $\leqslant 0.010\%$; arsenic 0.0003%; heavy metals $\leqslant 0.001\%$.

Standardized Protein Solution (Bovine Serum Albumin)

NCCLS Approved Standard ASC-1

Bovine serum albumin which meets the NCCLS requirements and a standardized protein solution which also meets the NCCLS requirements are available from the National Bureau of Standards as SRM No 926 and SRM No 927, respectively.

Storage: the standardized protein solution is packaged in sterile conditions under an inert gas in sealed glass ampoules. It must be stored refrigerated ($2\,°C$ to $8\,°C$). Stored ampoules must be brought to room temperature before opening and used within a working day after opening [1].

Reference

[1] Specifications for Standardized Protein Solution (Bovine Serum Albumin) National Committee for Clinical Laboratory Standards, Second edition 1979. 771 E. Lancaster Avenue, Villanova, PA. 19085.

Thiamine Hydrochloride

USP-RS; Cat. No 6560

Storage: the material is preserved in air-tight, light-resistant containers. Do not dry; determine the water content titrimetrically at the time of use.

Standard stock solution, 25 mg/l: 25 mg USP-thiamine-HCl-RS is transferred to a 1000 ml volumetric flask, observing precautions to avoid absorption of moisture in weighing the dried standard. The weighed material is dissolved in about 300 ml aqueous ethanol (1 in 5) adjusted to pH 4.0 with HCl, 3 mol/l, and the acidified dilute ethanol is added to volume.

The solution is stable for 1 month when stored refrigerated in a light-resistant container.

Standard solution, 0.2 mg/l: a portion of the standard stock solution is diluted quantitatively and stepwise with HCl, 0.2 mol/l, to obtain the concentration of 0.2 mg/l.

Purity: thiamine hydrochloride $98.0-102.0\%$ $C_{12}H_{17}ClN_4OS \cdot HCl$, calculated on the anhydrous basis (method: fluorescence); water $< 5.0\%$; ash $< 0.2\%$.

Thromboplastin, Human Brain

(for prothrombin time determination on blood plasma)

BCR No 147 (identical with BCT/099 [1]).

Lyophilized material from 1 ml aliquots of the aqueous extract of homogenized human brain tissue without calcium added.

Storage and stability: the evacuated rubber-capped vials should be stored at $-20\,°C$. The material is stable for 12 months.

Standard solution: reconstitution of the contents of one vial is performed by the addition of 1 ml of the phenolized water provided, which is stored at $+4\,°C$. The reconstituted sample is kept at room temperature and must be used within 2 h. Calibration of the BCR reference material No. 147 was performed against the WHO IRP 67/40 [2].

Purity: water $< 2.0\%$; haemoglobin concentration is below the detection limit.

References

[1] *J. M. Thomson, I. S. Chart,* The System for Laboratory Control of Oral Anticoagulant Therapy Using the British Comparative Thromboplastin (BCT). J. Med. Lab. Technol. *27*, 207 (1970).
[2] *E. A. Loeliger et al.,* Thromboplastin Calibration. Experience of the Dutch Reference Laboratory for Anticoagulant Control, Thromb. Haemostasis *42*, 1141 (1978).

Tris(hydroxymethyl)aminomethane
Tris(hydroxymethyl)aminomethane Hydrochloride

NBS-Nos: SRM 922 and SRM 923, resp.

Storage and stability: SRM 922 should be stored in a well-closed container at room temperature, protected from high temperatures (above $50\,°C$) and direct sunlight. Properly stored, the material is stable for at least 12 years.

SRM 923 is hygroscopic and should be stored in a desiccator in a well-closed container at room temperature. Exposure to higher temperatures (above 40 °C) and direct sunlight should be avoided. Properly stored, this material is stable for at least 5 years.

Standard buffer solution: this solution for the calibration of pH equipment is 0.01667 molal with respect to Tris and 0.0500 molal with respect to Tris-HCl. Such buffers are used in cases where phosphate buffers would cause undesirable side reactions or do not approximate to the desired variation of pH with temperature.

The pH of this solution as a function of temperature is as follows:

°C	0	5	10	15	20	25	30	35	37	40	45	50
pH	8.471	8.303	8.142	7.988	7.840	7.699	7.563	7.433	7.382	7.307	7.186	7.070

Preparation: transfer 7.800 g Tris-HCl and 1.999 g Tris to a 1 l volumetric flask. Dissolve and fill to the mark with distilled water (25 °C) which should be free from CO_2 and should have a conductivity $< 2 \times 10^{-6}\,Ohm^{-1}\,cm^{-1}$. If the solution is to maintain the assigned pH for a few weeks, the exclusion of CO_2 is essential.

Purity: Tris: 99.99 ± 0.02 mole % after drying; Tris-HCl: 99.69 ± 0.05 mole % (by coulometric assay).

Trypsin

Chemical Reference Substance; identical with FIP-Standard Reference Substance

The standard is prepared from bovine pancreas glands. It is a colourless, crystalline hygroscopic powder.

Stability: when stored in the dry state at 4 °C the shelf-life will be two years.

Standard enzyme solution: an amount consisting of approx. 50 FIP units of trypsin/ml is dissolved in HCl, 1 mmol/l.

Assay principle: the specific activity is determined using the hydrolysis of the synthetic substrate N-benzoyl-L-arginine ethyl ester under standard conditions at pH 8.0 and 25 °C. The base consumed necessary to neutralize liberated acid is recorded as a function of time.

Unit definition: one FIP unit of trypsin activity is contained in that amount of the standard preparation which hydrolyzes one micromole of the substrate per minute under the specified conditions.

Purity: specific activity not less than 40 FIP units/mg. Chymotrypsin activity < 5% (titration of N-acetyl-L-tyrosine ethyl ester at pH 8.0 and 25 °C). Residue on ignition at 600 °C for 24 h < 2.5%. Loss on drying *in vacuo* (60 °C, 4 h) < 2% of its weight.

Urea

NBS-No SRM 912a

Stability: refrigerated storage in a tightly-closed container to prevent change in moisture content is recommended. It should not be exposed to heat or direct sunlight. Under proper storage urea is stable for at least 5 years.

The standard urea nitrogen solution (20 mg/100 ml) is stable for three months when stored at +4 °C in an all-glass container.

Standard solution: a standard solution containing 20 mg urea nitrogen per 100 ml is prepared from 0.429 g urea in 1000 ml of ammonia-free distilled water.

Purity: 99.9 ± 0.1% (differential scanning colorimetry [1]); moisture 0.02 ± 0.003%; biuret 0.02 ± 0.02%; ash 0.001 ± 0.0007%; insoluble matter 0.0001 ± 0.00005%.

Reference

[1] *C. Plato, A. R. Glasgow Jr.,* Anal. Chem. *41*, 330 (1969).

Uric Acid (2,6,8-Trihydroxypurine)

NBS-No SRM 913

Stability: stored in well-closed containers at +30 °C or below, protected from heat and sunlight, the material is stable for at least 5 years.

The stock standard solutions (1 mg/ml) are stable for 3 months, when stored in an all-glass dark brown bottle at +4 °C.

Standard solution, 1 mg/ml: uric acid (SRM 913) dissolved in lithium carbonate (SRM 924) solution, containing formaldehyde and sulphuric acid [1, 2].

Purity: 99.7 ± 0.1% (paper-, thin-layer chromatography, emission spectrometry, neutron activation, UV absorption); volatile matter 0.014%; ash 0.057%.

References

[1] *W. T. Caraway,* Uric Acid, in: *David Seligson* (ed.), Standard Methods of Clinical Chemistry, Vol. *4,* Academic Press, New York, 1963, pp. 239–247.
[2] *S. Natelson,* Uric Acid, in: *Miriam Reiner* (ed.), Standard Methods in Clinical Chemistry, Vol. *1,* Academic Press, New York 1953, pp. 123–135.

Vanillino Mandelic Acid

cf. 4-Hydroxy-3-methoxy-D,L-mandelic Acid, p. 418.

Vitamin A

USP-RS; Cat. No 7160

Storage: the vitamin-A-containing capsules have to be preserved in air-tight light-resistant containers. Store in a cool, dry place. Portions remaining unused after opening individual capsules have to be discarded.

Standard solution: the contents of 1 capsule of USP-vitamin A-RS are dissolved in chloroform to make 25 ml. One capsule contains approximately 37500 USP units (one USP vitamin A unit is the specific biological activity of 0.3 μg of the *all-trans* isomer of retinol).

Purity: vitamin A capsules contain not less than 95.0% and not more than 120.0% of the labelled amount of vitamin A. The vitamin A assay procedure is described in [1].

Reference

[1] United States Pharmacopeia, USP XX, 1980, p. 933–934.

Vitamin B$_{12}$

cf. Cyanocobalamin (Vitamin B$_{12}$), p. 409.

2.3.2.2 Materials for Instrument Calibration

Absorbance Standard

cf. Crystalline Potassium Dichromate, p. 432; Glass Filters, p. 435; Liquid Absorbance Standards for UV and Visible Spectrophotometry, p. 436, Metal-on-quartz Filters for Spectrophotometry, p. 436, and 4-Nitrophenol, p. 436.

Clinical Laboratory Thermometer

NBS-No SRM 933

This SRM is intended for use as a primary calibrant, particularly in the area of clinical enzymology. It consists of three individually calibrated thermometers. Each thermometer, 180 ± 5 mm in length, has an auxiliary scale from − 0.20 to + 0.20 °C with 0.05 °C divisions. The main scales of the thermometers are 24.00 to 26.00 °C, 29.00 to 31.00 °C, and 36.00 to 38.00 °C with 0.05 °C divisions.

Calibrated points are 0 °C, and 25, 30, or 37 °C depending on the scale of the individual thermometer [1].

Reference

[1] *B. W. Mangum,* Clin. Chem. *20,* 670 (1974).

Clinical Laboratory Thermometer

NBS-No SRM 934

This SRM is intended for use as a primary calibrant. The thermometer has a length of 300 ± 5 mm and is individually calibrated at four points. It has an auxiliary scale from − 0.20 to + 0.20 °C with 0.05 °C divisions. The main scale extends from 24.00 to 38.00 °C in 0.05 °C divisions. The calibration points are 0, 25, 30, and 37 °C [1].

Reference

[1] *B. W. Mangum,* Clin. Chem. *20,* 670 (1974).

Crystalline Potassium Dichromate for Use as an Ultraviolet Absorbance Standard

NBS-No SRM 935

Solutions of known concentrations of this SRM in perchloric acid, 0.001 mol/l, are certified for their apparent specific absorbances at 23.5 °C for a 10 mm internal pathlength.

Apparent specific absorbance (kg $\times$ g^{-1} $\times$ cm^{-1})

Nominal concentration g $\times$ kg^{-1}	Wavelength and (bandpass) nm					Uncertainty
	235.0(1.2)	257.0(0.8)	313.0(0.8)	345.0(0.8)	350.0(0.8)	
0.040	12.304	14.318	4.811	10.603	10.682	± 0.020
0.080	12.390	14.430	4.821	10.601	10.701	± 0.020

Stability: solutions prepared in the concentration range of 0.02 to 0.1 g/kg and made according to the instructions given in [1] are stable for at least 6 months when stored at room temperature and protected from evaporation and light.

Reference

[1] *R. W. Burke, R. Mavrodineanu,* Certification and Use of Acidic Potassium Dichromate Solutions as an Ultraviolet Absorbance Standard, NBS Specific Publication 260-54 (1977), Office of Standard Reference Materials, National Bureau of Standards, Washington D.C. 20234, USA.

Crystalline Potassium Iodide
Crystalline Potassium Iodide with Attenuator

(heterochromatic and isochromatic, resp.; stray radiant energy standard for ultraviolet absorption spectrophotometry)

NBS-Nos: SRM 2032 and SRM 2033, resp.

Storage and stability: the materials should be stored in the original low actinic glass bottles and should be protected from exposure to light and humidity. The expected stability under proper storage is at least 3 years.

Solutions: solutions should be made in borosilicate glass containers using distilled water and transfer pipettes (*Pasteur* type) of the same glass and fitted with rubber bulbs. Fresh solutions should be made before each test.
   A 1% KI solution (c = 10 g/l) with 10 mm path length exhibits a sharp cut off in transmittance near 260 nm, i.e. it transmits more than 90% above 273 nm, but less than 0.01% below 258 nm. Therefore, with the monochromator set for a wavelength below 260 nm, any appreciable amount of light detected is heterochromatic stray-light which consists of wavelengths above the cut-off.

Purity: 99.8%.

Use of the Attenuator: this unit can be used to attenuate the incident radiation in the reference beam of the spectrophotometer in two steps by a total factor of about 100 [1]. Thus, in an instrument that cannot measure transmittances lower than 1%, the use of the attenuator will permit heterochromatic stray-light to be measured down to about 0.01%.

Reference

[1] "Estimating Stray Radiant Energy" ASTM Manual on Recommended Practices in Spectrophotometry, 3rd edition, 1916 Race St., Philadelphia, PA. 19103, USA 1969, pp. 94−105.

Cuvette

cf. Quartz Cuvette for Spectrophotometry, p. 438.

Didymium Glass Filter for Checking the Wavelength Scale of Spectrophotometers

NBS-Nos SRM 2009, SRM 2010, SRM 2013, SRM 2014

The SRMs noted above are identical in their general descriptions and show only slight differences from lot to lot.

These SRMs consist of rare earth oxides in a glass matrix. The filters are intended to be used in calibrating the wavelength scale in the visible spectral region (380 nm to 780 nm) for spectrophotometers having nominal bandwidths in the range 1.5 to 10.5 nm.

Fluorescence Calibrator

cf. Quinine Sulphate Dihydrate, p. 438.

Gallium Melting Point Standard

NBS-No SRM 1968

SRM 1968 should be used as a calibrant in a temperature-regulated bath. The cell containing the metallic gallium (approximately 25 g) consists of a Teflon body, a nylon well, and a nylon capstem. Filling, assembling, and sealing of the cell have been performed in a dry argon atmosphere.

Melting point 29.7723 ± 0.0007 °C (International Practical Temperature Scale of 1968 [1]).

Purity: 99.99999%.

Reference

[1] *D. D. Thornton,* The Gallium Melting Point Standards: A Determination of the Liquid-Solid Equilibrium Temperature of Pure Gallium on the International Practical Temperature Scale 1968, Clin. Chem. *23*, 719 – 724 (1977).

Glass Filters for Spectrophotometry

NBS-No SRM 930 D

This SRM is intended as a reference source for the verification of the transmittance and absorbance scales of spectrophotometers. It consists of three individual glass filters in their metal holders. The transmittance values (given for 440 nm, 465 nm, 546.1 nm, 590 nm, and 635 nm) have been measured against air at an ambient temperature of 23.5 °C.

The transmittance values are estimated to be accurate within ± 0.5% at the time of certification. Ageing of the glass may cause some filters to change transmittance by about ± 1% over a period of approximately one year from the date of calibration.

Reference

see description of SRM 2030.

Glass Filter for Transmittance Measurement

NBS-No SRM 2030

This SRM is intended as a reference source for one-point verification of the transmittance and absorbance scales of spectrophotometers at a given wavelength (465.0 nm) and measured transmittance (nominal 30%).

The transmittance value given was measured against air at an ambient temperature of 23.5 °C, and is estimated to be accurate within ± 0.5% at the time of certification. Ageing of the glass may cause some filters to change transmittance by about ± 1% over a period of approximately one year from the date of calibration. Improper storage or handling of the filters may cause changes [1].

Reference

[1] *R. Mavrodineanu, J. R. Baldwin,* Glass Filters as a Standard Reference Material for Spectrophotometry; Selection, Preparation, Certification, Use, NBS Spec. Publ. 260-51 (1975), Office of Standard Reference Materials, National Bureau of Standards, Washington, D.C. 20234, USA.

Holmium Oxide for Use as Wavelength Standard in Spectrophotometry and Fluorescence Spectrometry (planned SRM)

NBS-No SRM 2034

Planned SRM (issue projected in 1982)

Material: holmium oxide of established purity dissolved in perchloric acid.

Composition: holmium oxide is to be offered in sealed 10 mm quartz cuvettes.

Liquid Absorbance Standards for Ultraviolet and Visible Spectrophotometry

NBS-No SRM 931b

This SRM is certified as solutions of known net absorbance at specific spectral wavelengths. The material (three solutions and a blank) is applicable for calibrating those instruments that provide an effective spectral bandpass of 1.5 nm at 302 nm, <2 nm at 395 nm, <3.3 nm at 512 nm or <6.5 nm at 678 nm [1].

It is recommended that this material should not be used after three years from the date of purchase.

The uncertainties of the certified absorbance values, about 0.3, 0.6, and 0.9, amount to ± 0.002 to ± 0.003, to ± 0.003 to ± 0.004, and to ± 0.003 to ± 0.005, respectively.

Absorbances have been measured using 10.00 nm cuvettes (SRM 932) at 25 °C.

Reference

[1] *R. W. Burke, E. R. Deardorff, O. Menis,* J. Res. NBS *76A,* 469 – 482 (1972).

Metal-on-quartz Filters for Spectrophotometry

NBS-No SRM 2031

This SRM is intended for use in the verification of the transmittance and absorbance scales of spectrophotometers in the ultraviolet and visible regions. It consists of three individual filters in their metal holders and one empty filter holder.

The transmittance values for the filter with a nominal transmittance of 10% are estimated to be accurate within $\pm 1\%$, with nominal transmittance of 30% and 90% within $\pm 0.5\%$ [1].

Ageing of the material may cause some filters to change transmittance with time.

Reference

[1] *R. Mavrodineanu, J. R. Baldwin,* Metal-on-Quartz Filters as a Standard Reference Material for Spectrophotometry, NBS Spec. Publ. 260-68, U. S. Government Printing Office, Washington, D.C. 20402 (1979).

4-Nitrophenol

NBS-No SRM 938

Storage and stability: SRM 938 should be stored in the original low-actinic glass bottle at room temperature (30 °C or less). Refrigerated storage is recommended. It should not be exposed to heat, moisture, or direct sunlight. Under proper storage this material should be stable for at least 5 years.

Standard stock solution: approximately 140 mg of the SRM are weighed to the nearest 0.01 mg and transferred to a dry, one litre flask weighed to the nearest 0.1 g. Approximately one kg of either type I or type II Clinical Laboratory Reagent Water [1] is added, the crystals are dissolved and flask and contents are weighed to the nearest 0.1 g.

Working solution: a plastic syringe is used as a weigh-burette to weigh a 30 to 35 ml aliquot of the stock solution to the nearest 0.1 mg. Transfer to a one litre volumetric flask and reweigh the syringe. 10 ml of aqueous NaOH, 1 mol/l, are added and the flask is filled to the mark with the water. A blank solution is prepared using the same NaOH solution and water.

Purity: 99.75% (titration with alkali; nuclear magnetic resonance spectroscopy). Apparent specific absorbance ε_a at 401 nm and 23.5 °C is $13.148 \pm 0.033 \, l \times g^{-1} \times mm^{-1}$ (solution containing 4 to 5 mg/l of the SRM in aqueous sodium hydroxide, 0.01 mol/l).

Reference

[1] "Specifications for Reagent Water Used in Clinical Laboratory; NCCLS Approved Standard ASC-3", (1980), National Committee for Clinical Laboratory Standards, Villanova, PA., USA.

Potassium Fluoride

Standard for ion-selective electrode

NBS-No SRM 2203

Storage and stability: potassium fluoride is hygroscopic and therefore it is recommended to store SRM 2203 in a desiccator. No limitations on stability are given.

Standard solutions: before use the hygroscopic KF should be dried in two consecutive stages: initially for 2 h at 110 °C, followed by drying at 200 $\pm$ 10 °C for an additional 2 h. After drying, normal precautions should be exercised to prevent pickup of moisture during weighing. To prepare a 1.0 molal solution, 57.363 g KF (weight in air) is transferred to a 1 litre volumetric flask, dissolved and filled to the mark with distilled water at 25 °C. The distilled water should have a conductivity < 2×10^{-6} Ohm^{-1} $\times$ cm^{-1}. Appropriate dilution of this stock solution should be made to obtain standards in the concentration range of interest to the user. After preparation, standard solutions should be transferred to non-silicate storage containers to avoid fluorosilicate formation and a resulting decrease in fluoride ion activity.

Purity: 99.5% after drying. The material is not entirely free from traces of chloride, fluorosilicates, and heavy metals.

Potassium Iodide

cf. Crystalline Potassium Iodide, p. 433.

Quartz Cuvette for Spectrophotometry

NBS-No SRM 932

This SRM consists of a single accurately calibrated cuvette entirely made of fused silica that is issued for use in the production of accurate spectrophotometric data on liquids.
  Path length (nominal 10 mm) and parallelism are certified with an uncertainty of ± 0.0005 mm determined at 20 °C.

Quinine Sulphate Dihydrate

NBS-No SRM 936

This SRM is intended for use in the calibration of fluorescence spectrometers.

Stability: the material should be kept in its original bottle and stored in the dark at room temperature (30 °C or less). It should not be subjected to heat or direct sunlight. Under proper storage this material is stable for at least 3 years.

Standard stock solution: 0.1 mg/ml or 1.28 $\times$ 10^{-4} mol/l: 0.1 g SRM 936 is weighed into a 1000 ml volumetric flask, and diluted to the calibrated volume with $HClO_4$, 0.105 mol/l.

Working standard solution, 1 µg/ml or 1.28 $\times$ 10^{-6} mol/l: 10 ml of the stock solution are transferred to a 1000 ml volumetric flask and diluted to the calibrated volume with $HClO_4$, 0.105 mol/l.
  Both solutions are stored in the dark in well stoppered glass bottles. When stored as specified the stock standard solution is stable for 3 months, the working standard solution is stable for 1 month.

Ultraviolet absorption spectrum of SRM 936 in $HClO_4$, 0.105 mol/l:

250.0 nm $\varepsilon_{max} = 5699 \pm 9$ $l \times mol^{-1} \times mm^{-1}$

347.5 nm $\varepsilon_{max} = 1081 \pm 2$ $l \times mol^{-1} \times mm^{-1}$

on the side of a peak:

365.0 nm $\varepsilon_{abs} = 692 \pm 1$ $l \times mol^{-1} \times mm^{-1}$

Purity: quinine sulphate dihydrate $> 99.5\%$ (thin-layer chromatography); impurity (believed to be dehydroquinine sulphate dihydrate which has optical characteristics similar to the main component) 1.7% (HPLC); water $4.74 \pm 0.05\%$.

Stray light standard

cf. Crystalline Potassium Iodide, p. 433.

Temperature calibration

cf. Gallium Melting Point Standard, p. 434.

Thermometer

cf. Clinical Laboratory Thermometer, p. 432.

Wavelength calibration

cf. Didymium Glass Filter, p. 434, and Holmium Oxide, p. 435.

3 Evaluation of Experimental Data and Assessment of Results

3.1   Evaluation of Experimental Data

Hans Ulrich Bergmeyer and Marianne Graßl

The measuring techniques described in detail in the following volumes yield experimental values that cannot be compared directly with one another and cannot serve by themselves as a basis for conclusions. These values must first be transformed to experimental results, which are then expressed as conventional, understandable quantities. The following sections were written with photometric measurements in mind, but their content is also valid, in principle, for all other measuring techniques.

3.1.1   Experimental Data

The data provided by the measurements often cannot be used directly, but they must be corrected for one or more blank values. Frequently, they are meaningful only in combination with the results of other measurements (standard values, other reference magnitudes). Occasionally they are associated with large errors (for error theory, see p. 460) which are recognized by deviations from the expected course of the reaction. Two such phenomena are described below.

3.1.1.1   Non-constant End-point

In the determination of substance concentrations of substrates and coenzymes by the end-point method, the value of the physical parameter measured should not change after completion of the reaction. However, the absorbance occasionally changes slowly, i.e. it "creeps", for reasons that are not always apparent. Some of the causes are known. Thus the enzyme used may contain contaminating enzymes, which allow a slow reaction of other substances in the sample, or in NAD(P)H-dependent reactions the enzyme used may contain "NAD(P)H oxidase". An interfering reaction is therefore superimposed on the reaction to be measured. This may run in the same or in the opposite direction. Accordingly, instead of a constant end-point, there is an increase or decrease in the measured effect (Fig. 1, p. 443).

The additional change of absorbance, which often shows a linear variation with time, can easily be eliminated by graphical extrapolation to t_0 (the time of the start of the reaction). If no recorder is used, absorbance readings are taken at regular intervals, e.g. one minute, the values are plotted against time, and the linear portion of the resulting curve is extended back to time t_0. The point of intersection with the ordinate axis at time t_0 gives the correct end-point and hence the true absorbance change ΔA.

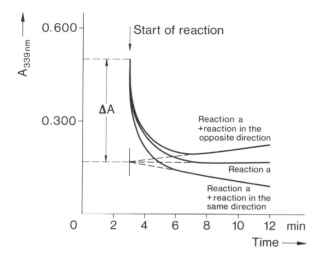

Fig. 1. End-point method for the determination of substance concentrations. Extrapolation when the end-point is not constant (schematic).

If the readings are taken or printed automatically at certain time intervals (e.g. every minute) the absorbance change (ΔA) caused by an interfering reaction can also easily be calculated. The ΔA of the "creep reaction" – related to the stated time interval – is deducted from the total ΔA observed. The principle may be demonstrated by an example:

A reaction is followed for 9 min after the start. After approx. 4 min a constant ΔA per min was observed. If A_1 is the absorbance read at the start of the reaction and A_2 the absorbance after 9 min

$$\Delta A_{\text{total}} = A_2 - A_1 = \Delta A_{\text{sample}} + \Delta A_{\text{"creep reaction"}}$$

and

$$\Delta A_{\text{sample}} = \Delta A_{\text{total}} - \Delta A_{\text{"creep reaction"}} .$$

The ΔA due to the "creep reaction" is calculated from the mean $\Delta A/\Delta t$ after the change of absorbance per unit time became constant (t in minutes). Then

$$\Delta A_{\text{"creep reaction"}} = (\Delta A/\Delta t) \times 9 .$$

Subtraction of the ΔA due to the "creep reaction" must take account of the direction of the reaction; i.e. a creep reaction in the same direction as the main reaction will lead to an overestimate of the true change, whereas a creep in the opposite direction will cause the uncorrected change to be underestimated.

3.1.1.2   Non-linear Reaction Curves

The form of the reaction curve is important in the determination of the catalytic activity of enzymes. By definition, the catalytic activity concentration of an enzyme is given by the rate of the reaction that it catalyzes, and in particular by the initial rate v_i.

If the reaction curve is non-linear, however, the reaction rate $\pm \Delta c/\Delta t$ or $\pm \Delta A/\Delta t$ depends on the position of the measurement interval on the time scale [1]. Any curvature indicates that the rate changes during the reaction, the direction of the change usually being a decrease as the reaction progresses. The greater the curvature, the more difficult it is to determine the initial rate accurately. Approximate values, which are usually adequate for practical purposes, can be obtained from several experimental readings taken shortly after the start of the reaction. These values may be extrapolated graphically to $t = 0$ (Fig. 2b).

To obtain accurate values for the initial rate, a tangent is drawn to the curve at $t = 0$ (Fig. 2a), using a mirror ruler. This is a surface mirror perpendicular to the plane of the graph; it is placed across the reaction curve and pivoted until the mirror image forms an unbroken continuation of the curve. A line drawn along the edge of the mirror is perpendicular to the tangent. If measurements are not carried out at regular intervals (in the "two-point method"), the error increases with increasing

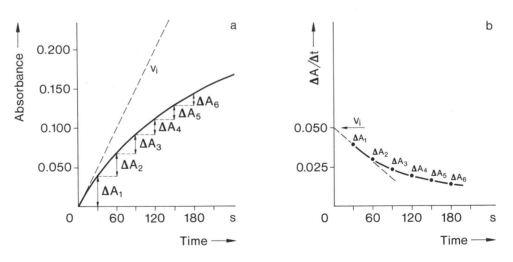

Fig. 2. Continuous measurement for the determination of catalytic activity of enzymes. Evaluation of the initial rate v_i when the reaction curve is non-linear. a) reaction curve with tangent at $t = 0$. b) Evaluation of v_i by extrapolation of the values for $\Delta A/\Delta t$ from the curve a versus $t = 0$. $t = 30$ s = constant.

curvature of the reaction curve. In the case of a curve with a lag phase (cf. Vol. I, p. 83) – assuming there is no possibility of avoiding this type of curve – an approximate value for the "initial rate" is calculated from the slope of the tangent to the point of inflection (cf. Vol. I, p. 139). Many of the photometers designed for routine clinical chemistry are now equipped with microcomputers and printers which indicate any significant deviation from the linearity. There are a number of instruments on which various methods are already programmed in such a way that the linear part of the reaction curve will be monitored. The corresponding data are used to calculate the final results.

3.1.1.3   Standard Curves

For a stoichiometric reaction, the experimental results can be calculated directly on the basis of the molar absorption coefficient, molar volume or other physical constants. If the reaction does not proceed to completion, standard curves must be constructed. It goes without saying that specified conditions of temperature, pH, and buffering must be maintained. Standard curves for determinations in biological material are best set up by adding the substance to be determined to the biological material, so that allowance may be made for any effects due to other compounds present in sample and due to the sample matrix.

Standard curves should pass through the origin in a plot of the unit of measurement against concentration. If they intersect the ordinate (on which the unit of measurement is plotted), the system already contains traces of the substance to be determined (or of one of the intermediates in the case of coupled reactions), or else the reagent blank has been wrongly prepared and has given too low a value. Intersection of the standard curve with the abscissa (on which the concentration or amount of the substance to be determined is plotted) is equivalent to intersection with the ordinate at negative values. This usually results from an excessively high reading for the reagent blank or from the presence of an impurity that reacts with the substance to be estimated so that the latter cannot be determined quantitatively.

Calibration points should lie on a straight line. When the curve is non-linear, the initial portion is often approximately linear: only this portion should be used. With non-linear standard curves, there is no linear relationship between the values measured and the concentrations. Curved standard plots can sometimes be linearized by using logarithmic, semi-logarithmic, reciprocal, semi-reciprocal, or other graphical presentations. Nowadays computer programs are available for curve fitting, e.g. specially for enzyme-immunoassays. Special care is necessary in the construction of standard curves for the determination of substrates by kinetic methods (cf. Vol. I, p. 182). Strict control of the conditions of measurement (particularly temperature, pH, and quantity of enzyme added) is very important. A single calibration point ("standard value") is sufficient if the "fixed-time" method can be used. The experimental data are referred to the standard (eqn. (d), p. 451).

3.1.1.4   Standards

Extremely pure standard substances are a prerequisite for use as a calibrator and for the construction of a standard curve. Because of the uncertainty associated with purity, the International Union of Pure and Applied Chemistry (IUPAC) has laid down the following classification of such substances (cf. [2]):

Grade A: atomic-weight standard

Grade B: ultimate standard. A substance which can be purified to virtually atomic-weight standard.

Grade C: primary standard. A commercially available substance of purity 100 ± 0.02%.

Grade D: working standard. A commercially available substance of purity 100 ± 0.05%.

Grade E: secondary standard. A substance of lower purity which can be standardized against primary (Grade C) material.

Because of the high quality of the standards and standard solutions, needed to ensure the accuracy and comparability of analytical results, various organizations have continued to discuss this topic and have developed definitions and requirements [3 – 5] for standards. The latest recommendations, intended e.g. especially for clinical chemistry, are given in the following.

Reference material: material of which the property(ies) is (are) exactly established to be used as basic substance for the calculation of analysis results, or for the calibration of an instrument ("calibrator"), or for the quality control of an analysis method (quality control material). The reference material may be supplied as a gas, liquid or solid substance, or as it may be easily prepared from defined materials [3, 4].

Certified reference material: reference material which is certified or accompanied by a certificate in which the values for the properties concerned are confirmed. They are distributed by public or private institutions generally acknowledged as technically competent [3, 4].

Standard: material or solution which will be compared with the sample to determine the concentration or the content of a component present in the sample (calibration standard) [5].

Primary standard: substance of known chemical composition and sufficient purity to be used in preparing a primary standard solution [5].

Secondary standard: substance of insufficiently known chemical composition and purity to be used in preparing a primary standard. The content can be determined indirectly by analysis only [6].

Primary standard solution: solution used as calibration standard in which the concentration is determined solely by dissolving a weighed amount of primary standard material in an appropriate solvent, and making to a stated volume or weight.

Secondary standard solution: solution used as a calibration standard in which the concentration or other quantity has been determined by an analytical method of stated reliability [5].

Tertiary standard solution: solution with an artificial matrix which takes into account the specific properties of the analysis method. It is prepared by dissolving a primary or secondary standard and defined additives in a suitable solvent and is used as calibration solution. The determination of the concentration is carried out with regard to the specific properties of the sample and the applied method [6]. Example: addition of detergents and other ions to calibrating solution for flame photometry.

Sample-resembling standard: standard of biological matrix which contains almost the same main components and by-products as the patient specimen to be investigated [6]. Example: standard based on human serum for the calculation of immunoglobulin fractions [7].

Combination standard: standard suitable for the calculation of several different components. Example: calibrator for Continuous Flow Systems.

Arbitrary standard: calibration standard containing an unknown quantity of specified substance. The content of the material concerned is assigned by agreement. Examples: international biological standards, immunoglobulin standards [5].

Internal standard: a substance, not normally present in the specimen and clearly distinguishable from the analyte, which is added in known amount to the sample, or to both the standard and the sample, for the purpose of correcting results for inaccuracy. Examples: calibrator for gas chromatography; addition of lithium to the diluent for flame photometry.

The requirements for and the application of calibration and control material (standard and standard solution) used for quality control in clinical chemistry have also been outlined by the Expert Panel on Nomenclature and Principles of Quality Control in Clinical Chemistry [8].

For standard materials cf. chapter 2.3, pp. 393 – 439.

3.1.2   Transformation of Data to Results

The result of a measurement should provide specific information on the sample under investigation. For this reason, the processed data should be referred to the measured parameters of the sample (e.g. volume or mass).

3.1.2.1   Definitions

Results for metabolites are usually expressed as a "substance concentration" or "mass concentration" (mol or g per unit volume of the sample), or else, with the unit weight of the sample as the reference quantity, as the "content" (mol per unit weight). IUPAC and IFCC propose "substance content", mol/kg, as a new derived kind of quantity [9]. (Substance content should not be confused with "molality", mol/kg, since this kind of quantity refers to the amount of substance of the component divided by the mass of the solvent). A term "mass content", kg/kg (corresponding to "mass concentration, kg/l) does not exist. The unit "mass fraction", w, with the demension one is used.

Bücher et al. [10] gave the following definitions for biological material (tissue): "The term metabolite content of a tissue should be used when the analytical value is referred to fresh weight of the tissue in question. No account is taken of the distribution of the metabolite within the tissue among the various types of cells and their intracellular compartments, the blood vessels, and the interstitial space. "Metabolite concentration" is used when the reference quantity is the volume of a given tissue compartment in which it is distributed".

The result of the determination of the catalytic activity of an enzyme is expressed by the activity unit referred to the weight or the volume. One unit (U) is the enzyme activity that converts 1 µmol of substrate in 1 min under optimal conditions. This international unit is replaced by the term katal (kat) as defined according to the SI (1 mol/s), cf. Vol. I, pp. 12, 13 and 105. Description of the catalytic activity of an enzyme in terms of the rate of conversion is always possible, but difficulties occasionally arise with the term "mol of substance converted", e.g. for some hydrolases that convert macromolecular substrates and for protein kinases where the molecular weight of the phosphorylated protein has not been established with sufficient accuracy. If an enzyme cannot be saturated with substrate (e.g. catalase), the concentration of the substrate should be given. For enzymes that catalyze definite equilibrium reactions, the substrate used for the measurement should be named. The terms "forward" and "reverse" reaction are not free from ambiguity. The activity of hydrolases, for example, is unambiguously defined only if one specifies which synthetic substrate (and there are many possibilities in some cases) was used (cf. Vol. I, p. 111). The temperature at which the measurements were carried out should also be stated, unless it is fixed by international agreement.

The activites of enzyme preparations are referred to the mass of enzyme protein. Exceptions may be made in some cases, however. For example β-fructosidase is only about 25% protein, the remainder being polysaccharide. Therefore, the activity is referred to the quantity of substance. The "catalytic activity content", formerly "specific catalytic activity" (kat/kg, mkat/g, U/g) is thus obtained.

The appropriate quantity for the catalytic activity of an enzyme per unit volume, corresponding to the substance concentration of metabolites, is the "catalytic activity concentration" (U/l, kat/l, µkat/l). The following symbols and units are equivalent:

Metabolites		*Enzymes*	
Substance concentration c	mol/l	Catalytic activity concentration b	kat/l; U/l
	mmol/l		mkat/l
Mass concentration ρ	g/l		µkat/l
	mg/l		nkat/l
Substance content n_c/m_s	mol/kg	Catalytic activity content z_c/m_s	kat/kg
	mmol/g		(U/g, U/mg)
Mass fraction w	1		mkat/g
	0.001		µkat/g

In general, therefore, the experimental data should be converted to concentrations or contents, and referred to specified reference quantities. Only then can one speak of a "result".

3.1.2.2   Formulae

The use of the basic kind of quantities of the SI (cf. Vol. I, pp. 10, 22) involves some changes in the symbols, quantities and units formerly used. For example, the base unit of length is the metre (m) and thousands or thousandths of it. The path length of a photometer cuvette is in general 10 mm (not 1 cm). Accordingly, the unit of the absorption coefficient also changes ($l \times mol^{-1} \times mm^{-1}$, instead of $l \times mol^{-1} \times cm^{-1}$).

For example, for NADH at 339 nm $\varepsilon = 6.3 \times 10^2 \, l \times mol^{-1} \times mm^{-1}$ instead of $6.3 \times 10^3 \, l \times mol^{-1} \times cm^{-1}$. For practical purposes it is proposed to use the path length $d = 10$ mm and the concentration c in mmol/l (μmol/ml). Then for NADH at 339 nm $\varepsilon \times d = 6.3 \, l \times mmol^{-1}$ applies.

The following symbols, kind of quantities and units are customary in this context. Throughout this book we use v, V for reaction rates and v, V for volumes (deviating from international recommendations). We also write T for transmission of light and T for thermodynamic temperature.

Symbols in radiometry are still not completely coherent with those in other laboratory sciences; also apparently no official symbols exist for various commonly used terms (cf. [11]).

General

c	substance concentration, mol/l
ρ	mass concentration, g/l
n_c/m_s	substance content, mol/kg
w	mass fraction, 1
MW	mass of one millimole or mole of substrate, mg/mmol, g/mol
V	assay volume, l
v	volume of sample used in assay, l
φ	volume fraction of sample in assay (incubation) mixture, v/V
t	time, s (min, h)
z	catalytic activity, kat, U
b	catalytic activity concentration, kat/l, U/l
z_c/m_s	catalytic activity content (specific catalytic activity), kat/kg, U/g
v_i	stoichiometric coefficient

Photometry, fluorimetry, luminometry

A	absorbance
ε	linear millimolar absorption coefficient, $l \times mmol^{-1} \times mm^{-1}$
I	light intensity
F	fluorescence intensity
d	light path, mm

For experimental data obtained by fluorimetric methods, F and ΔF are used instead of A and ΔA. In luminometry I and I_0 are used.

Radiometry

Y	number of counts measured
Z	decay rate, disintegrations per minute, dpm
C_b	background counting rate, cpm
C_p	sample (probe) counting rate, cpm
C_n	net counting rate, cpm
η	counting efficiency, cpm/dpm
X	specific radioactivity, Bq/mol (Ci/mol)
v_r	volume taken for scintillation counting, l
cpm	counts per minute
dpm	disintegrations per minute
2.22×10^{12}	factor for conversion of decay rate Ci to dpm (1 Ci $= 2.22 \times 10^{12}$ dpm)
3.7×10^{10}	factor for conversion of Ci to Bq (1 Ci $= 3.7 \times 10^{10}$ Bq)
2.7×10^{-11}	factor for conversion of Bq to Ci (1 Bq $= 2.7 \times 10^{-11}$ Ci)

Metabolites

Photometry, fluorimetry, luminometry

From *Lambert-Beer*'s law (cf. Vol. I, p. 283) follows

(a) $$c = \frac{\log I_0/I}{\varepsilon \times d} = \frac{A}{\varepsilon \times d} \quad \text{mmol/l} .$$

For chemical reactions this gives

$$c_1 - c_2 = \frac{A_1 - A_2}{\varepsilon \times d} ; \qquad \Delta c = \frac{\Delta A}{\varepsilon \times d} \quad \text{mmol/l}$$

and for complete conversion ($c_2 = 0$)

(a$_1$) $$c = \frac{\Delta A}{\varepsilon \times d} \quad \text{mmol/l} \qquad \text{(in the cuvette)} .$$

For the determination of the concentration of the analyte in the sample the ratio of assay volume : sample volume (V : v) or the volume fraction of sample in the assay (v/V) $= \varphi$), respectively, must be considered:

(b) $$c = \frac{\Delta A \times V}{\varepsilon \times d \times v} = \frac{\Delta A}{\varepsilon \times d \times \varphi} \quad \text{mmol/l} \qquad \text{(in the sample)}$$

(b_1) $$\rho = \frac{\Delta A \times V \times MW}{\varepsilon \times d \times v} = \frac{\Delta A \times MW}{\varepsilon \times d \times \varphi} \quad mg/l \qquad \text{(in the sample)} .$$

The substance content of the sample n_c/m_s (i.e. of the analyte in the material under investigation) is

(c) $$n_c/m_s = \frac{\Delta A \times V}{\varepsilon \times d \times v \times \rho_{sample}} = \frac{\Delta A}{\varepsilon \times d \times \varphi \times \rho_{sample}} \quad mmol/g$$

The mass fraction of the analyte in the sample (g/g) is

(c_1) $$w = \frac{\Delta A \times V \times MW}{\varepsilon \times d \times v \times \rho_{sample}} = \frac{\Delta A \times MW}{\varepsilon \times d \times \varphi \times \rho_{sample}} .$$

If two or more moles of light-absorbing reaction products (e.g. NADH) are formed or consumed per mole substrate that reacts, the corresponding stoichiometric coefficient v_i (cf. Vol. I, p. 10) also appears in the denominator of eqns. (b) and (c). If several moles of one substrate go to form the unit of substance on which the measurement is based, the corresponding factor appears in the numerator.

The result for the sample in relation to a standard is

(d) $$c = \frac{c_{sample \ (measured)}}{c_{standard \ (measured)}} \times c_{standard \ (weighed \ out)} \quad mmol/l$$

or

(d_1) $$c = \frac{A_{sample \ (measured)}}{A_{standard \ (measured)}} \times c_{standard \ (weighed \ out)} \quad mmol/l .$$

From the concentration of the substance in the sample solution (e.g. tissue extract), the content of the substance in the material under investigation is calculated by relating if to its mass concentration in the standard solution:

(e) $$n_c/m_s = \frac{c_{sample \ (measured)}}{\rho_{standard \ (weighed \ out)}} \quad \frac{mmol/l}{mg/l} = mmol/mg ,$$

(f) $$w = \frac{\rho_{sample \ (measured)}}{\rho_{standard \ (weighed \ out)}} \quad \frac{mg/l}{mg/l} = 1 .$$

Example

Determination of fructose 1,6-bisphosphate (molecular weight: 340) in rat liver. Enzymatic analysis with aldolase/triosephosphate isomerase/glycerophosphate dehydrogenase. Two moles of NADH are transformed per mole of fructose-1,6-P_2.

To prepare an "extract", 1 g of fresh liver was homogenized in 7.25 ml of $HClO_4$. With a value of 75% (w/w) for the fluid content of the liver, the volume of the extract is $7.25 + 0.75 = 8.00$ ml. To neutralize and remove perchlorate, 0.2 ml K_2CO_3 solution was added to 6 ml of the extract. The volume of the perchlorate-free extract is thus 6.2 ml. The dilution factor for the extract is $6.2/6.0 = 1.033$, and that for the tissue is $8 \times 6.2/6.0 = 8.267$. The experimental data must be multiplied by these values to express the results per 1 ml of acid extract or per 1 g of tissue.

The measured change in absorbance at 339 nm ($\varepsilon = 0.63 \, l \times mmol^{-1} \times mm^{-1}$) was $\Delta A = 0.120$; the volume of the assay solution was 3×10^{-3} l, the volume of sample was 1.5×10^{-3} l, and the light path was 10 mm. The concentration in the perchlorate-free sample used for the assay was:

according to eqn. (b)

$$c = \frac{0.120 \times 3 \times 10^{-3}}{0.63 \times 10 \times 1.5 \times 10^{-3} \times 2} = 0.0190 \, mmol/l$$

or according to eqn. (c)

$$\rho = \frac{0.120 \times 3 \times 10^{-3} \times 340}{0.63 \times 10 \times 1.5 \times 10^{-3} \times 2} = 6.46 \, mg/l \ .$$

Multiplication by the dilution factor gives the concentration in the acid extract:

$$c = 0.0190 \times 1.033 = 0.0196 \, mmol/l$$

$$\rho = 6.46 \times 1.033 = 6.67 \, mg/l \ .$$

The substance content of fructose-1,6-P_2 in the tissue is

$$n_c/m_s = 0.0190 \times 8.267 = 0.157 \, mmol/kg$$

or its mass fraction (g/g)

$$w = 6.46 \times 10^{-6} \times 8.267 = 5.4 \times 10^{-5} \ .$$

Radiometry

For calculation of the analyte concentration and content from the net counting rate C_n of sample and blank, the following relationships are valid.

Substance concentration

(g) $$c = \frac{\Delta C_n}{\eta \times 2.22 \times 10^{12} \times X \times \varphi \times v_r} \quad mol/l \text{ (in the sample)}$$

(units: C_n in cpm; X in Ci/mol or Bq/mol; $\varphi = v/V$ in l/l; v_r in l).

Mass concentration

(g₁) $$\rho = \frac{\Delta C_n \times MW}{\eta \times 2.22 \times 10^{12} \times X \times \varphi \times v_r}$$ g/l (in the sample)

(units as for eqn. (g); MW in g/mol).

Substance content

(h) $$n_c/m_s = \frac{\Delta C_n}{\eta \times 2.22 \times 10^{12} \times X \times \varphi \times v_r \times \rho_{sample}}$$ mol/g (in the sample)

(units as for eqn. (g); ρ_{sample} in g/l).

Mass fraction

(h₁) $$w = \frac{\Delta C_n \times MW}{\eta \times 2.22 \times 10^{12} \times X \times \varphi \times v_r \times \rho_{sample}}$$

(units as for eqn. (g₁); ρ_{sample} in g/l).

Enzymes

Photometry, fluorimetry, luminometry

For measurement of the catalytic activity z of enzymes the rate of the catalyzed substrate conversion per time unit is used, μmol/min or mol/s.

Concerning the stoichiometric coefficient v_1, cf. p. 451.

According to eqn. (a)

catalytic activity (conversion in mol* per time unit) is

(i) $$z = \frac{\Delta c \times V}{\Delta t} = \frac{\Delta A \times V}{1000 \times \varepsilon \times d \times \Delta t}$$ mol/s (kat)

(units: ε in l × mmol^{-1} × mm^{-1}; V of the assay volume in l; d in mm; t in s)

(i₁) $$z = \frac{\Delta c \times V}{\Delta t} = \frac{\Delta A \times V \times 1000}{\varepsilon \times d \times \Delta t}$$ μmol/min (U)

(units as for eqn. (i); t in min).

---

* $c \times V$ (mol/l) × l = mol.

The catalytic activity concentration in the sample is

(k) $$b = \frac{\Delta A \times V}{1000 \times \varepsilon \times d \times \Delta t \times v} = \frac{\Delta A}{1000 \times \varepsilon \times d \times \Delta t \times \varphi} \quad \text{mol} \times s^{-1} \times l^{-1} \text{ (kat/l)}$$

(units as for eqn. (i); v in l; $\varphi = v/V$ in l/l)

(k$_1$) $$b = \frac{\Delta A \times V \times 1000}{\varepsilon \times d \times \Delta t \times v} = \frac{\Delta A \times 1000}{\varepsilon \times d \times \Delta t \times \varphi} \quad \mu\text{mol} \times \text{min}^{-1} \times l^{-1} \text{ (U/l)}$$

(units as for eqn. (i$_1$); v in l; $\varphi = v/V$ in l/l).

The catalytic activity related to the mass of protein, catalytic activity content z_c/m_s (specific activity) is

(l) $$z_c/m_s = \frac{\Delta A \times V}{1000 \times \varepsilon \times d \times \Delta t \times v \times \rho_{\text{protein}}} = \frac{\Delta A}{1000 \times \varepsilon \times d \times \Delta t \times \varphi \times \rho_{\text{sample}}} \quad \text{kat/g}$$

(units as for eqn. (k); ρ_{protein} in g/l)

(l$_1$) $$z_c/m_s = \frac{\Delta A \times V \times 1000}{\varepsilon \times d \times \Delta t \times v \times \rho_{\text{protein}}} = \frac{\Delta A \times 1000}{\varepsilon \times d \times \Delta t \times \varphi \times \rho_{\text{protein}}} \quad \text{U/g}$$

(units as for eqn. (k$_1$); ρ_{protein} in g/l).

Example

Determination of the (specific) catalytic activity of an enzyme. The measurements were made at 339 nm; $\Delta A/\Delta t = 0.063/60$ s in a 3 ml assay mixture (V $= 3 \times 10^{-3}$ l). The volume of the sample was 2×10^{-4} l. The sample diluted one thousandfold for measurement contained 10 g of enzyme protein per litre.

In the assay mixture the catalytic activity is according to eqn. (i) or (i$_1$), respectively

$$z = \frac{0.063 \times 3 \times 10^{-3}}{10^3 \times 0.63 \times 10 \times 60} = 5 \times 10^{-10} \text{ kat} = 0.5 \text{ nkat}$$

or

$$z = \frac{0.063 \times 3 \times 10^{-3} \times 10^3}{0.63 \times 10 \times 1} = 0.03 \text{ U}.$$

The catalytic activity concentration in the sample solution according to eqn. (k) or (k$_1$), respectively, is

$$b = \frac{0.063 \times 3 \times 10^{-3}}{10^3 \times 0.63 \times 10 \times 60 \times 2 \times 10^{-4}} = 2500 \text{ nkat/l}$$

or

$$b = \frac{0.063 \times 3 \times 10^{-3} \times 10^3}{0.63 \times 10 \times 1 \times 2 \times 10^{-4}} = 150 \text{ U/l}.$$

Related to the mass of protein (dilution factor 1000), the catalytic activity content is

$$z_c/m_s = \frac{2500 \times 1000}{10} \frac{\text{nkat/l}}{\text{g/l}} = 250 \text{ µkat/g}$$

or

$$z_c/m_s = \frac{150 \times 1000}{10} \frac{\text{U/l}}{\text{g/l}} = 15 \text{ U/g}.$$

The result for the sample in relation to an enzyme standard solution is

(m) $\quad b = \dfrac{b_{\text{sample (measured)}}}{b_{\text{standard (measured)}}} \times b_{\text{standard (indicated)}} \quad$ nkat/l (U/l)

or simply

(m$_1$) $\quad b = \dfrac{A_{\text{sample (measured)}}}{A_{\text{standard (measured)}}} \times b_{\text{standard (indicated)}} \quad$ nkat/l (U/l)

Radiometry

According to eqns. (k) and (k$_1$), respectively, and using eqns. (g) and (h)

catalytic activity is

(n) $\quad z = \dfrac{\Delta C_n \times V}{\eta \times 2.22 \times 10^{12} \times 10^{-9} \times X \times v_r \times \Delta t} \quad$ nmol/s (nkat)

(units: C_n in cpm; X in Ci/mol or Bq/mol; v_r in l; t in s), (n$_1$) $$z = \frac{\Delta C_n \times V}{\eta \times 2.22 \times 10^{12} \times 10^{-6} \times X \times v_r \times \Delta t} \quad \mu\text{mol/min (U)}$$

(units as for eqn. (n); t in min).

The catalytic activity concentration in the sample is

(o) $$b = \frac{\Delta C_n}{\eta \times 2.22 \times 10^{12} \times 10^{-9} \times X \times \varphi \times v_r \times \Delta t} \quad \text{nkat/l}$$

(units as in eqns. (n), (n$_1$); $\varphi = v/V$ in l/l),

(o$_1$) $$b = \frac{\Delta C_n}{\eta \times 2.22 \times 10^{12} \times 10^{-6} \times X \times \varphi \times v_r \times \Delta t} \quad \text{U/l}$$

(units as in eqn. (o); t in min).

The catalytic activity related to the mass of protein, catalytic activity content z_c/m_s (specific activity), is

(p) $$z_c/m_s = \frac{\Delta C_n}{\eta \times 2.22 \times 10^{12} \times 10^{-9} \times X \times \varphi \times v_r \times \Delta t \times \rho_{protein}} \quad \text{nkat/g}$$

(units as for eqn. (o); $\rho_{protein}$ in g/l),

(p$_1$) $$z_c/m_s = \frac{\Delta C_n}{\eta \times 2.22 \times 10^{12} \times 10^{-6} \times X \times \varphi \times v_r \times \Delta t \times \rho_{protein}} \quad \text{U/g}$$

(units as for eqn. (o$_1$); $\rho_{protein}$ in g/l).

Statistics

For parameters and sample statistics, for mathematical evaluation of experiments (mean value, standard deviation, imprecision, inaccuracy) and for statistical evaluation of results, especially for regression analysis, cf. chapter 3.2.

3.1.2.3   Refering the Experimental Results to Biological Material

In the analysis of organ extracts, blood, serum, etc., the dilution resulting from deproteinization of samples must also be taken into account along with the fluid content of the sample. The following values give reasonable accuracy: blood 80% (w/w), tissue (liver, kidney, muscle, heart) 75% (w/w). Tissue samples are weighed out; a specific gravity of 1.06 is used for the conversion of blood volumes into mass.

Apart from the volume (in the case of serum, plasma, blood, urine, etc.), other reference quantities that may be used for biological material are the fresh weight, dry weight, total nitrogen, protein content, protein nitrogen; cell count, e.g. erythrocyte count; haemoglobin content, cytochrome c content, and dry weight of the cell-free sample solution.

Examples

Determination of the dilution factor for blood.

The specific gravity of blood is taken as 1.06, and its fluid content is taken as 80% (w/w). 2 ml of blood is deproteinized with 3 ml of perchloric acid and centrifuged, and 2.5 ml of the supernatant are neutralized with 1 ml K_2CO_3 solution. The blood is therefore diluted by the following factor:

$$F = \frac{2 \times 1.06 \times 0.8 + 3}{2} \times \frac{2.5 + 1}{2.5} = \frac{1.696 + 3}{2} \times \frac{3.5}{2.5} = 3.29 \,.$$

The experimental result obtained with the neutralized blood extract must be multiplied by this factor to obtain the content of the metabolite in the blood.

Calculation of the dilution factor for multi-stage assay mixtures.

In the determination of maltose in biological fluids, the various reaction steps have different pH-optima. The assay begins with an incubation: 0.5 ml sample with 0.5 ml acetate buffer and 0.02 ml α-glucosidase. After inactivation of the enzyme, the second step is the determination of the glucose in 0.2 ml of the incubation solution (V = 3.42 ml). The dilution factor is thus:

$$F = \frac{0.5 + 0.5 + 0.02}{0.5} \times \frac{3.42}{0.2} = 2.04 \times 17.1 = 34.88 \,.$$

The experimental result, after division by 2 (1 maltose $\triangleq$ 2 glucose), must be multiplied by this factor to obtain the maltose content in 1 ml sample.

To calculate the metabolite content of the cells of a tissue, the metabolite content of the blood in this tissue must be taken into account.

The mass fraction of blood in the tissue is determined according to *Bücher et al.* [10] from absorbance measurements at 578, 560, and 540 nm. Assuming that the proportion of oxyhaemoglobin (HbO_2) in the circulating blood and the tissue is approximately the same, it follows that the fraction of blood w in the tissue is

$$w = \frac{\Delta A_{HbO_2} \times F_1 \times d_1}{\Delta A'_{HbO_2} \times F_2 \times d_2}$$

where

ΔA_{HbO_2} is absorbance difference for tissue extract
$\Delta A'_{HbO_2}$ is absorbance difference for blood dilution
F_1 and F_2 are dilution factors
d_1 and d_2 are light paths of the cuvettes

ΔA_{HbO_2} and $\Delta A'_{HbO_2}$ are calculated [10] from the absorbance measurements at 578, 560, and 540 nm, according to the formula:

$$\Delta A_{HbO_2} \quad \text{or} \quad \Delta A'_{HbO_2} \; = \; (A_{578} - A_{560}) + [(A_{540} - A_{578}) \times 0.47].$$

If the metabolite concentration in the tissue sample is to be referred to the true volume of cellular fluid in order to give the physiological concentration, the result (in mmol/l of tissue extract) is multiplied by the following factor (*G. Michal,* unpublished):

$$F = \frac{V_{\text{after neutralization}}}{V_{\text{before neutralization}}} \times \frac{\dfrac{m_{\text{tissue}}}{\text{spec. gr.}} \times \varphi + V_d}{\dfrac{m_{\text{tissue}}}{\text{spec. gr.}} \times \varphi}$$

φ is the volume fraction of liquid in the tissue volume; m_{tissue} is tissue wet weight; specific gravity relates to the sample. The latter quantity may be taken as unity in most cases. V_d is volume of reagent solution (e.g. $HClO_4$) used for deproteinization.

A similar calculation for erythrocytes has been published by *Bürgi* [12] (cf. chapter 1.1.1.3, p. 13).

References

[1] *H. U. Bergmeyer,* Reaktionskinetische Untersuchungen zur Messung der Glykolyse, Biochem. Z. *323,* 163 – 180 (1952).
[2] Expert Panel on Nomenclature and Principles of Quality Control in Clinical Chemistry; Committee on Standards (IFCC): Provisional Recommendation on Quality Control in Clinical Chemistry. Part 1: General Principles and Terminology, Clin. Chim. Acta *63,* F25 – F38 (1975).
[3] ISO Guide 6 (1978): Mention of Reference Materials in International Standards.
[4] DIN 32811 (1979): Grundsätze für die Bezugnahme auf Referenzmaterialien in Normen.
[5] Expert Panel on Nomenclature and Principles of Quality Control in Clinical Chemistry; Committee on Standards (IFCC): Approved Recommendation (1978) on Quality Control in Clinical Chemistry, Part 1: General Principles and Terminology, J. Clin. Chem. Clin. Biochem. *18,* 69 – 77 (1980).
[6] *D. Stamm,* Reference Materials and Reference Methods in Clinical Chemistry, J. Clin. Chem. Clin. Biochem. *17,* 283 (1979).

[7] Expert Panel on Proteins; Committee on Standards (IFCC): The Human Serum Standard (Stage 2, Draft 1), Clin. Chim. Acta *98*, 179F–186F (1979).

[8] Expert Panel on Nomenclature and Principles of Quality Control in Clinical Chemistry; Scientific Committee (IFCC): Approved Recommendation (1979) on Quality Control in Clinical Chemistry, Part 3. Calibration and Control Materials, J. Clin. Chem. Clin. Biochem. *18*, 855–860 (1980).

[9] International Union of Pure and Applied Chemistry (IUPAC) and International Federation of Clinical Chemistry (IFCC): Approved Recommendation (1978), Quantities and Units in Clinical Chemistry, Clin. Chim. Acta *96*, 157F–183F (1979).

[10] *H. J. Hohorst, F. H. Kreutz, Th. Bücher,* Über Metabolitgehalte und Metabolit-Konzentrationen in der Leber der Ratte, Biochem. Z. *332*, 18–46 (1950).

[11] International Commission on Radiation Units and Measurements: ICRU Report 33 "Radiation Quantities and Units", 1980, 7910 Woodmont Ave., Washington D.C. 20014, USA.

[12] *W. Bürgi,* The Volume Displacement Effect in Quantitative Analysis of Red Blood Cell Constituents, J. Clin. Chem. Clin. Biochem. *7*, 458–460 (1969).

3.2   The Quality of Experimental Results

3.2.1   Quality Control

Horst Brettschneider and Manfred Glocke*

The aim of quality control is to keep the errors in quantitative analytical chemistry at a satisfactorily low level.

The extent of the errors in the analysis of biological specimens can, if not controlled, be large due to the analytical process or the limited stability of the biological specimens. The results of an inter-laboratory study of glucose [1] are given as an example and show unacceptably high deviations of the measured values from the expected value.

Because of the great importance of clinical chemistry for medical diagnosis, quality control was introduced early in this field [2, 3] and developed to a high standard of performance today. However, the principles of quality control can also be transferred to other applications of enzymatic analysis, such as food chemistry.

---

* This chapter is based on the contribution of *H. Büttner, E. Hansert* and *D. Stamm* in the 2nd edition.

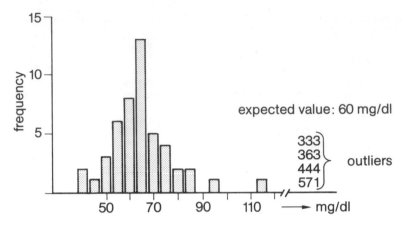

Fig. 1. Frequency distribution of glucose values

3.2.1.1   Errors in Quantitative Chemical Analysis

Types of errors

The results of quantitative chemical analyses are markedly susceptible to errors. It has become convenient in analytical laboratories to devide the errors into three main categories: gross, random and systematic errors.

Gross errors can be caused by confusing specimens, reagents and devices, reversing two digits when transcribing a reading, or by mistakes in the calculation of a result, etc. Gross errors are avoidable by a well-trained laboratory staff and an adequate laboratory organization.

Random errors occur when multiple analyses of a specimen are performed. The results obtained differ slightly from one another so that, on average, the plus or minus

error types		random error	systematic error	gross error
				used the wrong target
precision	optimal	bad	good	—
accuracy	optimal	good	bad	—

Fig. 2. Random, systematic and gross errors illustrated by example of shots at a bullseye with a rifle.

deviations from the mean values are equal. Random errors are caused by the imperfect repeatability of analysis. Examples of random errors are: reading a meniscus, variations in the incubation and measuring time, etc. The three types of errors can be illustrated by the example of shots at a bullseye [4].

Systematic errors can be due to the method, the reagents, the calibration standards, the pipettes, the measuring temperature, etc. The results always deviate from the expected value in the same direction, i.e. the values found are systematically too high or too low. Systematic errors are avoidable and can be traced back to a definite cause.

Picture A shows the bull's eye of a marksman who hits the target in an optimal way, i.e. without systematic error.

Picture B illustrates the results of a marksman who hits the target with relatively large random errors free from systematic errors.

Picture C demonstrates a systematic error of a marksman with an excellent random distribution.

Picture D shows no hits in the bullseye. Obviously the marksman makes a gross error.

Precision and accuracy

With regard to the pictures A to D, it is equally true to say that in case A the marksman shot at a target with a good precision (small random errors) and a good accuracy (no systematic errors), whereas in picture B precision is bad (large random errors) and accuracy good (no systematic errors) and so on (see Fig. 2). In the following the terms precision and accuracy are explained in detail.

Precision

The agreement between values from replicate measurements is defined as *precision*. The magnitude of the precision is expressed by the *imprecision*, characterized by the standard deviation (SD) or the coefficient of variation (CV) in percent:

(a) $$SD = \sqrt{\frac{\sum (x_i - \bar{x})^2}{n - 1}}$$

(b) $$CV = \frac{SD}{\bar{x}} \times 100$$

SD   standard deviation
CV   coefficient of variation in percent
x_i single value
$\bar{x}$ mean value of n replicate values
n number of values

The standard deviation and also the coefficient of variation, depend on:

– measuring conditions
– concentration of the analytes

With fixed concentration levels and measuring conditions two main types of imprecision can be distinguished, depending on the experimental designs:

– imprecision within a given laboratory (repeatability)
– imprecision between different laboratories (reproducibility)

Repeatability

Within a laboratory the best coefficient of variation is achieved within a single analytical run. The coefficient of variation always increases when the analyses are carried out in different runs on different days by the same analyst (see example: Fig. 3).

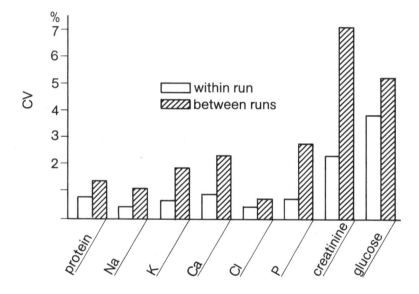

Fig. 3. Within-run and between-runs coefficients of variation

Reproducibility

The highest coefficient of variation of a method generally is seen when the specimen is analyzed by different analysts in various laboratories (see example: Fig. 4). This can be interpreted as showing that, in addition to the variances within the laboratories, systematic errors between the laboratories must be taken into account. This can be studied by an analysis of variance.

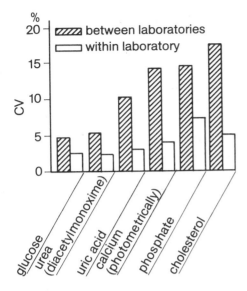

Fig. 4. Within-laboratory and between-laboratories coefficients of variation.

Accuracy

The agreement between the mean of replicate measurements and its expected value is defined as *accuracy*. The magnitude of the accuracy is expressed by the *inaccuracy*** (bias), characterized by the difference between the mean and the expected value:

(c) $d = \mu_0 - \bar{x}$

d inaccuracy (bias)
$\bar{x}$ mean of replicate measurements
μ_0 expected value

or the relative bias b in percent

(d) $b = \dfrac{d}{\mu_0} \times 100$.

Fig. 5 illustrates the mean classes of biases by their typical operational lines [5].

Constant bias: the measured means $\bar{x}$ (solid line) are each higher than the expected values μ_0 (dotted line) by a constant amount of bias, independent of the concentration of the analyte.

Proportional bias: the amount of inaccuracy (or bias) increases in direct proportion to the concentration of the analyte.

---

* Note: symbols in statistics must not be interchanged with symbols generally used in analysis.

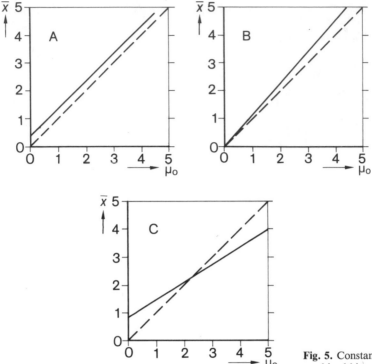

Fig. 5. Constant, proportional and combined biases.

combined bias: the figure shows a combination of constant and proportional in-accuracies (biases). It is a characteristic of combined biases to be zero at a definite concentration whereas at concentrations higher or lower than this the bias increases.

A bias caused by systematic error is, as already mentioned, avoidable. In practice, however, such a goal is often imperfectly achieved due to inherent weaknesses in the methods or measuring systems used.

A particular problem in the analysis of biological material is the non-specificity of many analytical procedures. The biases caused by non-specificity are only avoidable in many cases by changing the analytical principles. The determination of uric acid in serum is an example: the unspecific classical reduction methods with phosphotungstic acid yield results with a positive bias and should be replaced by the specific uricase methods.

Specimen errors

The errors dealt with so far occur in the analytical phase. However, for reliable results of analyses, potential specimen errors of the pre-analytical phase have also to be considered. Specimen errors include all sources of faults which may occur from the collection to the transport and storage of the specimen. Depending on the kind of biological material to be analyzed, specimen errors can be caused by a variety of factors, e.g. inhomogeneity of the analyte in the specimen, inadequate storage conditions, contamination, water loss due to evaporation, influence of light, etc. [6].

A typical example of a specimen error in clinical chemistry is the "haemolysis error". Lysis of the blood cells releases considerable amounts of enzymes into the plasma which results in an artificial increase of enzyme acitivities (Fig. 5).

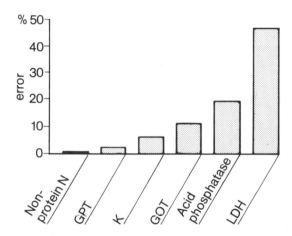

Fig. 6. Systematic errors caused by haemolysis.

3.2.1.2   Continuous Control of Quantitative Analysis

As was just discussed, chemical analyses are highly subject to errors. What can be done to avoid such errors so that reliable results can be obtained under the working conditions of a routine laboratory?

Practical experience has indicated that only by the use of particular analytical control procedures based on insight gained from error studies can results reliable within certain limits be guaranted.

A rather obvious method of control would be to compare each result with a second analysis based on an independent method as is done, for example, in forensic investigations. Such a total control, although theoretically convincing, cannot be realized in a practical way. The excessive costs in personnel, apparatus and reagents, together with the increased amount of specimen this would require, make such a control procedure out of the question. Duplicate determinations by the same method within given runs − often used for control purposes − fulfil this need to an insufficient degree only, since errors within run, i.e. under nearly the same working conditions, are usually small and do not represent the real error situation.

The only control procedures remaining are those based on samples, making use necessarily of statistical decision methods. Such procedures were first developed by *Shewhart* [7] for controlling the quality of factory products; he referred to them as methods of statistical quality control.

More recently they have been used successfully in analytical laboratories. Before the details of such methods are dealt with, general principles of "statistical quality control" will be considered.

Basic principles of quality control

Quality control of a process is based on the null hypothesis, that a measured value $\bar{x}$ is compatible with an expected value μ_0. The statistical test for the null hypothesis, $H_0: \bar{x} = \mu_0$, is carried out by calculating the difference between the mean value $\bar{x}$ and the expected value μ_0 divided by its standard error $\left(\dfrac{\sigma_0}{\sqrt{n}}\right)$ and compared with tabulated values for λ.

(e) $\xi = (\bar{x} - \mu_0) / \left(\dfrac{\sigma_0}{\sqrt{n}}\right)$

μ_0 expected value for the population
$\bar{x}$ mean of n replicate assay results
σ_0 standard deviation of the expected value
n number of replicate assay results
ξ observed value
λ critical (tabulated) value

If the observed ξ of the control sample is less than the tabulated λ there is no statistically significant difference between $\bar{x}$ and μ_0. The process is "under control". If the calculated ξ exceeds the tabulated λ, the difference becomes statistically significant and the process is "out of control".

The value for λ is a function of the degree of freedom and of a chosen probability (P). Usually in quality control a probability of $P = 1-\alpha$, with $\alpha = 0.01$ error risk, is chosen. This means that assuming a *Gaussian* distribution, 1 out of 100 values ($\alpha = 0.01$) are falsely rejected.

This error is called a statistical error of the first kind and can be designated as "false alarm of the control system". The statistical test just described can be carried out more simply and quickly by using a control chart [8]. Time or time-dependent quantities, i.e. number of the sample, are marked on the abscissa whereas the expected value μ_0 and the decision limits $\mu_0 \pm \lambda_{0.01} \dfrac{\sigma_0}{\sqrt{n}}$ are entered on the ordinate.

The value $\bar{x}$ of the control samples are entered successively on the diagram as shown in Fig. 7. If the distribution characterized by μ_0 and σ_0 is unchanged 99% ($\lambda_{0.01}$) of the

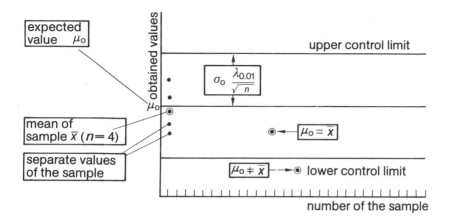

Fig. 7. Design of a control chart.

control values $\bar{x}$ can be expected to lie between the control limits. If a value falls outside the control limits, the process has gone "out of control", because the probability of this occurring is smaller than 1%.

The limitations of the control-chart technique must be recognized:

– The control limits are valid only for the *Gaussian* distribution. They will differ if the variable under test is differently distributed.

– The control limits involve not only statistical errors of the first kind (type I errors) but also statistical errors of the second kind (type II errors).

– The error-probability for type I error is fixed, e.g. at $\alpha = 0.01$. However, the error-probability of type II error depends in a complicated way on several factors, e.g. on the error risk (α) and the number of analyses in the run (n), and can be estimated by constructing characteristic operating curves (OC-curves) (Fig. 8).

As seen from Fig. 8, the probability of detecting a deviation (in σ units) of the sample mean $\bar{x}$ from the expected value μ_0 will increase with increasing number of analyses [9].

– Changes within the control limits are not recognized, but there are other criteria: According to the theory of runs [10] a process is likewise out of control if more than 7 successive values lie on one side of the μ_0-line or occur in increasing or decreasing order. Another approach for detecting trends is the cumulative sum or cusum chart [11]. In this technique the differences (either positive or negative) between $\bar{x}$ and μ_0 are calculated and consecutively added and entered in the cusum chart. Assuming that $\bar{x}$ is close to μ_0 and that positive and negative cusum increments will occur equally, the cusum values usually oscillate about zero: the method is considered to be in control.

When the method goes out of control, a high deviation of $\bar{x}$ from the expected value μ_0 or a series of small deviations in the same direction will cause the cusum values to depart from the vicinity of the zero line. The use of a V-mask helps to reveal the changed trend of the line.

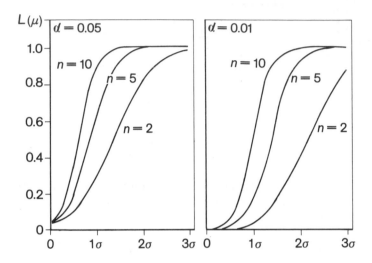

Fig. 8. Operating curves.

Application of statistical quality control to quantitative chemical analysis

Before introducing quality control in quantitative analytical chemistry the suitability of the control material, the analytical method and the control system should be considered.

Suitability of control specimen

For an effective control system, control specimens are needed which are as similar as possible to the specimens to be analyzed [12]. Only if the control specimens have nearly the same compositions and structure as the analytical specimens can an optimal control system be expected. This applies especially to biological material, e.g. serum, where the presence of protein affects pipetting and non-specific serum components may interfere with the test system. For these reasons aqueous control solutions are often ineffective and should only be used if for some reasons, such as lack of stability, a suitable control material is not available.

Suitability of the analytical method

The level of reliability of analytical results depends primarily on the choice of the analytical method.

The criteria for the selection of the analytical method are described elsewhere [13] and should be known before setting up a quality control system.

Suitability of the control system

The probability of type I errors can be kept as small as desired by the choice of λ. Usually $\lambda = 2.00$ ($\alpha = 0.05$) or $\lambda = 2.58$ ($\alpha = 0.01$) or $\lambda = 3.00$ ($\alpha = 0.0027$) are chosen. Type II errors on the other hand will increase with increasing values of λ.

With an increase of number of analyses per specimen, however, the probability of type II errors can be reduced.

Fig. 9 illustrates this with the example of quality control of protein analyses.

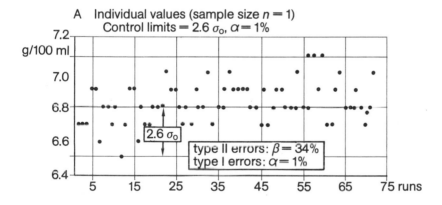

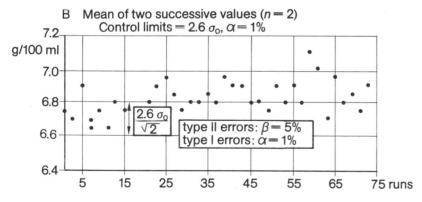

Fig. 9. Control chart of differing sample size.

Establishment of a quality control program

After finding a suitable control material and an analytical method, the first step in the establishment of a control system is the assessment of the provisional values for μ_0: $\bar{x}$ and σ_0: SD. For these purposes the control specimen is analyzed according to the experimental design, i.e. from day to day, at least $20-30$ times.

	x_i	$(x_i - \bar{x})$	$(x_i - \bar{x})^2$
1	193	0	0
2	190	3	9
3	202	9	81
4	200	7	49
5	197	4	16
6	187	6	36
7	188	5	25
8	193	0	0
9	185	8	64
10	197	4	16
11	190	3	9
12	195	2	4
13	188	5	25
14	192	1	1
15	200	7	49
16	202	9	81
17	190	3	9
18	192	1	1
19	198	5	25
20	189	4	16
n 20	$\sum (x_i)$ 3868		$\sum (x_i - \bar{x})^2$ 516

Mean value

$$\bar{x} = \frac{\sum x_i}{n}$$

$\bar{x} = 193{,}4$

$\bar{x} \sim 193 \text{ mg/dl}$

Standard deviation

$$SD = \sqrt{\frac{\sum (x_i - \bar{x})^2}{n-1}} = \sqrt{\frac{516}{19}}$$

$SD = 5{,}19$
$SD \sim 5 \text{ mg/dl}$

Coefficient of variation

$$CV = \frac{SD \times 100}{\bar{x}} \%$$

$CV = 2{,}68\%$
$CV \sim 2{,}7\%$

Control limits

$\bar{x} + 3\,SD = 208 \text{ mg/dl}$

$\bar{x} - 3\,SD = 178 \text{ mg/dl}$

Warning limits

$\bar{x} + 2\,SD = 203 \text{ mg/dl}$

$\bar{x} - 2\,SD = 183 \text{ mg/dl}$

Fig. 10. Calculation of mean value ($\bar{x}$) and standard deviation (SD).

The determination of cholesterol is used in Fig. 10 as an example of the calculation of the mean value ($\bar{x}$) and the standard deviation (SD). Results are given for a period of 20 days.

The steps in the calculation are as follows:

- compute the sum of the entries in column 1
- calculate the mean value by using the equation

$$\bar{x} = \frac{1}{n} \Sigma x_i$$

- calculate the difference between the mean $\bar{x}$ and the single values x_i (column 2)
- square the differences (column 3)
- add the entries in column 3, to obtain the sum of the squares
- find the standard deviation by means of the equation

$$SD = \sqrt{\frac{\Sigma (x_i - \bar{x})^2}{n - 1}} \ .$$

From the data in Fig. 10 the control limits $\mu_0 \pm \lambda \times \sigma_0$ for single determinations with $\mu_0 = \bar{x}$, $\lambda = 2$ and 3, $\sigma_0 = $ SD are computed:

warning limits: $x \pm 2\,SD = 193 \pm 10$
action limits: $\bar{x} \pm 3\,SD = 193 \pm 15$.

The mean value and the warning and action limits are entered on the control chart.

With the aid of the control chart it is now possible to recognize immediately whether the assay is in or out of control (Fig. 11).

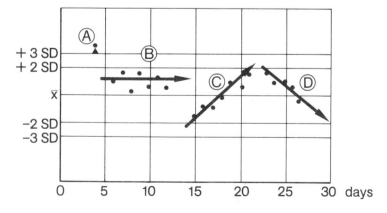

Fig. 11. Four criteria for the decision "analysis out of control".

The system is said to be *in control*, if:

– the values vary around the mean within the control limits
– from 20 values 1 of them can be found between the range
  ± 2 SD and ± 3 SD.

The system is said to be *out of control*, if

– the values lie outside of the ± 3 SD-limits (see Fig. 11 A)
– more than 7 values lie on one side of the mean or occur in increasing or decreasing order (see Fig. 11 B, C, D)

Precision can be controlled with the system described. The system does not allow a control of accuracy. The latter can be checked by analyzing a control specimen with a known value (accuracy control specimen), determined by weighing in the analyte or by value assignment by several independent laboratories [14]. Accuracy control should be carried out at least on two different points of the operational curve, in order to identify the nature of a possible systematic error.

3.2.1.3   Quality Control in Clinical Chemistry

The recommendations on quality control in clinical chemistry issued by the International Federation of Clinical Chemistry (IFCC) [15] define quality control in clinical chemistry as: "The study of those errors which are the responsibility of the laboratory and the procedures used to recognize and minimize them".

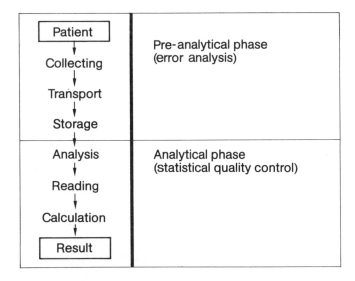

Fig. 12. The field of quality control in clinical chemistry.

Quality control has to monitor all sources of errors which arise within the laboratory between the receipt of the specimens and the output of results in the form of reports. In a wider sense quality control can be extended to the pre-analytical phase, which means control of collection, storage and transport of the specimen from the patient.

For practical reasons it is convenient to divide quality control into two major types: internal quality control (intra-laboratory quality control) and external quality control (inter-laboratory quality control). The term "external quality assessment" is now preferred to describe inter-laboratory schemes, since they are largely retrospective in nature, and the element of using the results for an immediate correction of the analytical process which is typical of control processes is not present.

The goal of the internal quality control is to keep precision and accuracy at an acceptable quality level, whereas the external quality control checks the reproducibility with other laboratories.

To achieve these goals a wide range of quality control schemes are available today. The selection of a more or less comprehensive quality control scheme depends on such factors as size and facilities of a laboratory. In the following some guidelines are given for the introduction of quality control in clinical laboratories.

Internal quality control

The first step in the establishment of an internal quality control program is to assess the statistical parameters for precision in an initial period.

Initial period

A control specimen with concentrations in the borderline range is included in at least 20 different runs of patient specimens, preferably on separate days. The observed control data are tested for their distribution and, when normal or symmetrical dispersion has been assured, the mean values ($\bar{x}$), the standard deviations (SD) and coefficients of variation (CV) are calculated for each constituent. If the precision has been found to be at an acceptably low level, the control limits ($\bar{x} \pm 3$ SD) and warning limits ($\bar{x} \pm 2$ SD) are computed and entered on a control chart for the continuous control period. There exist no simple rigorous criteria for acceptability of analytical performance from the clinical point of view today. From the analytical state of the art it can be recommended that the coefficient of variation from day to day for the determination of metabolites should be between 3 and 5% and for enzymes and hormones between 5 and 10%. The coefficient of variation may be greater than 10% in some exceptionally difficult analyses, e.g. of some hormones.

Continuous control periods

In the subsequent control period the values obtained for the control specimen are entered on the control chart.

As already shown, the control chart provides a rapid means of deciding whether the analytical process is in or out of control. Furthermore, trends, shifts and other types of irregularities can be detected easily. To utilize the control chart technique in an optimal sense, mean values, standard deviations and coefficients of variation should be updated monthly on the basis of all control values. The advantages of this procedure are more reliable calculations of the statistical parameters and recognition of long-term trends of precision and accuracy.

Accuracy should be monitored in addition to precision control. An effective program of accuracy control consists of two separate systems [16].

Open control: accuracy is monitored by the technicians themselves by operating with a known control specimen.

Blind control: accuracy is monitored by the head or supervisor of the laboratory; the control specimens are unknown to the technicians.

This dual system of accuracy control is essential, especially for manually operated tests, where the blind design obviates the analyst's conscious or unconscious bias.

With regard to the type and number of accuracy control specimens within and between runs, no general rules can be given. The inclusion of at least two control specimens with concentrations below and above the precision interval is recommended for accuracy control, and to detect changes in operational curves.

Additional procedures

It is worth noting that the correction of errors can often be simplified if the control system is extended to the continuous control of reagent blanks and standard solutions [17]. A further approach to refine the control system is the performance of instrumental checks; e.g. checks of absorbance, temperature or pipette volumes should be carried out on a daily, weekly or monthly basis [18]. Depending on the analytical system, drift or carryover within a run should be monitored also, by including several samples of the same control specimen or several control specimens of markedly different concentrations [19]. Some mechanized systems already have a software program that performs functional checks during the start-up phase.

External quality assessment

The major aim of external quality assessment is to supplement internal quality control by comparing the results of several laboratories which analyze the same specimens.

There are many clinical-chemistry survey programmes of varying scope available to laboratories organized by governmental, private and professional agencies.

In a typical programme each participant receives two unknown control specimens with different concentration levels. The specimens are analyzed and the results returned to the survey leader for computation. Thereafter each participant receives a tabulation of his own results compared to the mean values and standard deviations obtained by a group of selected reference laboratories. In addition often a scatter diagram or a frequency histogram is given to the participants to provide a better over-view.

Quality control specimens

Commercially-produced specimens are mostly used for quality control purposes today. In order to be effective, the control materials must meet three main requirements [12].

- the control specimen should be as similar as possible to the material being examined;
- the analytes in the control material have to be stable during storage and after reconstitution over a definite period of time;
- inter-vial variability in the concentration of the control specimen should be negligible compared to the imprecision of the analysis.

Two main types of control material are offered today: lyophilized and stabilized liquid preparations. To know the limitations of these materials and to allow proper interpretation of quality control data, characteristics and requirements of the two types are given in the following.

Lyophilized control specimens

Lyophilized control sera are produced from large pools of human or animal sera, which are supplemented with various analytes to achieve the desired concentration levels in the final product. The pool is dispensed into vials followed by chamber lyophilization for better stability. An alternative process is the spray-dry-freezing of the pool, followed by weighing in the control material.

The most critical production steps for assuring homogeneity over the total production lot are the dispensing and lyophilization procedures. In general, the final lot should have a vial-to-vial variability of less than 1 percent. To achieve this requirement well established filling and lyophilization procedures are needed. The lyophilized final products should be stable in storage at $+4°C$ at least for 2 years and after reconstitution for 2 days. The turbidity of the reconstituted control sera, due to denatured lipoproteins, should be kept at an acceptable level.

The similarity to patients' specimens is generally satisfactory for most quality control purposes.

However, in the case of the measurement of proteins, e.g. transferrin, immunglobulins and protein electrophoresis, it must be remembered that the protein composition of animal sera can vary markedly from that found in human sera. In these cases human-based control material is to be preferred.

Liquid control specimens

Liquid control specimens are produced in a very similar way to the lyophilized ones. In contrast to the lyophilized specimens, stability is achieved by addition of stabilizers e.g. glycerol, polyethyleneglycol, etc. The advantages of the liquid control specimens are readiness for use, clarity and homogeneity. On the other hand, however, it must be said that the high amount of stabilizer changes the matrix, so that similarity to patients' specimens is no longer present in some analytical procedures such as continuous flow analysis [20]. Further disadvantages of liquid control specimens available at present are the need for storage at very low temperature ($-20\,°C$) and their sensitivity to temperature variations.

As an alternative to the commercial control material, laboratories may prepare their own control material by pooling the remainder of specimens from patients. This is a cheap method of obtaining control specimens, but in many cases problems arise due to instability or interfering factors. The danger of infectious hepatitis should also be taken into consideration when working with surplus patient sera.

Control specimens with known values are needed for the control of accuracy. There are several concepts for defining target values [21].

– By weighing the pure analyte into a control specimen.
– As the assigned value (consensus value) obtained by different laboratories in a survey programme using the same method, after exclusion of outliers.
– As the assigned value obtained by a group of selected reference laboratories using a reference method ("reference method mean") or routine method with known bias ("method mean").

Today all three possibilities are used and have their respective advantages but also their limitations in practical use.

The weighing-in concept does not need a value assignment and money can be saved; however, the concept is only applicable to some low molecular-weight analytes. It is necessary also to ensure that the specimen matrix before spiking does not contain traces of the analyte in question.

The most reliable procedure for establishing assigned values is assignment a reference method and appropriate reference materials, under the responsibility of a group of selected reference laboratories. With the growing number of approved reference methods and reference materials this ought to be the method of choice for assignment of values in control specimens in the future. In cases where no reference method exists at present, the target value should be assigned by the best routine method available of known bias.

References

[1] *W. P. Belk, F. W. Sunderman,* A Survey of the Accuracy of Chemical Analyses in Clinical Laboratories, Am. J. Clin. Path. *17*, 853 (1947).
[2] *L. G. Whithby, F. L. Mitchell, D. W. Moss,* Quality Control in Routine Clinical Chemistry, Advances in Clinical Chemistry, Academic Press New York-London, *10*, 65 (1967).
[3] *H. Büttner,* Statistische Qualitätskontrolle in der klinischen Chemie, Z. Klin. Chem., *5*, 41 (1967).
[4] *H. Büttner,* Qualitätskontrolle im klin. chem. Routinelaboratorium, Z. Analyt. Chem. *212*, 109 (1965).
[5] *G. F. Grannis, B. E. Statland,* Monitoring the Quality of Laboratory Measurements, in: *J. B. Henry* (ed.), Clinical Diagnosis and Managements by Laboratory Methods, Vol. *2*, W. B. Saunders Comp., Philadelphia 1979, p. 2049.

[6] *W. G. Guder,* Einfluß von Probennahme, Probentransport und Probenverwahrung auf klinisch-chemische Untersuchungsergebnisse, Ärztl. Lab. *22*, 69 (1976).

[7] *W.A. Shewart,* Economic Control of Quality of Manufactured Product, van Nostrand, New York, 1931.

[8] *K. Stange,* Kontrollkarten für meßbare Merkmale, Springer Verlag, Berlin 1975.

[9] *R. Saracci,* Wieviel Mehrfachbestimmungen sind bei der Qualitätskontrolle klinisch-chemischer Tests erforderlich?, Ärztl. Lab. *23*, 187 (1977).

[10] *J. Wolfowitz,* On the Theory of Runs with some Applications to Quality Control, Ann. Math. Stat., *14*, 280 (1943).

[11] *R. W. H. Edwards,* Internal Analytical Quality Control using the Cusum Chart and Truncated V-mask Procedure, Ann. Clin. Biochem. *17*, 205 (1980).

[12] *D. Stamm,* Calibration and Quality Control Material, Z. Klin. Chem., Klin. Biochem. *12*, 137 (1974).

[13] *D. Stamm,* Selected Methods in Clinical Chemistry, J. Clin. Chem. Clin. Biochem. *17*, 277 (1979).

[14] *H. Passing, B. Müller, H. Brettschneider, W. Bablok, M. Glocke,* The Establishment of Assigned Values in Control Sera I, II, III, IV, J. Clin. Chem. Clin. Biochem. *19*, 1137 – 1179 (1981).

[15] The International Federation of Clinical Chemistry, Approved Recommendation (1978) on Quality Control In Clinical Chemistry, Part 1, General Principles and Terminology, J. Clin. Chem. Clin. Biochem. *18*, 69 (1980).

[16] *D. B. Tonks,* A Dual Program of Quality Control for Clinical Chemistry Laboratories, with a Discussion of Allowable Limits of Error, Z. Anal. Chem. *243*, 760 (1968).

[17] *W. Bürgi,* Aktuelle Probleme der Qualitätskontrolle in der klinischen Chemie, Med. Labor *27*, 105 (1974).

[18] *S. L. Inhorn,* Quality Assurance Practices for Health Laboratories, American Public Health Association, 1978.

[19] *M. Hjelm,* Quality Control of Automated Analytical Systems in Clinical Chemistry, Z. Anal. Chem. *243*, 781 (1968).

[20] The International Federation of Clinical Chemistry, Provisional Recommendation On Quality Control In Clinical Chemistry, Part 5, External Quality Control, Clin. Chem. *24*, 1213 (1978).

[21] *W. T. Pope, T. E. Curupher, G. F. Grannis,* An Evaluation of Ethylene Glycol-based Liquid Specimens for Use in Quality Control, Clin. Chem. *25*, 413 (1979).

3.2.2 Statistics as a Tool for Quality Control

Manfred Glocke and Horst Brettschneider*

3.2.2.1 Basic Concepts and Descriptive Statistics

Planning of experiments

Experimental design is one of the most important fields of mathematical statistics, because errors in the design of experiments cannot be eliminated later by statistical techniques, whatever sophisticated they are. By statistical planning and evaluation, it is

---

* This chapter is based on the contribution of *H. Büttner, E. Hansert* and *D. Stamm* in the 2nd edition.

possible to arrive at meaningful statements and rational decisions. Erroneous evaluations and interpretations are obtained when the conditions for certain statistical procedures are lacking. The laws of statistics must be taken into account in all phases of experimental work: in formulation of the problem (how accurate should the result be?), in the planning of the experiments (how many measurements are necessary?), in their performance (what methods and apparatus will be required?), and in the evaluation of the results (which statistical evaluation?).

To allow a definite statement to be made, a minimum number of experiments (samples*) must be performed. The number of experiments or the "sample size" means the number of measurements or analyses in well defined experimental units (sera, urines etc.). The number of samples depends on the precision required on the experimental result, on the range of biological variation in the material under investigation and on the Type I- and Type II-errors. The sample number must be particularly large when it is desired to determine small differences significantly. Doubling of the precision requires in many cases a 4-fold increase of the sample number. On the other hand, there is the danger with too few samples of failure to detect any differences which exist in reality.

The aim of the experimental design is based on the assumption that a limited number of experiments (samples) will lead to the deduction of a law that is valid even when the experiment is repeated an unlimited number of times on the same basic population. Conclusions concerning the population that are derived from a sample are permissible only when there is a random selection of the experimental units (sera, urines etc.). Each experimental unit of the population must basically have the same chance of being sampled in the random experiment, and this is often difficult to achieve.

The advantage of statistical planning and evaluation is the ability to generalize from sample findings to populations, but this advantage has to be paid for by predictable errors.

Descriptive evaluation of the experimental results

In descriptive statistics, experimental results are treated in three successive steps [8, 10].

---

* In analytical chemistry, "sample" refers to the portion of material on which an analysis is carried out, e.g. a small volume of blood serum removed from a single patient, or an extract of a portion of tissue. "Sample size" in these cases consequently usually refers to the volume of material to be analyzed. In statistics, the term "sample" is used to describe a number of items which are drawn from a larger group of items or from a population. The characteristics of the sample are used to infer the corresponding characteristics of the population from which it is drawn. "Sample size" in the statistical sense thus refers to the number of items which are drawn from the parent group or population.

It will be obvious that each of the items constituting a statistical sample may itself become a sample in the analytical sense; e.g. a number of blood specimens, together making up a statistical sample of a population of normal adults, may each become an analytical sample for the determination of some biochemical parameter such as blood glucose concentration.

- Tabulation of the results of observations, frequency tables.
- Graphical presentation: histogram, frequency distribution etc.
- Calculation of sample statistics (statistics of location and dispersion).

Tabulation of the observed results

All original data are organized synoptically in tabular form. This list is called the master list. It must contain complete and accurate information on all details of the experiment (worker, date, experimental arrangement, sequence of measurements, changes in the arrangement and/or performance of the experiments).

Frequency tables

In samples with $n > 20$ the investigator should, in addition to the tabulation of results, concentrate the data in frequency tables. To set up class intervals, a good working rule is to have $k = \sim \sqrt{n}$ class intervals. The limits of the class intervals should be chosen so that there is no ambiguity in assigning observed values to classes.

The absolute class frequency gives the number of values in the interval (class) concerned. The relative class frequency is obtained by division by the size of the sample. When this frequency is expressed as a function of the class means, the frequency distribution of the random samples subdivided into classes is obtained.

The cumulative frequencies are formed by the addition of relative class frequencies. They show the number of observations less than or equal to a specified class or given value.

Example:

Table 2 is obtained by grouping from a master list of 100 blood sugar values (Table 1).

Table 1. Analytical variability of the determination of glucose in serum (precision of the hexokinase/G6P-DH method, $n = 100$ single determinations on 100 days). Values in mmol/l.

4.37	4.13	4.46	4.45	4.73	4.20	3.64	4.15	3.96	3.61
3.71	4.05	3.70	4.21	4.35	3.89	3.67	4.50	4.16	4.07
4.31	4.08	3.65	4.23	4.00	4.58	4.30	4.42	3.90	4.36
3.93	4.14	3.76	3.82	4.16	4.17	4.30	4.18	4.56	3.93
4.60	3.94	4.32	4.55	4.40	4.55	4.82	4.56	4.19	4.52
4.38	4.36	3.69	4.04	4.40	3.98	4.38	4.00	4.08	4.16
4.10	3.84	4.03	4.58	4.20	4.37	4.58	4.14	4.88	4.21
4.11	4.31	4.27	4.31	4.62	3.80	4.08	4.05	3.85	4.21
3.87	4.05	3.94	4.23	3.95	4.32	4.03	3.91	4.18	4.23
3.84	4.03	3.56	3.81	3.93	4.28	4.03	3.74	4.27	4.72

Table 2. Class formation from the master list (Tab. 1).

Classes	Class means	Frequency			Cumulative frequency (relative)
		absolute		relative	
3.50 – 3.66	3.58	‖‖	4	0.04	0.04
3.67 – 3.83	3.75	𝅘𝅥‖‖	9	0.09	0.13
3.84 – 4.00	3.92	𝅘𝅥𝅘𝅥𝅘𝅥‖	16	0.17	0.30
4.01 – 4.17	4.09	𝅘𝅥𝅘𝅥𝅘𝅥𝅘𝅥‖	22	0.22	0.52
4.18 – 4.34	4.26	𝅘𝅥𝅘𝅥𝅘𝅥𝅘𝅥‖	24	0.21	0.73
4.35 – 4.51	4.43	𝅘𝅥𝅘𝅥‖‖	11	0.13	0.86
4.52 – 4.68	4.60	𝅘𝅥𝅘𝅥	10	0.10	0.96
4.69 – 4.85	4.77	‖‖	3	0.03	0.99
4.86 – 5.02	4.94	‖	1	0.01	1.00

If in an experiment two or more parameters (random samples) are investigated, all measured values are treated analogously.

Graphical presentation

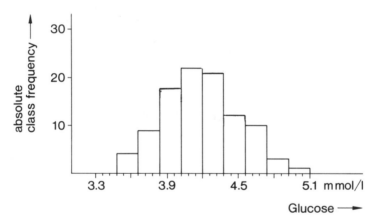

Fig. 1. Histogram of the random sample presented in Tab. 1.

The simplest form of graphical presentation is a histogram (Fig. 1). In many cases, this has the typical shape of a bell, which may be more or less biased toward the left or the right. Distributions with two or more peaks are more rare. The width of the distribution is a measure of the variance of the measured values. The peak indicates the position of the frequency maximum of the random samples (mode). Different choices of the origin of classes, limits and class interval may lead to completely different histograms, as shown in Fig. 2. Therefore, the specific choice of the classes is important for the resulting figure.

Normal distribution

In many samples it is of interest to determine whether the frequency distributions of the observed data follow specific theoretical distributions (Fig. 3). These distributions

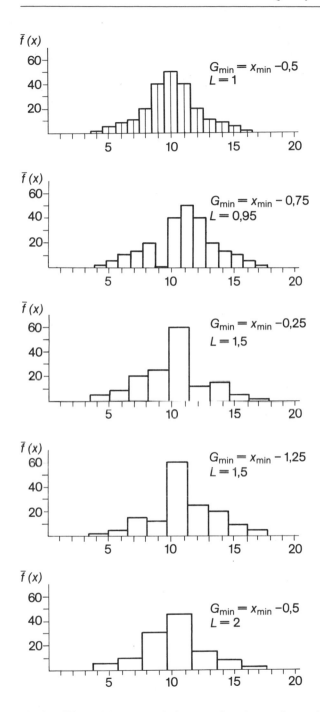

Fig. 2. Different histograms of the same data due to clumsy choice of class limits G_{min} and width L (possibility of bias)

are based on mathematical functional equations. An extraordinary role is played by the normal (*Gaussian*) distribution because many measurements are distributed normally and in other cases the underlying distributions, such as t-, F-distributions, lead to a normal distribution when *n* becomes large. The question whether measurements are normally distributed is important with respect to statistical tests which are

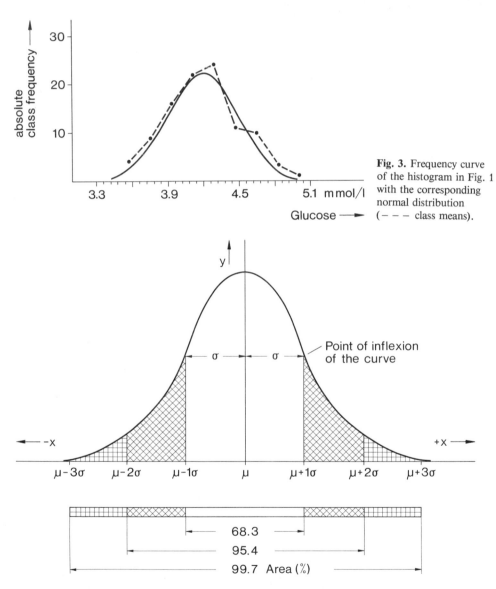

Fig. 3. Frequency curve of the histogram in Fig. 1 with the corresponding normal distribution (− − − class means).

Fig. 4. Normal distribution of a basic population (mean value μ and standard deviation σ) and relative areas of $\mu \pm 1\sigma$, 2σ, 3σ (in a random sample $\bar{x} \triangleq \mu$, $s \triangleq \sigma$).

based on normality; it can be answered by a statistical test of fit. Typical examples are methodological and technical errors.

The normal distribution is characterized by the fact that a definite percentage of values always lies within various multiples of σ, the population standard deviation (see Fig. 4 and Table 3 for explanation of symbols):

> 68.3% of all values lie within $\mu \pm 1\sigma$
> 95.4% of all values lie within $\mu \pm 2\sigma$
> 99.7% of all values lie within $\mu \pm 3\sigma$

The distance along the abscissa from the mean to points corresponding to either of the points of inflection of the two arms is the standard deviation σ.

The normal distribution and its modifications are of great importance in biology (e.g. in comparative statistics), because many statistical tests are based on normality. In addition, "non-parametric" test procedures are being used to an increasing extent. These have the advantage that the random samples need not exhibit any definite form of distribution. The disadvantage is the lower resolution of the corresponding tests with small numbers of samples ($n < 20$). A condition for the validity of the t-test that is often used in comparing mean values, is, for example, that the sample consists of "normally distributed" values. If this condition is not fulfilled, which is often the case, the use of the *t*-test may lead to false interpretations.

Calculation of sample statistics

Terms and symbols

Population and sample are characterized by parameters and statistics. Parameters belong to the population, statistics to the sample. The symbols differ accordingly.

Table 3. Parameters and statistics.

Term	Population Parameter	Sample Statistic
Sample size*	N	n
Mean	μ	$\bar{x}$
Variance	σ^2	s^2
Standard deviation	σ	s
Probability	P	–
Relative frequency	–	p
Correlation coefficient**	ρ	$r\,(\hat{\rho})$

* Number of samples analyzed for replicate measurements.

** If there is no specific symbol for the sample statistic, the symbol is simply provided with the sign $\wedge$. Symbols with $\wedge$ are therefore sample statistics.

Sample statistics

There are two different types of sample statistics, statistics indicating the location and statistics indicating the dispersion of the distribution.

1. Parameters of location

The mean value $\bar{x}$ of a random sample $x_1 \ldots x_n$ is defined as the arithmetic mean of values of the random sample:

(a) $\bar{x} = \dfrac{x_1 + x_2 \ldots + x_n}{n} = \dfrac{1}{n} \cdot \sum\limits_{i=1}^{n} x_i.$

and is an estimate of the true mean (μ) of the population to which the random sample belongs. It approaches this value more and more closely, the greater the number of observations. For $n \to \infty$, $\bar{x} \to \mu$.

The median $\bar{x}$ is the $(n + 1)/2$-th observation when the values are arranged in order of magnitude:

(b) $\bar{x} = \begin{cases} x_{(n+1)/2} & n \text{ odd} \\ \frac{1}{2}(x_{n/2} + x_{1+n/2}) & n \text{ even} \end{cases}$

that is, when n is an even number, the median is the mean of the $n/2$-th and the $\frac{n}{2} + 1$-th observations. One half of the observations is located below the median, one half above it. When the sample distribution is symmetrical, median and arithmetic mean are equal, in skew distributions there is a gap between both parameters. The advantage of the median is the fact that it is often more typical of all the measurements than is the mean since it is not affected by extreme values.

Besides the median there are other values which divide the observations into two groups, the so-called percentiles. A percentile P_p divides the observations into two groups such that p percent of the observations are less than P_p and (100-p) percent of the observations are greater than P_p. The percentile P_p can approximately be calculated by the $[p\,(n + 1)/100]$th observation in the sorted data array [8].

2. Parameters of dispersion

The variance shows how widely the sample values are scattered around the mean value. It is denoted by s^2:

(c) $s^2 = \dfrac{1}{n - 1} \sum\limits_{i=1}^{n} (x_i - \bar{x})^2$

To calculate the variance, the following relation derived from eqn. (c) is used:

(d) $s^2 = \dfrac{1}{n - 1} \left[\sum\limits_{i=1}^{n} x_i^2 - \dfrac{1}{n} \left(\sum\limits_{i=1}^{n} x_i \right)^2 \right]$

It gives an estimated value for the variance of the population σ^2. The expression $(n - 1)$ is termed the number of degrees of freedom.

The standard deviation is the name given to the square root of the variance:

$$\text{(e)}\quad s = \sqrt{\dfrac{\sum\limits_{i} (x_i - \bar{x})^2}{n - 1}}\ .$$

It has the same dimension as the measurements.

In practice, the quality of the analytical methods is characterized by the percentage error or coefficient of variation (CV); this is a specific measure of the imprecision of the measurement. It is the relative error:

$$\text{(f)}\quad CV = \dfrac{s \times 100}{\bar{x}}\ \%$$

and is expressed as a percentage.

The arithmetic mean $\bar{x}$ is the best estimated value for the mean of the population, but in common with the individual observations, it is itself affected by an error. The average error of the mean is:

$$\text{(g)}\quad s_{\bar{x}} = \dfrac{s}{\sqrt{n}}, \text{ the standard error of the mean.}$$

The accuracy of the mean increases in proportion to $1/\sqrt{n}$. If it were desired to reduce $s_{\bar{x}}$ by a power of 10, the number of measurements to be performed would have to be increased around a hundredfold.

Another parameter of dispersion is the interval between two percentile limits. For instance, the interval $P_{25} - P_{75}$ is called the interquartile and contains 50% of the observed data.

Confidence interval

The term confidence interval was defined in order to estimate the deviation of an approximate value of the sample (e.g. $\bar{x}$) from the true value (e.g. μ) of the population. The mean value $\bar{x}$ and the variance s^2 of a random sample are estimated values of the corresponding parameters of the basic population. The question is, how good are these approximate values of the samples?

Let the mean value $\bar{x}$ and the standard deviation of an approximately normally distributed random sample be given. When the values of the individual samples are shown as a frequency distribution (cf. p. 482), it is apparent that, for example, 95.4% of the values occur in the interval $\bar{x} \pm 2s$; in other words, the probability of observations lying within the interval $\bar{x} \pm 2s$ is roughly 95%. The same interpretation is valid for the interval $\bar{x} \pm 2s_{\bar{x}}$ as far as the mean μ is concerned. Consequently, this interval is called the 95% confidence interval for the mean value μ. The limiting values are called confidence limits. The corresponding confidence number is denoted by γ. In biology, it is usually chosen as 95% or 99%.

If in a statistical investigation a parameter is to be estimated, the corresponding confidence limits should always be determined as well. Without confidence limits the information about an estimated result is small.

This concept of confidence intervals is applied when control charts are used for internal quality control. Control charts usually contain boundaries in the $\pm 1s$, $\pm 2s$, $\pm 3s$ distance around the mean. Values which fall outside those limits only have a certain probability. According to these probabilities and the possible consequence, if a process is out of control, the investigator has to act (*Brettschneider* [11]).

The problem of outliers

Extremely high or low values within a series of measurements often raise the question whether they can be neglected under certain conditions. Errors of measurement, calculation, or a pathological case among the healthy subjects being investigated can lead to extreme values which must be deleted, since they arise from populations different from that of the random sample.

In all other cases such results should not be eliminated. Rules derived from statistical viewpoints, as for instance the elimination of values outside the $\pm 4s$-range, are often not applicable because those rules are bound to certain statistical assumptions which are not always fulfilled.

3.2.2.2   Statistical Evaluation of Experimental Results

The testing of hypotheses

The aim of inferential statistics is to test by a statistical procedure the validity of statistical hypotheses. A statistical hypothesis H_0 is a statement about a statistical population and usually is a statement about the values of one or more parameters of the population. For example: let the null hypothesis (H_0) be that two characteristic magnitudes (e.g. mean values) do not differ significantly. Let an alternative hypothesis (H_1) state that there is a significant difference between the two characteristic magnitudes. If H_1 is accepted, H_0 must automatically be rejected.

If the hypothesis that the mean value μ_1 and μ_0 of two distributions do not differ is valid, then $H_0: \mu_1 = \mu_0$. Possible alternatives are

$$\mu_1 > \mu_0$$
$$\mu_1 < \mu_0$$
$$\mu_1 \neq \mu_0$$

In the case of the hypotheses $\mu_1 > \mu_0$ and $\mu_1 < \mu_0$ one-sided differences are investigated (e.g. whether the mean value of one random sample is larger than that of the other). The alternative hypothesis $\mu_1 \neq \mu_0$ tests for deviations in both directions; the question is two-sided. The question of a one-sided or two-sided deviation is of considerable importance in testing for significance.

The decision whether to accept or reject hypotheses is made by comparing the experimentally determined characteristic magnitudes with "test distributions" which are valid under the null hypothesis. A test statistic is calculated from the experimental results, this statistic is compared with critical points. These critical points define the regions of acceptance and rejection. If the test statistic falls into the region of acceptance, H_0 cannot be rejected, i.e. it is accepted. If it falls into the region of rejection, H_0 must be rejected. But it should be pointed out that although we accept or reject a null hypothesis we have not proved or disproved the hypothesis.

In testing hypotheses, there are two types of errors which can be made. They are called Type I and Type II errors.

Type I error (α): the rejection of a hypothesis which is true

Type II error (β): the acceptance of a hypothesis which is false (Table 4)

Table 4. Definition of the types of errors associated with tests of hypotheses

Decision	True situation	
	Hypothesis is true	Hypothesis is false
Accept the hypothesis	No error	Type II error
Reject the hypothesis	Type I error	No error

If in a series of investigations the differences between populations are to be elucidated and the appropriate tests reveal no significant difference, the statement is made that "on the basis of the random sample used, no (statistical) difference can be ascertained (for the assumed probability of error)".

Significance

The range of the regions of acceptance and rejection are dependent on the choice of the Type I error.

Empirically, agreement has been reached on three levels of significance with the following probabilities of Type I error α:

$\alpha = 0.05$
     The null hypothesis (H_0) is erroneously
     rejected once in 20 measurements.

$\alpha = 0.01$

The null hypothesis (H_0) is erroneously rejected once in 100 measurements.

$\alpha = 0.001$

The null hypothesis (H_0) is erroneously rejected once in 1000 measurements.

When setting up an experiment to test a hypothesis it is desirable to minimize the probabilities of making these errors. But when the Type I error α gets smaller, the Type II error β gets larger (n fix).

So in selecting values for the two types of error, one should bear in mind the consequences of making these errors. The problem is that α can be chosen relatively arbitrarily, whereas β depends on the alternative hypothesis which cannot be known exactly. Normally, α and β become smaller as n gets larger.

Statistical tests

Statistical test procedures can be divided according to

– the underlying distribution, and

– the number of samples which are to be compared.

Those statistical tests which do not imply any specific distribution of the data are called distribution-free or non-parametric; the others are parametric tests [4, 5, 6].

In testing whether the mean values of experimental groups differ substantially (significantly, with a previously specified probability of error) the choice of test procedure must be governed by the nature of the random samples. The conditions for the applicability of the appropriate test must be satisfied absolutely, since otherwise conclusions may be drawn which are seriously in error.

Testing whether the mean value belongs to a normally distributed population with an unknown variance (1 sample, parametric)

The mean value $\bar{x}$ of a normally distributed random sample of size n (number of measurements) and variance s^2 is to be compared with the mean value μ of a normally distributed population of the same variance ($s^2 = \sigma^2$). The null hypothesis ($H_0: \mu = \mu_0$) that the two differ only by chance cannot be rejected with a probability of error α (e.g. 5%) when the value t_0 calculated according t_0 the test statistic

(h) $$t_0 = \frac{|\bar{x} - \mu_0|}{s_{\bar{x}}}$$

is smaller than the tabulated value t_τ of the t distribution for α and the number of degrees of freedom $f = n - 1$.

Example:

In $n = 30$ blood samples the protein content was found to be $\bar{x} = 55$ g/l, variance $s^2 = 7$ g/l. The normal value is $\mu_0 = 65$ g/l. With a probability of error $\alpha = 0.05$ are the observed deviations fortuitous? The value of the test statistic is

$$t_0 = \frac{|55 - 65|}{0.48} = 20.8 \,.$$

The tabulated value of the t distribution for $\alpha = 0.05$ (5%) and $f = n - 1 = 29$ degrees of freedom is $t_\tau = 2.05$ (Table 5). Since $t_0 > t_\tau$, the null hypothesis (H_0) must be rejected. Significant differences exist between the mean value of the sample and the assumed normal value.

Table 5. Tabulated values of the t-distribution

Degree of freedom f	Probability of errors $\alpha = 5\%$	$\alpha = 1\%$	Degree of freedom f	Probability of errors $\alpha = 5\%$	$\alpha = 1\%$
1	12.71	63.66	26	2.06	2.78
2	4.30	9.92	27	2.05	2.77
3	3.18	5.84	28	2.05	2.76
4	2.78	4.60	29	2.05	2.76
5	2.57	4.03	30	2.04	2.75
6	2.45	3.71	35	2.03	2.72
7	2.36	3.50	40	2.02	2.70
8	2.31	3.36	45	2.01	2.69
9	2.26	3.25	50	2.01	2.68
10	2.23	3.17	60	2.00	2.66
11	2.20	3.11	70	1.99	2.65
12	2.18	3.05	80	1.99	2.64
13	2.16	3.01	90	1.99	2.63
14	2.14	2.98	100	1.98	2.63
15	2.13	2.95	120	1.98	2.62
16	2.12	2.92	140	1.98	2.61
17	2.11	2.90	160	1.97	2.61
18	2.10	2.88	180	1.97	2.60
19	2.09	2.86	200	1.97	2.60
20	2.09	2.85	300	1.97	2.59
21	2.08	2.83	400	1.97	2.59
22	2.07	2.82	500	1.97	2.59
23	2.07	2.81	1000	1.96	2.58
24	2.06	2.80	∞	1.96	2.58
25	2.06	2.79			

t-Test for paired observations (2 samples, paired, parametric)

Some experiments yield as results paired data per experimental unit; that is, every experimental unit (serum, ...) is subjected two different treatments (for instance, two different chemical methods of determining a particular parameter may be used). The reason for designing an experiment in this way is the fact that most intermediary metabolites in one individual are subject to smaller fluctuations than the differences observed between different individuals. Pair analysis reduces the variability and the evaluation is thus made more effective.

The principle of the paired observations is to calculate the individual differences and to apply a statistical test to these differences. If no mean difference between the values of the control and test observations is to be supposed, positive and negative differences must cancel out.

Let x_i be the values of the control group, y_i of the test group, then the null hypothesis H_0: "The mean difference equals 0" will be tested against the alternative: "The mean difference will not be equal to 0".

The mean difference d, calculated from the individual differences $d_i = x_i - y_i$ is:

(i) $$\bar{d} = \frac{\Sigma d}{n} .$$

The standard deviation s_d is calculated from the individual differences, the standard deviation of the mean value of the differences is:

(j) $$s_{\bar{d}} = \frac{s_d}{\sqrt{n}} .$$

The statistic t_0 is calculated from the equation

(k) $$t_0 = |\bar{d}|/s_{\bar{d}} .$$

If the test magnitude t_0 is smaller than the tabulated value t_τ (Table 5), the null hypothesis (H_0) cannot be rejected, i.e. with a probability of Type I error of 5% no significant difference can be assumed between the x- and y-values.

Example

A new enzymatic procedure for the determination of uric acid is compared with the uricase method (reference method) on samples of human serum with different uric acid contents (Table 6). The test magnitude t_0 of the values in Table 6 is calculated from eqn. (i) to (k) as $t_0 = 0.0912$. Since for 19 degrees of freedom and $\alpha = 0.05$ (5%) t_0 is smaller than the tabulated value $t_\tau = 2.09$, the null hypothesis cannot be rejected on the basis of a probability of Type I error of 5%. No difference can be assumed between the analytical procedures.

Table 6. t-Test for paired data.

No.	x_i	y_i	$d = x_i - y_i$	d^2
1	2.9	2.8	-0.1	0.01
2	5.7	6.0	$+0.2$	0.09
3	3.3	3.7	$+0.4$	0.16
4	5.8	6.0	$+0.2$	0.04
5	4.4	4.3	-0.1	0.01
6	9.5	9.1	-0.4	0.16
⋮	⋮	⋮	⋮	⋮
20	11.3	11.0	-0.3	0.09
Σ	143.8	144.4	-0.1	1.21

Wilcoxon's test for paired observations (2 samples, paired, non-parametric)

Since the t-test for paired observations assumes the sample populations to be of known form (normal distribution) there is a non-parametric, i.e. distribution free, analogue in the matched-pairs design: the *Wilcoxon* signed rank test [8].

The procedure is as follows:

the individual differences are ranked without regard to sign, that is, the absolute values of the differences are ranked. Then the sign is assigned to each rank and the sum of positive and negative ranks is obtained. The smaller rank sum of both sums is compared with the critical values tabulated. If the observed rank sum is less than or equal to the tabulated value for the chosen significance level, then the hypothesis that no differences exist between the groups is rejected. As in the parametric case, first the individual differences are found and thus the 2-sample problem is reduced to a 1-sample problem. The difference between the t-test and *Wilcoxon*'s test is that the latter has slightly less statistical power, but is more widely applicable because its use is not restricted to normal distributions.

t-Test for the comparison of the mean values of two random samples (2 samples, parametric)

Two normally distributed mutually independent random samples with random variances $(s_1)^2$ and $(s_2)^2$ and sample sizes n_1 and n_2 are to be compared with one another. Let the null hypothesis (H$_0$) be that their mean values μ_1 and μ_2 differ only by chance with the probability of error α (H$_0$: $\mu_1 - \mu_2 = 0$). The null hypothesis (H$_0$) is to be confirmed if, for α and the number of degrees of freedom $f = n_1 + n_2 - 2$, the calculated value t_0 for the test statistic (t) is smaller than the tabulated value t_τ (Table 5) which is the critical value of the t-distribution.

The calculation is performed from the following equations:

(l) $$t = \frac{|\bar{x}_1 - \bar{x}_2|}{s_d} \times \sqrt{\frac{n_1 \times n_2}{n_1 + n_2}}$$

(m) $$s_d = \sqrt{\frac{s_1^2 \times (n_1 - 1) + s_2^2 \times (n_2 - 1)}{n_1 + n_2 - 2}} \ .$$

Example

In two groups of rats average increases in weight per week of $\bar{x}_1 = 18$ and $\bar{x}_2 = 24$ g were found in $n_1 = 20$ and $n_1 = 32$ experiments, resp. The variances were $s_1^2 = 4$ and $s_2^2 = 6$. Do the two experimental groups differ if a probability of error $\alpha = 5\%$ is assumed ($H_0: \mu_1 - \mu_2 = 0: \alpha = 0.05$)?

For the test statistic we calculate

$$t_0 = \frac{|18 - 24|}{\sqrt{5.24}} \times \sqrt{\frac{20 \times 32}{20 + 32}} = 9.20 \ .$$

The tabulated value of the *t*-distribution for a probability of error $\alpha = 5\%$ and $f = n_1 + n_2 - 2 = 50$ degrees of freedom is $t_0 = 2.01$. Since $t_0 > t_\tau$, the null hypothesis (H_0) must be rejected, i.e. significant differences can be assumed between the two mean values assuming a probability of Type I error of 5%.

U-test for the comparison of two random samples (2 samples, non-parametric)

A suitable non-parametric test for the above mentioned situation is the U-test by *Mann-Whitney-Wilcoxon*. Supposed the *n* experimental units are divided at random into a group of n which will receive a new treatment and a control group of m which will be treated by the standard method. At the termination of the study the observations of both groups are put together and ranked according to the value of the observations the smallest value getting the rank 1 etc. Then the ranks are added up according to each group so that each group has a rank sum. The null hypothesis of no effect of treatment (that is, equality of both groups) is rejected if in this ranking the one rank-sum is much higher than the other one. The term much higher is defined by a tabulated critical value, which is dependent on the chosen level of significance. Using the rank-sum U-test and applying it to the original 52 data of the above mentioned example yields the same decision in respect to the rejection of H_0.

The difference from the t-test for two groups lies in the fact that not the actual values of the observations, but the rank are used. This represents a certain loss of information which on the other hand is balanced by the unlimited usefulness of this test because it is distribution-free.

The extension from 2-sample problems to k-sample comparisons ($k \geq 3$) leads to analysis of variances if normal distributions can be assumed. If the assumption are not fulfilled, the H-test by *Mann-Whitney-Wilcoxon* or *Friedman*'s rank analysis may be applied. Since distribution-free test procedures have a wider range of applications their use is recommended.

3.2.2.3   Curve-fitting and the Method of Least Squares

In many experiments the relationship between two variables is to be represented, for example the dependence of the absorbance of a coloured solution on the concentration. One begins with the collection of corresponding values. For the above example, the values of the concentration are denoted by x_1, x_2, ... x_n and the corresponding absorbances by y_1, y_2, ... y_n, and the pairs of values (x_1, y_1), (x_2, y_2) ... (x_n, y_n) are plotted in a system of coordinates. The question arises whether the scattered pairs of values reflect a certain relationship between the two variables, and if this dependence can be described by a curve or a straight line (Fig. 5).

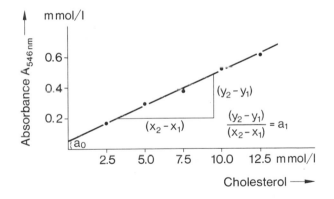

Fig. 5. Regression analysis of the relationship between concentration and absorbance in the enzymatic determination of cholesterol.

Example

The relationship between concentration and absorbance in an enzymatic determination of cholesterol is to be tested on the basis of measured values and their graphical representation (Fig. 5). Manual curve-fitting shows that a linear relationship can be assumed between the two variables.

Non-linear relationship can sometimes be linearized by plotting the results on a semi-logarithmic or logarithmic scale.

Linear regression

Regression means movement backward, namely to the origin. If in an analytical process, for example a photometric procedure, the concentrations of the samples (random variable x) are associated with absorbance (random variable y), it is necessary to determine whether a higher concentration is or is not associated, on average, with a greater absorbance, whether this relationship is linear (regresses), and what average concentration corresponds to a given absorbance. The calculation of regression is used to answer the second question. The simplest linear fitting of a curve is a straight line with the equation:

(n) $y = a_0 + a_1 x$.

The additive term a_0 defines the point of intersection of the straight line with the y-axis (the intercept).

The constant a_1 defines the slope of the regression line, which is characterized by

(o) $\dfrac{y_2 - y_1}{x_2 - x_1}$ (Fig. 5) .

The method of least squares

In order to avoid subjective evaluation in the fitting of straight lines or curves to the measured values, it is necessary to agree on what is understood by the best fitting.

A measure of the quality of the fitting of the curve to the given data is the sum of the squares of the vertical deviations $(d_1^2 + d_2^2 + \ldots + d_n^2)$, when the vertical distance between the observed value and the straight line is called d. The best fitted curve is that for which the sum of the squares of the deviation is a minimum (best fitting curve, least-squares fit).

The least squares regression (least squares line)

The constants a_0 and a_1 of the straight line equation $y = a_0 + a_1 \times x$ are calculated from

(p) $a_0 = \dfrac{\sum y - a_1 \sum x}{n}$

(q) $a_1 = \dfrac{n \sum xy - \sum x \sum y}{n \sum x^2 - (\sum x)^2}$

By calculation according to equations (p) and (q), the data in Table 6 yield the equation $y = 0.051 + 0.045x$. The correlation coefficient is calculated from equation (s); it is $r = 0.998$.

Correlation Analysis

Whereas the regression function indicates the type of the relationship, correlation analysis investigates the degree of dependence between two or more random variables [1, 7]. The most important measure of dependence is the correlation coefficient ρ. An estimate of ρ on the basis of a random sample is the correlation coefficient r; it can assume values between $+1$ and -1. If $r = 1$ or $r = -1$, a functional relationship exists between x and y. If $r = 0$, no relationship exists between x and y; the variables are uncorrelated (Fig. 6).

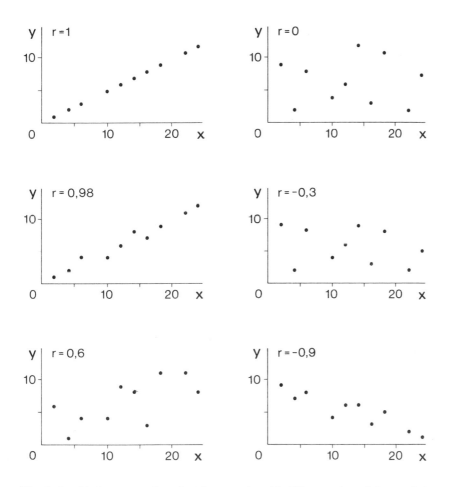

Fig. 6. Graphical representation of random samples with different values of the correlation coefficient r. From: *Erwin Kreyszig,* Statistische Methoden und ihre Anwendungen. Vandenhoek & Ruprecht, Göttingen, 1968.

Calculation

Let a random sample of a two-dimension xy-population be given. The random sample consists of n pairs of values $(x_1 y_1)$, $(x_2 y_2)$, ... $(x_n y_n)$. Then $\bar{x} = (x_1 + x_2 + \ldots + x_n)/n$ is the mean of the x values in the random sample. The variance of these figures is s_x^2 resp. s_y^2.

The product of the two deviations of the observed values (x, y) from the corresponding mean $(\bar{x}, \bar{y})$ is taken as a measure of the covariances (s_{xy}) of the random sample. This is calculated from

(r) $$s_{xy} = \frac{1}{n-1} \sum_{i=1}^{n} (x_i - \bar{x})(y_i - \bar{y}) .$$

The quotient $r = \dfrac{s_{xy}}{s_x s_y}$

is called the correlation coefficient of the random sample. The following formula is more suitable for calculation:

(s) $$r = \frac{\sum x_i \times y_i - \dfrac{1}{n} \sum x_i \sum y_i}{\sqrt{\left[\sum x_i^2 - \dfrac{1}{n}(\sum x_i)^2 \right]\left[\sum y_i^2 - \dfrac{1}{n}(\sum y_i)^2 \right]}}$$

Obviously, the concept of correlation is not restricted to linear relationships but can be generally used. A wide field of applications of linear regressions is the graphical comparison of different methods of clinical chemistry where it is very often used besides other statistical tools which are designed for the correct statistical comparison.

Two of these procedures are especially for comparing two chemical methods.

- The standardized principle component method [3],

- a newly developed statistical method which is distribution-free [9].

As in the comparison of different groups there are distribution-free regression and correlation analyses which use the principle of the ranks of the observations as input variables. They are known as *Kendall's* τ and *Spearman's* rank correlation coefficient, respectively [2].

References

[1] *N. B. Chapman, J. Shorter,* Correlation Analysis in Chemistry, Recent Advances, Plenum Press, New York and London (1978).
[2] *P. J. Cornbleet, M. C. Shea,* Comparison of Product Moment and Rank Correlation Coefficients in the Assessment of Laboratory Method – Comparison Data, Clin. Chem. *24*, 857–891 (1978).

[3] *U. Feldmann, B. Schneider, H. Klinkers,* A Multivariate Approach for the Biometric Comparison of Analytical Methods in Clinical Chemistry, J. Clin. Chem. Clin. Biochem. *19,* 121 – 137 (1981).

[4] *J. D. Gibbons,* Non-Parametric Statistical Inference, McGraw Hill, New York (1971).

[5] *M. Hollander, D. A. Wolfe,* Non-Parametric Statistical Methods, J. Wiley & Sons, New York (1973).

[6] *E. L. Lehmann,* Non-Parametrics: Statistical Methods Based on Ranks, McGraw Hill, New York (1975).

[7] *D. L. Massart, A. Dijkstra, L. Kaufmann,* Evaluation and Optimization of Laboratory Methods and Analytical Procedures, Elsevier, Amsterdam, Oxford and New York (1978).

[8] *B. Ostle,* Statistics in Research, The Iowa University Press, Iowa, 2nd edition (1966).

[9] *H. Passing, W. Bablok,* A New Biometrical Procedure for Testing the Equality of Measurements from Two Different Analytical Methods, Application of Linear Regression Procedures for Method Comparison Studies in Clinical Chemistry, Part I, J. Clin. Chem. Clin. Biochem. *21,* 709 – 720 (1983).

[10] *L. Sachs,* Statistische Auswertungsmethoden, Springer, Berlin (1969).

Appendix

1   Symbols, Quantities, Units and Constants

Units and symbols in ⟨ ⟩ should not be used.

m	Metre, m	g	Gram, g
mm	Millimetre, 10^{-3} m	mg	Milligram, 10^{-3} g
μm	Micrometre, 10^{-6} m	μg	Microgram, 10^{-6} g
nm	Nanometre, 10^{-9} m	ng	Nanogram, 10^{-9} g
		pg	Picogram, 10^{-12} g
l	Litre, 10^{-3} m^3	fg	Femtogram, 10^{-15} g
ml	Millilitre, 10^{-3} l	ag	Attogram, 10^{-18} g
μl	Microlitre, 10^{-6} l		

h	Hour	kg	Kilogram, 10^3 g
min	Minute	Mg	Megagram, 10^6 g
s	Second		

t	Time, s (h, min)
Δt	Interval between measurements, s (h, min)

t	Temperature, °C
T	Temperature, K
K	Kelvin

V	Volume (usually volume of assay mixture), ml, l
v	Volume (usually volume of sample in assay mixture), ml, l
φ	Volume fraction of sample in assay mixture
	Mass of substance, g

c	Substance concentration, mol/l
ρ	Mass concentration, g/l
n_c/m_s	Substance content, mol/kg
w	Mass fraction
⟨ppm	Parts per million⟩
⟨ppb	Parts per billion⟩

%	Percentage
% (v/v)	Percentage, volume related to volume
% (v/w)	Percentage, volume related to weight
% (w/v)	Percentage, weight related to volume
% (w/w)	Percentage, weight related to weight

M_r	Molecular weight, relative molecular mass

v	Rate of reaction, $\text{mol} \times l^{-1} \times s^{-1}$ ($\mu\text{mol} \times \text{ml}^{-1} \times \text{min}^{-1}$)
V	Maximum rate of reaction, $\text{mol} \times l^{-1} \times s^{-1}$
$\dot{\xi}$	Rate of conversion, $\text{mol} \times s^{-1}$ ($\mu\text{mol} \times \text{min}^{-1}$)
v_i	Stoichiometric coefficient

kat	Katal, $\text{mol} \times s^{-1}$
U	International unit (for enzymes), $\mu\text{mol} \times \text{min}^{-1}$
mU	International milliunit, 10^{-3} U
kU	International kilounit, 10^{3} U
MU	International megaunit, 10^{6} U
Inh.U	Inhibitor unit, $\mu\text{mol} \times \text{min}^{-1}$

z	Catalytic activity, U, kat
b	Catalytic activity concentration, U/l, kat/l
z_c/m_s	Catalytic activity content (specific catalytic activity), U/g, kat/kg

ε	Linear molar absorption coefficient, $l \times \text{mol}^{-1} \times \text{mm}^{-1}$
d	Light path, mm
A	Absorbance
F	Fluorescence intensity
I	Light intensity
T	Transmission

Bq	Becquerel, s^{-1}
$\langle\text{Ci}$	Curie, $s^{-1}\rangle$
cpm	Counts per minute, min^{-1}
dpm	Disintegrations per minute, min^{-1}
C	Counting rate (cpm) of the radioactive product
η	Counting efficiency
X	Specific radioactivity, $Bq \times \text{mol}^{-1}$ $\langle\text{Ci} \times \text{mol}^{-1}\rangle$
Y	Number of counts measured
Z	Decay rate, disintegrations per minute, dpm

k	Reaction constant
K	Equilibrium constant
K'	Apparent equilibrium constant
K_m	Michaelis constant, mol/l
K_I	Inhibitor constant, mol/l
pH	Negative logarithm of the hydrogen ion concentration
pK	Negative logarithm of the dissociation constant, $-\log K$

$[\alpha]_D^{20}$	Specific rotation (sodium D-line at $20\,^\circ\text{C}$)
sp.gr.	Specific gravity at $20\,^\circ\text{C}$ relative to water at $4\,^\circ\text{C}$
g	Acceleration due to gravity; $9.81\ \text{m} \times s^{-2}$
rpm	Revolutions per minute, min^{-1}

J	Joule, $m^2 \times kg \times s^{-2}$
R	Gas constant, $8.312\ J \times mol^{-1} \times K^{-1}$
h	*Planck's* constant
v	Frequence of emitted light
$\bar{x}$	Mean value
s, SD	Standard deviation
RSD	Relative standard deviation
⟨CV	Coefficient of variation⟩

2 Abbreviations for Chemical and Biochemical Compounds

It is unavoidable with the numerous abbreviations in use that one abbreviation occasionally is used for different compounds. In such cases the correct meaning can be obtained from the text. Only unequivocal abbreviations are used in the book without further explanation.

A	Adenosine
ABTS®	2,2'-Azino-bis(3-ethylbenzthiazoline) sulphonate
Acetoacetyl-CoA	Acetoacetyl coenzyme A
Acetyl-CoA	Acetyl coenzyme A
AChE	Acetylcholinesterase
AcP	Acid phosphatase
ADA	Adenosine deaminase
ADH	Alcohol dehydrogenase
ADP	Adenosine 5'-diphosphate
ADPglucose	Adenosine 5'-diphosphoglucose
ALAT (ALT, AlaAT)	Alanine aminotransferase
Alcohol-OD	Alcohol oxidase
ALD	Aldolase
AK	Acetate kinase
AK	Adenylate kinase
Ammediol	2-Amino-2-methyl propane-1,3-diol
AMP	Adenosine 5'-monophosphate
A-2-MP	Adenosine 2'-monophosphate
A-3-MP	Adenosine 3'-monophosphate
A-3,(2)-MP	Adenosine 3'(2')-monophosphate
A-3:5-MP (cAMP)	Adenosine 3':5'-monophosphate, cyclic
AOD	Amino acid oxidase
AP	Alkaline phosphatase
Ap$_5$A	Diadenosine 5'-pentaphosphate
APAD	Acetylpyridine-adenine dinucleotide
APADH	Acetylpyridine-adenine dinucleotide, reduced
ASAT (AST; AspAT)	Aspartate aminotransferase
Ascorbate-OD	Ascorbate oxidase
ATP	Adenosine 5'-triphosphate
ATPase	Adenosine 5'-triphosphatase
BAEE	Benzoyl-L-arginine ethyl ester
BAPNA	N-Benzoyl-arginine-4-nitroanilide
Benzoyl-CoA	Benzoyl coenzyme A
BES	Bis(2-hydroxyethylamino)ethanesulphonic acid
Bicine	N,N-Bis(2-hydroxyethyl)glycine

BMTD	6-Benzamido-4-methoxy-3-toluidinediazonium chloride
BSA	Bovine serum albumin
C	Cytidine
CDP	Cytidine 5'-diphosphate
CDPglucose	Cytidine 5'-diphosphoglucose
CE	Cholesterol esterase
Cellosolve	Ethylene glycol monomethyl ether
CHA	Cyclohexylammonium
ChE	Cholinesterase
ChOD	Cholesterol oxidase
CK	Creatine kinase
CL	Citrate lyase
CMP	Cytidine 5'-monophosphate
C-2-MP	Cytidine 2'-monophosphate
C-2:3-MP	Cytidine 2':3'-monophosphate, cyclic
C-3-MP	Cytidine 3'-monophosphate
C-3,(2)-MP	Cytidine 3'(2')-monophosphate
CoA	Coenzyme A
COX	Cytochrome c oxidase
CS	Citrate synthase
CTP	Cytidine 5'-triphosphate
Cyt-c	Cytochrome c
d	deoxy (prefix)
DAP	Dihydroxyacetone phosphate
DEA	Diethanolamine
DEAE	Diethylaminoethyl
DFP	Diisopropyl fluorophosphate
DM-POPOP	2,2'-(1,4-Phenylene)bis(4-methyl-5-phenyloxazole)
DMSO	Dimethylsulphoxide
DNase	Deoxyribonuclease
DNP	Dinitrophenylhydrazine
DTNB	5,5'-Dithiobis(2-nitrobenzoic acid)
DTT	Dithiothreitol
EDTA	Ethylenediamine tetraacetate
EGTA	Ethyleneglycol-bis(β-aminoethyl ether) N,N'-tetraacetic acid
FAD	Flavin-adenine dinucleotide
FDH	Formate dehydrogenase
FDNB	1-Fluoro-2,4-dinitrobenzene
FH_2	Dihydrofolate
FH_4	Tetrahydrofolate

FiGlu	N-Formimino-L-glutamate
FMN	Flavin mononucleotide
F-1-P	Fructose 1-phosphate
F-1,6-P$_2$	Fructose 1,6-bisphosphate
F-2,6-P$_2$	Fructose 2,6-bisphosphate
F-6-P	Fructose 6-phosphate
F-6-PK	Fructose-6-phosphate kinase
GABAse	Mixture of GAB-AT and SS-DH
GAB-AT	γ-Aminobutyrate aminotransferase
Gal-DH	Galactose dehydrogenase
Gal-OD	Galactose oxidase
GAP	Glyceraldehyde phosphate
GAPDH	D-Glyceraldehyde-3-phosphate dehydrogenase
GDH	L-Glycerol-3-phosphate dehydrogenase (glycerol-1-phosphate dehydrogenase; α-glycerophosphate dehydrogenase)
GDP	Guanosine 5′-diphosphate
GlDH	L-Glutamate dehydrogenase
GMP	Guanosine 5′-monophosphate
G-2-MP	Guanosine 2′-monophosphate
G-3-MP	Guanosine 3′-monophosphate
G-3,(2)-MP	Guanosine 3′(2′)-monophosphate
G-3:5-MP (cGMP)	Guanosine 3:5-monophosphate, cyclic
GMPK	GMP kinase
GOD	Glucose oxidase
GOT	Glutamate-oxaloacetate transaminase/Aspartate aminotransferase
G-1-P	Glucose 1-phosphate
G-1,6-P$_2$	Glucose 1,6-bisphosphate
G-6-P	Glucose 6-phosphate
G6P-DH	Glucose-6-phosphate dehydrogenase
GPT	Glutamate-pyruvate transaminase/Alanine aminotransferase
GR	Glutathione reductase
GSH	Glutathione, reduced
GSSG	Glutathione, oxidized
γ-GT	γ-Glutamyltransferase
GTP	Guanosine 5′-triphosphate
Hb	Haemoglobin
Hepes	N-2-Hydroxyethyl piperazine-N′-2-ethanesulphonic acid
HK	Hexokinase
HMG-CoA	3-Hydroxy-3-methylglutaryl coenzyme A
I	Inosine
IDP	Inosine 5′-diphosphate

Ig	Immunoglobulin
IMP	Inosine 5′-monophosphate
INT	2-(4-Iodophenyl)-3-(4-nitrophenyl)-5-phenyltetrazolium chloride
ITP	Inosine 5′-triphosphate
LDH (L-LDH)	L-Lactate dehydrogenase
D-LDH	D-Lactate dehydrogenase
MDH	L-Malate dehydrogenase
MES	2-(N-Morpholino)ethanesulphonic acid
4-Met-um	4-Methylumbelliferon (4-methylumbelliferyl)
MK	Myokinase/adenylate kinase
Mops	3-(N-Morpholino)propanesulphonic acid
4-NA	4-Nitroaniline
NAC	N-Acetylcysteine
NAD	Nicotinamide-adenine dinucleotide
NADH	Nicotinamide-adenine dinucleotide, reduced
NADP	Nicotinamide-adenine dinucleotide phosphate
NADPH	Nicotinamide-adenine dinucleotide phosphate, reduced
NBT	Nitro-BT-tetrazolium salt; 2,2′-bis(4-nitrophenyl)-5,5′-diphenyl-3,3′-(dimethoxy-4,4′-diphenylene)ditetrazolium chloride
NBTH	N-Methyl-2-benzothiazolone hydrazone
NDPK	Nucleoside diphosphate kinase
NMN	Nicotinamide mononucleotide
NMPK	Nucleoside monophosphate kinase
NP	Nucleoside phosphorylase
4-NP	4-Nitrophenol (4-Nitrophenyl)
4-NP-G_1	4-Nitrophenyl α-D-glucopyranoside
4-NP-G_2	4-Nitrophenyl [α-D-glucopyranosyl-(1 → 4)-α-D-glucopyranoside]/4-Nitrophenyl α-D-maltoside
4-NP-G_3	4-Nitrophenyl {di[α-D-glucopyranosyl-(1 → 4)]-α-D-glucopyranoside}/4-Nitrophenyl α-D-maltotrioside
4-NP-G_4	4-Nitrophenyl {tri[α-D-glucopyranosyl-(1 → 4)]-α-D-glucopyranoside}/4-Nitrophenyl α-D-maltotetraoside
4-NP-G_5	4-Nitrophenyl {tetra[α-D-glucopyranosyl-(1 → 4)]-α-D-glucopyranoside}/4-Nitrophenyl α-D-maltopentaoside
4-NP-G_6	4-Nitrophenyl {penta[α-D-glucopyranosyl-(1 → 4)]-α-D-glucopyranoside}/4-Nitrophenyl α-D-maltohexaoside
4-NP-G_7	4-Nitrophenyl {hexa[α-D-glucopyranosyl-(1 → 4)]-α-D-glucopyranoside}/4-Nitrophenyl α-D-maltoheptaoside
4-NPgal	4-Nitrophenyl galactoside
4-NPgluc	4-Nitrophenyl glucoside

4-NPman	4-Nitrophenyl mannoside
4-NPP	4-Nitrophenyl phosphate
NTA	Nitriloacetate
OMP	Orotidine 5'-monophosphate
PALP	Pyridoxal 5'-phosphate
PAMP	Pyridoxamine 5'-phosphate
PDE	Phosphodiesterase
PEP	Phosphoenolpyruvate
6-PGDH	6-Phosphogluconate dehydrogenase
PGK	3-Phosphoglycerate kinase
PGluM	Phosphoglucomutase
PGM	Phosphoglycerate mutase
P_i	Inorganic phosphate
Pipes	Piperazine-N,N'-bis(2-ethanesulphonic acid)
PK	Pyruvate kinase
PL-A	Phospholipase A
PL-C	Phospholipase C
PL-D	Phospholipase D
PMS	Phenazine methosulphate
POD	Peroxidase
POPOP	2,2'-(1,4-Phenylene)bis(5-phenyloxazole)
PPase	Pyrophosphatase, inorganic
PP_i	Inorganic pyrophosphate
PPO	2,5-Diphenyloxazole
PRPP	5-Phospho-α-D-ribose 1-diphosphate
PTA	Phosphotransacetylase
PyDC	Pyruvate decarboxylase
RNA	Ribonucleic acid
rRNA	Ribosomal ribonucleic acid
tRNA	Transfer ribonucleic acid
RNase	Ribonuclease
SOD	Superoxide dismutase
SS-DH	Succinate-semialdehyde dehydrogenase
ST	Sialyltransferase
SUPHEPA	N-Succinyl-L-phenylalanine-4-nitroanilide
T	Ribosylthymidine
dT	Thymidine
TA	Transaldolase
TAT (TyrAT)	Tyrosine aminotransferase
dTDP	Thymidine 5'-diphosphate

dTDPglucose	Thymidine 5'-diphosphoglucose
TCA	Trichloroacetic acid
TEA	Triethanolamine
THF	Tetrahydrofolic acid
TIM	Triosephosphate isomerase
TPP	Thiamine pyrophosphate
Tricine	N,N,N-Tris(hydroxymethyl)methylglycine
Tris	Tris(hydroxymethyl)aminomethane
dTTP	Thymidine 5'-triphosphate
U	Uridine
UDP	Uridine 5'-diphosphate
UDPG (UDPglucose)	Uridine 5'-diphosphoglucose
UDPAG	Uridine 5'-diphospho-N-acetylglucosamine
UDPGA	Uridine 5'-diphosphoglucuronate
UDPGal	Uridine 5'-diphosphogalactose
UDPG-DH	UDPglucose dehydrogenase
UDPG-PP	UDPglucose pyrophosphorylase
UMP	Uridine 5'-monophosphate
U-2-MP	Uridine 2'-monophosphate
U-2:3-MP	Uridine 2':3'-monophosphate, cyclic
U-3-MP	Uridine 3'-monophosphate
U-3,(2)-MP	Uridine 3'(2')-monophosphate
UTP	Uridine 5'-triphosphate
X	Xanthosine
XOD	Xanthine oxidase

3   Formulae

This topic is the subject of chapter 3.1.2 "Transformation of Data to Results" in the present volume and, therefore, it is not repeated here. However, in volume III and in each of the subsequent volumes of the series, the contents of chapter 3.1.2 are reproduced in Appendix 3, "Formulae".

4 Absorption Coefficients of NAD(P)H

The absorption curves and absorption coefficients depend on the temperature, pH, and the ionic strength of the solution. The best-investigated cases [1–4] are NADH and NADPH. At $\lambda = 334$ nm the temperature dependence of ε is approximately zero. The value of ε falls with rising temperature at wavelengths $\lambda > 334$ nm (including the maximum of the absorption curve) and increases at $\lambda < 334$ nm; the absorption maximum is shifted accordingly.

Table 1. Molar decadic absorption coefficients ($l \times mol^{-1} \times mm^{-1}$) for β-NADH and β-NADPH (measured in triethanolamine/HCl buffer, 0.1 mol/l; pH = 7.6) [1].

	°C	Hg 334 nm	339 nm	340 nm	339.85 nm[+]	Hg 365 nm
β-NADH	25	6.176×10^2	no measurement	$6.317 \times 10^{2[++]}$	6.292×10^2	3.441×10^2
		$6.182 \times 10^{2*}$	cf. Table 2	–	$6.298 \times 10^{2*}$	$3.444 \times 10^{2*}$
β-NADH	30	$6.187 \times 10^{2*}$	–	–	–	$3.427 \times 10^{2*}$
β-NADPH	25	$6.178 \times 10^{2*}$	–	–	–	$3.532 \times 10^{2*}$
β-NADPH	30	$6.186 \times 10^{2*}$	–	–	–	$3.515 \times 10^{2*}$

[+] Checking the photometer revealed 339.85 nm to be correct, instead of 340 nm.
[++] Acc. to [2] in Tris buffer, 0.1 mol/l; pH 7.8.
* Values were not corrected for beam convergence or intrinsic absorbance of the oxidized coenzyme.

The above mentioned investigations [1–4] on NAD(P)H show:

– ε is different for NADH and NADPH,
– ε is temperature-dependent,
– ε depends on the pH and the ionic strength of the solution to be measured,
– ε cannot be determined sufficiently accurately at 37 °C because of the instability of the coenzyme,
– the absorption maximum of NAD(P)H is not located at exactly 340 nm; 339 nm can be taken as a first approximation,
– the absorption maximum is temperature-dependent,
– the differences in the value of ε due to the factors mentioned above are smallest at Hg 334 nm.

All of the influences mentioned above lead to deviations of the value of ε, not exceeding 0.5% at Hg 334 nm. Consequently, it is best, since practically independent of the conditions of measurement, to perform measurements at this wavelength. Moreover, the values of ε are identical here for both coenzymes.

However, the values of ε at 25 °C and 30 °C at 340 nm (or 339 nm) and Hg 365 nm are also sufficiently close for practical purposes ($<1\%$ error at 340 (339) nm; about 2% at Hg 365 nm) and are independent of the other conditions of measurement. For measurement at Hg 365 nm, however, one must distinguish between the values of ε for NADH and NADPH.

In a routine laboratory it would not be practical to use the exact values of the absorption coefficients given above. It would then be necessary, under certain circumstances, to determine the exact value of ε for each experiment. The figures recommended for practical purposes in the routine laboratory vary, according to temperature and other measurement conditions, by less than 1 to 2%, which is within the limits of error attainable for routine enzymatic analyses.

Molar decadic absorption coefficients ($1 \times mol^{-1} \times mm^{-1}$) for NADH and NADPH at temperatures of 25 °C and 30 °C [3] are for practical use:

	Hg 334 nm *(334.15 nm)*	*340 nm* *(339 nm)*	*Hg 365 nm* *(365.3 nm)*
NADH	6.18×10^2	6.3×10^2	3.4×10^2
NADPH	6.18×10^2	6.3×10^2	3.5×10^2

References

[1] *J. Ziegenhorn, M. Senn, T. Bücher,* Molar Absorptivities of β-NADH and β-NADPH, Clin. Chem. *22,* 151 – 160 (1976).
[2] *R. B. McComb, L. W. Bond, R. W. Burnett, R. C. Keech, G. N. Bowers, jr.,* Determination of the Molar Absorptivity of NADH, Clin. Chem. *22,* 141 – 150 (1976).
[3] *H. U. Bergmeyer,* Neue Werte für die molaren Extinktions-Koeffizienten von NADH und NADPH zum Gebrauch im Routine-Laboratorium, J. Clin. Chem. Clin. Biochem. *13,* 507 – 508 (1975).
[4] *Th. Bücher, G. Lüsch, H. Krell* in: *G. Anido, E. J. van Kampen, S. B. Rosalki, M. Rubin* (eds.), Temperature Dependence of Difference-Absorption Coefficients of NADH Minus NAD$^+$ and NADPH Minus NADP$^+$ in the Near Ultraviolet, Quality Control in Clinical Chemistry, Walter de Gruyter, Berlin, New York 1975, pp. 301 – 310.

5 Numbering and Classification of Enzymes

Extract of Enzyme Nomenclature*

1. Oxidoreductases

1.1 Acting on the CH-OH group of donors

    1.1.1   With NAD^+ or $NADP^+$ as acceptor
    1.1.2   With a cytochrome as acceptor
    1.1.3   With oxygen as acceptor
    1.1.99 With other acceptors

1.2 Acting on the aldehyde or oxo group of donors

    1.2.1   With NAD^+ or $NADP^+$ as acceptor
    1.2.2   With a cytochrome as acceptor
    1.2.3   With oxygen as acceptor
    1.2.4   With a disulphide compound as acceptor
    1.2.7   With an iron-sulphur protein as acceptor
    1.2.99 With other acceptors

1.3 Acting on the CH-CH group of donors

    1.3.1   With NAD^+ or $NADP^+$ as acceptor
    1.3.2   With a cytochrome as acceptor
    1.3.3   With oxygen as acceptor
    1.3.7   With an iron-sulphur protein as acceptor
    1.3.99 With other acceptors

1.4 Acting on the CH-NH$_2$ group of donors

    1.4.1   With NAD^+ or $NADP^+$ as acceptor
    1.4.2   With a cytochrome as acceptor
    1.4.3   With oxygen as acceptor
    1.4.4   With a disulphide compound as acceptor
    1.4.7   With an iron-sulphur protein as acceptor
    1.4.99 With other acceptors

---

* Enzyme Nomenclature, Recommendations (1978) of the Nomenclature Committee of the International Union of Biochemistry, Published for the International Union of Biochemistry by Academic Press, Inc., New York, United Kingdom Edition Academic Press, Inc., London (Ltd.) (1979), p. 19–26.

1.5   Acting on the CH-NH group of donors

1.5.1   With NAD$^+$ or NADP$^+$ as acceptor
1.5.3   With oxygen as acceptor
1.5.99 With other acceptors

1.6   Acting on NADH or NADPH

1.6.1   With NAD$^+$ or NADP$^+$ as acceptor
1.6.2   With a cytochrome as acceptor
1.6.4   With a disulphide compound as acceptor
1.6.5   With a quinone or related compound as acceptor
1.6.6   With a nitrogenous group as acceptor
1.6.7   With an iron-sulphur protein as acceptor
1.6.99 With other acceptors

1.7   Acting on other nitrogenous compounds as donors

1.7.2   With a cytochrome as acceptor
1.7.3   With oxygen as acceptor
1.7.7   With an iron-sulphur protein as acceptor
1.7.99 With other acceptors

1.8   Acting on a sulphur group of donors

1.8.1   With NAD$^+$ or NADP$^+$ as acceptor
1.8.2   With a cytochrome as acceptor
1.8.3   With oxygen as acceptor
1.8.4   With a disulphide compound as acceptor
1.8.5   With a quinone or related compound as acceptor
1.8.7   With an iron-sulphur protein as acceptor
1.8.99 With other acceptors

1.9   Acting on a heam group of donors

1.9.3   With oxygen as acceptor
1.9.6   With a nitrogenous group as acceptor
1.9.99 With other acceptors

1.10 Acting on diphenols and related substances as donors

1.10.1 With NAD$^+$ or NADP$^+$ as acceptor
1.10.2 With a cytochrome as acceptor
1.10.3 With oxygen as acceptor

1.11 Acting on hydrogen peroxide as acceptor

1.12 Acting on hydrogen as donor

    1.12.1 With NAD^+ or $NADP^+$ as acceptor
    1.12.2 With a cytochrome as acceptor
    1.12.7 With an iron-sulphur protein as acceptor

1.13 Acting on single donors with incorporation of molecular oxygen (oxygenases)

    1.13.11 With incorporation of two atoms of oxygen
    1.13.12 With incorporation of one atom of oxygen (internal monooxygenases or internal mixed function oxidases)
    1.13.99 Miscellaneous (requires further characterization)

1.14 Acting on paired donors with incorporation of molecular oxygen

    1.14.11 With 2-oxoglutarate as one donor, and incorporation of one atom each of oxygen into both donors
    1.14.12 With NADH or NADPH as one donor, and incorporation of two atoms of oxygen into one donor
    1.14.13 With NADH or NADPH as one donor, and incorporation of one atom of oxygen
    1.14.14 With reduced flavin or flavoprotein as one donor, and incorporation of one atom of oxygen
    1.14.15 With a reduced iron-sulphur protein as one donor, and incorporation of one atom of oxygen
    1.14.16 With reduced pteridine as one donor, and incorporation of one atom of oxygen
    1.14.17 With ascorbate as one donor, and incorporation of one atom of oxygen
    1.14.18 With another compound as one donor, and incorporation of one atom of oxygen
    1.14.99 Miscellaneous (requires further characterization)

1.15 Acting on superoxide radicals as acceptor

1.16 Oxidizing metal ions

    1.16.3 With oxygen as acceptor

1.17 Acting on -CH$_2$ groups

    1.17.1 With NAD^+ or $NADP^+$ as acceptor
    1.17.4 With a disulphide compound as acceptor

1.97 Other oxidoreductases

2. Transferases

2.1   Transferring one-carbon groups

2.1.1   Methyltransferases
2.1.2   Hydroxymethyl-, formyl- and related transferases
2.1.3   Carboxyl- and carbamoyltransferases
2.1.4   Amidinotransferases

2.2   Transferring aldehyde or ketonic residues

2.3   Acyltransferases

2.3.1   Acyltransferases
2.3.2   Aminoacyltransferases

2.4   Glycosyltransferases

2.4.1   Hexosyltransferases
2.4.2   Pentosyltransferases
2.4.99   Transferring other glycosyl groups

2.5   Transferring alkyl or aryl groups, other than methyl groups

2.6   Transferring nitrogenous groups

2.6.1   Aminotransferases
2.6.3   Oximinotransferases

2.7   Transferring phosphorus-containing groups

2.7.1   Phosphotransferases with an alcohol group as acceptor
2.7.2   Phosphotransferases with a carboxyl group as acceptor
2.7.3   Phosphotransferases with a nitrogenous group as acceptor
2.7.4   Phosphotransferases with a phosphate group as acceptor
2.7.5   Phosphotransferases with regeneration of donors (apparently catalysing intramolecular transfers)
2.7.6   Diphosphotransferases
2.7.7   Nucleotidyltransferases
2.7.8   Transferases for other substituted phosphate groups
2.7.9   Phosphotransferases with paired acceptors

2.8   Transferring sulphur-containing groups

2.8.1   Sulphurtransferases
2.8.2   Sulphotransferases
2.8.3   CoA-transferases

3. Hydrolases

3.1   Acting on ester bonds

   3.1.1   Carboxylic ester hydrolases
   3.1.2   Thiolester hydrolases
   3.1.3   Phosphoric monoester hydrolases
   3.1.4   Phosphoric diester hydrolases
   3.1.5   Triphosphoric monoester hydrolases
   3.1.6   Sulphuric ester hydrolases
   3.1.7   Diphosphoric monoester hydrolases
   3.1.11 Exodeoxyribonucleases producing 5′-phosphomonoesters
   3.1.13 Exoribonucleases producing 5′-phosphomonoesters
   3.1.14 Exoribonucleases producing other than 5′-phosphomonoesters
   3.1.15 Exonucleases active with either ribo- or deoxyribonucleic acids and producing 5′-phosphomonoesters
   3.1.16 Exonucleases active with either ribo- or deoxyribonucleic acids and producing other than 5′-phosphomonoesters
   3.1.21 Endodeoxyribonucleases producing 5′-phosphomonoesters
   3.1.22 Endodeoxyribonucleases producing other than 5′-phosphomonoesters
   3.1.23 Site-specific endodeoxyribonucleases: cleavage is sequence-specific
   3.1.24 Site-specific endodeoxyribonucleases: cleavage is not sequence-specific
   3.1.25 Site-specific endodeoxyribonucleases: specific for altered bases
   3.1.26 Endoribonucleases producing 5′-phosphomonoesters
   3.1.27 Endoribonucleases producing other than 5′-phosphomonoesters
   3.1.30 Endonucleases active with either ribo- or deoxyribonucleic acids and producing 5′-phosphomonoesters
   3.1.31 Endonucleases active with either ribo- or deoxyribonucleic acids and producing other than 5′-phosphomonoesters

3.2   Acting on glycosyl compounds

   3.2.1   Hydrolysing *O*-glycosyl compounds
   3.2.2   Hydrolysing *N*-glycosyl compounds
   3.2.3   Hydrolysing *S*-glycosyl compounds

3.3   Acting on ether bonds

   3.3.1   Thiolether hydrolases
   3.3.2   Ether hydrolases

3.4   Acting on peptide bonds (peptide hydrolases)

   3.4.11 α-Aminoacylpeptide hydrolases
   3.4.12 Peptidylamino-acid or acylamino-acid hydrol
   3.4.13 Dipeptide hydrolases
   3.4.14 Dipeptidylpeptide hydrolases

6. Ligases (synthetases)

6.1   Forming carbon-oxygen bonds

  6.1.1   Ligases forming aminoacyl-tRNA and related compounds

6.2   Forming carbon-sulphur bonds

  6.2.1   Acid-thiol ligases

6.3   Forming carbon-nitrogen bonds

  6.3.1   Acid-ammonia (or amine) ligases (amide synthetases)
  6.3.2   Acid-amino-acid ligases (peptide synthetases)
  6.3.3   Cyclo-ligases
  6.3.4   Other carbon-nitrogen ligases
  6.3.5   Carbon-nitrogen ligases with glutamine as amido-*N*-donor

6.4   Forming carbon-carbon bonds

6.5   Forming phosphate ester bonds

6   Atomic Weights

Lanthanides and actinides are incomplete

International atomic weights approved by the commission on atomic weights and isotopic abundances of IUPAC, 1983.

Element	Symbol	Atomic number	Atomic weight	Element	Symbol	Atomic number	Atomic weight
Aluminium	Al	13	26.9815	Iridium	Ir	77	192.22
Antimony	Sb	51	121.75	Iron	Fe	26	55.847
Argon	Ar	18	39.948	Krypton	Kr	36	83.80
Arsenic	As	33	74.9216	Lanthanum	La	57	138.9055
Barium	Ba	56	137.33	Lead	Pb	82	207.2
Beryllium	Be	4	9.0122	Lithium	Li	3	6.941
Bismuth	Bi	83	208.980	Lutetium	Lu	71	174.967
Boron	B	5	10.811	Magnesium	Mg	12	24.305
Bromine	Br	35	79.904	Manganese	Mn	25	54.9380
Cadmium	Cd	48	112.41	Mercury	Hg	80	200.59
Caesium	Cs	55	132.905	Molybdenum	Mo	42	95.94
Calcium	Ca	20	40.078	Neodymium	Nd	60	144.24
Carbon	C	6	12.011	Neon	Ne	10	20.179
Cerium	Ce	58	140.12	Nickel	Ni	28	58.69
Chlorine	Cl	17	35.453	Niobium	Nb	41	92.9064
Chromium	Cr	24	51.996	Nitrogen	N	7	14.0067
Cobalt	Co	27	58.9332	Osmium	Os	76	190.2
Copper	Cu	29	63.546	Oxygen	O	8	15.9994
Dysprosium	Dy	66	162.50	Palladium	Pd	46	106.42
Erbium	Er	68	167.26	Phosphorus	P	15	30.9738
Europium	Eu	63	151.96	Platinum	Pt	78	195.08
Fluorine	F	9	18.9984	Potassium	K	19	39.0983
Gadolinium	Gd	64	157.25	Praseodymium	Pr	59	140.908
Gallium	Ga	31	69.723	Rhenium	Re	75	186.207
Germanium	Ge	32	72.59	Rhodium	Rh	45	102.9055
Gold	Au	79	196.9665	Rubidium	Rb	37	85.4678
Hafnium	Hf	72	178.49	Ruthenium	Ru	44	101.07
Helium	He	2	4.0026	Samarium	Sm	62	150.36
Holmium	Ho	67	164.930	Scandium	Sc	21	44.9559
Hydrogen	H	1	1.00794	Selenium	Se	34	78.96
Indium	In	49	114.82	Silicon	Si	14	28.0855
Iodine	I	53	126.9045	Silver	Ag	47	107.8682

Element	Symbol	Atomic number	Atomic weight	Element	Symbol	Atomic number	Atomic weight
Sodium	Na	11	22.9898	Titanium	Ti	22	47.88
Strontium	Sr	38	87.62	Tungsten	W	74	183.85
Sulphur	S	16	32.066	Uranium	U	92	238.0289
Tantalum	Ta	73	180.9479	Vanadium	V	23	50.9415
Tellurium	Te	52	127.60	Xenon	Xe	54	131.29
Terbium	Tb	65	158.9254	Ytterbium	Yb	70	173.04
Thallium	Tl	81	204.383	Yttrium	Y	39	88.9059
Thorium	Th	90	232.0381	Zinc	Zn	30	65.39
Thulium	Tm	69	168.9342	Zirconium	Zr	40	91.224
Tin	Sn	50	118.710				

Index